Lecture Notes in Mechanical Engineering

Lecture Notes in Mechanical Engineering (LNME) publishes the latest developments in Mechanical Engineering—quickly, informally and with high quality. Original research reported in proceedings and post-proceedings represents the core of LNME. Volumes published in LNME embrace all aspects, subfields and new challenges of mechanical engineering. Topics in the series include:

- Engineering Design
- Machinery and Machine Elements
- Mechanical Structures and Stress Analysis
- Automotive Engineering
- Engine Technology
- Aerospace Technology and Astronautics
- Nanotechnology and Microengineering
- Control, Robotics, Mechatronics
- MEMS
- Theoretical and Applied Mechanics
- Dynamical Systems, Control
- Fluid Mechanics
- Engineering Thermodynamics, Heat and Mass Transfer
- Manufacturing
- Precision Engineering, Instrumentation, Measurement
- Materials Engineering
- Tribology and Surface Technology

To submit a proposal or request further information, please contact the Springer Editor of your location:

China: Ms. Ella Zhang at ella.zhang@springer.com
India: Priya Vyas at priya.vyas@springer.com
Rest of Asia, Australia, New Zealand: Swati Meherishi at swati.meherishi@springer.com
All other countries: Dr. Leontina Di Cecco at Leontina.dicecco@springer.com

To submit a proposal for a monograph, please check our Springer Tracts in Mechanical Engineering at https://link.springer.com/bookseries/11693 or contact Leontina.dicecco@springer.com

Indexed by SCOPUS. All books published in the series are submitted for consideration in Web of Science.

Jyotirmay Banerjee · Rupesh D. Shah ·
Ramesh K. Agarwal · Sushanta Mitra
Editors

Recent Advances in Fluid Dynamics

Select Proceedings of ICAFFTS 2021

Springer

Editors
Jyotirmay Banerjee
Department of Mechanical Engineering
S. V. National Institute of Technology
Surat, India

Rupesh D. Shah
Department of Mechanical Engineering
S. V. National Institute of Technology
Surat, India

Ramesh K. Agarwal
Department of Mechanical Engineering
and Materials Science
Washington University
Saint Louis, MO, USA

Sushanta Mitra
Waterloo Institute for Nanotechnology
University of Waterloo
Ontario, ON, Canada

ISSN 2195-4356 ISSN 2195-4364 (electronic)
Lecture Notes in Mechanical Engineering
ISBN 978-981-19-3378-3 ISBN 978-981-19-3379-0 (eBook)
https://doi.org/10.1007/978-981-19-3379-0

This Springer imprint is published by the registered company Springer Nature Singapore Pte Ltd.
The registered company address is: 152 Beach Road, #21-01/04 Gateway East, Singapore 189721, Singapore

About This Book

This book presents selected proceedings of the International Conference on Advances in Fluid Flow and Thermal Sciences (ICAFFTS 2021) and summarizes the modern research practices in fluid dynamics and fluid power. The content of the book involves advanced topics on droplet deposition, oscillating flows, wave breaking, spray structure and its atomization and flow patterns in mini- and micro-channels. Technological concerns relevant to erosion of steam turbine blade due to droplets, influence of baffle cut and baffle pitch on flow regime, bubble formation and propagation in pool boiling, design optimization of flow regulating valves are included in the book. In addition, recent trends in small-scale hydropower plant and flow stability issues in nanofluids, solar water heating systems and closed loop pulsating heat pipes are discussed. Special topics on airflow pattern in railway coach and vortex tube are also included. This book will be a reliable reference for academicians, researchers and professionals associated with advanced applications of "Fluid Dynamics and Fluid Power."

Contents

About the Editors

Dr. Jyotirmay Banerjee is Professor in Department of Mechanical Engineering at Sardar Vallabhbhai National Institute of Technology Surat. His research interests include computational fluid dynamics, multiphase flow and heat transfer and phase change applications. His has authored a monograph published by Tech. science press, USA under CREST (Contemporary Research in Emerging Science and Technology) and a book on "Conduction and Radiation", published by Narosa Publishing house, New Delhi. He has published more than 80 peer reviewed journal papers, two of which are "Editor's pick" in "Physics of Fluids" published by American Institute of Physics. He was a Guest Editor for Special Issue on Fluid Mechanics and Fluid power published by Sadhana, Indian Academy of Science, Springer. He has completed several sanctioned research projects funded by Science and Engineering Research Board (SERB), Board of Research Nuclear Studies (BRNS), Aeronautical Research and Development Board (ARDB) and Consultancy Assisted Research Services (CARS) sanctioned by Defence Research and Development Organization (DRDO). He has rendered major consultancy services to several industries including THDC HPP, Uttaranchal Jal Vidyut Nigam Ltd. (UJVNL) and Koteshwar Hydroelectric Project.

Dr. Rupesh D. Shah is presently affiliated to Sardar Vallabhbhai National Institute of Technology Surat as Associate Professor in Mechanical Engineering Department. He received his Ph.D. degree from SVNIT, Surat in 2014 and specializes in Gas Turbine Combustors. His research area includes combustion, inverse diffusion flame, porous media combustion and heat transfer augmentation. He has authored more than 20 papers in journals of repute and conferences in the field of inverse diffusion, gas turbine combustors and hydrogen energy.

Prof. Ramesh K. Agarwal is the William Palm Professor of Engineering in the department of Mechanical Engineering and Materials Science at Washington University in St. Louis, USA. From 1994 to 2001, he was the Sam Bloomfield Distinguished Professor and Executive Director of the National Institute for Aviation Research at Wichita State University in Kansas. From 1978 to 1994, he was the Program

Director and McDonnell Douglas Fellow at McDonnell Douglas Research Laboratories in St. Louis. Dr. Agarwal received Ph.D. in Aeronautical Sciences from Stanford University in 1975, M.S. in Aeronautical Engineering from the University of Minnesota in 1969 and B.S. in Mechanical Engineering from Indian Institute of Technology, Kharagpur, India in 1968. Over a period of 45 years, Prof. Agarwal has worked in Computational Fluid Dynamics (CFD) and its applications to fluid flow problems in mechanical and aerospace engineering and in energy and environment. In past fifteen years, the focus of his research has been on renewable and clean energy technologies. In renewable energy, he has worked on energy from wind, solar and biomass. In clean energy technologies, he has pioneered chemical looping combustion and Carbon, Capture, Utilization and Sequestration (CCUS). He is the author and coauthor of over 600 publications. He has given many plenary, keynote and invited lectures at various national and international conferences worldwide in over sixty countries. He is a Fellow of 25 professional societies including American Institute of Aeronautics and Astronautics (AIAA), American Society of Mechanical Engineers (ASME), Institute of Electrical and Electronics Engineers (IEEE), Society of Automotive Engineers (SAE), American Association for Advancement of Science (AAAS), American Physical Society (APS), and UK Energy Institute among others. He has received many prestigious honors and national/international awards from various professional societies and organizations for his research contributions including the AIAA Reeds Aeronautics Award, SAE Medal of Honor, ASME Honorary Membership and Honorary Fellowship from Royal Aeronautical Society.

Sushanta Mitra is the Executive Director of the Waterloo Institute for Nanotechnology and a Professor in Mechanical and Mechatronics Engineering at the University of Waterloo. His research interests are in the fundamental understanding of fluid transport in micro and nano-scale confinements with applications in energy, water, and bio-systems. For his contributions in science and engineering, he is an elected Fellow of several professional bodies, including the Canadian Academy of Engineering, the American Physical Society, the Royal Society of Chemistry, and the American Association for the Advancement of Science.

Evaporation of Sessile Droplets Placed Adjacent to Each Other on a Solid Surface

Prathamesh G. Bange, Manish Kumar, and Rajneesh Bhardwaj

1 Introduction

Understanding sessile droplet evaporation on a surface is crucial for applications such as surface cooling, surface coating [1], disease diagnosis [2] and bio-sensing of protein [3]. The difference in liquid vapour concentration around the liquid–gas interface and the surrounding atmosphere drives the evaporation. The mechanism governing the evaporation here is diffusion. Droplet evaporates two modes: constant contact radius (CCR) and constant contact angle (CCA). In CCR mode, the droplet evaporates with a constant contact radius, and droplet height decreases with time. The contact line remains pinned to the substrate. In CCA mode, contact angle remains constant and contact radius starts to recede as the droplet evaporates. Evaporation flux is larger near the contact line compared to the top of the droplet and to maintain the spherical shape of the droplet, and the liquid mass loss at the contact line is filled from the region around the apex of the droplet.

While a single droplet evaporation is much studied, few studies reported the evaporation of neighbouring droplets and its influence on the evaporation process. For example, Pradhan and Panigrahi [4] studied the drying pattern of evaporation of two adjacent droplets. Non-uniform deposit patterns and asymmetric convection were observed. The deposit pattern near the contact line of the two droplets is weak compared to the rest of the contact line. Carles and Cazabat [5] reported motion of the PDMS droplet towards the trans-decaline droplet when those were placed together. Carrier et al. [6] compared the evaporation of a single droplet of 10 μl with the closely spaced array of ten droplets, each having a volume of 10 μl and a bigger droplet with a volume of 100 μl. They found that if several droplets surround the droplet, the evaporation rate reduces due to the saturation of liquid vapour concentration in the

P. G. Bange · M. Kumar · R. Bhardwaj (✉)
Department of Mechanical Engineering, Indian Institute of Technology Bombay, Mumbai 400076, India
e-mail: rajneesh.bhardwaj@iitb.ac.in

J. Banerjee et al. (eds.), *Recent Advances in Fluid Dynamics*, Lecture Notes in Mechanical Engineering, https://doi.org/10.1007/978-981-19-3379-0_1

atmosphere above the droplets. Edwards et al. [7] used an interferometric technique to determine the accurate measurement of droplet evaporation rate. They recorded evaporation rates of random two-dimensional arrays of droplets and concluded that the evaporation rate depends upon the droplet placement in the array. Shaikeea and Basu [8] studied the evaporation rate of pair of droplets on a hydrophobic surface. They found asymmetry in the left and right contacts angle due to different evaporation rates in the droplet evaporation. In a follow-up study, Shaikeea and Basu [9] observed a single toroid flow inside the droplet due to the interlocking of space between the two droplets, evaporating on a hydrophobic surface. The flow alters to double toroid as the distance between droplets increases as the evaporation happens.

The above literature survey shows that the evaporation of two or more than two sessile droplets placed adjacent to each other has not fully explored yet. In particular, the role of the gap between the droplets on the evaporation is not clear yet. Therefore, the objective of the present work is to study the effect of the gap between droplets on the evaporation time of the droplet.

2 Experimental Details

The set-up utilized in the present study is shown in Fig. 1. We used a pipette to generate microliter water droplets, and they were gently placed on the glass slide (Sigma-Aldrich Inc). The slides were thoroughly cleaned isopropanol, deionized water and blown dry using nitrogen gas. We employed a high-speed camera (MotionPro, Y-3 Classic) with a long-distance objective (Qioptiq Inc.) to visualize the droplet evaporation from the side. A white LED lamp served as a light source for the optical visualization. One frame per second was used for the slow evaporation process. The focal length of the optical set-up was 9.5 cm. A resolution of 1280 × 184 pixels was used for the images collected during the experiments. The calibration of the set-up was 14 μm/pixel. We used in-house MATLAB® (www.mathworks.com) image processing codes for obtaining droplet dimensions from recorded images. The ambient temperature was maintained at 25 ± 2 °C during the experiments. The relative humidity was recorded as 40 ± 4%.

Fig. 1 Experimental set-up used in the present study

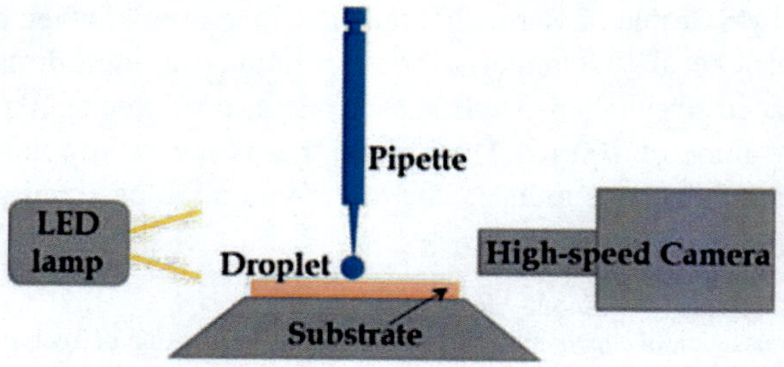

3 Results and Discussion

Figure 2 shows schematic for droplet array arrangement in the experiments performed. The evaporation of three cases was considered: 1, 2 and 3 sessile droplets placed adjacent to each other. We vary distance (d) between the droplets to study the effect of droplets proximity on droplet evaporation time. Figure 3 presents the temporal evolution of evaporation of water droplets on a glass substrate. In all cases, the droplet substrate evaporates in two distinct stages, first in CCR mode (approximately 80% droplet initial volume) and later in mixed mode, where droplet contact radius and contact angle change as the droplet volume evaporates. We consider only 80% of evaporation volume for the analysis. The evaporation times to evaporate 80% of the initial volume are 12.74, 14.54 and 15.5 min for one droplet, two droplets and three droplets adjacent to each other. The increase in the evaporation time suggests that the neighbouring droplet influences the evaporation of the droplet and delays the evaporation process.

Further, we have varied distance between the droplets in three-droplet configuration. Figure 1 represents the normalized evaporation time (t/t_f) against the distance between the droplets. The measured evaporation time is normalized with the theoretical evaporation time (t_f) for a single droplet evaporation, reported by Popov [10]. The latter time is given by

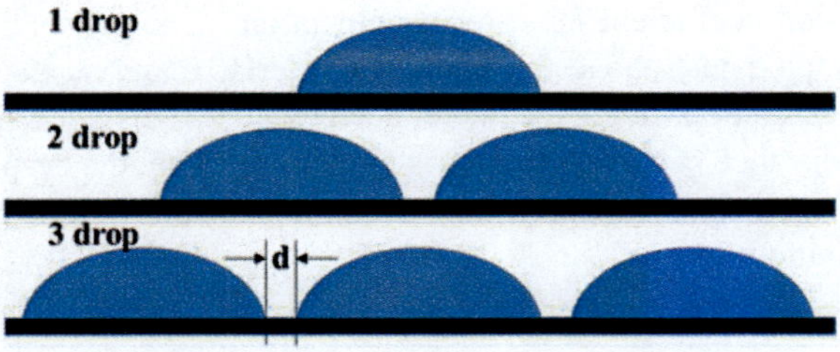

Fig. 2 Droplet array arrangements in the current study

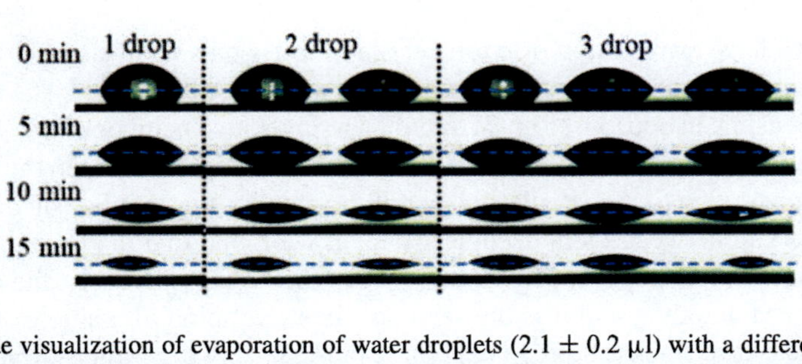

Fig. 3 Side visualization of evaporation of water droplets (2.1 ± 0.2 µl) with a different array of droplets. The evaporation time are 12.74 min, 14.54 min and 15.5 min in one droplet, two droplets and three droplets configuration, respectively

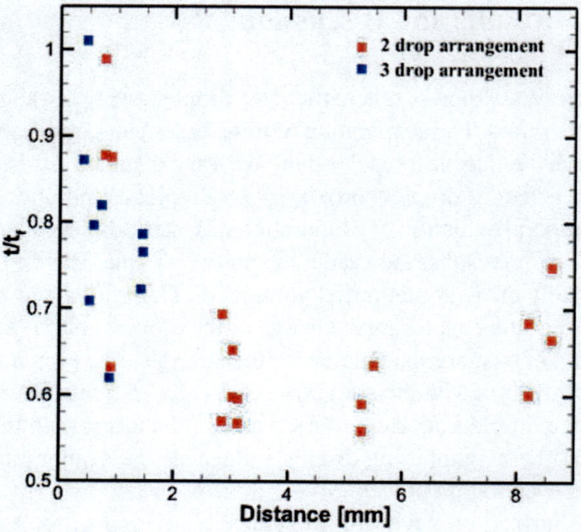

Fig. 4 Variation of normalized evaporation time against the distance or gap between the droplets in different experimental runs

$$t_f = \frac{\pi \rho R_i^2 \theta_i}{16D(n_s - n_\infty)}$$

where ρ is the density of a liquid, R_i is an initial wetted radius, θ_i is the initial contact angle, D is diffusion coefficient of water vapour in air, n_s is concentration of the saturated water vapour just above the liquid–air interface, and n_∞ is the ambient water vapour concentration. In Fig. 4, t/t_f reduced by 40% if distance changes from 1 to 3 mm. If the distance between droplets is more than 3 mm, the evaporation time does not change significantly. Clearly, the smaller distance (1 mm) corresponds to the longest evaporation time.

4 Conclusions

In closure, we have evaporation of multi-sessile droplets on a solid surface. Microliter droplets of water on a glass surface were considered. We used side visualization of the droplet using an optical set-up. We studied a single, two neighbouring droplets and three neighbouring droplets. We have found that the distance between the neighbouring droplets affects the droplet evaporation rate. As the number of droplets increases and gap between them becomes smaller, we found that the time required for evaporation increases. As the distance between the droplet increases, the evaporation time does not alter, and it is the same as the evaporation of a single droplet. These findings have implications in design of bio-assays, where multi-droplets in arrays evaporate on a surface.

References

1. Dugas V, Broutin J, Souteyrand E (2005) Droplet evaporation study applied to DNA chip manufacturing. Langmuir 21(20):9130–9136
2. Brutin D, Sobac B, Loquet B, Sampol J (2011) Pattern formation in drying drops of blood. J Fluid Mech 667:85–95
3. Wen JT, Ho C-M, Lillehoj PB (2013) Coffee ring aptasensor for rapid protein detection. Langmuir 29(26):8440–8446
4. Pradhan TK, Panigrahi PK (2015) Deposition pattern of interacting droplets. Colloids Surf A 482:562–567
5. Carles P, Cazabat AM (1989) Spreading involving the Marangoni effect: some preliminary results. Colloids Surf 41:97–105
6. Carrier O, Shahidzadeh-Bonn N, Zargar R, Aytouna M, Habibi M, Eggers J, Bonn D (2016) Evaporation of water: evaporation rate and collective effects. J Fluid Mech 798:774–786
7. Edwards AM, Cater J, Kilbride JJ, Le Minter P, Brown CV, Fairhurst DJ, Ouali FF (2021) Interferometric measurement of co-operative evaporation in 2D droplet arrays. Appl Phys Lett 119(15):151601
8. Shaikeea AJD, Basu S (2016) Insight into the evaporation dynamics of a pair of sessile droplets on a hydrophobic substrate. Langmuir 32(5):1309–1318
9. Shaikeea AJD, Basu S (2016) Evaporating sessile droplet pair: Insights into contact line motion, flow transitions and emergence of universal vaporisation pattern. Appl Phys Lett 108(24):244102
10. Popov YO (2005) Evaporative deposition patterns: spatial dimensions of the deposit. Phys Rev E 71(3):036313

Natural Convection of CMC/Water Mixture and Alumina Nanoparticles in a Cavity with Two Isoflux Heaters

D. S. Loenko and M. A. Sheremet

Nomenclature

a	The beginning of the local heater at X-axis
b	The end of the local heater at X-axis
d	Diameter, (m)
D_{ij}	The components of the strain rate tensor
K	Coefficient of flow density, ($Ns^n \cdot m^{-2}$)
n	Power-law index
q_1, q_2	Constant heat fluxes for left and right sources, respectively, ($W \cdot m^{-2}$)
T_c	Temperature of a cold vertical walls, (K)
T_{fr}	The freezing point of base fluid, (K)
X, Y	Dimensionless Cartesian coordinates
U, V	Dimensionless velocity components in X, Y - directions, respectively

Greek Symbols

λ	Thermal conductivity, ($W \cdot m^{-1} \cdot K^{-1}$)
β	Heat expansion factor, (K^{-1})
ΔT	Temperature drop, (K)
Λ	Relative coefficient of thermal conductivity
Θ	Dimensionless temperature
$\overline{\nu}$	Effective kinematic viscosity, ($m^2 \cdot s^{-1}$)
τ	Dimensionless time

D. S. Loenko (✉) · M. A. Sheremet
Laboratory on Convective Heat and Mass Transfer, Tomsk State University, 634050 Tomsk, Russia
e-mail: d.s.loenko@mail.tsu.ru

J. Banerjee et al. (eds.), *Recent Advances in Fluid Dynamics*, Lecture Notes in Mechanical Engineering, https://doi.org/10.1007/978-981-19-3379-0_2

τ_{ij}	Components of the deviatoric part of the stress tensor
Ψ	Dimensionless stream function
Ω	Dimensionless vorticity

Subscripts

avg	Average
bf	Base fluid
nf	Nanofluid
p	Particles
r	Relative

1 Introduction

In recent years, liquids consisting of base media with the addition of nanoparticles have received particular interest from researchers [1, 11]. Nanofluids have brought an innovative approach to many technical industries [15]. Such popularity is associated with the peculiarity of nanoparticles to significantly increase the thermal conductivity of the base medium [8]. In addition, due to nanosize and larger relative surface area of nanoparticles, nanofluids usually exhibit better stability, higher mobility, and lower pressure drop [5]. Thus, the simulation of flows of such fluids is used in relation to completely different problems. Several recent works on this topic will be presented below.

Li et al. [11] investigated the anisotropic natural convective heat transfer of a water-based Fe_2O_3 nanofluid, which exhibits pseudoplastic properties. The liquid is in a square cavity, the left wall of which has an isothermal heater, and the right one is kept at a low temperature. The process is also influenced by the magnetic field. The results showed that heat transfer enhances with increasing Rayleigh number and decreasing Hartmann number. The influence of a magnetic field on natural convective heat and mass exchange of Cu–Al_2O_3/water nanofluid contained in a cavity with several heat sources on the bottom wall is studied by Roy [16]. The problem was solved by the finite difference method and was confirmed by experimental and numerical results. The author found that convective heat and mass transfer is intensified due to an increase in the number of heat sources, the angle of the magnetic field and the Rayleigh number. Numerical modeling of the real configuration of the rib and its natural convective cooling using CuO/water nanofluid was carried out by Hejria and Malekshah [7]. The results confirmed that the average Nusselt number increases with an increase in the volume fraction of nanoparticles. It has also been found that thinner fins cause better thermal performance. Natural convection of Cu/water nanofluid in a square cavity with cooling and heating cylinders is numerically studied by Cao et al. [3]. The effects of magnetic field, Brownian motion of nanoparticles and thermophoresis are estimated. The authors found that an increase in the Rayleigh number and the volume fraction of nanoparticles enhance the heat transfer in the cavity. The effect of a porous cavity on natural convection of a Cu/water nanofluid was modeled by Cho [4]. The cavity has wavy top and bottom walls, as well as a partially heated left wall. The results showed that an increase in the Darcy number and Rayleigh number leads to an intensification of convective heat transfer. A numerical analysis of natural convection of alumina-water nanofluid in H-shaped cavity with V-shaped partition is carried out by Keramat et al. [9]. The upper rib and the entire lower wall act as a heater, while the side walls act as cooling ones. The authors found that installing a baffle on the bottom rib improves heat transfer. Nia et al. [14] investigated the influence of the Rayleigh number and the volume fraction of nanoparticles on the natural convection of Cu-water nanofluid in L-shaped enclosure with a baffle. Various configurations of the baffle were also investigated. The study showed that for low Rayleigh numbers an addition of a baffle always enhances natural convection, and a longer baffle is most effective under all conditions.

The presented brief analysis of studies on this topic characterizes the popularity and prospects of such works. Therefore, the purpose of this study is mathematical modeling of a pseudoplastic fluid natural convection in a square cavity with two heat-generating lower wall sections.

2 Problem Description

The geometric region of the solution is a closed square cavity (Fig. 1). The vertical walls are kept at a constant low temperature T_c. The horizontal walls are completely insulated. On the bottom wall of the cavity, there are two heat sources with different heat flux densities q_1 and q_2. The cavity is filled with a pseudoplastic nanofluid consisting of a CMC/water mixture and Al_2O_3 nanoparticles. Chemical attributes of the used nanosuspension are shown by Maleki et al. [13].

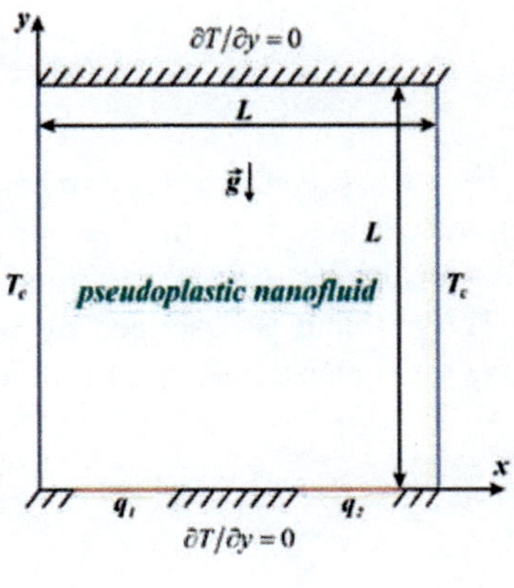

Fig. 1 Domain of interest

A nanosuspension is a power-law liquid that satisfies the Boussinesq approximation. To describe the non-Newtonian nature of the flow, the Ostwald-de-Waele power-law has been used Khezzar et al. [10]:

$$\tau_{ij} = 2\mu_{nf} D_{ij} \tag{1}$$

Here $D_{ij} = \frac{1}{2}\left(\frac{\partial u_i}{\partial x_j} + \frac{\partial u_j}{\partial x_i}\right)$ are the components of the strain rate tensor; u_i, u_j are the velocity components corresponding to the coordinates x_i, x_j; μ_{nf} is the effective coefficient of the nanosuspension viscosity.

The effective viscosity coefficient of the nanosuspension was calculated based on the Corcione model [6]:

$$\frac{\mu_{\mathrm{nf}}}{\mu_{\mathrm{bf}}} = \frac{1}{1 - 34.87\left(\frac{d_{\mathrm{p}}}{d_{\mathrm{bf}}}\right)^{-0.3}\phi^{1.03}} \tag{2}$$

Here ϕ is the concentration of solid particles, d_{bf} is the equivalent diameter of the base fluid molecule, d_{p} is the diameter of nanoparticles. In this case, to determine the effective coefficient of viscosity μ_{bf}, the following correlation was used $\mu_{\mathrm{bf}} = K(2D_{\mathrm{kl}}D_{\mathrm{kl}})^{\frac{n-1}{2}}$, where K is the coefficient of flow density; n is an indicator of fluid behavior, which are presented in the work of Khezzar et al. [10]. Taking into account that the index of liquid behavior is $n < 1$, let us note its correspondence to the pseudoplastic properties of the medium.

The physical characteristics of the nanosuspension were obtained using the following relationships [2]:

$$\begin{aligned} \rho_{\mathrm{nf}} &= \rho_{\mathrm{bf}}(1-\phi) + \rho_{\mathrm{p}}\phi \\ (\rho\beta)_{\mathrm{nf}} &= (\rho\beta)_{\mathrm{bf}}(1-\phi) + (\rho\beta)_{\mathrm{p}}\phi \\ (\rho c)_{\mathrm{nf}} &= (\rho c)_{\mathrm{bf}}(1-\phi) + (\rho c)_{\mathrm{p}}\phi \end{aligned} \tag{3}$$

Thermal conductivity was calculated using also the Corcione model [6]:

$$\frac{\lambda_{\mathrm{nf}}}{\lambda_{\mathrm{bf}}} = 1 + 4.4\mathrm{Re}^{0.4}\mathrm{Pr}^{0.66}\left(\frac{T}{T_{\mathrm{fr}}}\right)^{10}\left(\frac{\lambda_{\mathrm{p}}}{\lambda_{\mathrm{bf}}}\right)^{0.03}\phi^{0.66} \tag{4}$$

Here Re is the Reynolds number, Pr is the Prandtl number, T is the temperature of nanosuspension, T_{fr} is the freezing point of base fluid, λ_{p} is the thermal conductivity of nanoparticles, λ_{bf} is the thermal conductivity of the base fluid.

The problem is described by a system of non-stationary differential equations using the dimensionless non-primitive variables stream function and vorticity:

$$\frac{\partial^2\Psi}{\partial X^2} + \frac{\partial^2\Psi}{\partial Y^2} = -\Omega \tag{5}$$

$$\begin{aligned}&\frac{\partial\Omega}{\partial\tau}+\frac{\partial\Psi}{\partial Y}\frac{\partial\Omega}{\partial X}-\frac{\partial\Psi}{\partial X}\frac{\partial\Omega}{\partial Y}\\&\quad=H_1(\phi)\left(\frac{Ra}{\Pr}\right)^{\frac{n-2}{2}}\left[\nabla^2\left(\overline{M}\Omega\right)+S_\Omega\right]+H_2(\phi)\frac{\partial\Theta}{\partial X}\end{aligned}\tag{6}$$

$$\begin{aligned}&\frac{\partial\Theta}{\partial\tau}+\frac{\partial\Psi}{\partial Y}\frac{\partial\Theta}{\partial X}-\frac{\partial\Psi}{\partial X}\frac{\partial\Theta}{\partial Y}\\&\quad=\frac{H_3(\phi)}{\sqrt{Ra\cdot\Pr}}\left[\frac{\partial}{\partial X}\left(\Lambda\frac{\partial\Theta}{\partial X}\right)+\frac{\partial}{\partial Y}\left(\Lambda\frac{\partial\Theta}{\partial Y}\right)\right]\end{aligned}\tag{7}$$

Dimensionless viscosity $\overline{M}$ and source term S_Ω are

$$\overline{M}=\left[4\left(\frac{\partial^2\Psi}{\partial X\partial Y}\right)^2+\left(\frac{\partial^2\Psi}{\partial Y^2}-\frac{\partial^2\Psi}{\partial X^2}\right)^2\right]^{\frac{n-1}{2}}$$

$$S_\Omega=2\left[\frac{\partial^2\overline{M}}{\partial X^2}\frac{\partial^2\Psi}{\partial Y^2}+\frac{\partial^2\overline{M}}{\partial Y^2}\frac{\partial^2\Psi}{\partial X^2}-2\frac{\partial^2\overline{M}}{\partial X\partial Y}\frac{\partial^2\Psi}{\partial X\partial Y}\right]$$

Governing Eqs. (5)–(7) contain the following dimensionless complexes: Rayleigh number $\mathrm{Ra}=g\beta\Delta TL^3/(\overline{\nu}\alpha)$, Prandtl number $\Pr=\overline{\nu}/\alpha$, and $\Lambda=\lambda_{nf}/\lambda_{bf}$ thermal conductivity ratio. The effective kinetic viscosity coefficient is determined as follows $\overline{\nu}=(K/\rho)^{\frac{1}{2-n}}\cdot L^{\frac{2(1-n)}{2-n}}$. Additional dimensionless complexes are

$$H_1(\phi)=\frac{\mu_{nf}}{\mu_{bf}}\frac{\rho_{bf}}{\rho_{nf}}=\frac{1}{1-34.87\left(\frac{d_p}{d_{bf}}\right)^{-0.3}\phi^{1.03}}\cdot\frac{1}{\left(1-\phi+\phi\rho_p/\rho_{bf}\right)}$$

$$H_2(\phi)=\frac{(\rho\beta)_{nf}}{(\rho\beta)_{bf}}\frac{\rho_{bf}}{\rho_{nf}}=\frac{1-\phi+\phi(\rho\beta)_p/(\rho\beta)_{bf}}{1-\phi+\phi\rho_p/\rho_{bf}}$$

$$H_3(\phi)=\frac{(\rho c)_{bf}}{(\rho c)_{nf}}=\frac{1}{1-\phi+\phi(\rho c)_p/(\rho c)_{bf}}$$

The initial and boundary conditions for the formulated system of differential Eqs. (5)–(7) in dimensionless form look as follows, where $q_r=q_1/q_2$ is the heat flux ratio:

$$\begin{aligned}&\tau=0\rightarrow\Psi=\Omega=0,\ \Theta=0.5,\\&\tau>0\rightarrow\\&X=0\ X=1,\ 0\le Y\le 1,\ \Psi=0,\ \frac{\partial\Psi}{\partial X}=0,\ \Theta=0;\end{aligned}$$

$$Y = 0, \left| \begin{array}{l} 0 \le X \le 0.2, \\ 0.4 \le X \le 0.6, \\ 0.8 \le X \le 1 \end{array} \right. \quad \Psi = 0, \ \frac{\partial \Psi}{\partial Y} = 0, \ \frac{\partial \Theta}{\partial Y} = 0;$$

$$Y = 0, \ 0.2 \le X \le 0.4, \ \Psi = 0, \ \frac{\partial \Psi}{\partial Y} = 0, \ \frac{\partial \Theta}{\partial Y} = -\frac{\lambda_{\mathrm{bf}}}{\lambda_{\mathrm{nf}}};$$

$$Y = 0, \ 0.6 \le X \le 0.8, \ \Psi = 0, \ \frac{\partial \Psi}{\partial Y} = 0, \ \frac{\partial \Theta}{\partial Y} = -\frac{\lambda_{\mathrm{bf}}}{\lambda_{\mathrm{nf}}} q_{\mathrm{r}};$$

$$Y = 1, \ 0 \le X \le 1, \ \Psi = 0, \ \frac{\partial \Psi}{\partial Y} = 0, \ \frac{\partial \Theta}{\partial Y} = 0$$

The system of unsteady differential Eqs. (5)–(7) with auxiliary functions and the additional restrictions was solved employing the finite difference technique [12]. The time derivatives were approximated with the first order of accuracy, and the space derivatives were discretized with the second order of accuracy. The elliptic equation for the stream function (5) was discretized by central differences. The resulting set of algebraic equations was numerically worked out using the successive over relaxation algorithm. The relaxation parameter was determined based on a numerical experiment. The convective terms of the parabolic equations for vorticity (6) and temperature (7) were discretized by the "donor cells" difference scheme, which is also called the second upstream scheme. The diffusion terms were approximated employing central differences. The obtained two-dimensional equations were reduced to the system of one-dimensional equations using the locally one-dimensional Samarskii method. The final system of linear algebraic equations was resolved by the Thomas technique.

The solution method was tested on the model problem of natural convection of a non-Newtonian fluid in a differentially-heated cavity. Comparison of the results is presented in Fig. 2, which demonstrates good agreement between them.

In the course of the work, the grid parameters influence on the calculation results has been validated. Figure 3 demonstrates the values of the average *Nu* and average Θ at the left local energy source. The values of the average *Nu* were calculated using the following expression: $Nu_{\mathrm{avg}} = \frac{1}{(b-a)} \int_a^b \left(\frac{1}{\Theta_{Y=0}} \right) \mathrm{d}X$ where *a* and *b* are the beginning and end of the left local heater at *X*-axis. Grids with 100 × 100 and 150 × 150 subdivisions provide an independent solution, so a uniform rectangular grid of 100 × 100 elements was chosen to reduce the computation time.

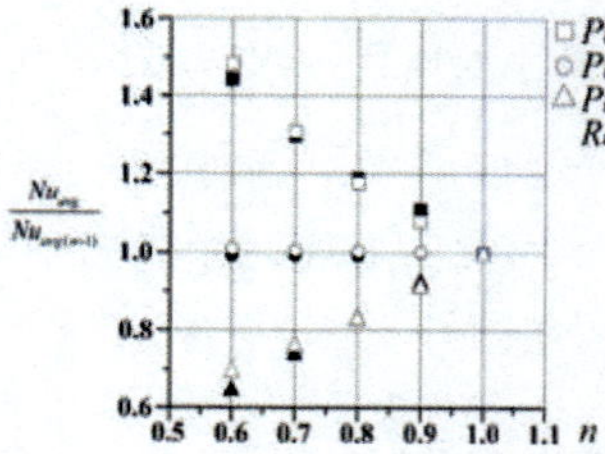

Fig. 2 Comparison with work of Turan et al. [17] (data of Turan et al. [17]—black points and obtained results—white points)

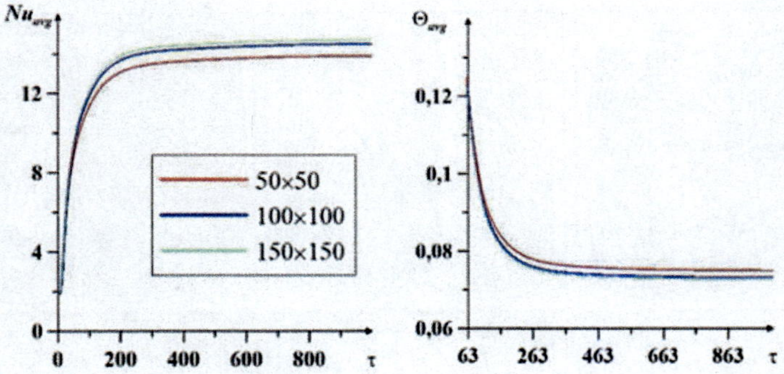

Fig. 3 Influence of grid parameters on the Nu_{avg} and Θ_{avg} at $\phi = 0.01$, $q_r = 0.5$, $Ra = 10^5$ for CMC/water (0.1%) + Al_2O_3

3 Results

The study was devoted to mathematical modeling of natural convective heat transfer of a pseudoplastic fluid with the addition of nanoparticles in a closed square cavity with two heat sources. In the course of the study, the analysis of the influence of the nanoparticles material, the nanoparticles volume fraction $\phi = 0, 0.005, 0.01$, the CMC volume fraction in water (0–0.3%), as well as the relative density of the heat flux $q_r = 0.5, 1, 2$ was carried out. The results were analyzed by constructing streamlines and isotherms depending on the above parameters. Integral characteristics were also plotted for both heating elements. Some results will be presented below.

Figure 4 shows the dependence of the average Nusselt number and average temperature for the left source on the nanoparticles material. For this study, the following nanoparticles were used: Al_2O_3, Cu, CuO, TiO_2. It can be seen from this figure that the most intense convective heat transfer mechanism corresponds to aluminum oxide (red line) and titanium oxide (black line), since in this case the average Nusselt

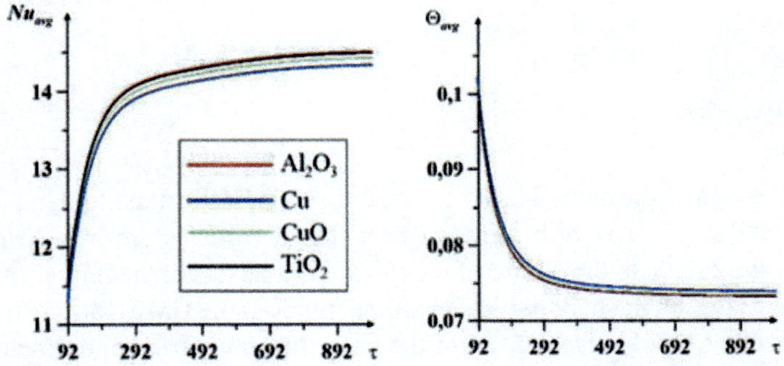

Fig. 4 Influence of the nanoparticles material on the average Nusselt number Nu_{avg} and average temperature Θ_{avg} at $\phi = 0.01$, $q_r = 0.5$

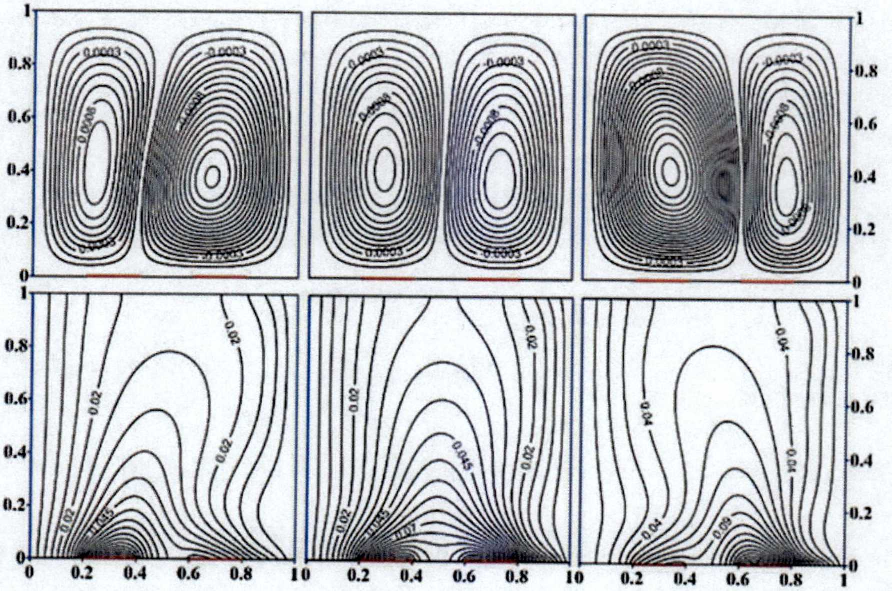

Fig. 5 Effect of q_r on the distribution of streamlines Ψ and isotherms Θ for CMC/water (0.1%) + Al_2O_3 at $\phi = 0.01$

number has the maximum values. It is also seen that in the case of using Al_2O_3 and TiO_2, the average temperature of the source has minimum values.

Figure 5 shows the effect of the relative density of the heat flux q_r on the distribution of streamlines Ψ and isotherms Θ in the cavity. Streamlines reflect the formation of two convective cells in the cavity. In cases of $q_r = 0.5$ and $q_r = 2$, one of the cells is more intense, and the other one is weaker. In case of $q_r = 1$, the cells have the same intensity. Also, in the case of $q_r = 0.5$ and $q_r = 2$, the isotherms tend to a less hot source. In case of $q_r = 1$, a two-dimensional thermal plume is formed in the center of the cavity, reflecting the temperature stratification inside the enclosure.

4 Conclusions

This work investigates the natural convection of a pseudoplastic nanofluid in a square closed cavity. Two sources with different heat flux densities were located on the lower wall of the cavity. In the course of the study, the analysis of the influence of the nanoparticles material, the nanoparticles volume fraction, the CMC volume fraction in water, as well as the relative density of the heat flux was carried out. Streamlines and isotherms were plotted depending on the varied parameters. The average Nusselt number on the surface of these two sources and average temperatures have been examined.

Acknowledgements The work was supported by the Russian Science Foundation (Project No. 17-79-20141).

References

1. Aboud ED, Rashid HK, Jassim HM, Ahmed SY, Khafaji SOW, Hamzah HK, Ali FH (2020) MHD effect on mixed convection of annulus circular enclosure filled with non-newtonian nanofluid. Heliyon 6:e03773
2. Ali FH, Hamzah HK, Egab K, Arici M, Shahsavar A (2020) Non-newtonian nanofluid natural convection in a U-shaped cavity under magnetic field. Int J Mech Sci 186(105887)
3. Cao Y, Ayed H, Jarad F, Togun H, Alias H, Issakhov A, Dahari M, Wae-hayee M, Ouni MHEl (2021) MHD natural convection nanofluid flow in a heat exchanger: effects of Brownian motion and thermophoresis for nanoparticles distribution. Case Stud Therm Eng 28(101394)
4. Cho CC (2020) Effects of porous medium and wavy surface on heat transfer and entropy generation of Cu-water nanofluid natural convection in square cavity containing partially-heated surface. Int Commun Heat Mass Transf 119(104925)
5. Choi SUS (1995) Enhancing thermal conductivity of fluids with nanoparticles. Am Soc Mech Eng Fluids Eng Div FED. 231:99–105
6. Corcione M (2011) Empirical correlating equations for predicting the effective thermal conductivity and dynamic viscosity of nanofluids. Energy Convers Manag 52:789–793
7. Hejria S, Malekshah EH (2021) Cooling of an electronic processor based on numerical analysis on natural convection and entropy production over a dissipating fin equipped with copper oxide/water nanofluid with Koo-Kleinstreuer-Li model. Therm Sci Eng Prog 23(100916)
8. Jou RY, Tzeng SC (2006) Numerical research of nature convective heat transfer enhancement filled with nanofluids in rectangular enclosures. Int Commun Heat Mass Transf 33(6):727–736
9. Keramat F, Dehghana P, Mofarahiab M, Lee CH (2020) Numerical analysis of natural convection of alumina–water nanofluid in H-shaped enclosure with a V-shaped baffle. J Taiwan Inst Chem Eng 111:63–72
10. Khezzar L, Siginer D, Vinogradov I (2012) Natural convection of power law fluids in inclined cavities. Int J Therm Sci 53:8–17
11. Li MG, Zheng C, Feng F, Chen X, Wu WT (2021) Natural convection and anisotropic heat transfer of shear-thinning ferro-nanofluid in partially heated rectangular enclosures under magnetic field. Therm Sci Eng Prog 25(100992)
12. Loenko D, Shenoy A, Sheremet M (2020) Influence of the chamber inclination angle and heat-generating element location on thermal convection of power-law medium in a chamber. J Numer Methods Heat Fluid Flow 31(1):134–153
13. Maleki H, Safaei MR, Alrashed AAAA, Kasaeian A (2019) Flow and heat transfer in non-Newtonian nanofluids over porous surfaces. J Therm Anal Calorim 135:1655–1666
14. Nia SN, Rabiei F, Rashidi MM, Kwang TM (2020) Lattice Boltzmann simulation of natural convection heat transfer of a nanofluid in a L-shape enclosure with a baffle. Results Phys 19(103413)
15. Putra N, Roetzel W, Das SK (2003) Natural convection of nano-fluids. Heat Mass Transf 39(8–9):775–784
16. Roy NC (2021) MHD natural convection of a hybrid nanofluid in an enclosure with multiple heat sources. Alexandria Eng J
17. Turan O, Sachdeva A, Chakraborty N, Poole RJ (2011) Laminar natural convection of power-law fluids in a square enclosure with differentially heated side walls subjected to constant temperatures. J Nonnewton Fluid Mech 166:1049–1063

A Numerical Study of Film Cooling on NASA-C3X Vane by Forward and Reverse Injection

Sumit Kumar, Kuldeep Singh, and Dushyant Singh

Nomenclature

D	Hole diameter, m
GT	Gas turbine
PR	Pressure ratio, $\frac{P_{\text{sec}}}{P_{\text{ms}}}$, non-dimensional
Re	Reynolds number is defined by mainstream flow
T	Temperature, K
U	Velocity, m/s

Symbols

α	Film cooling hole angle, °
η	Effectiveness, $\frac{T_{\text{ms}}-T_{\text{w}}}{T_{\text{ms}}-T_{\text{sec}}}$, non-dimensional
ρ	Fluid density, kg/m^3
μ	Fluid dynamic viscosity, Pa s
C_{p}	Specific heat, J/kg °C
K	Thermal conductivity, W/m k

S. Kumar · D. Singh
National Institute of Technology Manipur Imphal, Lamphelpat, Manipur 795004, India

K. Singh (✉)
Gas Turbine and Transmissions Research Centre, University of Nottingham, Nottingham NG7 2TU, UK
e-mail: kuldeep.singh@nottingham.ac.uk

J. Banerjee et al. (eds.), *Recent Advances in Fluid Dynamics*, Lecture Notes in Mechanical Engineering, https://doi.org/10.1007/978-981-19-3379-0_3

Subscript

c Coolant
ms Mainstream
p Pressure side
s Suction side

1 Introduction

Turbine entry temperature (TET) of the modern gas turbine engine has increased drastically to improve the power output and thermal efficiency of the GT [6]. The structural rigidity of the components in the pathway of hot gases are compromised. Despite of the advancement in the materials till date, no material can sustain such a high temperature without adequate cooling [13]. Film cooling is a common technique to cool turbine blades and key components in the pathways of hot gases. In this technique, coolant is injected from the cooling holes made on the surface which is to be cooled. The injected coolant makes a film onto the test plate and avoid hot gases directly touching the metallic surface. Film cooling has been investigated since last seven decades by conducting experimental and numerical study. Review of some of the relevant studies to the current work is presented here. Hylton et al. [5] experimentally studied film cooling of NASA C3X vane with and without coolant injection at the leading edge. They found the influence of parameters such as Re No., Mach No., wall to gas and coolant to gas temperature ratio, turbulence, coolant to gas pressure ratio of coolant to gas in actual engine conditions. Considerable cooling benefits were achieved because of the coolant injection. Narzary et al. [8] experimentally investigated the film cooling on GT blade using pressure sensitive paint technology. They studied blowing ratio (1.2, 1.7 and 2.2) on suction side and blowing ratio (1.1, 1.4 and 1.8) on pressure side, average coolant density (1.0, 1.5 and 2.5), turbulence intensity maintained was (4.2 and 10.5%). Three foreign gases (oxygen, Argon, carbon dioxide) were considered for coolant density study. Results suggests that PSP is very powerful technique for clear film effectiveness contours on pressure side higher effectiveness was observed as compared to suction side with increment in the blowing ratio. Camci and Arts [3] experimentally analysed the film cooling on a GT blade. Film injection was done on suction side, pressure side and leading edge side. The effect of mass weight ratio, turbulence intensity and coolant to hot stream temperature ratio was studied. According to this study, the coolant to free stream temperature ratio has a considerable effect on the convection heat transfer coefficient.

Drost-Bolcs [4] experimentally studied the film cooling on NGV airfoil using the transient liquid crystal technique on pressure and suction side. They investigated Reynolds no (0.52e + 06, 1.02e + 06, 1.45e + 06), Exit Mach no (0.33, 0.62, 0.8), density ratio (1.65), blowing ratio of 0.25–2.3 on suction side, 0.55–7.3 on pressure side. Single compound angle row was studied in pressure side, single and double both were studied on the suction side. Higher density coolant is more effective on suction and pressure side. In comparison to single row and double row, double row provides better effectiveness on suction side. Nathan et al. [9] experimentally studied the showerhead side effectiveness of the blade and used IR camera to measure the temperature of the blade. This study compared the adiabatic and overall effectiveness of the showerhead region. In this study two model are studied one with C3X blade without internal jet impingement and other with jet impingement. Adiabatic film cooling effectiveness enhance up to a momentum flux of 6.7. After momentum flux of 0.76 film starts to detach from the surface. Overall film effectiveness was higher when compared to adiabatic film effectiveness due to the additional jet impingement.

Baghel et al. [2] numerically studied the film cooling on airfoil using standard $k - \varepsilon$ turbulence model was used on suction and pressure side with single hole on each side. Pressure ratio studied was 1.1–1.2 and density ratio maintained was 2.0. As pressure ratio was increased at constant blowing rate film effectiveness increased on both sides of the blade, although more spreading was observed on the pressure side. Ke and Wang [7] numerically investigated the pulsed film cooling on NASA C3X vane using SST $k - \omega$ turbulence model on suction, pressure and leading edge portion. The effect of blowing ratio and Strouhal number was studied. At the leading edge effectiveness increases as blowing ratio or Strouhal no increases whereas on the suction side reverse trend were reported. Shetty et al. [11] numerically investigated the film cooling on GT blade by using RANS turbulence model $k - \varepsilon$. The effect of blowing ratio, forward and backward injection orientation of cooling hole was investigated. The centreline film cooling effectiveness decrease on both pressure and suction side due to backward injection but the lateral spreading of the coolant increases. It was also noted that the film effectiveness on suction side increases with higher blowing ratio but decreases for the forward injection. Singh et al. [14] conducted experimental and numerical study on the film cooling of a flat plate with reverse and forward injection film effectiveness and coverage of coolant increases. The problem of kidney-vortex formation was mitigated with the reverse injection. These vortices are mainly responsible for enhancing mixing of the injected fluid with the hot stream fluid. However, film cooling studies on the turbine vane with the reverse injection is not available in the literature. Hence, in the present study influence of reverse and forward injection holes on the film cooling is presented for the operating conditions of a GT.

2 Problem Description

In the current study, film cooling of NASA C3X vane is studied. The geometric parameters of the investigated vane are documented in a report by Hylton et al. [5]. This vane is exposed to a temperature of 1500 °C. Air is used as a coolant to keep the temperature of the vane within the safe limit. Cooling air is injected at a temperature of 500 °C through two rows of film cooling holes arranged inline on the suction and

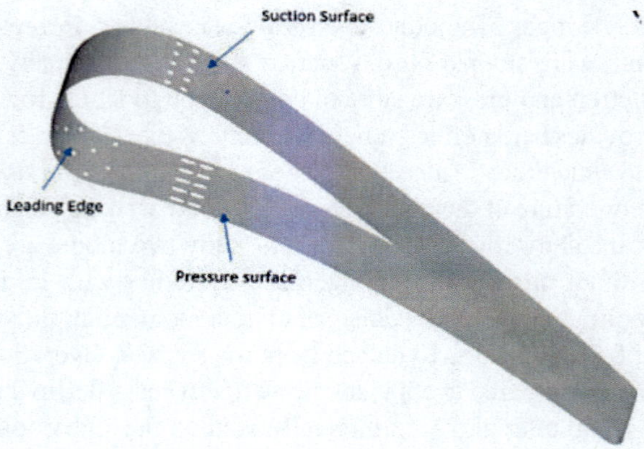

Fig. 1 NASA C3X vane with cooling holes on various portion

Table 1 Investigated cooling hole configurations

Sr. No.	Angle of injection	Hole orientation
1	$\alpha_p = 30°, \alpha_s = 35°$	Forward hole
2	$\alpha_p = 30°, \alpha_s = 35°$	Reverse hole
3	$\alpha_p = 45°, \alpha_s = 50°$	Forward hole
4	$\alpha_p = 45°, \alpha_s = 50°$	Reverse hole
5	$\alpha_p = 60°, \alpha_s = 75°$	Forward hole
6	$\alpha_p = 60°, \alpha_s = 75°$	Reverse hole

pressure portion. The arrangement of film holes on the leading edge is kept staggered following Hylton et al. [5], as shown in Fig. 1. Cooling air is injected at a pressure ratio of 1.15 times to the mainstream pressure which is maintained at 285,130 Pa. The investigated cases are listed in Table 1.

3 Mathematical Modelling

The conservation equations such as: mass, momentum and energy are solved with time averaged for the computational domains in three-dimensional assuming that the flow is incompressible steady, and statistically turbulent. The conservation equations are given as:

$$\frac{\partial(\rho u_i)}{\partial x_i} = 0 \tag{1}$$

$$\frac{\partial(\rho u_j u_i)}{\partial x_j} = -\frac{\partial P}{\partial x_j} + \frac{\partial}{\partial x_j}\left[\mu\left(\frac{\partial u_i}{\partial x_j} + \frac{\partial u_j}{\partial x_i}\right) - \rho\overline{u_i' u_j'}\right] \tag{2}$$

$$\frac{\partial(\rho u_j T)}{\partial x_j} = \frac{\partial}{\partial x_j}\left[\frac{\mu}{\Pr}\left(\frac{\partial T}{\partial x_j}\right) - \rho\overline{T'u'_j}\right] \tag{3}$$

where P, T, u are represent of pressure, temperature and the mean velocity components whilst u' and T' nomenclature are represent of the fluctuating components of velocity and temperature. The term $-\rho\overline{u'_i u'_j}$ is the Reynolds stress and the term of $C_p\overline{T'u'_j}$ represents the specific turbulent heat fluxes. To solve Eqs. (2–3), $-\rho\overline{u'_i u'_j}$ and $C_p\overline{T'u'_j}$ terms have to be modelled. The Boussinesq hypothesis is considered to obtained Reynolds stresses, as given in Eq. (4).

$$-\rho\overline{u'_i u'_j} = \mu_t\left(\frac{\partial u_i}{\partial x_j} + \frac{\partial u_j}{\partial x_i}\right) - \frac{2}{3}\rho k \delta_{ij} \tag{4}$$

where μ_t and k are represent turbulent viscosity and turbulent kinetic energy. Turbulent heat flux is solved by simple eddy diffusivity which is described in Eq. (5).

$$\rho\overline{T'u'_j} = -\frac{\mu_t}{\Pr_t}\frac{\partial T}{\partial x_j} \tag{5}$$

where $\Pr_t$ is turbulent Prandtl number.

In order to solve Eqs. (2)–(3), Boussinesq hypothesis Eq. (4) and simple eddy diffusivity model Eq. (5) are considered. To obtain the turbulent quantities and turbulent viscosity, such as $k - \varepsilon$ are considered to obtain by length scale and velocity. $\sqrt{k}$ is considered as velocity scale in $k - \varepsilon$ models, whilst $\frac{k^{3/2}}{\varepsilon}$ is represent length scales of $k - \varepsilon$ models in the mixing length hypothesis to calculate the eddy viscosity. The Realisable $k - \varepsilon$ turbulent model is considered in the present study work.

3.1 Boundary Conditions

The numerical domain investigated is shown in Fig. 2. Temperature and pressure at the inlet of the computational domain is specified whereas pressure outlet is specified at the outlet of the domain. Temperature and pressure condition is given at each plenum. Front and back side of the computational domain are specified as a periodic boundary and all other surfaces are specified as non-slip wall boundaries.

3.2 Fluid and Thermal Properties

In both primary and secondary stream, Air is taken as working fluid. As the temperature difference is very large so the variation of fluid properties with temperature is

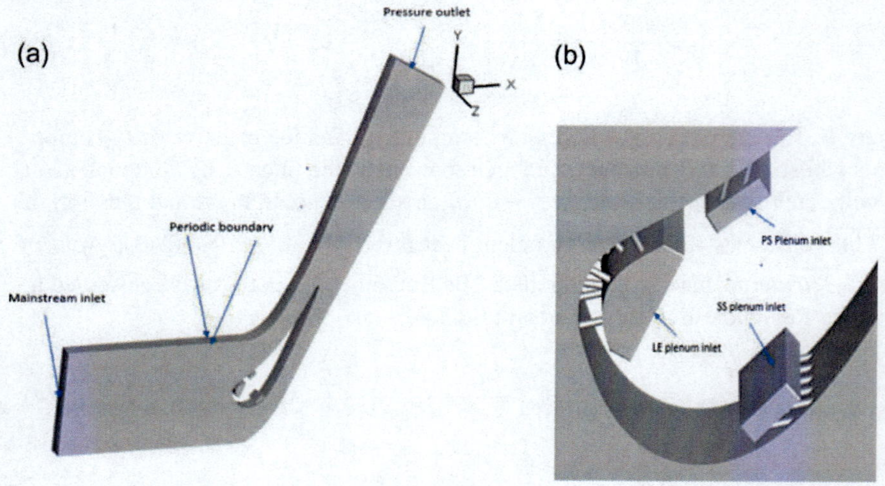

Fig. 2 **a** Computational domain **b** enlarged view showing cooling holes with plenum arrangement

considered. Air is assumed to be an ideal gas for the current analysis. Fourth order polynomials variations of C_p, k and μ of fluid and thermal properties was considered by Singh et al. [12]. These equations are valid in the range of 100–2300 K. Present study involves the temperature variation between 773 and 1773 K and polynomial equations proposed by Singh et al. [12] are applicable to the current study as well.

3.3 Solution Procedure

The current simulation is carried out in ANSYS FLUENT [1] to solve the energy, momentum and continuity equation. The solution is considered as converged when residuals of fluid flow equation fall below 10^{-5} and that for energy below 10^{-8}. Simple scheme is considered for velocity coupling and pressure.

3.4 Grid Sensitivity Study

For the present study structured hexahedral mesh considered for simulations. 1 ANSYS ICEM-CFD 18.1 software package is used for generating grids. The grid independence test is done at the pressure ratio of 1.15 and temperature ratio of 3 for the forward holes. Grids near the wall are placed such that the wall $y+$ is less than 1.

A typical mesh consider for the present analysis is shown in Fig. 3. The grid sensitivity study is carried out for optimal grid through varying mesh size from 2 million cells to 4.5million cells. The results obtained from these grids are shown in Fig. 4.

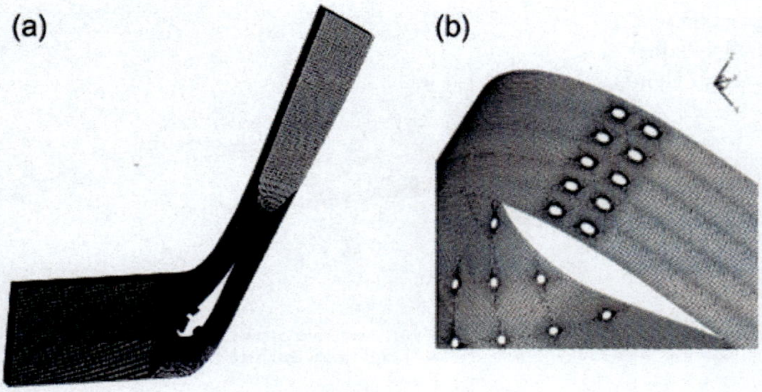

Fig. 3 **a** Typical mesh of whole domain **b** mesh of the blade surface showing the filholes

Fig. 4 Grid sensitivity study

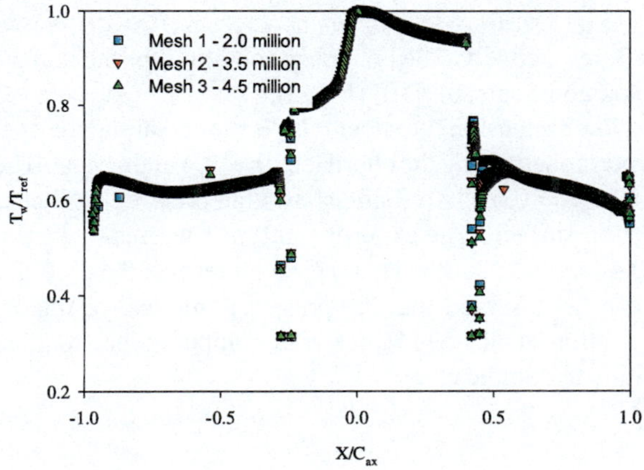

Non-dimensional temperature is plotted along the span of the vane which is non-dimensionalised by the cord length of the vane. The non-dimensional temperature obtained from mesh with 2 million cells slightly deviates from the other two meshes. However, the results obtained from the mesh with cell size 3.5 and 4.5 million are identical and hence mesh with 3.5 million cells was selected for further studies.

3.5 *Validation*

The experimental results of Hylton et al. [5] for the test case of mesh 3 of 4.5 million is used for validation of numerical model. The pressure and temperature of the mainstream were 285,130 Pa and 701 K corresponding to the experimental conditions. Pressure ratio used for the plenum inlet on suction side, pressure side, leading edge side is 1.051, 1.048 and 1.05. The temperature ratio is 0.85, 0.86 and

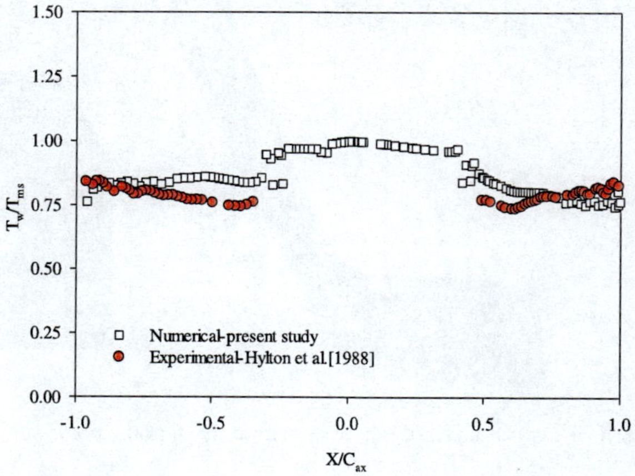

Fig. 5 Numerical present results compared with the experimental data of Hylton et al. [5]

0.83, respectively. The boundary condition for the radial channels are taken from the Prapamonthon et al. [10] (Fig. 5).

Non-dimensional temperature is plotted along the span of the vane which is non-dimensionalised by the chord length (C_{ax}) of the vane. The stagnation point is marked at $X/C_{ax} = 0$, negative side shows the pressure surface and positive side shows the suction surface. The experimental measurements were not available for the region $-0.45 < X/C_{ax} < 0.5$. This region represents the part of vane near the leading edge. It can be observed that the present numerical results are in good agreement with of Hylton et al. [5]. Hence, the computational methodology can be utilised for parametric studies.

4 Results and Discussion

4.1 Effect of Injection Angle on the Pressure Side

In current study, on pressure side injection angle of film hole is varied from 30° to 60°. The variation of lateral local effectiveness (η) on the pressure side of the vane is presented in Fig. 6 for forward and reverse injection.

The lateral local effectiveness is plotted at a downstream distance $X/C_{ax} = 0.31$. The value of pressure ratio and temperature ratio is kept 1.15 and 3. It is seen from Fig. 6 that the lateral local effectiveness with the reverse injection is higher in all the injection angle studied. The flat profile of lateral local effectiveness shows the uniform spreading of the cooling air in the lateral direction for the reverse injection. On the contrary of this, a wavy film profile for the forward injection indicates that the cooling air is skewed towards the centreline of the cooling hole. It is also observed that the film cooling effectiveness is better for the injection angle $\alpha_P = 30°$. At the higher injection angle, cooling air penetrates in the mainstream that promotes mixing

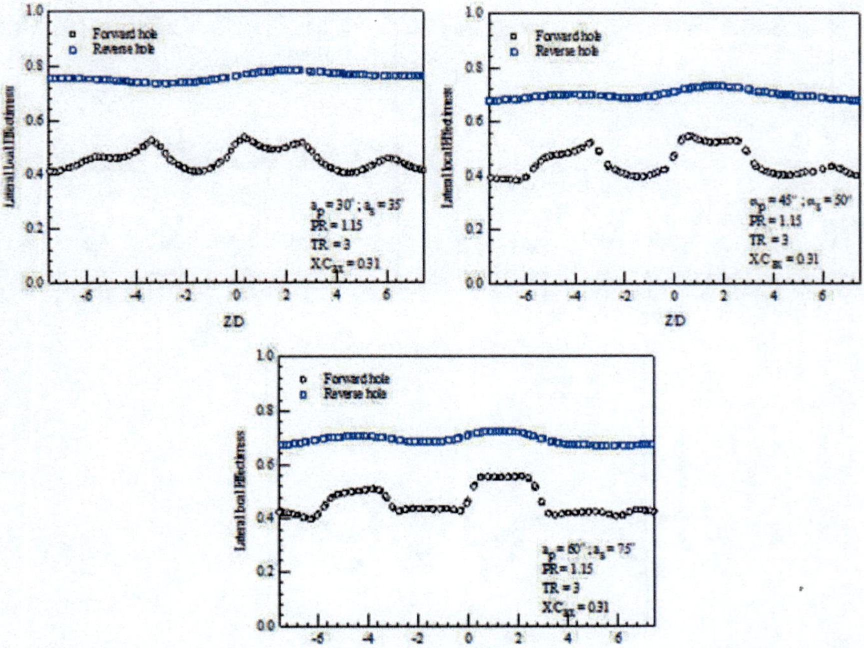

Fig. 6 Variation of lateral average effectiveness for forward and reverse injection on the pressure side of vane

of cooling air with the hot mainstream. Consequently, effectiveness on the surface of the vane decreases with the increasing injection angle.

4.2 Effect of Injection Angle on the Suction Portion

In the present study, inclination angle of the film hole is studied from 35° to 75° on the suction portion of the vane. This should be noted that the injection angle on the pressure side and suction side are different. The curvatures of vane on the pressure side and suction side are different. If the inclination of cooling holes is kept same for pressure and suction side then either suction side will have shallow injection or pressure side will have higher injection angle. It is shown by Fig. 6 that the higher injection angles diminish film cooling. So does the shallow injections because of the friction effects [12]. Hence, in the present study hole inclination angle are kept different for pressure and suction side. The lateral local effectiveness is plotted in Fig. 7 at a downstream distance $X/C_{ax} = 0.48$ in both the forward and reverse injections. The investigated hole inclinations angles are $\alpha_S = 35°$, 50° and 75°. It is observed in Fig. 7 that the lateral local effectiveness for the reverse injection is better for all the studied cooling hole inclination angles as compared to the forward hole. This trend is identical to the pressure side cooling of the vane.

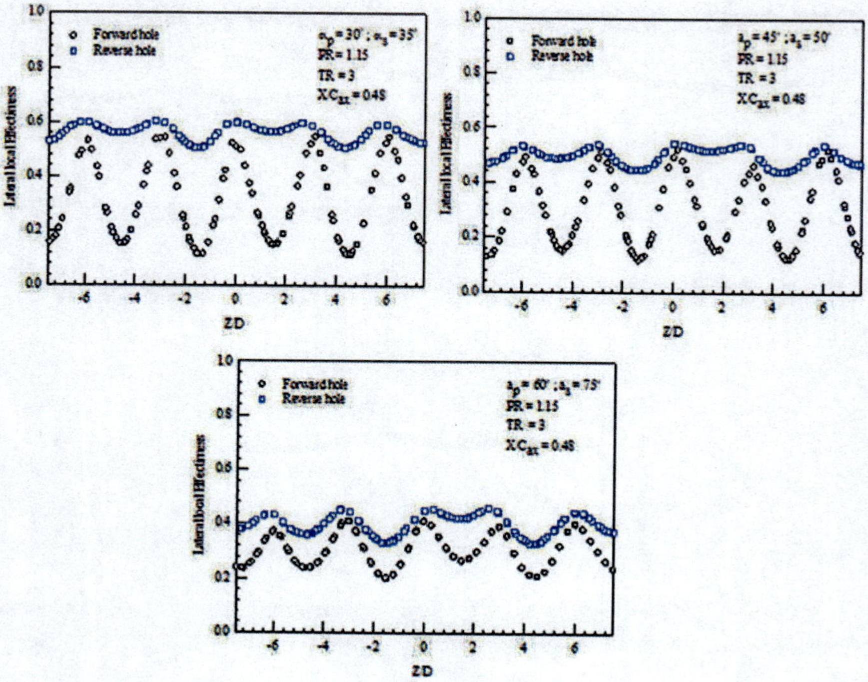

Fig. 7 Lateral average effectiveness variation for forward and reverse injection on the pressure side of vane

Moreover, the profile of lateral local effectiveness exhibits some wavy characteristics even for the reverse injection. The waviness in the profile of forward injection has also been magnified on the suction surface. The interaction of the hotstream air on the pressure and suction side is different which is responsible for the wavy film effectiveness profile. It can also be noticed that the magnitude of the lateral local effectiveness is higher for the lowest injection angle, i.e. $\alpha_S = 35°$. The temperature contours are shown on surface of the vane in Fig. 8. These contours confirm the skewed spread of the cooling air from the forward injection whereas almost uniform spread of the cooling air is observed on the vane surface with the reverse injection.

Fig. 8 Temperature contours on the surface of vane with the **a** forward and **b** rever injection

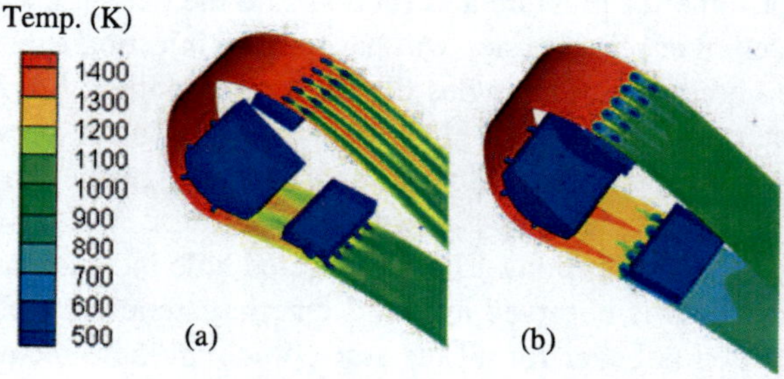

It can also be observed from Fig. 8 that the leading edge area without the cooling holes is exposed directly to high temperature and film cooled region are maintained at the lower temperature. Thus, film cooling is an effective technique to protect the surfaces exposed to very high temperature and reverse injection gives better cooling as compared to the forward injection.

5 Conclusions

In this numerical study, film cooling on a turbine vane (NASA-C3X) is investigated numerically. coolant is injected in both the forward and the reverse direction to the mainstream flow. The effect of inclination angle on the effectiveness is investigated for both reverse and forward injection by varying inclination angle from 30° to 60° on pressure and from 35° to 75° for the suction side of the vane. Film cooling effectiveness downstream of the cooling holes are compared for both the reverse and forward injection at the investigated cooling hole inclination angle. Based on the presented results, it can be concluded that the reverse injection angle provides better cooling as compared to the forward injection. The inclination angle of the film cooling hole also plays a significant role. Higher injection angle diminishes cooling effectiveness.

References

1. ANSYS Academic Research (2018) ANSYS fluent theory guide. ANSYS Help Syst
2. Baghel V, Chandel S, Reddy RS (2013) Effect of pressure ratio on film cooling of turbine aerofoil using CFD. Univ J Mech Eng 1(4):122–127. https://doi.org/10.13189/ujme.2013.010403
3. Camci C, Arts T (1985) Experimental heat transfer investigation around the film-cooled leading edge of a high-pressure gas turbine rotor blade. J Eng Gas Turbines Power 107(4):1016–1021. https://doi.org/10.1115/1.3239805
4. Drost U, Bolcs A (1998) Investigation of detailed film cooling effectiveness and heat transfer distributions on a gas turbine airfoil. Am Soc Mech Eng (Paper), (GT)
5. Hylton LD, et al (1988) The effects of leading edge and downstream turbine vane heat transfer. NASA Cr-182133 1–172
6. Ito S, Goldstein RJ, Eckert ERG (1978) Film cooling of a gas turbine blade. J Eng Gas Turbines Power 100(3):476–481. https://doi.org/10.1115/1.3446382
7. Ke Z, Wang J (2015) Numerical investigations of pulsed film cooling on an entire turbine vane. Appl Therm Eng. Elsevier Ltd. 87:117–126. doi:https://doi.org/10.1016/j.applthermaleng.2015.05.022
8. Narzary DP, et al (2011) Influence of coolant density on turbine blade film-cooling using pressure sensitive paint technique. J Turbomach 134(3). doi: https://doi.org/10.1115/1.4003025
9. Nathan ML et al (2013) Adiabatic and overall effectiveness for the showerhead film cooling of a turbine vane. J Turbomach 136(3):1–9. https://doi.org/10.1115/1.4024680
10. Prapamonthon P, et al (2018) Investigation of cooling performances of a non-film-cooled turbine vane coated with a thermal barrier coating using conjugate heat transfer. Energies 11(4). doi:https://doi.org/10.3390/en11041000

11. Shetty S, Li X, Subbuswamy G (2017) HT2012-58469 numerical simulation on gas turbine film cooling of curved surface, pp 1–8
12. Singh K, Premachandran B, Ravi MR (2015) A numerical study on the 2D film cooling of a flat surface. Num Heat Transf Part A Appl 67(6):673–695. doi:https://doi.org/10.1080/10407782.2014.949131
13. Singh K, Premachandran B, Ravi MR (2016) Experimental and numerical studies on film cooling of a corrugated surface. Appl Thermal Eng Elsevier Ltd 108(July):312–329. https://doi.org/10.1016/j.applthermaleng.2016.07.093
14. Singh K, Premachandran B, Ravi MR (2017) Experimental and numerical studies on film cooling with reverse/backward coolant injection. Int J Thermal Sci Elsevier Masson SAS 111:390–408. https://doi.org/10.1016/j.ijthermalsci.2016.09.027

Modelling and Analysis of AC System in Bus Considering Indian Drive Condition

Saurabh, Swastik Acharya, and Avanish Kumar Dubey

Nomenclature

A	Apparent direct normal solar flux at the outer edge of the earth's atmosphere
H	Passenger height
Pa	Air pressure
Ps	Saturation pressure at temperature T
W	Passenger weight
X	Humidity ratio

Greek Symbols

Φ	Relative humidity
θ	Angle between surface normal and position of the sun in sky

1 Introduction

As technology and society evolve, people's living standards improve, and higher comfort conditions are required. Therefore, the air conditioning is expected to be given so much weightage, mainly in residential buildings, offices, cars, buses, trucks, trains, etc. Hegar et al. [1] described that the fuel consumption increases due to air conditioning units installed inside the bus and determined it at various vehicle speed

Saurabh · S. Acharya (✉) · A. K. Dubey
Motilal Nehru National Institute of Technology, Allahabad, India
e-mail: swastik@mnnit.ac.in; swastik.acharya8@gmail.com

J. Banerjee et al. (eds.), *Recent Advances in Fluid Dynamics*, Lecture Notes in Mechanical Engineering, https://doi.org/10.1007/978-981-19-3379-0_4

rates. Suh et al. [2] developed an energy management control system to achieve high efficiency. They integrated the heat pump with the air conditioning system and recovered waste heat from the drive motor. They also concluded that the energy consumption for heating mode is less than 20%, and the cooling mode is less than 25% of the total power consumption. Eda et al. [3] converted a conventional bus into an electric bus (Waseda electric buses) and tested it on public roads to determine the effect of environmental conditions on performance and energy cost. He concluded that the CO_2 emission is reduced 28–42% and the energy cost to operate the bus reduces from 57 to 64%. He modified the heating system and found that energy consumption is further reduced by 11% and observe that if vehicle weight reduces, the power consumption is also reduced.

From the above literature survey, it is concluded that most of the studies yet have to consider the effect of the fluctuation of the vehicle speed on the calculation of heating load. As the speed of the bus would be higher, the heating load becomes less, and simultaneously, the power consumption for the compressor becomes less. Moreover, the AC system analysis in the Indian environmental condition has not been studied yet, which needs to be explored. Thus, in the present work, we focus on the analysis of the AC system in an electric bus considering the Indian environmental and drive condition. The input heating loads such as solar load, ventilation load, and ambient load have been calculated depending on various environmental conditions and the speed of the vehicle at three different places. Finally, the load on the compressor of the AC system has been evaluated considering the input heating load inside the cabin. By knowing the fluctuation of the heating load and required power consumption during a day, the amount of fuel supply to run the compressor could be optimized to enhance the vehicle's driving range.

2 Problem Description

Figure 1 shows the graphical representation of a bus [4] from the side and top view. The main surfaces like the glass windows and doors which have high transmissivity to the sun rays are considered for the analysis.

The heating loads in the present study are metabolic heat load, solar heat load (direct, diffused, and reflected), ventilation heat load, ambient heat load, and blower heat load. The metabolic heat load mainly depends on the metabolism process inside the human body which is kept constant in the present study. The direct solar heat load mainly depends on the apparent normal solar heat flux (A) (W/m^2), altitude angle (β), and the apparent atmospheric extinction coefficient (B). The altitude angle (β) depends on the position and time of the particular location. The apparent direct solar flux (A) and the apparent atmospheric extinction coefficient (B) have been plotted against several days for a year in the modelling section. The diffuse heat load is a function of the vehicle's surface area. The solar heat diffuses through the exposed surface into the bus. The reflected heat load is the solar radiation emitted from ground and diffuses into the vehicle body. The ventilation heat load depends on the

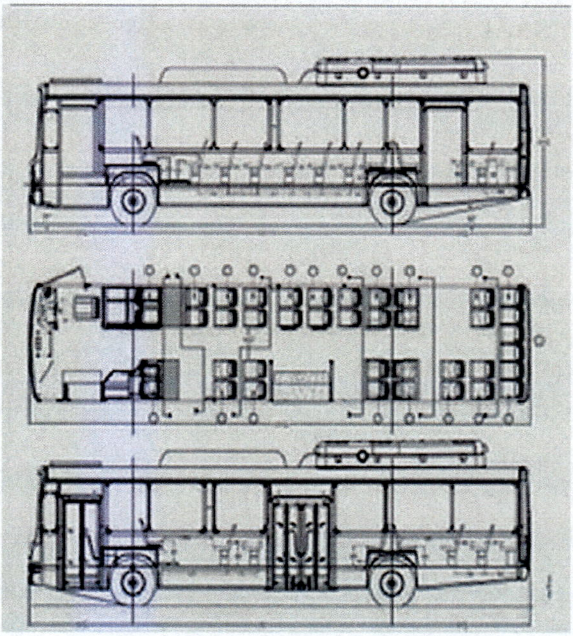

Fig. 1 Graphical representation of a bus

temperature of the fresh ambient air coming into the cabin of the bus for ventilation purposes. The more the ambient temperature of air more is the ventilation heat load. The ambient heat load is generated due to the higher temperature of the ambient air than that of the cabin temperature. Since we are considering situation with the initial cabin temperature more than the ambient temperature, the ventilation load would be coming out negative as heat is losing from the cabin. Generally, the ambient heat load is a vital function of the velocity of the vehicle. As the vehicle's speed increases, the heat transfer coefficient increases, and the heating load on the vehicle decreases. Therefore, the total heating load becomes a function of the vehicle speed and the ambient conditions in different cities (Indore, Lucknow, and Kolkata). The dimensions of the glass windows, metal roof, floor, and other bus elements could have been found in specification of Tata Starbus Ultra Electric 9/12 eV Bus [4]. The vehicle speed mainly depends on the road condition, traffic, and other road parameters. Therefore, the drive cycle data (fluctuation of the vehicle speed with respect to time (second)) have been recorded for various regions. The fluctuation of the total heating load and power consumption to compressor has been evaluated against the drive cycle data. The location, altitude, year, date, time, and ambient temperature considered in the present work have been shown in Table 1. The properties of glass and metal surface could be found in [5].

Table 1 Latitude, month, date, temperature, and time considered for Indore location

Location	Latitude	Month	Date of year	Comfort temperature	Time
Indore	22.7196°N	June	166 (25th June 2019)	23 °C	14:00
Kolkata	22.5726°N	July	202 (30th July 2019)	23 °C	14:00
Lucknow	26.8467°N	July	202 (30th July 2019)	23 °C	14:00

3 Mathematical Modelling

Various heat loads on the bus have been modelled by adopting the heat balance method over the control surface around the bus, which are presented in a systematic manner in this section. By using those mathematical equations, we can perform an energy analysis for both the cabin and component of the AC system. Here, the objective is to find the total heating load on the AC system and to evaluate the performance of each of its components with the input parameters.

3.1 Cabin Model

The power consumption of air conditioning system depends upon the amount of total heat load. Total heat load is obtained from cabin model data. For this, we have to focus on the sources of heat load inside the vehicle. Various sources cause an increase or decrease in heat load. Sources for the total heat load inside bus cabin are

Metabolic heat load: Due to human movement, metabolic heat is generated inside the body. It is transferred to the cabin through respiration and is evaluated by [5],

$$Q_{\text{meta}} = \sum_{\text{driver}} (Md * A_{\text{Du}}) + \sum_{\text{passenger}} (Mp * A_{\text{Du}}) \tag{1}$$

Dubois area (A_{Du}) [6] is the approximate surface area of a human body calculated by

$$A_{\text{Du}} = 0.202\, W^{0.425} H^{0.725} \tag{2}$$

Solar heat load: Solar heat load increases the thermal energy of space, structure, or object due to incident solar radiation. Different types of solar load are

(a) *Direct heat load*: Solar load acting on the vehicle directly. This load [7] is calculated by

$$Q_{\text{direct}} = \sum_{\text{surface}} (I_{\text{direct}} * S * \tau * \cos\theta) \tag{3}$$

τ is transmissivity of glass. Direct heat gain is the function of altitude angle which is based on the position and time of a particular location

$$I_{\text{direct}} = \frac{A}{\exp\left(\frac{B}{\sin \beta}\right)} \tag{4}$$

β is the angle of altitude that depends on latitude and the current time of driving location (Table 1).

The values of A, B, and C are calculated as the cosine function of day number (n) to be valid for all the year days [6, 8]:

$$A = 1162.12 + 77.0323 * \cos\left(n * \frac{360}{365}\right) \tag{5}$$

$$B = 0.171076\text{–}0.0348944 * \cos\left(n * \frac{360}{365}\right) \tag{6}$$

$$C = 0.0897334\text{–}0.0412439 * \cos\left(n * \frac{360}{365}\right) \tag{7}$$

where n = day number of the year starting from 1st January.s

The term $\sin \beta$ in Eq. 4 could be evaluated [6, 8] as

$$\sin \beta = \sin \Phi . \sin \delta + \cos \Phi . \cos \delta . \cos \omega \tag{8}$$

where Φ, δ, and ω represent the latitude, declination, and the hour angle. Declination angle (δ) changes with the tilting of the Earth on its rotational axis over year around the sun.

$$\delta = 23.45 \sin\left(360\left(\frac{284 + n}{365}\right)\right) \tag{9}$$

Hour angle (ω) is the angle between the observer's celestial meridian as well as the celestial object's hour circle, measured west from the meridian. It is evaluated by

$$\omega = 15 * (t - 12) \tag{10}$$

where t = time in an hour. It varies from $-180°$ to $+$ss$180°$.

(b) *Diffused heat load*: Heat load due to indirect radiation of sunlight on any surface or vehicle surface. Diffused heat load [6] is the function of the surface area of the vehicle, and its surface element transmissivity is evaluated by

$$Q_{\text{diffused}} = \sum_{\text{surface}} (I_{\text{diffused}} * S * \tau) \tag{11}$$

where Q_{diffused} is a diffused heat load measured in term of (watts) W, I_{diffused} is a diffused heat gain per unit area, W/m^2, S is a surface area in term of (metre square) m^2 and τ is a surface element transmissivity.

$$I_{\text{diffused}} = C * I_{\text{direct}} * \left(\frac{1+\cos\sum}{2}\right) \tag{12}$$

where C is a dimensionless constant and $\sum$ represents the angle of inclination of the metal or glass surface with horizontal.

(c) *Reflection heat load*: Reflected radiation load generated by solar radiation impacts the vehicle's body surfaces that reflected from ground. It is evaluated by

$$Q_{\text{reflected}} = \sum_{\text{surface}} (I_{\text{reflected}} * S * \tau) \tag{13}$$

where $Q_{\text{reflected}}$ is a reflected heat load which have unit (watts) W, $I_{\text{reflected}}$ is a reflected heat gain per unit area having unit of (watt per metre square) W/m^2, S is a surface area having unit of m^2.

$I_{\text{reflected}}$ [6] is evaluated by

$$I_{\text{reflected}} = (I_{\text{direct}} + I_{\text{diffused}}) * \rho * \left(\frac{1-\cos\sum}{2}\right) \tag{14}$$

where ρ is a ground reflectivity coefficient.

Ventilation heat load: To maintain the air quality, some external air enters the cabin. The ventilation heat gains due to entering of the external air [7] can be evaluated as

$$Q_{ven} = m_{ven} * (eo - ei) \tag{15}$$

where Q_{ven} is a ventilation heat load, m_{ven} is the mass flow rate of external air, eo and ei represent the enthalpy of the ambient air and cabin, respectively, which are calculated as

$$e = 1006\,T + (2.5 \times 10^6 + 1770\,T) * X \tag{16}$$

where T denotes as air temperature in degree Celsius (°C) and X is a humidity ratio (gm of water/gm of dry air).

Humidity ratio [6] is defined as a function of relative humidity

$$(\Phi),\ X = 0.62198\left(\frac{\Phi Ps}{Pa - \Phi Ps}\right) \tag{17}$$

Ambient heat load: Ambient heat load Q_{amb} [7] is generated because of the difference between the ambient air and cabin air.

$$Q_{\text{amb}} = \sum_{\text{Surface}} SU(T_{\text{amb}} - T_i) \tag{18}$$

where U (W/m^{2}K) is the overall heat transfer coefficient. T_{amb} is the ambient temperature, T_i is the average cabin temperature (K), and R is the thermal resistance. The overall heat transfer coefficient is calculated by

$$U = \frac{1}{R}; \quad R = \frac{1}{ho} + \frac{\lambda}{k} + \frac{1}{hi} \tag{19}$$

ho and hi (W/m^{2}K) are the outside and inside convection coefficient, respectively, λ (m) is the thickness of surface element, and k (W/m K) is the thermal conductivity.

Convection heat transfer coefficient [5] as a function of vehicle speed (v) and is calculated by

$$h = 0.6 + 6.64\sqrt{v} \tag{20}$$

This load is coming negative when the AC system is on and becomes positive when AC system runs for minimum 10 min to bring the cabin temperature down the ambient temperature. The same thing happens with the ventilation load.

Blower heat load: Blower heat load is the function of blower wattage and load factor. It is inversely proportional to efficiency. As the efficiency increases, the heat generated by the blower motor decreases.

$$Q_{\text{blower}} = \frac{\text{Blower wattage} * \text{ Load factor}}{\text{Efficiency}} \tag{21}$$

3.2 Component Model

The concerned parts are placed outside the cabin or inside the air conditioning unit in the component model. These parts include an evaporator, blower subsystem, compressor, condenser, and condenser fan unit.

An evaporator is a type of heat exchanger in which heat transfer occurs between the refrigerant and the cabin air. As per the energy balance equation at the evaporator

$$Q = \dot{m}_{\text{cabin}} * Cp_{\text{a}} * \Delta T_{\text{cabin}} = \dot{m}_{\text{ref}} * Cp_{\text{ref}} * \Delta T_{\text{ref}} \tag{22}$$

where

Q = total heat load, Cp_a = specific heat of air, Cp_{ref} = specific heat of refrigerant, $\dot{m}_{cabin}$ = mass flow rate of cabin air, $\dot{m}_{ref}$ = mass flow rate of refrigerant.

An electric blower is to extract the hot air from the cabin and send it back to the cabin after being cooled. In the air conditioning system, the number of blowers depends on the size of the electric vehicle and the amount of heat load needed to be extracted. The amount of power required for each blower depends on the amount of mass flow rate of air handled by the same blower. This could be found out from the blower performance as specified by the manufacturer. The total mass flow rate through the blowers could be computed as

$$\dot{m}_{blowertotal} = \frac{Q}{Cp_{air}(T_{cabin} - T_{ambient})} \tag{23}$$

Therefore, for each blower, the mass flow rate is $\frac{\dot{m}_{blowertotal}}{n}$, where n = number.

Work done by condenser fan = amount of heat released from the coolant (Q_{cond}) = Total heat load + Work done by compressor

$$Q_{cond} = Q_{total} + W_{comp} \tag{24}$$

$$\dot{m}_{cond} = \frac{Q_{cond}}{Cp_a * \Delta T} \tag{25}$$

By knowing the mass flow rate of air by the condenser fan, the electric current could be obtained from the fan's performance as specified by the manufacturer.

4 Validation

Since it is challenging to solve all the mathematical equations using pen and paper with greater accuracy, we solved them using MATLAB Simulink software. Such software helps in carrying out complex mathematical computing with negligible error. There are two different types of models. The first is the cabin model based on passenger, temperature, humidity, number of times door opening-closing, etc. The second one is the component model, which contains the evaporator, blower, condenser, and compressor. The cabin model is integrated with the component model to study the performance of each component of AC system with respect to the input heat load depending on various input parameters. Since the validation of the implementation of the mathematical model is very much needed, we compared the result of Suh et al. [2] with the present simulation. Suh et al. have considered the weather condition of Korea for their analysis. They plotted the compressor work and coefficient of performance (COP) as a function of outdoor temperature and condenser flow rate (m^3/min). We have performed the same analysis but in Indian weather conditions and obtained the results which lie in the same order as Suh et al. [2]. Table 2 shows

Table 2 A comparison of the present result with Suh et al. [2]

Q (Suh et al.)	Q (present analysis)	% Error
27 kW	**26.01 kW**	**0.037**
W_{Comp} (Suh et al.)	W_{Comp} (present analysis)	% Error
10 kW	**10.43 kW**	**4.3**
COP (Suh et al.)	COP (present analysis)	% Error
2.9	**3.022**	**4.20**

the comparison of the present results with Suh et al. with an error of less than 5%, which justifies the accuracy of the current work.

5 Results and Discussion

In this section, we intend to discuss the total heat load in a cabin of a bus and the component performance of the AC system against the input parameters. For this purpose, both the cabin and component model are integrated together to evaluate the power consumption, total heat load, COP, and mass flow rate of the blower. The speed at various times has been represented in the drive cycle plotted by taking real-time data in a bus at different India locations. The results are obtained keeping other conditions like ambient temperature, outside humidity, number of passengers as constant.

5.1 Variation of Heating Load, Power Consumption, COP of AC System Over Time (Minimum 6 h Duration) in Indore, Kolkata, and Lucknow Driving Condition

The specification of the driving condition is mentioned in Table – 1. The drive cycle in Indore, Kolkata, and Lucknow has been represented in Fig. 2a, b, and c.

The drive cycle shows the vehicle's dynamic response in terms of the fluctuation of velocity with time during the journey. The speed of the vehicle mainly depends upon the traffic condition, number of stoppages, road condition, etc., of the particular city. The fluctuation of velocity significantly impacts the input heat load, which simultaneously alters the pumping power and COP of the refrigeration system. Hence, the fluctuation of the total heat load, blower mass flow rate, pumping power, and COP of the refrigeration system has been evaluated and produced in terms of the graphs for discussions.

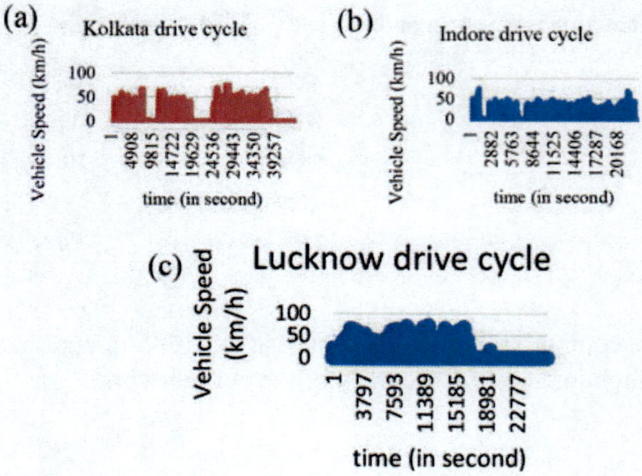

Fig. 2 Drive cycle data—**a** Indore, **b** Kolkata, and **c** Lucknow

5.1.1 Fluctuation of Total Heat Load with Time

The total heating load depends on many factors such as the number of passengers, location of the city, solar radiative flux, and drive cycle. Since we are taking the factors like weather conditions and altitude angle of a particular city as constant, the fluctuation in the total heat load arises due to the fluctuation of the vehicle's velocity. With an increase in vehicle speed, the heat load decreases due to the wind flowing outside the vehicle cools the outer surface by means of the convection process. Thus, a vehicle with a higher speed throws away more heat, reducing the cabin temperature. In Indore, the average velocity of the vehicle is around 30–40 km/h, whereas the average velocity of the vehicle clocks around 50–70 km/h for Kolkata and Lucknow. Figure 3 shows that the total heat load lies in the range of 31.45 kW to 31.85 kW in Indore. Similarly, the total heating load varies from 25.05 to 26 kW in Kolkata and 25.1 to 26.05 kW in Lucknow. The total heat load in Kolkata and Lucknow is marginally less compared to Indore owing to both geographical location and higher bus speed. From Figs. 2 and 3, it can be noticed that the fluctuation in heating load is very less whereas the fluctuation of velocity ranges from 0 to 70 km/h. It concludes

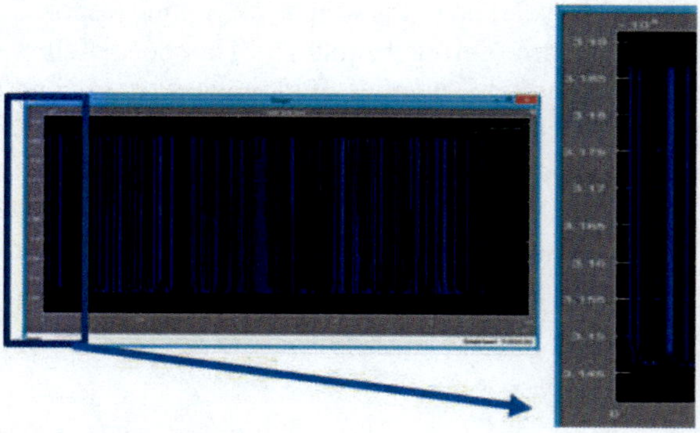

Fig. 3 Fluctuation of heating load as a function of time of the bus for Indore drive conditions

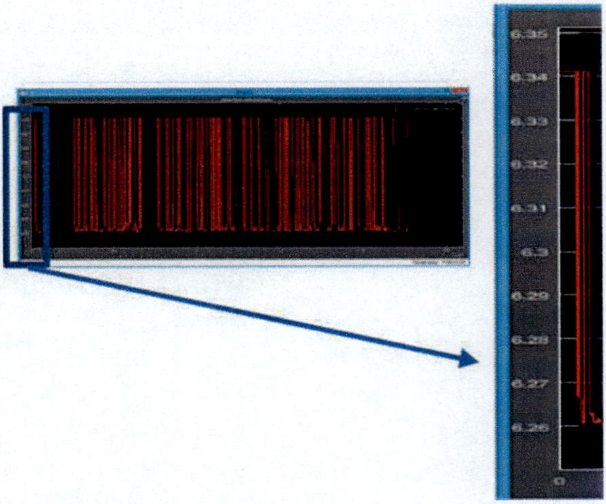

Fig. 4 Fluctuation of mass flow for a bus in Indore drive conditions

that the velocity has a little impact on the heating load when it is compared to the solar load.

5.1.2 Evaluation of Blower Mass Flow Rate

In air conditioning system, the blower is used to force the air in and out of the cabin. With an increase in the heating load inside the cabin, the blower has to do more work, and the mass flow of air has to be increased. The mass flow rate has been evaluated based on the cabin heat load. Figure 4 shows that the fluctuation of the mass flow rate in the blower lies between 6.26 and 6.34 kg/s in Indore drive conditions. At the same time, the blower mass flow rate in Kolkata and Lucknow varies between 4.95 and 5.18 kg/s and 4.98 and 5.18 kg/s.

5.1.3 Fluctuation of Power Consumption

Power consumption of air conditioning system depends upon the total heat load. If the heat load inside the cabin increases, more work is needed to run the compressor and blower. Heating load due to solar radiation is significantly higher than other factors and substantially impacts the power consumption. The speed has a minor effect on evaluating the absolute value of the power consumption. However, the fluctuation in power consumption is seen because of the velocity fluctuation which can be illustrated in Fig. 5. It can be observed that the power consumption fluctuates between 13.7 and 13.95 kW in Indore. For Kolkata and Lucknow drive conditions, it fluctuates between 9.85 and 10.34 kW and 9.86 and 10.45 kW.

Fig. 5 Fluctuation of power consumption for a bus in Indore drive conditions

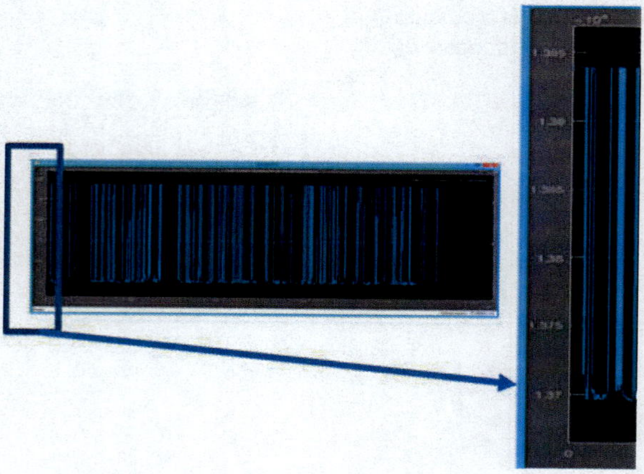

5.1.4 Evaluation of COP of the AC Unit

Coefficient of performance (COP) of the air conditioning unit is defined as the ratio between the amount of heat extracted from the cabin and the amount of work required to be done on the compressor. The more work to heat load ratio, the lesser the COP is. Mainly, it indicates the performance of the compressor unit used in the bus. If it is less, more electricity is required to run the compressor, which discharges the battery faster. In such a situation, a replacement of the AC unit is required. Figure 6 shows the fluctuation of COP during the journey of a bus in Indore drive condition. In such a condition, the COP varies between 2.63 to 2.65. However, the COP varies between 3.02 to 3.118 in Kolkata, 3.02 and 3.118 in Lucknow.

Fig. 6 Fluctuation of COP for a bus in Indore drive conditions

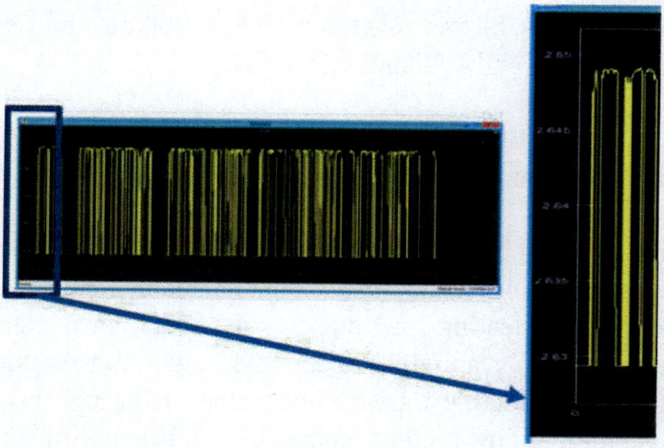

6 Conclusions

For this research work, a 9 m electric bus by TATA Motors is considered. This research work aims to estimate the fluctuation of the thermal load inside the vehicle and calculate the COP of the air conditioning system.

Based on the results, the following conclusions can be drawn:

1. The heat balance method estimates the total heat load generated inside the cabin, including the metabolic, solar, ambient, ventilation, and blower heat load.
2. Simulation has been carried out by keeping the ambient temperature constant and vehicle speed variable. The air conditioning system's heat load for electric bus lies between 25 and 27 kW depending upon four different test conditions, and the total power consumption by the air conditioning system lies between 9.8 and 12 kW based on different test conditions.
3. Engineers can use these simulation results to design a more efficient air conditioning system such that its power consumption by the battery can be minimized and the operating range of the vehicle can be increased.
4. Model predictive control can be applied in the electric bus to estimate the effect of fluctuation in passengers along with a change in ambient conditions.

Acknowledgements The author would like to thank TATA Motors Limited, Pune, India, for providing the technical support required to carry out the experiments/study/research.

References

1. Hegar M, Kolda M, Kopecka M, Rajtmajer V, Ryska A (2013) Bus HVAC energy consumption test method based on HVAC unit behavior. Int J Refrig 36(4):1254–1262
2. Suh I, Lee M, Kim J, Oh S, Won J (2015) Design and experimental analysis of an efficient HVAC (heating, ventilation, air-conditioning) system on an electric bus with dynamic on-road wireless charging. Energy 81:262–273
3. Eda K, Kanaga R, Yang W, Hirota T, Kamiya Y, Daisho Y (2015) Development and performance evaluation of advanced electric bus transportation system. World Electr Veh 7(3):349–356
4. Tata Starbus Ultra Electric 9/12 EV Bus I Price, Features & Specifications. [online] Available at: <https://www.buses.tatamotors.com/products/brands/starbus-ultra/starbus-ultra-electric-9-12-ev/. Accessed 22 May 2020
5. Fayazbakhsh M, Bahrami M (2013) Comprehensive modelling of vehicle air conditioning loads using heat balance method. SAE Technical Paper, No. 2013-01-1507
6. Vaghela J, Kapadia R (2014) The load calculation of automobile air conditioning system. IJEDR 2(1). ISSN: 2321-9939
7. Abdulsalam O, Santoso B, Aries D (2015) Cooling load calculation and thermal modelling for vehicle by MATLAB. Int J Innov Res Sci Eng 4(5)
8. ASHRAE Handbook of Fundamental (1988) American society of heating, refrigerating, and air conditioning. Atlanta, GA

Numerical Investigation of Erosion of a Steam Turbine Blade Due to the Impact of Condensed Water Droplets

Nagula Venkata Anirudh and Vipul M. Patel

Nomenclature

c	Speed of sound
c_s	SPEED of sound in solid (m/s)
E_s	Youngs modulus (GPa)
P	Pressure (GPa)
v_0	Impact velocity (m/s)

Greek Symbols

ρ	Density (kg/m^3)
Ψ	Stream function (m^2/s)
σ	Stress (N/m^2)
Σ	Non-dimensional stress

1 Introduction

Steam turbine blades are prone to be eroded by the water droplets that are formed by the condensation of steam at the final stages of the stator blades. The water droplets that are deposited are dragged along with the steam flow towards the trailing edge where it gets collected to form larger droplets. These droplets can have the diameter

N. V. Anirudh (✉) · V. M. Patel
Mechanical Engineering Department, Sardar Vallabhbhai National Institute of Technology Surat, Gujarat 395 007, India
e-mail: p19tm011@med.svnit.ac.in

J. Banerjee et al. (eds.), *Recent Advances in Fluid Dynamics*, Lecture Notes in Mechanical Engineering, https://doi.org/10.1007/978-981-19-3379-0_5

Fig. 1 Schematic of development of shock wave

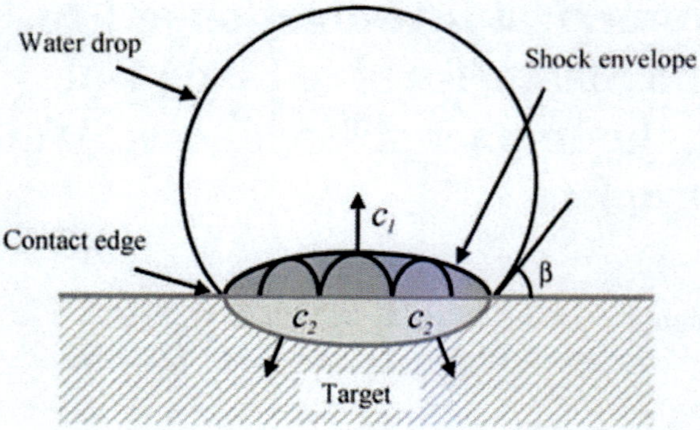

up to 1 mm. When these droplets are torn by the trailing edge, they get scattered and broken into smaller droplets. As the steam expands below saturation line in different stages, it condenses into little droplets of water which lead to nucleation making these water droplets bigger. These water droplets impact the turbine blades at high speeds and cause erosion onto the blade damaging it and lowering its performance over time. Due to this impact, a pressure wave is produced inside the water droplet. This pressure wave inside the water droplet propagates back and forth inside the droplet itself at the speed of sound. If the contact velocity is higher than the velocity of the shock wave inside the water, this wave remains in contact with the solid surface as shown in Fig. 1.

As the shockwave propagates through the water droplet at the speed of sound, the pressure inside it keeps increasing until the edge of the shockwave detaches the surface of the solid. This is the point where lateral jetting occurs and is the point where the maximum pressure inside the water droplet is seen. This pressure is higher than the water hammer pressure and depends on the velocity of impact. Since the water droplets impact the surface of the material at high impact velocities, compression and non-uniform distribution pressure distribution inside the water droplet occurs. This pressure is what we call water hammer pressure.

$$p = \rho \times c \times v \tag{1}$$

As shown in Fig. 2, since the volume of the water is highest at the centre of the surface where the water hits the surface of the material, the maximum pressure is generated at this point of the surface. The water droplet oscillates back and forth inside the droplet itself resisting the change of shape before it performs a radial flow while still in the collision of the surface of the material. This resistance is directly proportional to the impact velocity of the water droplet causing water hammer effect onto the surface of the material. As the shockwave of the pressure wave fronts detach from the contact surface during the radial flow, lateral jetting is produced within the water droplet with the velocities higher than the initial velocity of impact. During the lateral jetting, the pressures inside the water droplet reach its maximum which is much higher than the water hammer pressure of the droplet. This pressure is the

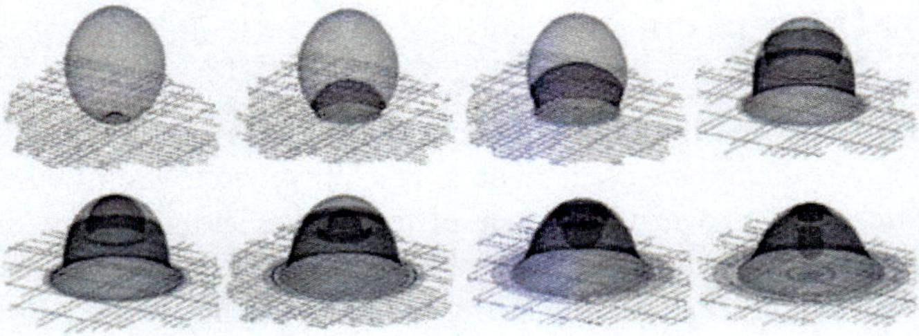

Fig. 2 Stages of droplet impact on solid surface [1]

most dangerous of all the other pressures as the lateral jetting pressures may reach up to three times more than that of the water hammer pressures.

As the blade surface gets constantly shot on by the water droplets at high speeds, the stresses induced onto the surface of the blade by these droplets are high which causes water droplet erosion. To reduce these stresses, metal coating is done on the surface of the blade. Coating for water droplet erosion is done with the metal that has a lower Young's modulus in order to reduce stress.

In the past couple of decades, several efforts have been made to quantify the water droplet erosion. Naib [1] did numerical analysis on water droplet erosion (WDE) for two different metals and compared them. It is seen that the titanium alloy had a better resistance towards WDE when compared to the steel alloy. It is also seen that the impact velocity, impact diameter and impact angle play a major role in determining WDE. Ahmad et al. [2] performed an experimental analysis on the water droplet erosion and observed that the impact velocity is the main factor for determining the peak pressure and stress while Chidambaram et al. [3] carried out three-dimensional numerical analysis of the water droplet erosion and came out with the same conclusion as observed by Ahmad et al. [2]. Li et al. [4] performed one-dimensional numerical analysis of water droplet erosion mechanism and their results showed that the higher the impact velocity results into greater peak pressure and higher stress on the blade. Marzbali [5] in his thesis performed two-dimensional numerical analysis to determine the peak pressures and stresses on the blade and found that the peak pressure occurs not at the instant of impact but after a brief moment and the peak pressure occurs inside the droplet and not at the interface. Huang et al. [6] did a numerical analysis on a 2D axisymmetric model of the water droplet erosion and compared the maximum stresses and pressures for different coating thicknesses. It is seen that the pressures and stresses on the blade are higher for a non-coated blade when compared to a coated blade.

In the present work, a one-dimensional numerical model is developed to study water droplet erosion. This work mainly focuses on comparing the pressure and stress curves of the rigid blade impact model with the elastic blade impact model. Moreover, pressure and stress curves of the blade material 1Cr13 (which has a Young's modulus of 200 GPa) and a coated blade (which has a Young's modulus of 100 GPa) are compared to appreciate the effect of coating of steam turbine blade.

2 Problem Description

In the present work, a one-dimensional numerical analysis is carried out on the water droplet erosion over the turbine blade to obtain the pressure curves inside the liquid droplet and stress curves over the surface of the blade. The governing equations consist of two parts one of liquid regime and the other of solid regime. Both are done for linear and nonlinear model. These governing equations are derived from the famous Navier–Stokes equations.

Governing equations. To simplify the Navier–stokes equations to arrive at the governing equations for the present problem, the momentum and viscosity terms are removed. This can be justified by the following reasons. (i) Since, the order of magnitude of the shear viscosity coefficient is low when compared to the density, the viscosity force tensor can be ignored. (ii) Since the shockwave travels at the speed of sound, the momentum cannot travel faster than the shockwave and hence can also be ignored.

Applying these conditions to the Navier–stokes equations, they can be simplified to the following equations.

$$\frac{\partial \Delta\rho}{\partial t} + \nabla.(\rho V) = 0 \tag{2}$$

$$\frac{\partial \rho V}{\partial t} + \nabla.(\Delta p I) = 0 \tag{3}$$

For a fluid, $\Delta\rho$ can be written as

$$\Delta\rho = \frac{\partial \rho}{\partial p}\Delta p = \frac{1}{c^2}\Delta p \tag{4}$$

Substituting Eq. 4 in Eqs. (2) and (3), the following equations can be obtained.

$$\nabla^2(p) = \frac{1}{c^2}\frac{\partial^2(p)}{\partial t^2} \tag{5}$$

$$\nabla^2(V) = \frac{1}{c^2}\frac{\partial^2(V)}{\partial t^2} \tag{6}$$

By introducing the stream function into the equations, Eqs. (5) and (6) can be merged into one equation using the following equations.

$$V = \nabla\psi \tag{7}$$

$$p = -\rho\frac{\partial(\Psi)}{\partial t} \tag{8}$$

The merged Eq. (9) represents a wave equation for the liquid regime.

$$\nabla^2\psi = \frac{1}{c^2}\frac{\partial^2(\psi)}{\partial t^2} \tag{9}$$

The speed of sound in equation will determine the linearity of the equations. If the value of 'c' is taken constant throughout the process, the equation will be turned into a linear partial differential equation.

The equations used for calculating pressure, density and speed of sound are as follows.

$$p = -\rho\frac{\partial(\Psi)}{\partial t} \tag{10}$$

$$p = A\rho^{\eta - B} \tag{11}$$

$$c = \sqrt{\eta\frac{p + B}{\rho}} \tag{12}$$

Equation (11) is known as Tait equation of state of water where A, B and η are constants with values being $A = 1.0147663 \times 10^{-19}$, $B = 2.858987 \times 10^8$, $\eta = 7.15$ [7].

To obtain the linear peak pressures and stresses, Eq. (12) is to be omitted from the set of equations meaning, the speed of sound throughout the water droplet is constant.

If the blade is taken as an elastic blade, the governing equation for solid regime is also added with the liquid regime. Under 1D coordinate system, the elastic solid also obeys the wave equation:

$$\nabla^2 U_s = \frac{1}{c_s^2}\frac{\partial^2(U_s)}{\partial t^2} \tag{13}$$

Boundary conditions. At the interface, the pressure that is induced by the droplet is absorbed by the blade in terms of stress so by equating these two terms, a boundary condition for the solid regime can be obtained as shown in Eq. (14).

$$\frac{\partial(U_s)}{\partial x}|_{x=0} = \frac{p_s|_{x=0}}{E_s} = -\frac{\rho}{E_s}\frac{\partial\psi}{\partial t}|_{x=0} \tag{14}$$

For the elastic blade model, the deformation velocity is to be incorporated to the liquid regime boundary condition which is Eq. (7). By doing so, the same transforms into Eq. (15).

Table 1 Boundary conditions

Coupled equations	Liquid	$\frac{\partial^2(\Psi)}{\partial x^2} = \frac{1}{c^2}\frac{\partial^2(\Psi)}{\partial t^2}$
		$\Psi\|_{t=0} = 0$
		$\frac{\partial \Psi}{\partial x}\|_{x=0} = v_0 - \frac{\partial U_s}{\partial t}\|_{x=0}$
	Solid	$\frac{\partial^2(U_s)}{\partial x^2} = \frac{1}{c_s^2}\frac{\partial^2(U_s)}{\partial t^2}$
		$U_s\|_{t=0} = 0$
		$\frac{\partial(U_s)}{\partial x}\|_{x=0} = -\frac{\rho}{E_s}\frac{\partial \psi}{\partial t}\|_{x=0}$

$$\frac{\partial \Psi}{\partial x}|_{x=0} = v_0 - \frac{\partial U_s}{\partial t}|_{x=0} \tag{15}$$

Initialization of the problem solution is done by substituting the initial values of stream function and deformation to be zero. Table 1 gives a general idea of all the boundary conditions and governing equations that are used to analyse the problem.

These boundary conditions are used to get the pressure curves inside the water droplet and stress curves on the surface of the blade using FDM discretization.

The discretization is done using 2nd order central difference method with 1st degree accuracy.

$$\frac{\partial^2 \emptyset}{\partial x^2} = \frac{\emptyset_{i+1}^n - 2 \times \emptyset_i^n + \emptyset_{i-1}^n}{\Delta x^2} \tag{16}$$

$$\frac{\partial^2 \emptyset}{\partial t^2} = \frac{\emptyset_i^{n-1} - 2 \times \emptyset_i^n + \emptyset_i^{n+1}}{\Delta t^2} \tag{17}$$

The discretized forms of the wave equations are as follows.

$$\psi_i^{n+1} = c^2 \times \Delta t^2 \times \left(\frac{\psi_{i-1}^n - 2 \times \psi_i^n + \psi_{i+1}^n}{\Delta x^2}\right) - \left(\psi_i^{n-1} - 2 \times \psi_i^n\right) \tag{18}$$

$$U_i^{n+1} = c_S^2 \times \Delta t^2 \times \left(\frac{U_{i-1}^n - 2 \times U_i^n + U_{i+1}^n}{\Delta x^2}\right) - \left(U_i^{n-1} - 2 \times U_i^n\right) \tag{19}$$

Methodology. The following steps are followed to construct a code for the problem at hand.

1. The initial values of pressure and displacement are given zero as the water droplet is yet to be disturbed by the impact with the surface.
2. The density and sound speed are to be calculated from Eqs. (11) and (12), respectively. (For linear wave model, speed of sound is to be omitted out of the calculations as it is to be kept constant in the stream function wave equation.)
3. Stream function (ψ) is to be calculated using the wave Eq. (18) which can be discretized using Eqs. (16) and (17) over the grid using the value of c obtained from the previous step.

4. Pressure is to be calculated from Eq. (10) using the density and stream function obtained from the previous steps. (For rigid, stop the 1st time step here and go to the next time step and redo the steps 1 to 4 until the pressure difference between consecutive time steps reach specific threshold.)
5. The pressure at this time step is to be taken as the boundary condition for the solid region as in Eq. (14).
6. Using these values of displacement (U), the displacements at the next time step are to be calculated using the displacement wave Eq. (19)
7. This displacement is to be used to calculate the velocity at which the deformation occurs at the interface
8 This velocity is used as the boundary condition of the liquid region (15).
9 Move to the next time step.
10. These steps 2–9 are to be repeated until the desired threshold pressure difference between the consecutive time steps is reached.

The solution methodology adopted in the present work is summarized in the flowchart, as shown in Fig. 3.

The schematic of the blade and the surroundings are shown in Fig. 4.

3 Results and Discussions

When the water droplet hits the surface of the blade, it creates a shockwave that travels at the speed of sound inside the water droplet. The incoming shockwave when breaks, creates a pressure that is more than the water hammer pressure due to the lateral jetting. The pressure at the interface has a different value when compared to the pressure at inside the droplet. As the time progresses, wave propagates into the droplet while varying the pressure inside it. This can be seen in Fig. 5a, b.

The peak pressures obtained for rigid impacts at two different velocities, namely 10 m/s and 100 m/s are compared with the work of Li et al. [4] and the corresponding plots are shown in Figs. 6 and 7, respectively.

Figure 8a, b shows the variation of the pressures and stresses against time on the interface for all the models for the impact velocities of 10 m/s and 100 m/s, respectively

From Fig. 8a, b, it is seen that all the peak pressures for the linear model are lower than that of the nonlinear model. This can be attributed to the fact that in the nonlinear model, the compressibility of the liquid causes the liquid to get harder which intern induces higher pressure onto the surface of the blade. All the peak pressures on the interface of the blade are higher for the rigid blade when compared to the elastic blade as the blade gets deformed with a deformation speed v_s which cushions the relative velocity of impact which leads to lower pressure action on the surface.

The peak pressure occurs at the time period of 0.2 nanoseconds at the interface and then decreases in an ever-damping wave pattern as shown in Figs. 4 and 5. This phenomenon is also seen in the finding of Marzbali [5].

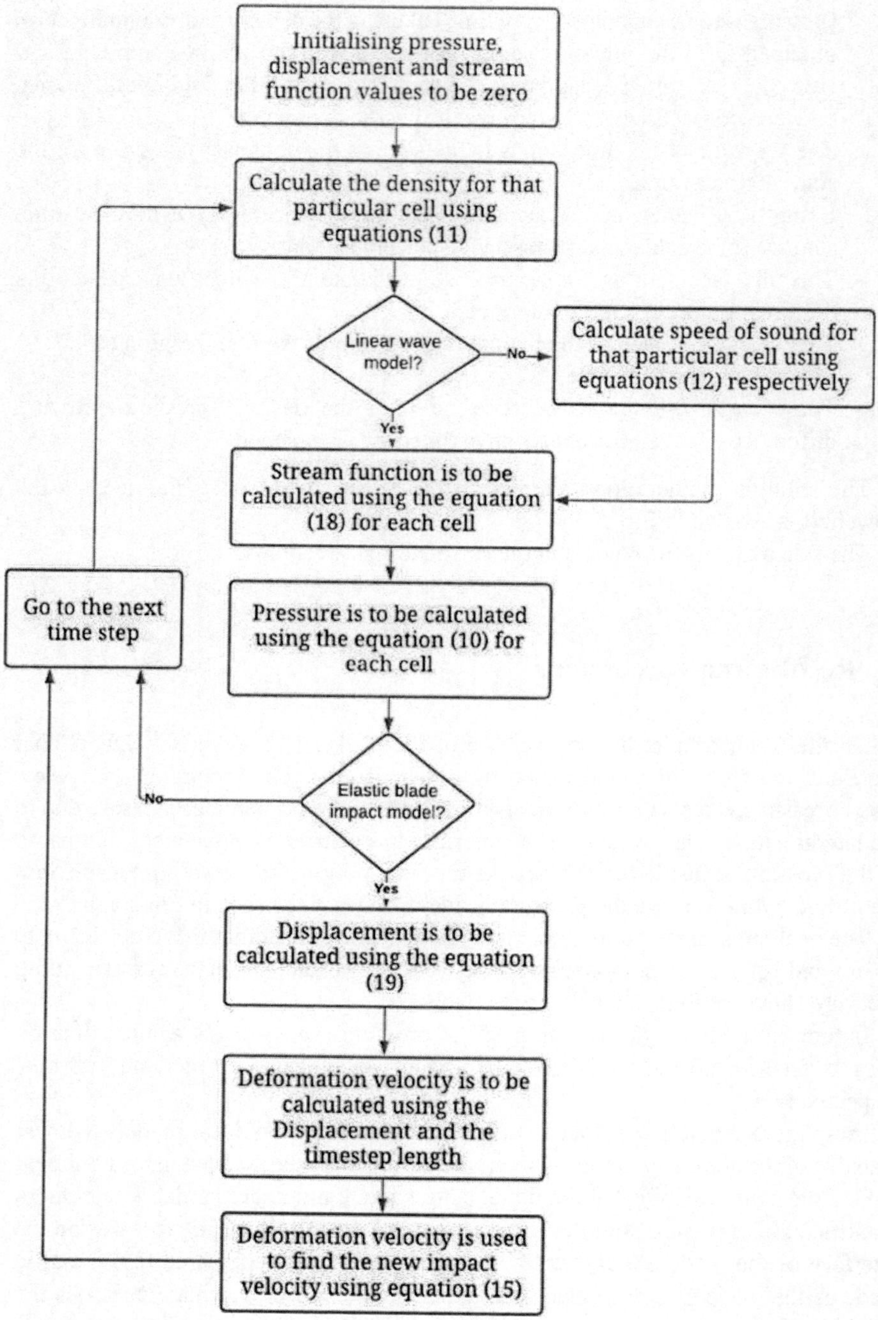

Fig. 3 Methodology flowchart

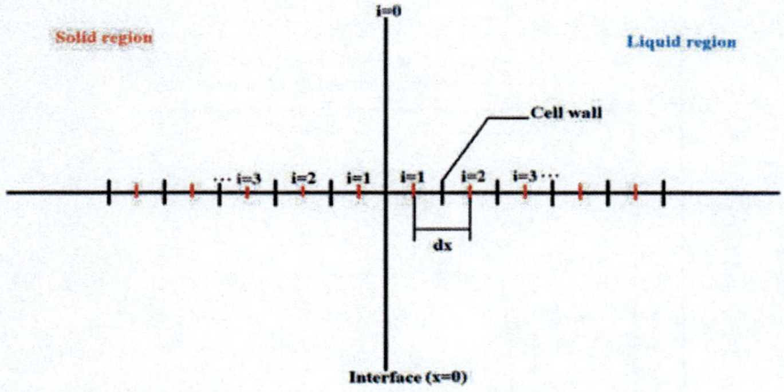

Fig. 4 Schematic of the blade domain

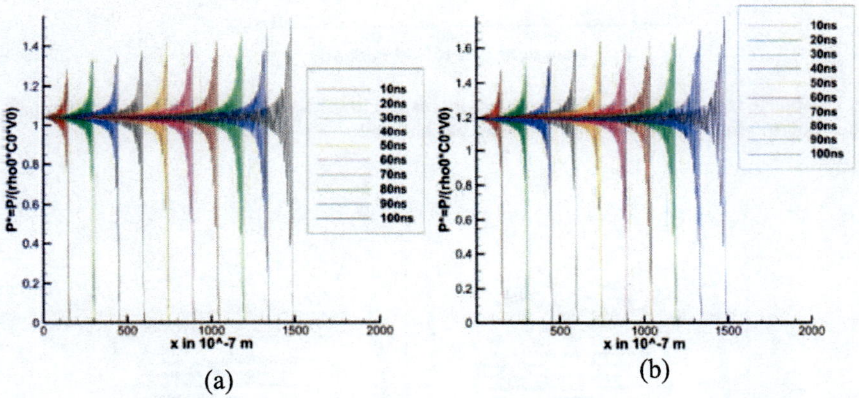

Fig. 5 Pressure distributions inside the liquid droplet at different time steps for the impact velocities of 10 m/s (**a**) and 100 m/s (**b**) for the nonlinear wave model

Fig. 6 Change in non-dimensional pressure with respect to the time at the interface for the impact velocity of 10 m/s obtained from a nonlinear wave model

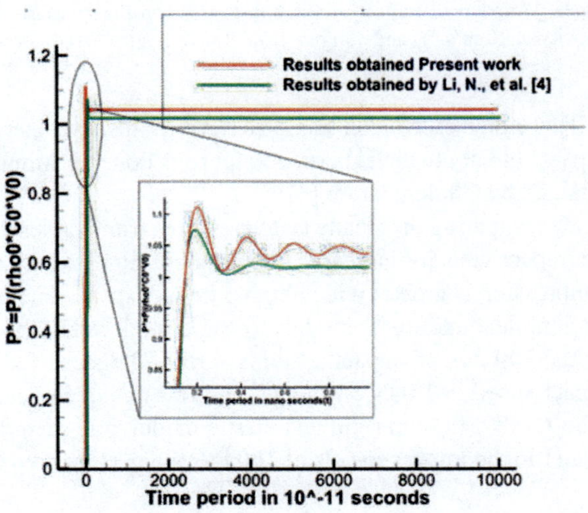

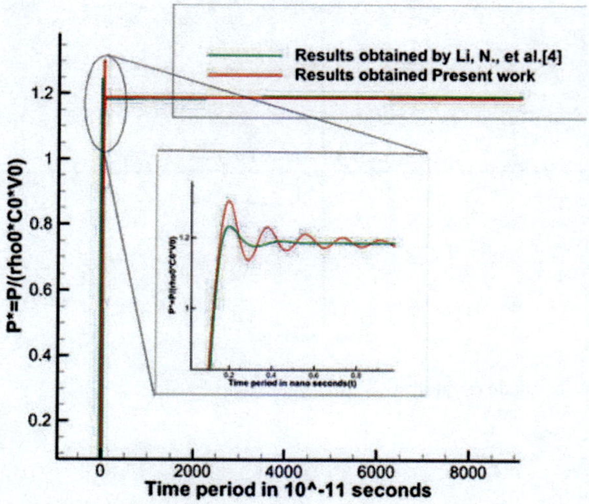

Fig. 7 Change in non-dimensional pressure with respect to the time at the interface for the impact velocity of 100 m/s obtained from a nonlinear wave model

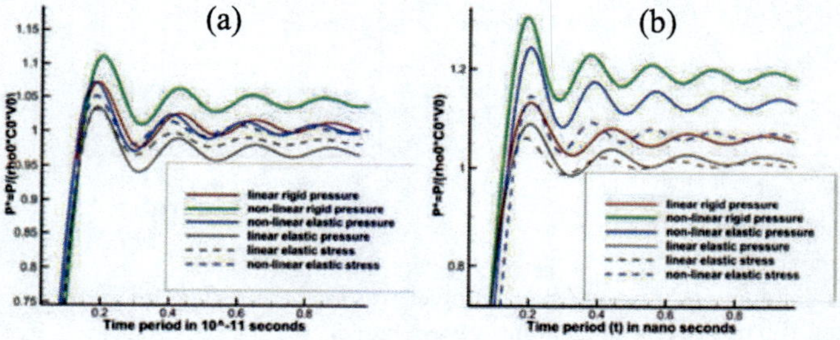

Fig. 8 Variation of impact pressure and stress on the interface w.r.t time period for the impact velocity of 10 m/s (**a**) and 100 m/s (**b**)

The effect of erosion has also seemed to have increased with increase in the impact velocity in the experimental results done by Ahmad [2] and numerical results obtained by Chidambaram [3].

By comparing the steady state pressures from Tables 2 and 3, it can be said that as the impact velocity increases, the effect of nonlinearity also increases as the liquid compression is greater with a higher impact speed. The peak pressure ratio between the nonlinear and linear models for rigid model at 10 m/s impact speed is 1.040 and for the 100 m/s of impact speed is 1.150. The same for elastic model at 10 m/s of impact speed is 1.0357; whereas, for 100 m/s of impact speed, the ratio turns out to be 1.1487. In both rigid and elastic model, it is seen that the ratio of pressures is greater in the impact speeds of 100 m/s when compared to 10 m/s.

Table 2 Results of the linear solid–liquid impact model

Model of solid	Rigid		Elastic	
Impact velocity (m/s)	10	100	10	100
Deformation speed (v_s)	0	0	0.438	4.547
Peak pressure at the interface (P^*)	1.112	1.302	1.072	1.251
Peak stress at the interface (Σ)	0	0	1.051	1.148
Steady state pressure (P^*)	1.042	1.192	1.021	1.124
Steady state stress (Σ)	0	0	1.018	1.064

Table 3 Results of the nonlinear solid–liquid impact model

Model of solid	Rigid		Elastic	
Impact velocity (m/s)	10	100	10	100
Deformation speed (v_s)	0	0	0.423	4.235
Peak pressure at the interface (P^*)	1.069	1.132	1.035	1.089
Peak stress at the interface (Σ)	0	0	1.031	1.061
Steady state pressure (P^*)	0.989	1.057	0.972	1.023
Steady state stress (Σ)	0	0	0.985	1.011

At higher impact speeds, the pressure at the interface is higher which causes the deformation velocity to increase making the blade's elasticity become more significant. From Table 2, it is observed that for the nonlinear model, at an impact velocity 10 m/s, the ratio between the deformation velocity v_s and the impact velocity v_0 $(v_s/v_0) = 0.0438$ and at 100 m/s, the ratio turns out to be 0.04547. This proves that the deformation effect is higher at higher impact speeds.

Coated blade. Figure 9a, b shows the results of the non-dimensional pressure and stress progression with respect to time for the metal coated elastic blade with the Young's modulus of 100 GPa [8].

It can be seen that the peak pressure and peak stress at the interface for the coated blade surface are lower than that of the uncoated blade surface. Young's modulus of

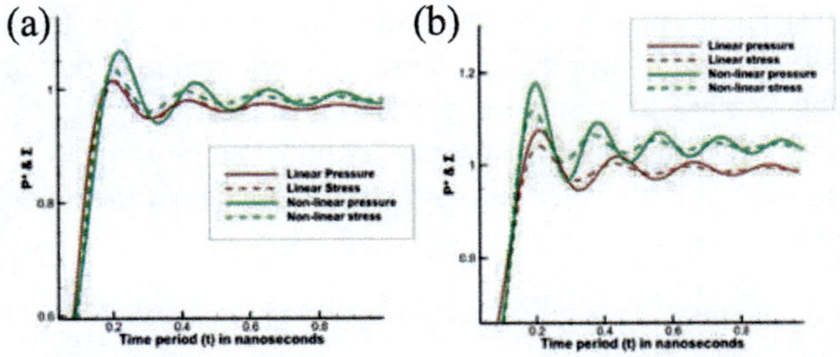

Fig. 9 Pressure and stress variations w.r.t time period for the coated blade with Young's modulus of 100 GPa for impact speeds of 10 m/s (**a**) and 100 m/s (**b**)

Table 4 Percentage change in pressure and stress in all scenarios for coated blade

Model	Velocity (m/s)	Pressure (%)	Stress (%)
Linear	10	1.98	1.54
	100	3.184	3.002
Nonlinear	10	2.144	2.015
	100	3.845	3.452

the coated surface changes the elasticity of the blade surface which in turn changes the amount of deformation that occurs in it. This causes the relative impact velocity to be lowered even further which causes the decrease in peak pressure and stress at the surface of the blade.

As seen in Table 4, the variation in both pressure and stress is higher for the higher impact speeds. This is because the property of elasticity highly affects the pressure and with the increase in velocity, the pressure difference is much higher as the deformation speed is more for lower Young's modulus material.

4 Conclusions

In this work, the one-dimensional numerical simulation of erosion of the steam turbine blade due to the condensed water droplets is carried out. For this, a nonlinear acoustic wave equation for the liquid phase is coupled with the elastic solid wave equation for solid phase. Since the viscous forces is insignificantly less, its effect is completely ignored. The key observations made during this work are as follows.

1. The nonlinear wave model has a high-steady state and peak pressures which signifies the importance of the nonlinear property which is the variation of speed of sound.
2. This work also simulates the shockwave travel inside the droplet while also showing that the elasticity of the solid affects the pressure and stress acted in the liquid droplet and the solid, respectively.
3. The analysis is done on the 1Cr13 material turbine blade and a coated turbine blade there by comparing these two results to accurately show the significance of erosion resistance coating onto the blade. This work mainly focuses on the pressure inside the water droplet and the stress induced due to that pressure on the turbine blade while changing the droplet impact speed.
4. This work also proves that the impact velocity plays a major role in determining the impact pressure of the water droplet onto the surface of the blade.

References

1. Naib S (2015) Modelling the influence of water droplet impacts on steam turbine blade surfaces, MS Thesis, Institute of Materials Testing, Materials Science and Strength of Materials (IMWF), University of Stuttgart
2. Ahmad M, Casey M, Sürken N (2009) Experimental assessment of droplet impact erosion resistance of steam turbine blade materials. Wear 267(9–10):1605–1618
3. Chidambaram PK, Kim HD (2018) A numerical study on the water droplet erosion of blade surfaces. Comput Fluids 164:125–129
4. Li N, Zhou Q, Chen X, Xu T, Hui S, Zhang D (2008) Liquid drop impact on solid surface with application to water drop erosion on turbine blades, Part I: nonlinear wave model and solution of one-dimensional impact. Int J Mech Sci 50(10–11):1526–1542
5. Marzbali M (2017) Numerical analysis of high-speed droplet impingement on elastic and rigid substrates. Doctoral dissertation, Concordia University
6. Huang S, Zhou Q, Li N, Teng Z, Yin H (2017) The effect of coating layer in liquid–solid impact problem. Int J Mech Sci 128:583–592
7. Macdonald JR (1966) Some simple isothermal equations of state. Rev Mod Phys 38(4):669
8. Tushinsky LI, Kovensky I, Plokhov A, Sindeyev V, Reshedko P (2002) Mechanical properties of coatings. In: Coated metal. Berlin: Springer, pp 85–131

CFD Analysis of Ventilation of Indian Railway 2 Tier AC Sleeper Coach

Jay S. Kachhadiya, Mukul Shukla, Swastik Acharya, and S. K. Singh

Nomenclature

ACH	Air Change per Hour
ASHRAE	American Society of Heating, Refrigerating and Air conditioning Engineers
CFD	Computational Fluid Dynamics
CAD	Computer-Aided Design
DNS	Direct Numerical Solution
LES	Large Eddy Simulation
LHB	Linke Hofmann Busch
RDSO	Research Design and Standards Organization
RANS	Reynolds Averaged Navier Stokes

1 Introduction

Railways are the most preferred transportation mode in India due to their low cost, good speed and comfort, particularly for the long journey. Due to the present COVID situation, the ventilation system of closed air conditioned rail coaches needs to be re-evaluated to reduce the chances of airborne infection. The current ventilation layout is not designed to remove the infectious particles effectively, as it is designed to provide uniform temperature and velocity to the passengers. To modify the ventilation system,

J. S. Kachhadiya (✉) · M. Shukla · S. Acharya
Department of Mechanical Engineering, MNNIT Allahabad, Prayagraj, India
e-mail: kachhadiyajay123@gmail.com

S. K. Singh
EDR, RDSO Lucknow, Lucknow, India

J. Banerjee et al. (eds.), *Recent Advances in Fluid Dynamics*, Lecture Notes in Mechanical Engineering, https://doi.org/10.1007/978-981-19-3379-0_6

it is necessary to analyse the flow field of the various layouts for temperature and velocity distribution inside the cabin. The air conditioning system is installed without any knowledge of the flow pattern required in the cabin. This gives rise to the problems of uneven air distribution in the coach compartments and uncomfortable passenger ride. Various parameters including the location, number and geometry of the inlet and outlet ports in the compartment affect the airflow distribution and heat transfer. These parameters can cause discomfort to the passengers as due to insufficient air conditioning the human body is unable to dissipate the waste thermal energy to the surroundings.

There are various studies on the airflow simulation with the help of CFD in aircraft cabin [7, 8, 11, 12, 16–18, 21], train seater coach [3, 4, 20], hospital ward [5, 10, 12]. The main focus of these studies is to simulate the airflow field inside the cabin with new ventilation layouts [7, 16, 21] or to study the contamination concentration in the cabin [8]. There are few studies on the airflow analysis of different ventilation layouts of railway coaches [1] and hospital ward [6]. Most of these studies are on seater coaches/cabins, but very few studies on the airflow simulation of the sleeper coaches are reported [1].

No previous work on the airflow simulation of the Indian railway's sleeper coaches is available making our study novel. It is very important to model the airflow characteristics inside the cabin to plan for further modification in the ventilation system to improve the ride experience of the passenger or in a pandemic situation, to redesign the ventilation system to reduce the transmission of airborne infection.

This study is about exploring the different ventilation layouts that can be used as an alternative. The layouts are analysed for the temperature and velocity distribution inside the cabin so that they can be checked against the thermal comfort standards of the ASHRAE.

2 Problem Description

This study was carried out on the 2 tier AC sleeper coach model of Indian Railways. Due to the repeatable layout of cabins throughout the coach, we have used only one cabin for analysis. The dimensions of the cabin were provided by RDSO Lucknow and the CAD model of original scale was prepared by simplifying the smaller features to reduce the simulation computational cost. The dimensions of the cabin are taken as $1.9 \times 3.07 \times 2.7\ \text{m}^3$ (Fig. 1a). The cabin geometry consists of the walls, partitions between the cabin, seats, inlet ports, outlet ports, manikins and gallery. A total of six manikins are seated inside the cabin on the lower berths (two manikins per berth) as shown numbered in Fig. 1b (Table 1).

The manikin geometry is simplified to reduce the mesh size and the dimensions are taken from anthropometric data [15]. The mouth area was kept at about $4\ \text{cm}^2$ [9]. The manikins were modelled as a box-like structure with realistic dimensions as smaller features do not affect the airflow inside the cabin significantly. The mouth surface normal is assumed to be at 30° to the horizontal plane.

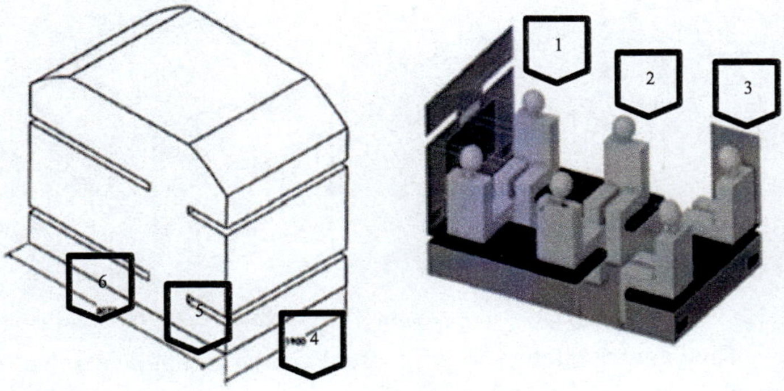

Fig. 1 **a** Train cabin geometry and **b** six manikins seated in the cabin

Table 1 Dimensions of the fluid domain geometry

	Length	Width	Height
Cabin	1900	3070	2719
Manikin	533	551	1200
Inlet	250	180	
Outlet	250	180	
Mouth	4 cm^2		

A total of 12 layouts with different inlet and outlet port positions have been used to analyse the airflow distribution. The possible positions to place inlet and outlet ports can be on sidewalls and roof. As the particles need to be quickly removed from the cabin the inlet port should be placed such that it pushes the particles away from the manikins and towards the outlet port. To make this possible we have used two inlet port positions, one with an inlet port at the centre as it is used in the current ventilation system in Indian railways and, the other with 3 inlet ports on the roof, two directly above the manikins on the lower berth and third in between the gallery area between the two cabins Fig. 2.

The outlet ports are placed at six possible positions including three positions on the roof, one above the window and below the seat is generally used and then pairing the two configurations we get another three layouts Fig. 3.

The velocity of cold air coming in the cabin from inlet ports and inlet port area. The inlet velocity was calculated assuming the ventilation rate of 12 ACH. For the first inlet layout, we have used a velocity of 0.475 m/s whilst in the second inlet layout we have used a velocity of 0.317 m/s due to the presence of 3 inlet ports. The inlet air temperature is kept as 296 K (23 °C) with the turbulent intensity of 5% and the outlet condition is assumed to be atmospheric and the walls adiabatic. The manikins are considered as a wall with a constant and uniform temperature of 310 K.

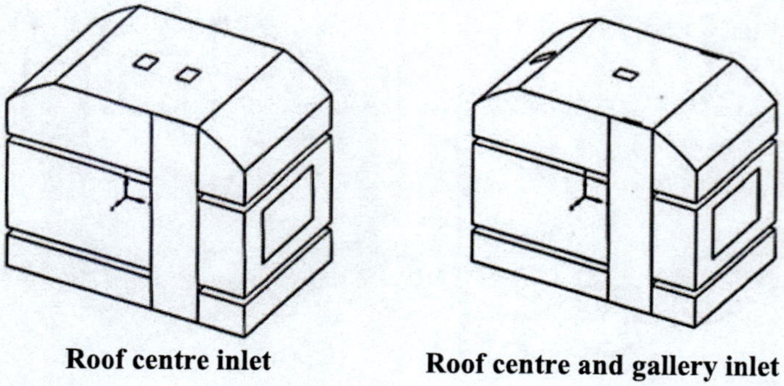

Fig. 2 Inlet port layouts

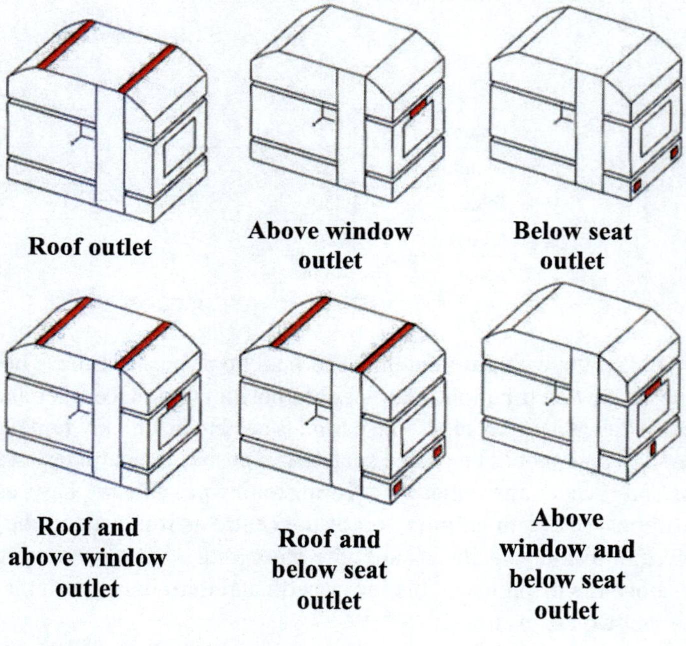

Fig. 3 Outlet port layouts

The gallery wall is considered as a periodic boundary condition due to the repeating cabin layout.

2.1 Numerical Formulation and Methodology

As the flow entering the cabin through the inlet port has a Reynolds number value of 1.87×10^6, we have used the turbulence model. Various turbulence models like DNS, LES, RANS and Reynolds stress model are being used in the literature [13]. But due to limitations of computational power, the RANS model was selected in this research. The realizable k-epsilon model was selected after validation tests, as it better predicts the velocity.

The steady-state, 3-dimensional Navier-Stokes equations including gravity were solved with pressure-velocity coupling achieved by using a coupled algorithm which is good when the mesh quality is poor and time steps are large. The energy equations were solved to account for temperature variations. The transport equations were discretized using a second-order upwind scheme. The numerical analysis was performed using FLUENT the commercially available CFD code, based on the finite volume method. Over each control volume, the coupled algorithm was used to iteratively solve these governing equations. As we are solving turbulence for steady-state, so the temporal terms will be removed and we are left with the continuity (1), momentum (2) and energy (3) equations.

$$\frac{\partial u_i}{\partial x_i} = 0 \tag{1}$$

$$u_j \frac{\partial u_i}{\partial x_j} = -\frac{1}{\rho}\frac{\partial \rho}{\partial x_i} + \frac{\partial}{\partial x_j}\left(2\vartheta S_{ij} - \overline{u_j^{'} u_i^{'}}\right) \tag{2}$$

$$u_j \frac{\partial \theta}{\partial x_j} = -\frac{1}{\rho e_p}\frac{\partial}{\partial x_j}\left(k\frac{\partial \theta}{\partial x_j} - \overline{u_j^{'}\theta^{'}}\right) \tag{3}$$

where ρ is the density of fluid, $\vec{u}$ is the velocity field, θ is the energy field, S_{ij} is strain vector and $u_j^{'}, \theta^{'}$ are the fluctuating components of turbulence flow. To solve for the turbulent components, the transport equations (4 and 5) are introduced.

$$\frac{\partial(\rho k u_j)}{\partial x_i} = \frac{\partial}{\partial x_j}\left(\left(\mu + \frac{\mu_t}{\sigma_k}\right)\frac{\partial k}{\partial x_j}\right) + G_k + G_b - \rho\varepsilon - Y_M \tag{4}$$

$$\frac{\partial(\rho \varepsilon u_i)}{\partial x_i} = \frac{\partial}{\partial x_j}\left(\left(\mu + \frac{\mu_t}{\sigma_\varepsilon}\right)\frac{\partial \varepsilon}{\partial x_j}\right) + \rho C_1 S\varepsilon - \rho C_2\frac{\varepsilon^2}{k + \sqrt{\varepsilon v}} + C_{1\varepsilon}\frac{\varepsilon}{k}C_{3\varepsilon}G_b \tag{5}$$

In the RANS equations, the instantaneous velocity has two components: time averaged and fluctuating. Then the Navier-Stokes equation is time averaged, which gives us Eqs. 1–5. The standard k-ε model differs from the realizable k-ε model in formulating the eddy viscosity and dissipation rate equation. In the standard model,

Fig. 4 (from left) Coarse, medium and fine-meshed fluid domain used for grid independence test

C_{μ} varies with turbulent kinetic energy (k), turbulent dissipation rate (ε), mean strain and rotation rates. The transport equation for turbulence kinetic energy (4) is the same for both the models, whilst the dissipation rate Eq. (5) is different.

The value of constants used in the Realizable k-ε model are, $C_{1\varepsilon} = 1.44$, $C_2 = 1.9$, $\sigma_k = 1.0$ and $\sigma_{k\cdot} = 1.2$. The terms μ and μ_t represent the viscosity and turbulent viscosity whilst G_k, G_b and Y_m are the terms for generation of turbulent kinetic energy due to mean velocity gradient, due to buoyancy and contribution of fluctuating dilation to overall dissipation rate, respectively. σ_k and σ_ε are the turbulent Prandtl numbers for k and ε, respectively.

Grid Independence Test: The unstructured mesh with tetrahedral cells was generated in the computational domain for spatial discretization of the governing equations. For a grid independence test the mesh size was varied with a reduction factor of 0.71. The average velocity and temperature inside the cabin were noted down against the mesh size. The final mesh size was selected when the change in average velocity was less than 0.1% (Fig. 4).

As the mesh size decreases, the number of cells increases, increasing the computational time. Therefore, after a grid independence test, an optimum mesh size of 59 mm was selected for further simulations, resulting in 82.3% mesh quality and a total of 2,85,056 cells.

Validation: The present result was validated with the data from two different experimental setups for velocity [19] and temperature [17]. The present result has been obtained using Laminar, RNG k-ε, Realizable k-epsilon and SST k-ω models. Figure 5a shows that the Realizable k-epsilon model predicts the results better with a maximum error of 10%. Therefore, we selected the realizable k-ε model for further simulations.

3 Results and Discussion

Figure 6 shows the temperature distribution inside the cabin, having the inlet port placed at the centre of the roof. The contours in the first column has been shown on the horizontal plane at a height of the mouth of manikins. The circular holes represent the cross-sectional view of the manikin's head.

Fig. 5 Comparison of simulated (a) temperature and (b) velocity with experimental data [17, 19]

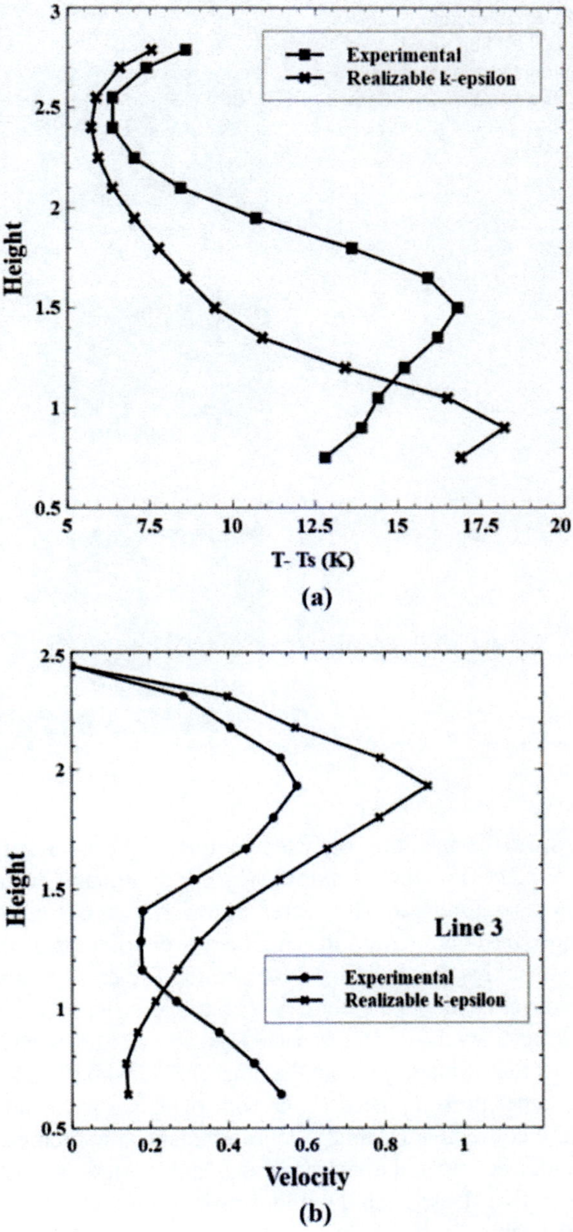

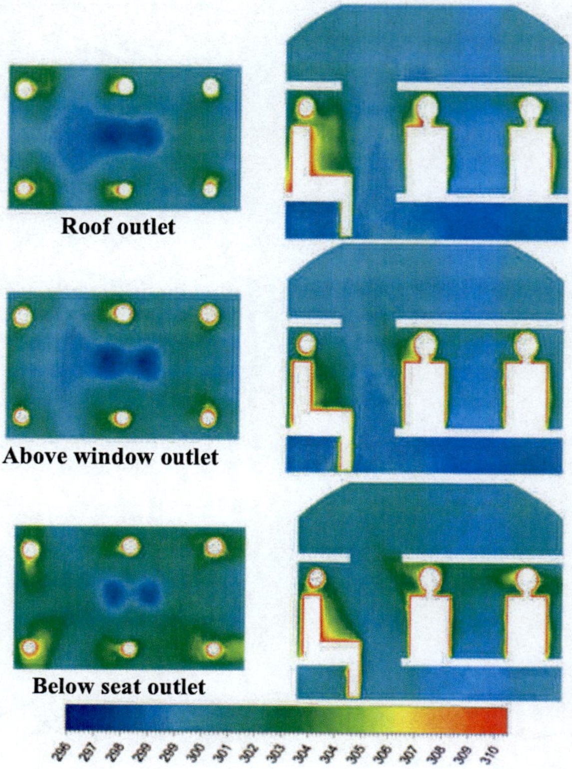

Fig. 6 Temperature contour on a (left) horizontal plane cutting the mouth of the seated manikins and (right) vertical plane cutting manikins

Since the manikin's body temperature is greater than the inlet air, the heat transfer takes place from the manikins to air. The air remains hot adjacent to the manikins and its temperature gradually decreases towards the centre of the cabin. The contours on the second column show the temperature distribution on the vertical plane cut through the manikins seated on the lower berth. The cold stream from the inlet port moves downward and hits the floor. After hitting, it tries to move upwards. However, it gets obstructed by the lower berth and fills the space below the seat. Thus, the temperature below the seat is lower than the other region which can be illustrated in Fig. 6. When the outlet port is placed at the roof or above the window, the upward movement of the convective current is more efficient in reducing the temperature around the manikins compared to other layouts. In the below seat outlet, a significant amount of the cold air leaves cabin without recirculating which leads to a higher temperature in the cabin.

Figure 7 shows the temperature distribution inside the cabin having inlet ports at the roof centre and the gallery. There are three inlet ports on roof one at the centre, one at the window side and the last at the gallery such that it is between two cabins (Fig. 2). It is observed that such a layout is very effective in heat transfer which results in a lower temperature around the manikins compared to the previous layout (Fig. 6).

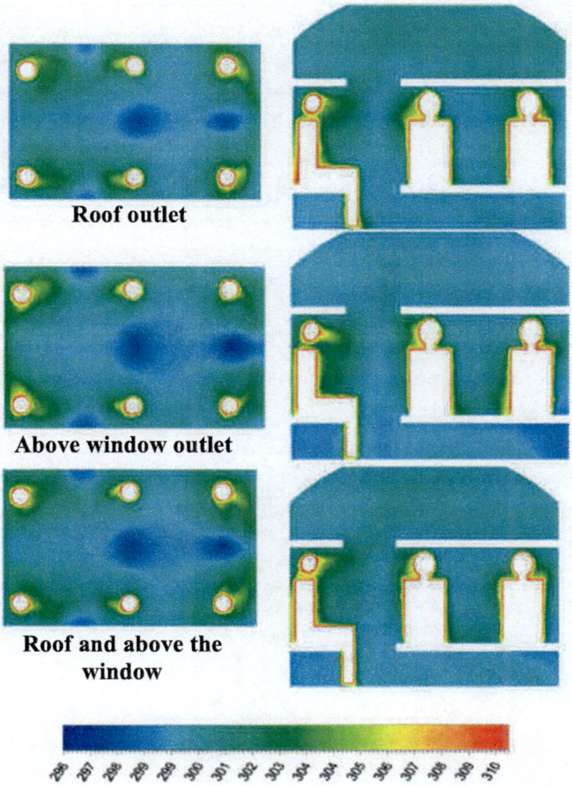

Fig. 7 Temperature contour on a (left) horizontal plane at the mouth level of the seated manikins and (right) vertical plane through manikins

Whilst comparing the different layouts for the outlet port, the lowest temperature inside the cabin is achieved for the window outlet due to better circulation of cold air as evident from Fig. 7 (second column, third row).

The layouts with either roof or above window outlet have lowest temperature around the manikins. In case of roof centre and gallery inlet, due to the three inlet ports on the roof, the temperature insisde the cabin is more uniform than the other layouts. The temperature around the side lower berth manikins also decreased thus improving the thermal comfort in the cabin.

The local temperature and velocity have been plotted along Line 1 and Line 2 (Fig. 8) which are drawn at the height of the mouth of manikins. Figures 9 and 11 show the variation of the temperature along Line 1 for the roof centre inlet. The temperature variation in the vertical direction is 2–4° (Fig. 6) and in the horizontal direction 5–7 degrees (Fig. 9). The lowest temperature is recorded in the layout of the roof outlet because the air moves upward to reach the outlet. During this time, it removes most of the heat from the cabin. The same phenomenon happens in the layout of the above window outlet. However, the opposite scenario arises for the roof and below seat outlets. The maximum temperature reaches 305 K for such a layout.

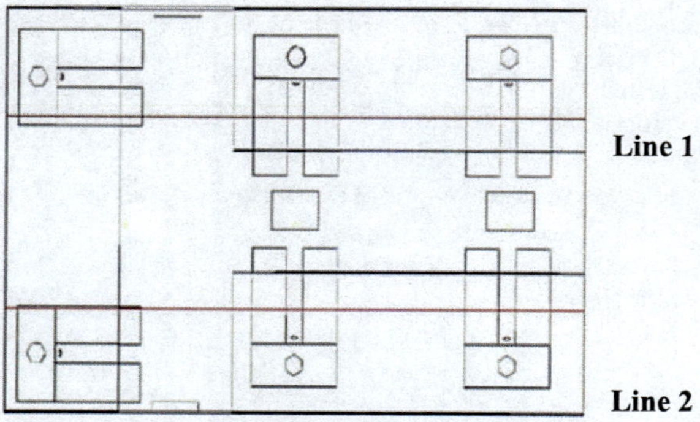

Fig. 8 Reference lines drawn for the extraction of local velocity and temperature values

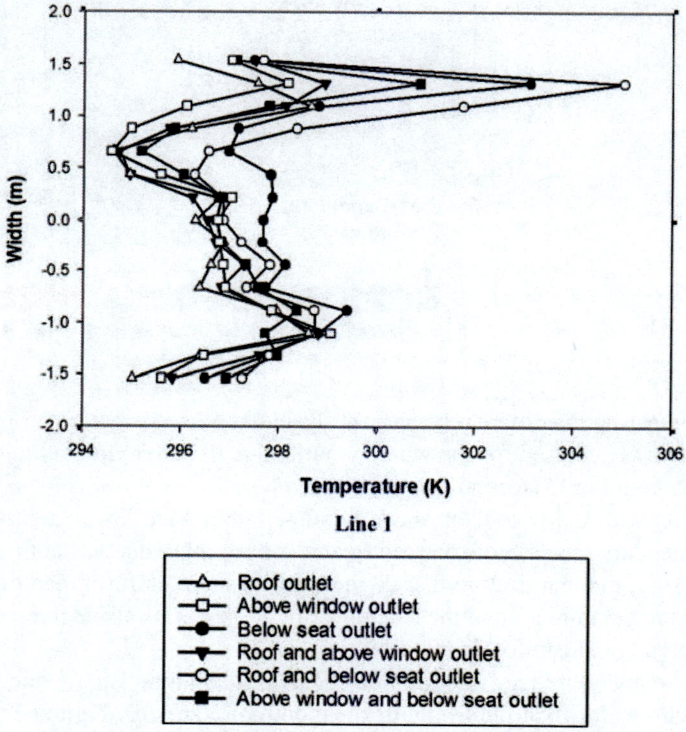

Fig. 9 Temperature variation along line 1 in all six layouts with roof centre inlet

Variation of temperature in the horizontal direction is almost the same in the case of roof outlet and above window outlet due to nearly similar flow pattern.

Figure 10 shows the velocity distribution of air along Line 1. The velocity is maximum in the roof outlet due to better recirculation of air inside the cabin. In the below seat outlet, the inlet air heads towards the outlet straightforwardly without

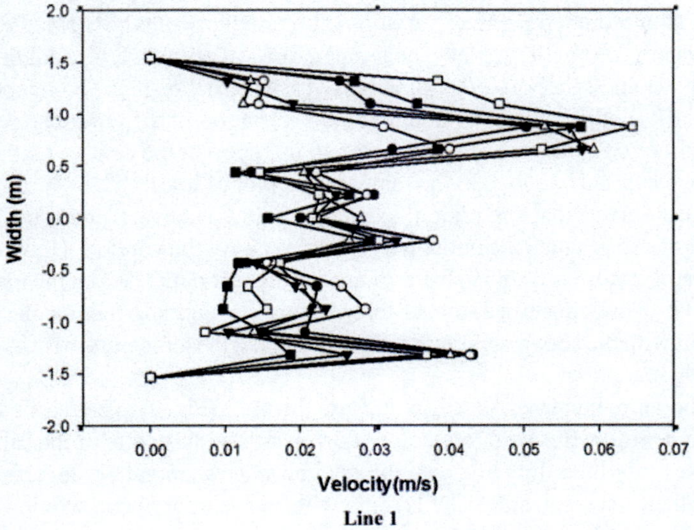

Fig. 10 Velocity variation along line 1 in all 6 layouts with roof centre inlet

any hindrance. Thus, smaller air circulation occurs around the manikins, resulting in minimum velocity (Fig. 9). In the case of the above window outlet, the velocity becomes maximum near the mouth of the manikins due to the presence of the outlet port just over the manikin's head.

The variation of the velocity in the horizontal direction is about 0.07 m/s (Fig. 10).

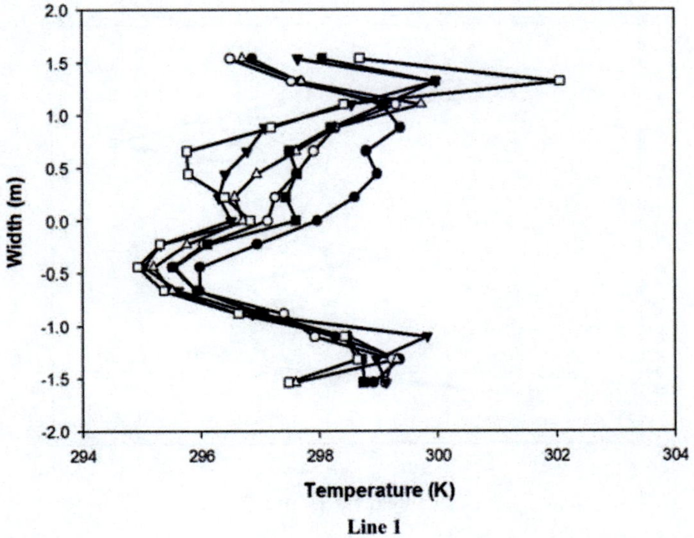

Fig. 11 Temperature variation along line 1 in all 6 layouts with roof centre and gallery inlet

Figure 9 shows the temperature variation along the horizontal lines in the layout of the roof centre and gallery inlet. The temperature variation in a vertical direction is 2–5° (Fig. 7) and in a horizontal direction, it is 6–7° (Fig. 11). The temperature remains at 301 K adjacent to the mouth and the variation in the temperature in the horizontal direction is due to the presence of an inlet port at the centre. As there are two inlets, we can see two dips in the temperature plot of line1.

It can be observed that the region near the window is covered with the lowest temperature of 295 K when the outlet port is placed above the window. However, the maximum temperature is found in the roof and below seat outlet. When the manikins on the side berth are concerned, the air temperature surrounding them is the lowest for the layout with the above window outlet. In contrast, the temperature is maximum for the below seat outlet.

The temperature increases as we go up from the floor (Fig. 7). The reason behind this is the cold airflow from the roof comes downwards at the centre of the cabin and then disperses in all the directions, so the cool air enters the gallery area from the bottom and then rises upwards. The layouts with roof and/or above window outlet has the maximum velocity inside the cabin due to the path of the air covering almost the whole cabin.

The peaks in the velocity plot (Fig. 12) are due to the inlet port position directly above it. The velocity at the mouth level of manikins is 0.02–0.04 m/s which is unnoticeable.

The average temperature and velocity in all the layouts are given in Table 2. In the roof centre inlet, the velocity is slightly higher, but the average temperature in all the layouts lies between 295 and 297 K (22–24 °C). The velocity in all the layouts lies between 0.03 and 0.05 m/s. The minimum average temperature and maximum

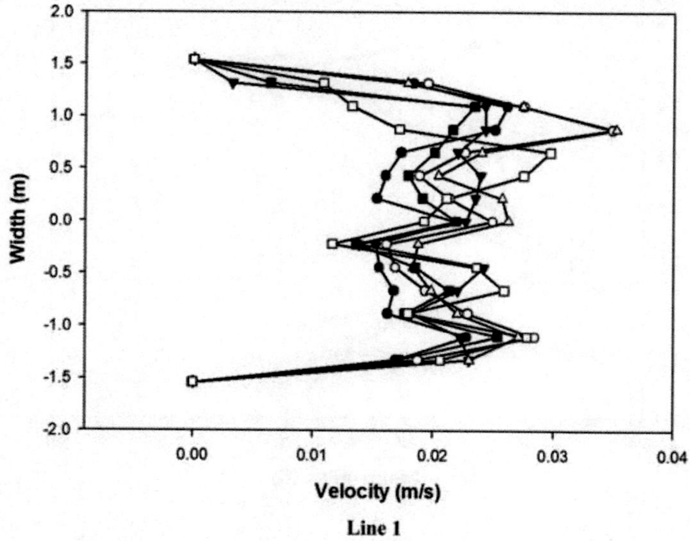

Fig. 12 Velocity variation along line 1 in all 6 layouts with roof centre and gallery inlet

Table 2 Average temperature and velocity in all the ventilation layouts

Outlet	Roof centre inlet		Roof centre and gallery inlet	
	Avg. temp. (°C)	Avg. velocity (m/s)	Avg. temp. (°C)	Avg. velocity (m/s)
Roof	22.59	0.05	23.27	0.034
Above window	22.49	0.051	22.54	0.035
Below seat	24	0.046	23.85	0.032
Roof and above window	22.71	0.05	22.96	0.034
Roof and below seat	23.37	0.049	23.43	0.033
Above window and below seat	22.84	0.047	23.28	0.033

average velocity are observed in the case of the above window outlet for both the inlet layouts. The air temperature and velocity for human comfort vary between 23–26 °C and 0.2–0.8 m/s [2]. The average temperature in all the layouts satisfies this thermal comfort condition, but the average velocity falls below the thermal comfort condition.

The temperature in the roof centre inlet and above window and/or roof outlet remains around 295 K (22 °C), which is below the thermal comfort condition. Whilst in the roof centre and gallery inlet, the temperature is within 23–24 °C, which is well within the thermal comfort condition. So by implementing three inlet ports, the overall distribution of the cold air gets enhanced, which results in increasing the temperature and decreasing the velocity.

4 Conclusions

From this study, it is concluded that when the inlet port is at the centre of the roof, the above window outlet is considered as the best layout. In such a layout, the average temperature is lower and the average velocity higher compared to others. This layout has better air distribution with higher velocity, which helps in providing greater comfort to the passengers. It is also concluded that the above window outlet has more uniform and lower temperature field inside the cabin with a maximum velocity of 0.05 m/s for the roof centre inlet and 0.035 m/s for the roof centre and gallery inlet. The below seat outlet is the worst layout because of having the highest temperature and lowest velocity in both the inlet conditions.

The above window, roof, and roof and above window outlets with the roof centre inlet are three favourable ventilation layouts since they provide the lowest average temperature and highest average velocity. The inlet and outlet layouts should be set such that the air has to travel maximum distance before leaving through the outlet.

The longer the air travels in the cabin the higher the heat transfer and the more effective the ventilation system. In both inlet layouts, the above window and/or roof outlet layouts performes well and can be used for the ventilation purpose in practise. The air velocity in this case is lower than the comfort velocity, so velocity needs to be increased to reach comfort conditions.

This study can be used to modify the ventilation layout of AC coaches to improve the thermal comfort, air quality or to remove infectious particles. The study can be further extended by analysing the effect of the shape of inlet port, temperature and velocity at the inlet on the thermal comfort in the cabin. With the help of more powerful computational resources, a more realistic and robust model can be developed to study the different ventilation systems. The study can also be extended in modifying the current ventilation system in light of COVID by analysing the effect of change in the ventilation system on the dispersion of the cough particles inside the cabin for a variety of coaches (work in progress).

Acknowledgements We are thankful to Mr. Sanjeev Garg from RDSO Lucknow for providing us all the technical information needed for this study. Thanks also go to the MHRD GOI for financially supporting the lead author's studentship.

References

1. Aliahmadipour M, Abdolzadeh M, Lari K (2017) Air flow simulation of HVAC system in compartment of a passenger coach. Appl Therm Eng 123:973–990. https://doi.org/10.1016/j.applthermaleng.2017.05.086
2. ASHRAE (2020) Standard 55-thermal environmental conditions for human occupancy
3. Berlitz T, Matschke G (2002) Interior air flow simulation in railway rolling stock. Proc Inst Mech Eng Part F J Rail Rapid Transit 216(4):231–236. https://doi.org/10.1243/095440902321029181
4. Chow WK (2002) Ventilation of enclosed train compartments in Hong Kong. Appl Energy 71(3):161–170. https://doi.org/10.1016/S0306-2619(02)00008-9
5. Correia G, et al (2020) Airborne route and bad use of ventilation systems as non-negligible factors in SARS-CoV-2 transmission. Med Hypotheses 141. doi:https://doi.org/10.1016/j.mehy.2020.109781
6. Deepthi UG, Kenneth RM (2017) Computational fluid dynamics study on the influence of an alternate ventilation configuration on the possible flow path of infectious cough aerosols in a mock airborne infection isolation room. Physiol Behav 176(3):139–148. doi:https://doi.org/10.1080/23744731.2016.1222212
7. Dygert RK, Dang TQ (2010) Mitigation of cross-contamination in an aircraft cabin via localized exhaust. Build Environ 45(9):2015–2026. https://doi.org/10.1016/j.buildenv.2010.01.014
8. Giaconia C, Orioli A, Di Gangi A (2013) Air quality and relative humidity in commercial aircrafts: anexperimental investigation on short-haul domestic flights. Build Environ 67:69–81. https://doi.org/10.1016/j.buildenv.2013.05.006
9. Gupta JK, Lin CH, Chen Q (2009) Flow dynamics and characterization of a cough. Indoor Air 19(6):517–525. https://doi.org/10.1111/j.1600-0668.2009.00619.x
10. Hathway EA et al (2011) CFD simulation of airborne pathogen transport due to human activities. Build Environ 46(12):2500–2511. https://doi.org/10.1016/j.buildenv.2011.06.001

11. Li J et al (2018) PIV experimental research on gasper jets interacting with the main ventilation in an aircraft cabin. Build Environ 138(April):149–159. https://doi.org/10.1016/j.buildenv.2018.04.023
12. Li Y et al (2005) Role of air distribution in SARS transmission during the largest nosocomial outbreak in Hong Kong. Indoor Air 15(2):83–95. https://doi.org/10.1111/j.1600-0668.2004.00317.x
13. Liu W et al (2012) State-of-the-art methods for studying air distributions in commercial airliner cabins. Build Environ 47(1):5–12. https://doi.org/10.1016/j.buildenv.2011.07.005
14. Liu W et al (2013) Evaluation of various categories of turbulence models for predicting air distribution in an airliner cabin. Build Environ 65:118–131. https://doi.org/10.1016/j.buildenv.2013.03.018
15. NASA, man system integration standards (no date) Anthropometry and biomechanics. Available at: https://msis.jsc.nasa.gov/sections/section03.htm
16. Pang L et al (2013) Evaluation of an improved air distribution system for aircraft cabin. Build Environ 59:145–152. https://doi.org/10.1016/j.buildenv.2012.08.015
17. Rees SJ, Haves P (2013) An experimental study of air flow and temperature distribution in a room with displacement ventilation and a chilled ceiling. Build Environ 59:358–368. https://doi.org/10.1016/j.buildenv.2012.09.001
18. Sze To GN et al (2009) Experimental study of dispersion and deposition of expiratory aerosols in aircraft cabins and impact on infectious disease transmission. Aerosol Sci Technol 43(5):466–485. https://doi.org/10.1080/02786820902736658
19. Topp C, et al (2003) Influence of geometry of thermal manikins on room airflow, pp 1–6
20. Zhang L, Li Y (2012) Dispersion of coughed droplets in a fully-occupied high-speed rail cabin. Build Environ 47(1):58–66. https://doi.org/10.1016/j.buildenv.2011.03.015
21. Zhang T, Chen Q (Yan) (2007) Novel air distribution systems for commercial aircraft cabins. Build Environ 42(4):1675–1684. doi:https://doi.org/10.1016/j.buildenv.2006.02.014

Numerical Study of Double Wall Oscillating Lid Driven Cavity

Dintakurthi Yaswanth and Ranjith Maniyeri

Nomenclature

U	Non-dimensional velocity in x-direction
V	Non-dimensional velocity in y-direction
C	Non-dimensional concentration
t	Non-dimensional time
T	Time period of oscillation
ω	Non-dimensional frequency
Re	Reynolds number
Sc	Schmidt number

1 Introduction

Lid driven cavity has been a benchmark problem in computational fluid dynamics (CFD) since it provides an excellent understanding not only in the vortex flow but also in the mixing of fluids in a cavity. Fluid mixing is mainly used in chemical industries. The most general method used in mixing is the usage of stirrers, but an oscillating lid driven cavity provides a faster and better way of mixing because it makes more disturbances in the flow regime.

Mendu and Das [9] have studied about the effect of Reynolds number (Re) and oscillating frequency (ω) in an oscillating lid driven cavity when the top wall is oscillating using lattice Boltzmann method (LBM) and found that with the increase in Re for a fixed ω promotes more eddies. Indukuri and Maniyeri [8] have done the

D. Yaswanth · R. Maniyeri (✉)
Biophysics Laboratory, Department of Mechanical Engineering, National Institute of Technology Karnataka, Surathkal, Mangalore, Karnataka 575025, India
e-mail: mranji1@nitk.edu.in

J. Banerjee et al. (eds.), *Recent Advances in Fluid Dynamics*, Lecture Notes in Mechanical Engineering, https://doi.org/10.1007/978-981-19-3379-0_7

numerical simulation using finite volume method (FVM) in an oscillating lid driven cavity for parallel and antiparallel wall motion for different Reynolds numbers and different frequency. They found that optimum frequency for effective mixing as $2\pi/6$. Bhopalam et al. [4] studied about fluid behavior in two-sided oscillating lid driven cavity for parallel and antiparallel wall oscillations using LBM by varying parameters like Re, ω and speed ratio.

Bettaibi et al. [3] conducted numerical study of thermo-solutal mixed convection in rectangular cavity using D_2Q_9 MRT-LBM model and found their method is effective and accurate. Ghaffarpasand [5] studied about the thermo-solutal mixed convection with consideration of soret and dufour effects and found these effects are insignificant in forced convection dominated flow. Benghein et al. [2] studied about steady state thermo-solutal convection in a square cavity. Rao et al. [11] investigated about inertial effect on chaotic advection and mixing in lid driven cavity at finite Re. Huang et al. [7] have done experiments on mixing of two miscible fluids in a lid driven cavity and studied about the effect of density differences and viscosity of two miscible fluids. Nishimura and Kunitsugu [10] have studied mass transfer and fluid mixing in cavities with time periodic lid velocity using Galerkin finite element method and found that the mixing enhances with increase of amplitude of lid velocity when the top wall of the cavity is oscillating.

Encouraged from previous studies, the present work is intended to develop a two-dimensional computational model to simulate parallel and antiparallel wall oscillating square lid driven cavity to explore mixing phenomena by solving mass, momentum and concentration equations. In the present study we simulate the oscillating lid driven cavity for various Schmidt number (Sc) for fixed Re and ω. A staggered grid system is employed using FVM for discretization. The discretized non-dimensional continuity, Navier-Stokes and concentration equations are solved using a SIMPLE algorithm.

2 Problem Description

A two-dimensional computational setup of double wall oscillating lid driven cavity is shown in Fig. 1. The horizontal walls of the cavity are oscillating while the vertical walls are stationary ($U = 0$, $V = 0$). The velocity of the top oscillating wall in the x-direction is $U_o \cos(\omega t)$. In contrast, the velocity of the bottom wall is given $\pm U_o \cos(\omega t)$, the direction of the wall velocity depends on the parallel or antiparallel cases. Here U_o is maximum lid velocity and can be referred to as the amplitude of oscillation. We assume fluid flow is incompressible, Newtonian and laminar.

The non-dimensional continuity, Navier-Stokes and concentration equations are given by

$$\frac{\partial U}{\partial X} + \frac{\partial V}{\partial Y} = 0 \tag{1}$$

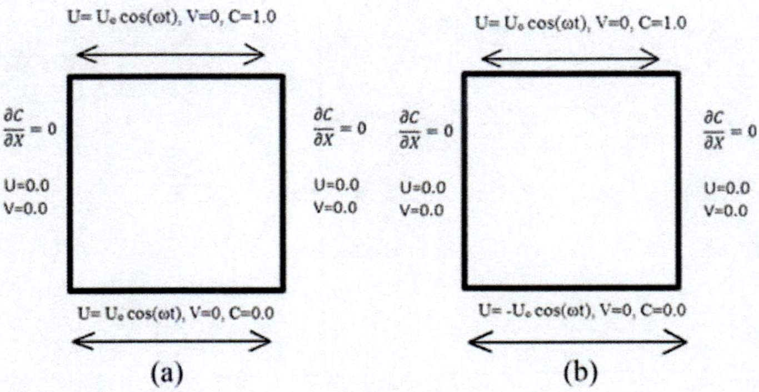

Fig. 1 Geometric representations of the oscillating lid driven cavity **a** parallel wall oscillation **b** antiparallel wall oscillation

$$U\frac{\partial U}{\partial X} + V\frac{\partial U}{\partial Y} = \frac{1}{\mathrm{Re}}\left[\frac{\partial^2 U}{\partial X^2} + \frac{\partial^2 U}{\partial Y^2}\right] \tag{2}$$

$$\frac{\partial V}{\partial t} + U\frac{\partial V}{\partial X} + V\frac{\partial V}{\partial Y} = \frac{1}{\mathrm{Re}}\left[\frac{\partial^2 V}{\partial X^2} + \frac{\partial^2 V}{\partial Y^2}\right] \tag{3}$$

$$\frac{\partial C}{\partial t} + U\frac{\partial C}{\partial X} + V\frac{\partial C}{\partial Y} = \frac{1}{\mathrm{Re\ Sc}}\left[\frac{\partial^2 C}{\partial X^2} + \frac{\partial^2 C}{\partial Y^2}\right] \tag{4}$$

The boundary conditions for the cavity are given by
For left and right wall $\frac{\partial C}{\partial X} = 0$, $U = 0$, $V = 0$
For bottom wall $C = 0.0$, $U = \pm\, U_o \cos(\omega t)$, $V = 0$
For top wall $C = 1.0$, $U = U_o \cos(\omega t)$, $V = 0$
For simulations we consider $U_o = 1.0$, $\omega = 2\pi/6$, Re $= 2000$ (based on Indukuri and Maniyeri [8]), time step $dt = 10^{-3}$ and Sc is varied to see its effect on flow and concentration behavior.

3 Results and Discussion

The numerical simulations are executed by constructing a code in FORTRAN. The developed code is validated first with the results of Ghia et al. [6] for the optimum grid size of 128 × 128 for finite wall motion (top lid moving with constant velocity) with Re = 100. Centreline U and V velocities and streamline plot are shown in Figs. 2 and 3 which are in good agreement with the results of Ghia et al. [6] serving the validation.

In addition, the numerical model is used to generate the concentration and streamline plots for the case of Re = 100 and Sc = 10 when top wall moves with constant

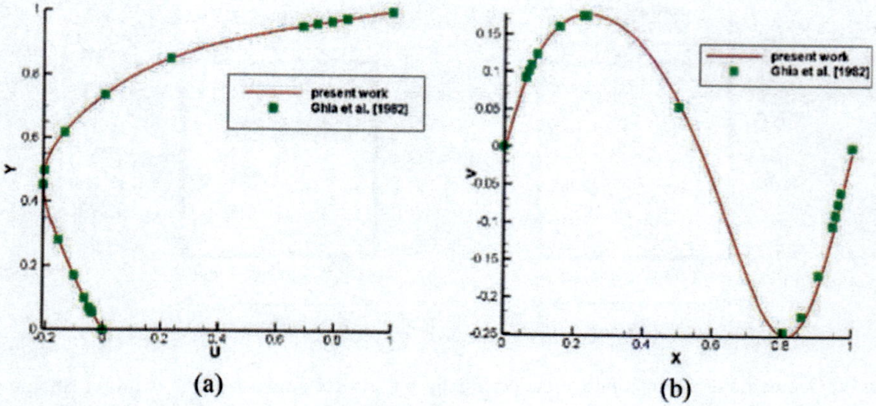

Fig. 2 Validation of the present work with Ghia et al. [6] for finite wall motion at Re = 100 **a** variation of centreline U velocity along y-axis **b** variation of centreline V velocity along x-axis

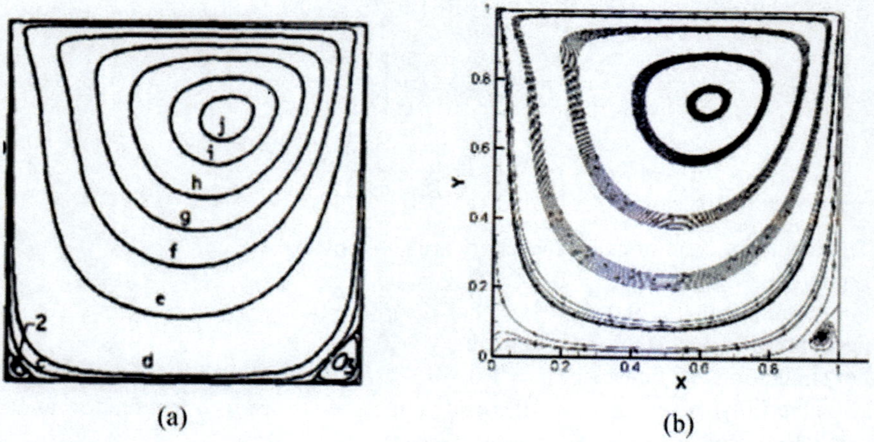

Fig. 3 Comparison of streamline plot of **a** Ghia et al. [6] and **b** present work

velocity. The results are compared with that of Al-Amiri et al. [1] and shown in Fig. 4. A good matching can be observed.

Now, simulations are performed for parallel and antiparallel wall oscillations as depicted in Fig. 1. For antiparallel wall oscillation (i.e., the top and bottom walls are oscillating in the opposite direction) and parallel wall oscillation (i.e., both the top and bottom walls are oscillating in the same direction), Re = 2000 is kept as constant and Sc is varied as Sc = 0.5, 1.0, 5.0.

Figure 5 represents the isoconcentration lines of antiparallel oscillation for Re = 2000 and Sc of 0.5, 1.0, 5.0 and time of 0.3T, 0.4T, 0.5T. Two vortices are formed one at the left top and other at right top of the cavity for all the cases but their intensity is different. For Sc = 5.0 number of isoconcentration lines are closer to the vortex

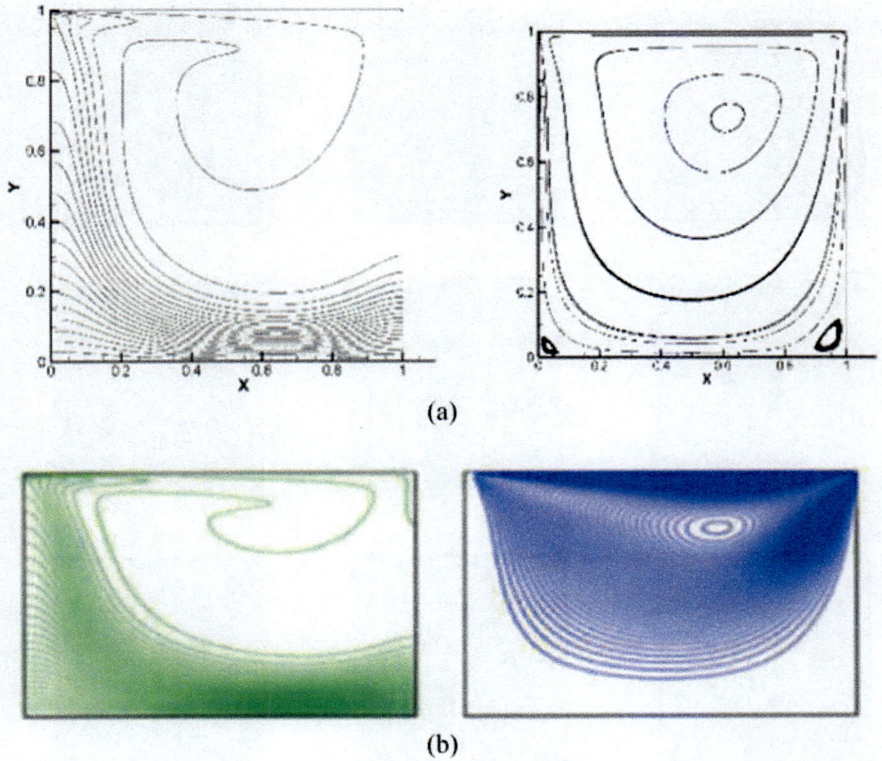

Fig. 4 Isoconcentration lines (left) and streamline (right) for Re = 100 Sc = 10 when top wall moving with constant velocity **a** present work **b** Al-Amiri et al. [1]

while for the Sc = 0.5 number of iscoconcentration lines are widely spread over the cavity and provides better mixing. For 0.5T there exist a small vortex at top left of the cavity for all Sc and its intensity increases with increase in Sc.

Figure 6 represents the streamline plot of antiparallel oscillation for Re = 2000. It is observed with increase in the time period from 0.3T to 0.5 T a small eddy is developing at the top left and bottom right of the cavity. The streamline plot is similar for all the Sc, so it can be inferred that the variation of Sc does not affect the velocity of fluid in the cavity.

Figure 7 represents isoconcentration lines for parallel oscillation for Re = 2000. Two primary vortices are formed one at the top left and other at the top right of the cavity for all the Sc. For higher Sc the isoconcentration lines are more closer at the vortex and no isoconcentration lines are formed at the bottom half of the cavity. For lower Sc the isoconcentration lines are widely spread over the top half of the cavity and slightly at the bottom half of the cavity.

Figure 8 shows the streamline plot of parallel wall oscillation for Re = 2000. Four primary vortices cover the complete cavity and two small secondary vortices formed at right middle of the cavity. With increase in time period a small tertiary

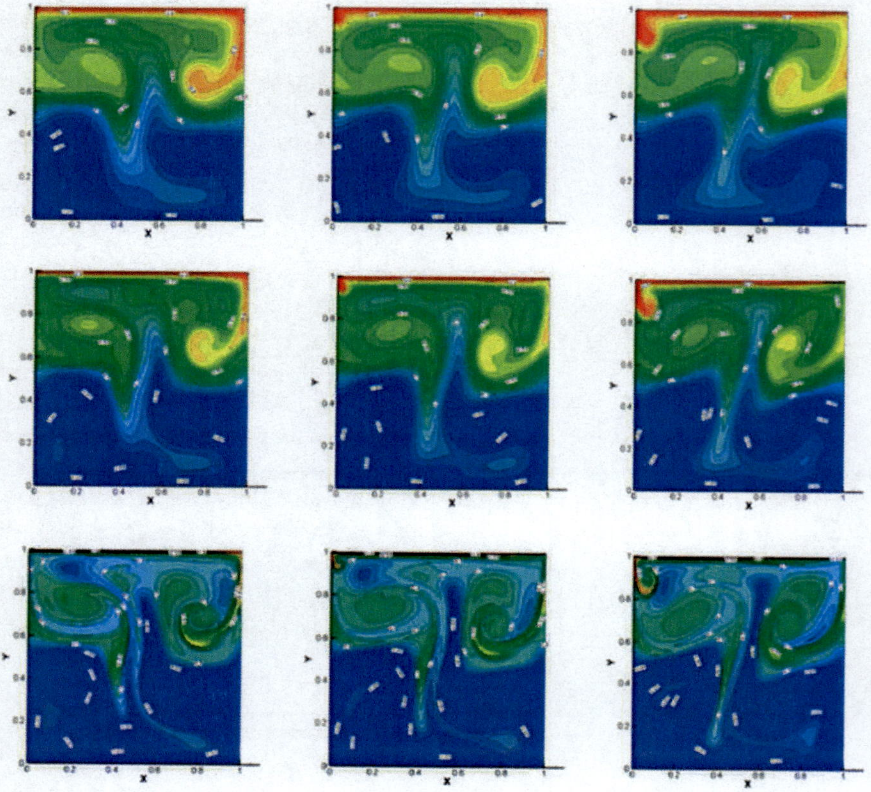

Fig. 5 Iscoconcentration lines of antiparallel oscillation for Re = 2000 by varying Sc from top to bottom (0.5, 1.0, 5.0) and time period left to right (0.3T, 0.4T, 0.5T)

vortex is developing at the left of the cavity. It is also noticed that with increase in Sc the streamline pattern is not changed. Hence, it can be believed that Sc does not influence the velocity of fluid in the cavity in parallel case also.

Comparing the streamline plot for parallel and antiparallel wall oscillations, the streamlines in parallel wall oscillation have four primary vortices. They do not intersect or interact with each other. In contrast, in antiparallel wall oscillation, the top right vortex and bottom left vortex are connected so that there is a chance to transfer the concentration to bottom of the cavity so that the mixing is better in antiparallel wall oscillation than parallel wall oscillation.

From Fig. 9 the variation of centreline U velocity for different Sc number are exactly matching for both parallel and antiparallel wall oscillation and we can conclude that Sc does not affect the velocity of the fluid. Since Sc is the ratio of momentum diffusivity to mass diffusivity and for low Sc mass diffusion dominates and provides better mixing for both parallel and antiparallel wall oscillations compared with momentum diffusion.

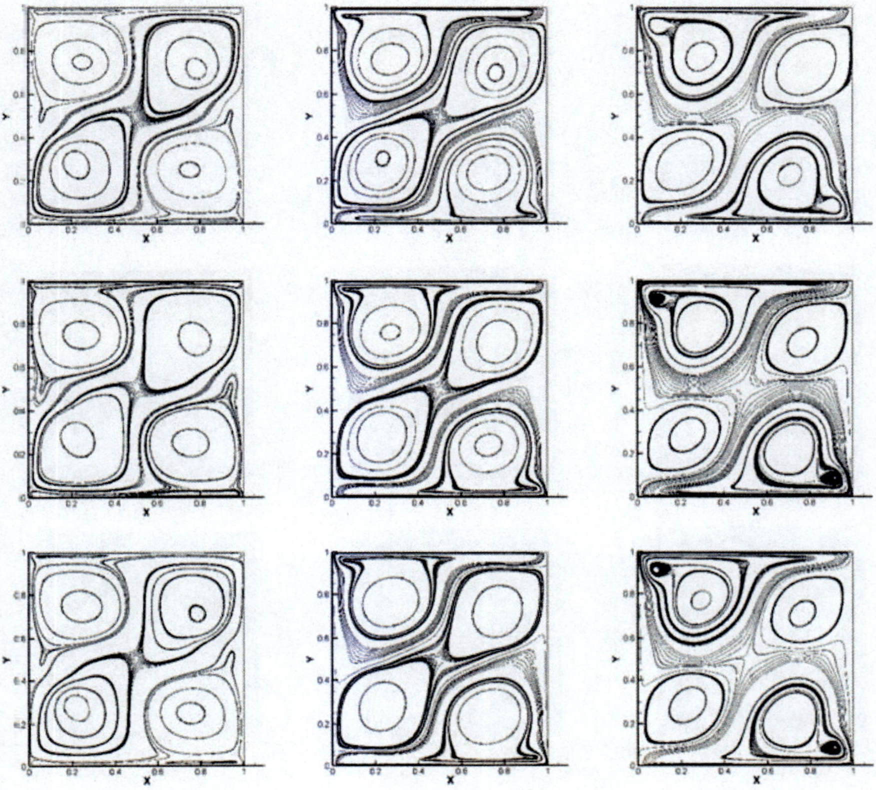

Fig. 6 Stream line plot of antiparallel oscillation for Re = 2000 by varying Sc from top to bottom (0.5, 1.0, 5.0) and time period left to right (0.3T, 0.4T, 0.5T)

4 Conclusions

A two-dimensional computational model is developed to study the mixing phenomenon in a double wall oscillating lid driven cavity considering both parallel and antiparallel wall oscillations. The finite volume method is adopted to discretize the governing equations and SIMPLE algorithm is used to carry out numerical simulations. A FORTRAN code is developed and validated and results are in good agreement with previous work. Streamlines and isoconcentration lines are plotted for fixed Re, ω and different Sc for both parallel and antiparallel wall oscillations. It is found that low Sc provides better mixing. The antiparallel wall oscillation provides good mixing than parallel wall oscillation because when we compare the streamline plots, the streamlines are connected diagonally whereas in parallel wall oscillations the top and bottom halves streamlines are different and the concentration does not carry from top to bottom. Hence low Sc and antiparallel wall oscillations provides good mixing.

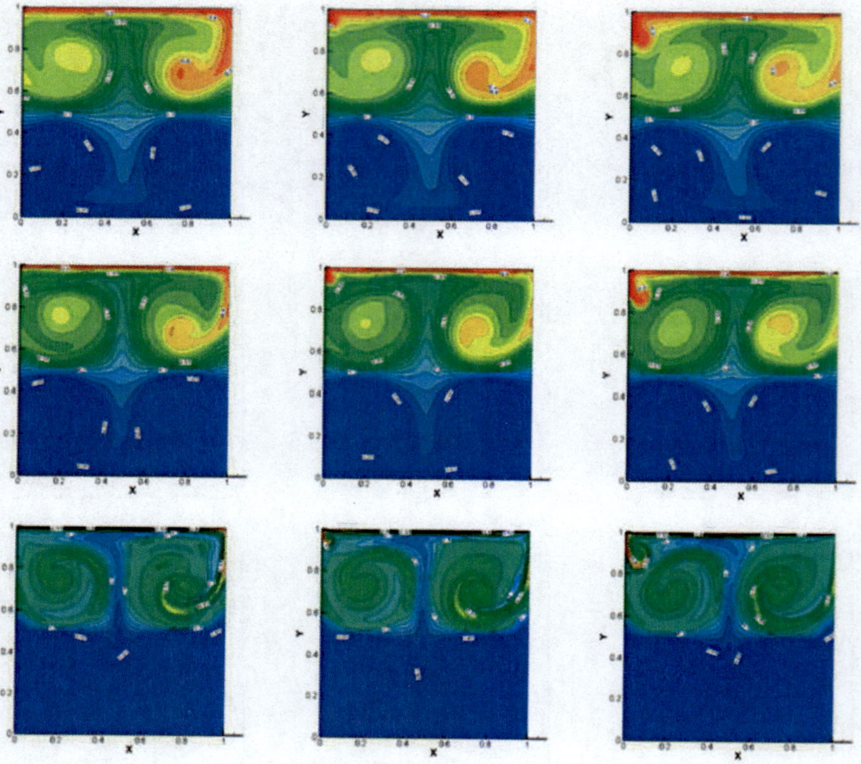

Fig. 7 Iscoconcentration lines of parallel oscillation for Re = 2000 by varying Sc from top to bottom (0.5, 1.0, 5.0) and time period left to right(0.3T, 0.4T, 0.5T)

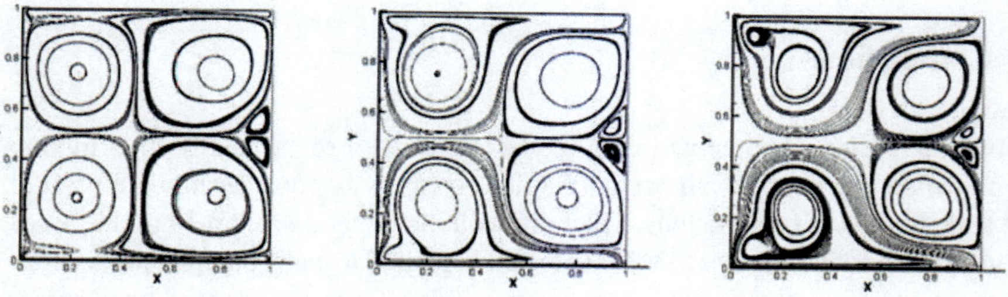

Fig. 8 Stream line plot of parallel oscillation for Re = 2000 by varying Sc from top to bottom (0.5, 1.0, 5.0) and time period left to right(0.3T, 0.4T, 0.5T)

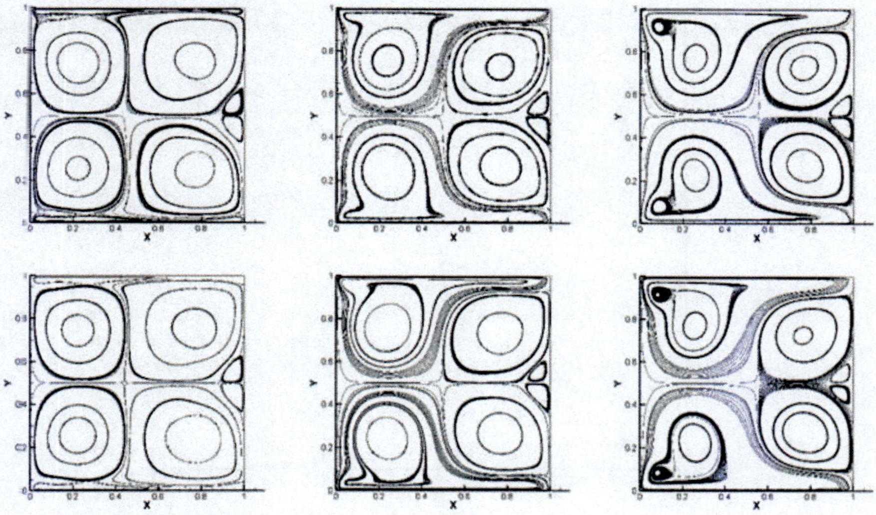

Fig. 8 (continued)

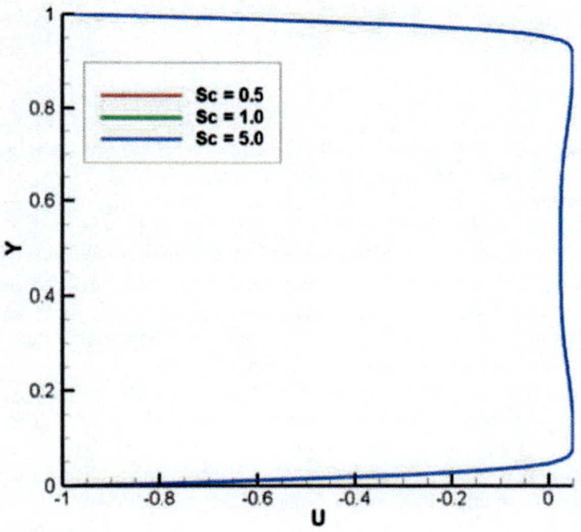

(a)

Fig. 9 Variation of centreline U velocity along the y-axis for different Sc number for Re = 2000 **a** parallel wall oscillation **b** antiparallel wall oscillation at $t = 0.5$T

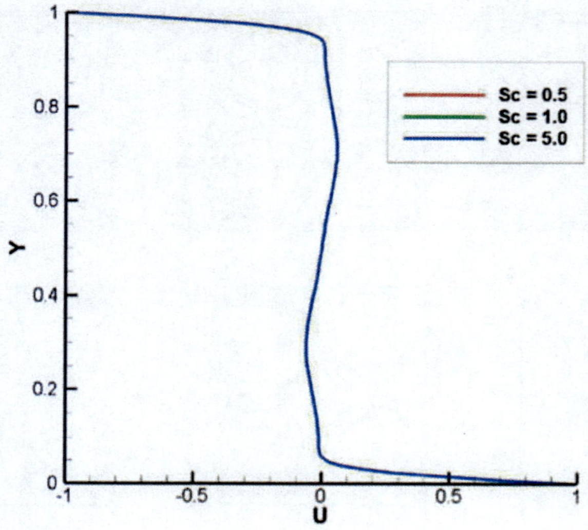

(b)

Fig. 9 (continued)

References

1. Al-Amiri AM, Khanafer KM, Pop I (2007) Numerical simulation of combined thermal and mass transport in a square lid-driven cavity. Int J Therm Sci 46:662–671
2. Benghein C, Haghighat F, Allard F (1992) Numerical study of double -diffusion natural convection in a square cavity. Int J Heat Mass Transfer 35(4):833–846
3. Bettaibi S, Kuznik F, Sediki E (2016) Hybrid LBM-MRT model coupled with finite difference method for double-diffusive mixed convection in rectangular enclosure with insulated moving lid. Physica A 444:311–326
4. Bhopalam SR, Perumal DA, Yadav AK (2021) Computational appraisal of fluid flow behaviour in two-sided oscillating lid-driven cavities. Int J Mech Sci 196
5. Ghaffarpasand O (2018) Characterization of unsteady double-diffusive mixed convection flow with soret and dufour effects in a square enclosure with top moving lid. J Heat Mass Transfer Res 5:51–68
6. Ghia U, Ghia KN, Shin CT (1982) High-Re solutions for incompressible flow using the Navier-stokes equationa and a multigrid method. J Comput Phys 48(3):387–411
7. Hung F, Wang D, Li Z, Gao Z, Derksen JJ (2019) Mixing process of two miscible fluids in a lid-driven cavity. Chem Eng J 362:229–242
8. Indukuri JV, Maniyeri R (2018) Numerical simulation of oscillating lid driven square cavity. Alexandria Eng J 57:2609–2625
9. Mendu SS, Das PK (2013) Fluid flow in a cavity driven by an oscillating lid—a simulation by lattice Boltzmann method. Eur J Mech B Fluids 39:59–70
10. Nishimura T, Kunitsugu K (1997) Fluid mixing and mass transfer in two-dimensional cavities with time periodic lid velocity. Int J Heat Fluid Flow 18:497–506
11. Rao P, Duggleby A, Stremler MA (2012) Mixing analysis in a lid driven cavity flow at finite Reynolds number. J Fluids Eng 134

Thermal and Fluid Property Analysis of Biodiesel Produced from Waste Cooking Oil Using Sodium Methoxide and Calcium Oxide

Abhishek Patel, Hinal Vachhani, and Yash Patel

1 Introduction

Industrialization and commercial advancements have changed the global energy scenario in recent years. The focus has been shifted towards meeting the needs for every sector. A rise in modernization is being observed in the developed as well as the developing nations. This, in turn, has burdened the energy requirements around the globe. The basic living standards for a major part of the society has increased, thus stressing more onto the already depleting resources. A direct reflection of this scenario is evident over the drastic reduction in availability of conventional fuel reserves. Although a lot of technological advancements have been made to squeeze out every ounce of resources present, it does not suffice and is certainly not adequate in today's situation [1–3].

With an over-exploited non-renewable energy sector, various groups have already worked upon focusing the efforts on the aspect of sustainable future. Thus, in recent decades, a lot of work focuses on renewable energy sources. Although methods to tap the wind, hydro and solar power have been in full swing, these come with their own geological limitations which cannot be overcome by humans [4–6].

A shift of focus is now seen towards utilization of biomass and the agro-products, either the produce itself or the residues. The residues have a lot of potential but come with their own drawbacks. This has led to a minimal amount of work being done commercially on agriculture residue-based fuels [7–9]

It can be clearly noted that there is an increase in oil production in India. This is in view of the rise in demand and supply. By this, we can infer that a lot of used edible oil is also being generated. Thus, channelling the efforts towards conversion

A. Patel (✉) · H. Vachhani · Y. Patel
Bachelor of Chemical Engineering, Department of Chemical Engineering, G H Patel College of Engineering and Technology, Anand, Gujarat, India
e-mail: abhishekpatel345@gmail.com

J. Banerjee et al. (eds.), *Recent Advances in Fluid Dynamics*, Lecture Notes in Mechanical Engineering, https://doi.org/10.1007/978-981-19-3379-0_8

Table 1 Bio oil production data in 10^8 kgs [10, 11]

Oil year (November–October)	Production of oilseeds*	Total availability of edible oils
2015–16	252.5	234.8
2016–17	312.76	254.16
2017–18	314.59	249.72
2018–19#	315.22	259.22
2019–20##	334.23	240.04

*Ministry of agriculture
**Directorate general of commercial intelligence and statistics (Ministry of Commerce)
#Based on final estimates (declared by the Ministry of Agriculture on 18.02.2020)
##Based on fourth advance estimates (declared by the Ministry of Agriculture on 19.08.2020)

of waste edible oils into energy sources would lead us closer towards meeting the ever-increasing energy needs (Table 1).

Various methods can be listed to produce biofuel from waste edible oil such as thermal cracking, transesterification, microemulsions with solvents like ethanol and others [12–14]. These processes include some being catalytic while others being non-catalytic processes. The catalytic process for production seems more useful since more output is observable at lower temperature and pressure conditions. The catalysts can be differentiated into being heterogeneous or homogeneous catalysts [15, 16].

It is obvious that not only the energy production is required, but also the qualitative analysis is important to know the feasibility of any new product. The global pollution effects as well as the severity of concerns regarding material handling, transportation and application make it necessary for analysing the properties. The thermal and fluid properties of the biofuel would make such concerns clear.

Biodiesel as biofuel proves to be a better option for approach. The easy to produce, lower emissions and alternative to conventional diesel makes it a viable alternative. Since India already has a vast source of waste edible oil, using it will give a positive push towards energy security since India has been largely dependent on imports for energy needs. Biodiesel has near similar emissions as compared to conventional diesel. The ease of direct usage in engines and added lubricating properties make biodiesel a good option. Analysis of various other properties indicates that biodiesel can be the go-to fuel [17–19].

The advantages though very reasonable, the challenges cannot be overlooked. The policies related to biodiesel inclusion in the energy sector are very inconsistent in India. The problems such as choking of injectors in engines due to higher viscosity, a higher engine wear factor and reduced power output are a point of debate.

The use of edible oil for biodiesel production raises a concern of it competing with the more important factor of national food security. Moreover, competing with the food sector would mean a rise in deforestation, which the nation cannot afford as of now [20].

Table 2 Waste cooking oil properties

Sr No.	Properties	Values
1	Kinematic viscosity (@40 °C, cSt)	43.34
2	Heating value (MJ/kg)	32.8
3	Density (@40 °C, g/cm^3)	0.8962
6	Pour point	3
7	Cloud point	7
9	Moisture content (wt. %)	0.19
10	Acid value (mg KOH/g oil)	3.2

Along with the challenges come the opportunities and their solutions knocking the door. Overcoming such impediments would result in a clear path towards a sustainable energy sector. Along with these, since many research groups have recently focused on the following catalysts, this study is also using these catalysts for a better retrospection.

2 Materials and Methods

2.1 Waste Cooking Oil

After filtering out the unnecessary cooking debris, the cooking vegetable oil used in the frying of onion and potato fritters was collected and stored in a glass container at room temperature. The cooking oil used in the preparation of these fritters was purchased at the market's lowest price from a nearby supermarket. The physiochemical parameters of waste vegetable oil were determined individually and given in Table 2. A pycnometer and viscometer were used to estimate density and viscosity, respectively, while acid-base titration procedures were used to assess FFA and acid levels [21].

2.2 Catalysts

Sodium methoxide (CH_3ONa). An exothermic reaction occurs at room temperature between sodium and methanol, yielding in the metal salt solution sodium methoxide, for which an equation is presented below.

$$2Na + 2CH_3OH \rightarrow 2CH_3ONa + H_2 \quad (1)$$

The resultant is a saccharomyces that can then be employed as a source of sodium methoxide, which serves as a homogeneous alkali catalyst in this case. Sodium methoxide appears to be a white solid upon further refining (evaporation and heating). Solid sodium methoxide, on the other hand, has a high air instability and is promptly deteriorated. The molecular weight, melting point, boiling point and specific gravity are 54.02, 127 °C, 350 °C and 1.1, respectively. Sodium methoxide dissolves readily in water, ethanol, and methanol, but not as readily in hydrocarbons.

Calcium oxide (CaO). Calcium oxide is produced as a byproduct of the thermal decomposition of calcium carbonate, along with carbon dioxide. The following is an equation for this.

$$CaCO_3 \rightarrow CaO + CO_2 \tag{2}$$

Once obtained through the aforementioned process, it appears to be a white solid. In this research project, it acts as a heterogeneous alkali catalyst. Calcium oxide has a molecular weight of 56.08, a melting point of 2572 °C, a boiling point of 2850 °C and a specific gravity of 3.34. In terms of solubility, calcium carbonate is easily soluble in both water and glycerol.

2.3 Biodiesel Production

Waste vegetable oil (WVO) was first acid catalysed with 0.10 N sulphuric acid and methanol. This process was performed to remove all superfluous water and high FFA content from the WVO, as well as to minimize oil loss due to saponification. In a temperature-controlled magnetic plate stirrer, 500 ml WVO, 0.8 ml Sulphuric acid and 100 ml methanol were agitated. For 1 h, the stirrer's revolution and reaction temperature are kept at 400 rpm and 60 °C, respectively. The mixture was then allowed to settle for an hour, forming two distinct layers: a methanol layer on top and a blend of methyl ester and unreacted triglycerides on the bottom [21]. A separating funnel was used to separate these. The bottom layer then was collected and employed in the alkali transesterification reaction.

In this study, the methanol: Oil ratio was 5:1, with a homogeneous alkali catalyst (sodium methoxide (CH_3ONa) concentration of 3.5% wt. and a heterogeneous alkali catalyst (calcium oxide (CaO)) concentration of 4% wt. used in two separate trials [22, 23]. The reaction temperature for the experiments with sodium methoxide was optimized to 60 °C while for the experiments with calcium oxide catalyst, it was optimized to 100 °C. Thereafter, 100 ml of WVO, determined methanol concentration and sodium hydroxide catalyst are placed in a magnetic stirrer and reacted at 400 rpm for 5 h. Once the reaction was completed, and it was poured into a separating funnel and allowed to settle for another hour to produce distinct layers. The top layer composed of crude biodiesel, while the bottom layer comprised of glycerol. The crude biodiesel, along with catalysts and methanol, is achieved by removing the glycerol from the bottom. Methanol was retrieved from the crude by vacuum distilling it at 80 °C. Then after, the biodiesel is washed with hot water (55 °C) until its pH is neutral, and then dried with anhydrous sodium sulphate [24–26]. As a direct consequence, biodiesel is finally available.

3 Result and Discussion

The waste cooking oil that was procured was analysed for its properties. These have been given in Table 2.

In order to compare the two different catalyst activities over the waste cooking oil, an optimum catalyst percentage was derived by comparing the amount of catalyst used with the biodiesel yield. This was graphically represented in Fig. 5. The optimum catalyst amounts were found to be 3.5 wt.% and 4 wt.% for sodium methoxide and calcium oxide, respectively. In a similar fashion, the optimum temperature for the biodiesel production process was determined by comparing the yield of biodiesel samples obtained with the temperature of the process. It was found that the biodiesel produced by using sodium methoxide had best yield at 60 °C while that produced by using calcium oxide as catalyst has the peak yield at 100 °C.

The data obtained from the process of finding optimum catalyst percentage and the optimum temperature was further used to compare the yields obtained by using the two catalysts sodium methoxide and calcium oxide, keeping reaction time as the common basis of 5 h. Biodiesel yield was calculated using the following equation.

Fig. 1 Calcium oxide (Left) and sodium methoxide (Right) (catalysts)

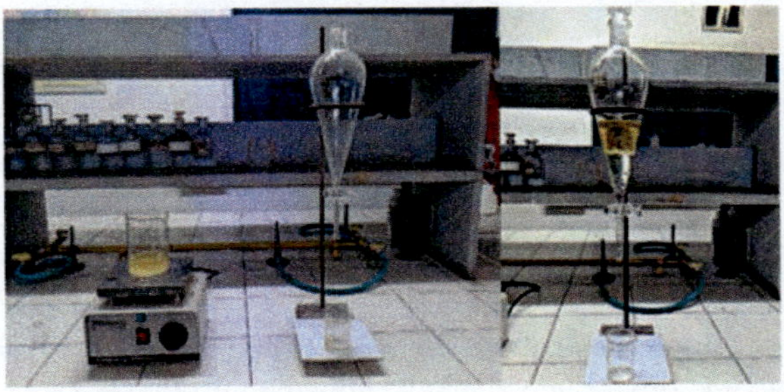

Fig. 2 Experimental setup including hotplate magnetic stirrer and separating funnel

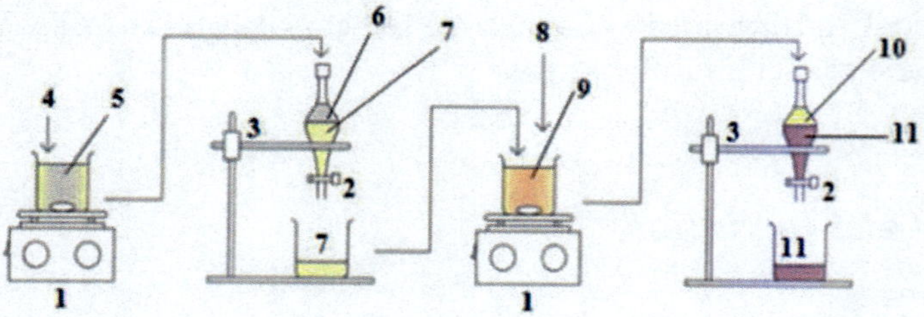

Fig. 3 Basic flow diagram of biodiesel production. **1** Temperature-controlled magnetic stirrer, **2** Separating funnel, **3** Holder, **4** Mix of WVO, H_2SO_4 and CH_3OH, **5** Reaction mix of four, **6** CH_3OH, **7** Methyl ester and Unreacted triglycerides, **8** Catalyst, **9** Reaction mix of seven and eight, **10** Biodiesel and **11** Glycerine

Fig. 4 Biodiesel 1(Sodium methoxide) (Left) and biodiesel 2 (Calcium oxide) (Middle) compared with standard diesel (Right)

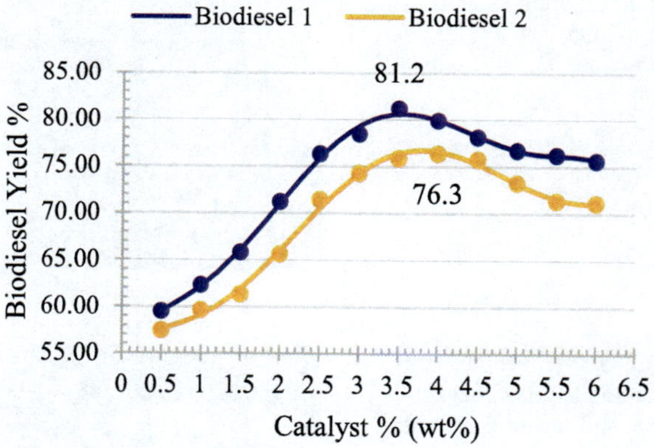

Fig. 5 Optimum catalyst versus biodiesel yield %

$$\text{Biodiesel Yield \%} = \frac{\text{Mass of Biodiesel obtained}}{\text{Mass of WVO used}} \tag{3}$$

The biodiesel samples that were obtained were analysed and compared for their yield and properties. Planned comparisons (Tables 3 and 4) reveal that properties of the biodiesel highly depend upon the reaction conditions and the catalyst used. As per the experimentation a higher yield was observed in case of biodiesel made using

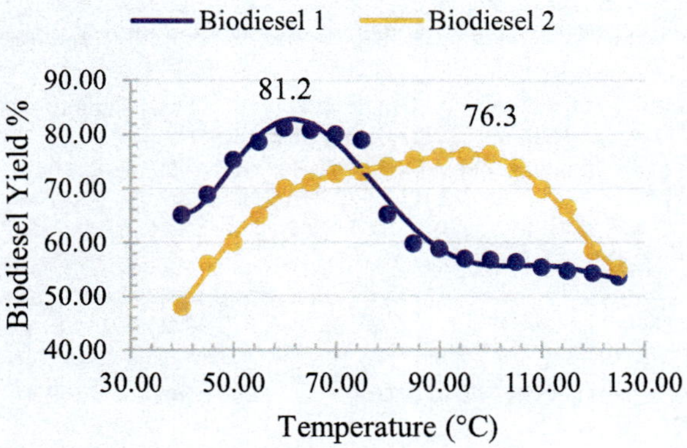

Fig. 6 Temperature versus biodiesel yield %

Table 3 Reaction conditions

Sr. No.	Trials/runs	Oil used	Catalyst used	Catalyst Amt. (wt. %)	Temperature (°C)	Time (h)	Biodiesel yield (wt. %)
1	1	Waste cooking oil	Sodium methoxide	3.5	60	5	81.2
2	2	Waste cooking oil	Sodium methoxide	3.5	60	5	79.9
3	1	Waste cooking oil	Calcium oxide	4	100	5	76.3
4	2	Waste cooking oil	Calcium oxide	4	100	5	75.8

sodium methoxide catalyst as compared to calcium oxide catalyst. The heating values when compared show that the biodiesel obtained by sodium methoxide catalysis was higher than that of the biodiesel obtained from calcium oxide catalysis. It was also inferred from the analysis that there was no drastic change in other properties as listed in Table 4.

It may be attributed to the high degree of interaction due to similar phases of the catalyst and WVO, thus the yield obtained for sodium methoxide is higher than that of calcium oxide.

A gradual shift from homogeneous catalyst to heterogeneous catalyst has been seen recently. Previous studies outline the fast reaction rates of the homogeneous catalyst as well as mild reaction conditions with easier handling, high selectivity and many more features. However, there is a spike in post-production cost especially in

Table 4 Comparison between properties of biodiesel attained from various catalysts with standard diesel properties

Sr. No.	Properties	Biodiesel 1 (sodium methoxide)	Biodiesel 2 (calcium oxide)	Diesel sample	ASTM D6571
1	Kinematic viscosity (@40 °C, cP)	4.4	4.1	3.4	4–6
2	Heating value (MJ/Kg)	34.02	30.83	44.6	−33 to −36
3	Density (@40 °C, g/cm^3)	0.8536	0.8409	0.79	–
4	Flash point (°C)	158	162	84	100–170
5	Fire point (°C)	163	168	91	–
6	Pour point (°C)	−3	1	15	−15 to 10
7	Cloud point (°C)	4	7	19	−3 to 12
8	Cetane no.	46.4	42.9	49	48–60
9	Carbon residue (wt. %)	0.24	0.2	0.15	– -
10	Acid value (mg KOH/g)	0.56	0.47	0.2	0.8 max
11	Sulphur content (wt. %)	0.02	0.02	0.10	0–15 ppm

purification and separation that leads us to think of alternative ways to make biodiesel. Overall, these findings are in accordance as reported by other researchers [27–30].

4 Conclusion

Several biodiesel production technologies have been proposed. Nevertheless, transesterification is considered the best option among all known procedures. In the transesterification reaction, a primary catalyst, such as alike and varied catalysts, is utilized. In terms of biodiesel production rate, the employment of homogeneous catalysts has been proven to be promising. However, it is not without flaws. The homogeneous catalyst-based transesterification reaction costs a lot. Furthermore, the existence of unreacted compounds necessitates further downstream processing. Owing to these limits, the creation of effective catalysts was required, which was achieved by heterogeneous catalysts. The above catalysts drew significant attention from the scientific world across the planet because of their numerous benefits over homogeneous catalysts, such as the output of relatively pure glycerol, the overall lack of alkaline catalyst neutralizing step and the absence of the need to replace the spent catalyst. For these merits, many current research groups are interested in heterogeneous catalysts. However, various parameters impact the solid catalyst's reactivity; the most important is the molar ratio of alcohol to oil, type of reactor and temperature. The optimal catalyst dosages of 4 g for calcium oxide and 3.5 g for sodium methoxide reflect this. Moreover, the optimal temperature for ultimate yield when using Biodiesel 1 (sodium methoxide) and Biodiesel 2 (calcium oxide) was established to be 60 °C and 100 °C, respectively. Consequently, the average yield for sodium methoxide at 3.5 g was revealed to be 80.55%, while that for calcium oxide at 4 g was shown to be 76.05%, demonstrating that choosing these variables at an ideal level is an important step. As a result, scientists are looking to nanotechnology and other catalysts for fresh hope.

Acknowledgements The authors are grateful to G. H. Patel College of Engineering and Technology, Vallabh Vidyanagar, Gujarat for their immense support and the Department of Chemical Engineering GCET for providing the required resources. This study would not have been possible without the guidance of Dr. Kaushik Nath, Head of Chemical Engineering Department, GCET and Vinay Patel, Associate Professor.

References

1. Arto I, Capellán-Pérez I, Lago R, Bueno G, Bermejo R (2016) The energy requirements of a developed world. Energy Sustain Dev 33:1–13
2. Dyatlov SA, Didenko NI, Ivanova EA, Soshneva EB, Kulik SV (2020) Prospects for alternative energy sources in global energy sector. In: IOP conference series: earth and environmental science, Vol 434, no 1. IOP Publishing, pp 012014
3. Ahmad T, Zhang D (2020) A critical review of comparative global historical energy consumption and future demand: the story told so far. Energy Rep 6:1973–1991
4. Li L, Lin J, Wu N, Xie S, Meng C, Zheng Y, Wang X, Zhao Y (2020) Review and outlook on the international renewable energy development. Energy Built Environ
5. Xu X, Wei Z, Ji Q, Wang C, Gao G (2019) Global renewable energy development: Influencing factors, trend predictions and countermeasures. Resour Policy 63:101470
6. Barbir F, Veziroğlu TN, Plass HJ Jr (1990) Environmental damage due to fossil fuels use. Int J Hydrogen Energy 15(10):739–749
7. Shyam M (2002) Agro-residue-based renewable energy technologies for rural development. Energy Sustain Dev 6(2):37–42
8. Akram MK, Talukdar IA, Khan MS (2020) Agro residual biomass conversion: a step towards pollution control and sustainable waste management. In: Smart cities—opportunities and challenges. Springer, Singapore, pp 887–892
9. Bhatia RK, Ramadoss G, Jain AK, Dhiman RK, Bhatia SK, Bhatt AK (2020) Conversion of waste biomass into gaseous fuel: present status and challenges in India. BioEnergy Research 13:1046–1068
10. https://dfpd.gov.in/oil-division.htm
11. https://www.nfsm.gov.in/StatusPaper/NMOOP2018.pdf
12. Laksmono N, Paraschiv M, Loubar K, Tazerout M (2013) Biodiesel production from biomass gasification tar via thermal/catalytic cracking. Fuel Process Technol 106:776–783

13. Karpagam R, Jawaharraj K, Gnanam R (2021) Review on integrated biofuel production from microalgal biomass through the outset of transesterification route: a cascade approach for sustainable bioenergy. Sci Total Environ 766:144236
14. Espinosa EAM, Rodríguez RP, Sierens R, Verhelst S (2016) Emulsification of waste cooking oils and fatty acid distillates as diesel engine fuels: an attractive alternative. Int J Sustain Energy Plan Manage 9:3–16
15. Rizwanul Fattah IM, Ong HC, Mahlia TMI, Mofijur M, Silitonga AS, Rahman SMA, Ahmad A (2020) State of the art of catalysts for biodiesel production. Front Energy Res 8:101
16. Mardhiah HH, Ong HC, Masjuki HH, Lim S, Lee HV (2017) A review on latest developments and future prospects of heterogeneous catalyst in biodiesel production from non-edible oils. Renew Sustain Energy Rev 67:1225–1236
17. Mishra VK, Goswami R (2018) A review of production, properties and advantages of biodiesel. Biofuels 9(2):273–289
18. Yusuf NNAN, Kamarudin SK, Yaakub Z (2011) Overview on the current trends in biodiesel production. Energy Convers Manage 52(7):2741–2751
19. Lin L, Cunshan Z, Vittayapadung S, Xiangqian S, Mingdong D (2011) Opportunities and challenges for biodiesel fuel. Appl Energy 88(4):1020–1031
20. Abd Manaf IS, Embong NH, Khazaai SNM, Rahim MHA, Yusoff MM, Lee KT, Maniam GP (2019) A review for key challenges of the development of biodiesel industry. Energy Convers Manage 185:508–517
21. Khoobbakht G, Kheiralipour K, Rasouli H, Rafiee M, Hadipour M, Karimi M (2020) Experimental exergy analysis of transesterification in biodiesel Production. Energy 196:117092
22. Metre AV, Nath K (2015) Super phosphoric acid catalyzed esterification of palm fatty acid distillate for biodiesel production: physicochemical parameters and kinetics. Polish J Chem Technol 17(1)
23. Metre A, Nath K (2015) Palm fatty acid distillate based biodiesel: characterization and emission analysis
24. Zhao H, Cao Y, Orndorff W, Cheng YH, Pan WP (2012) Thermal behaviors of soy biodiesel. J Therm Anal Calorim 109(3):1145–1150
25. Elkady MF, Zaatout A, Balbaa O (2015) Production of biodiesel from waste vegetable oil via KM micromixer. J Chem
26. Kamel DA, Farag HA, Amin NK, Zatout AA, Fouad YO (2019) Utilization of Ficus carica leaves as a heterogeneous catalyst for production of biodiesel from waste cooking oil. Environ Sci Pollut Res 26(32):32804–32814
27. Lawer-Yolar G, Dawson-Andoh B, Atta-Obeng E (2021) Synthesis of biodiesel from tall oil fatty acids by homogeneous and heterogeneous catalysis. Sustain Chem 2(1):206–221
28. Singh B, Jain S, Gangil B (2021) Effectiveness of homogeneous and heterogeneous catalyst on biodiesel yield: a review. Adv Clean Energy Technol 375–385
29. Kiss FE, Jovanović M, Bošković GC (2010) Economic and ecological aspects of biodiesel production over homogeneous and heterogeneous catalysts. Fuel Process Technol 91(10):1316–1320
30. Pasae Y, Tangdilintin S, Bulo L, Allo EL (2020) The contribution of heterogeneous and homogeneous catalysts towards biodiesel quality. J Phys Conf Ser 1464(1):012054 (IOP Publishing)

Experimental Investigation on Use of Activated Alumina and Molecular Sieve 13× In Heatless Desiccant Air Dryer

A. J. D'souza and P. K. Brahmbhatt

Nomenclature

M	Total water adsorbed, kg
t	Temperature, °C
P	Pressure, MPa
c	Drying cycle, hours
W	Water content of gas, kg/10^6 Std m^3
V	Volume, m^3
D	Bed Diameter, m
Q	Gas flow rate, 10^6 Std m^3/day

Subscripts

B	Bed
V	Vessel
g	Gas

A. J. D'souza (✉)
Gujarat Technological University, Ahmedabad, India
e-mail: alicedsz08@gmail.com

P. K. Brahmbhatt
Government Engineering College, Dahod, India

J. Banerjee et al. (eds.), *Recent Advances in Fluid Dynamics*, Lecture Notes in Mechanical Engineering, https://doi.org/10.1007/978-981-19-3379-0_9

1 Introduction

Pneumatics is widely used in industrial applications especially in the field of automobile, aerospace, construction, mining, agriculture and chemicals. Pneumatics involves the utilization of compressed air to perform mechanical operations [16]. Compressors are widely used for different applications of pneumatics. They compress atmospheric air or gas, and the atmospheric air contains water droplets in saturated form. These concentrated water droplets in saturated form can be harmful to instrumentation, air system infrastructure, as well as the end product [1]. Hence it is desirable to have moisture free compressed air. One of the ways to remove moisture from compressed air is by the use of air dryers. Compressed air dryers are designed on the basis of required dew point which signifies the temperature at which air will saturate with moisture. Moment the temperature of the air decreases to or lesser than the dew point, condensation is bound to happen. It is suggested as dew point decreases, moisture present in the air reduces [10]. Moisture removal from compressed air is the prime work of an air dryer. Air dryer are broadly classified on the basis of the integration of the technology used to achieve the moisture removal, which is namely, refrigeration, regenerative, deliquescent and membrane [16]. Refrigerant dryers tend to have high popularity when dry air is needed at +2 °C. The high temperature compressed air enters the refrigerant dryer where its temperature is reduced. This reduction in temperature causes the saturated water droplets to condense which are then removed, thus, leaving the compressed air dryer. However, this dryer tends to have limitation of dew point temperature which cannot go below +20 °C [16]. Hence in applications where temperature characteristically is in the −40 °C to −100 °C dew point range can be addressed by desiccant dryers [15]. Desiccant dryers can be further classified into, single tower deliquescent and twin tower regenerative. Twin tower regenerative closed loop desiccant dryers consist of the following types: Blower Heater Non-Purge (BHNP) with/without water cooling pumps, Compressed air Heater Purge (CHP), Blower Heater Purge (BHP) and Pressure Swing Heater-less (PSH) which are characteristically used in industries where requirement of extremely dry air is essential in the range of pressure dew point −40 °C to −100 °C. This extremely dry air is used in sensitive electronics, food manufacturing, pharmaceuticals and hospital surgical air [1]. The distinguishing feature in these dryers is the regeneration method used to replenish the saturated desiccants. Methods consist of using selective supplementary equipment such as blower, heater, purged compressed air or sometimes combination of these equipment for moisture removal. Twin tower regenerative desiccant dryers as the name suggest consist of two identical towers filled with solid adsorbents. Moisture laden compressed air is passed through the first tower, where moisture is adsorbed by the solid adsorbents. Once the tower reaches its saturation, the moisture laden compressed air is rerouted to the second tower, whilst the first tower regenerates. This cycle continues to give dry compressed air [1]. Regeneration method employed dominates the type of dryer. The type of desiccant, regeneration temperature, purge intake humidity and desiccant bed operation temperature all influence the regeneration process' efficacy [14]. Solid adsorbents in granule form for example silica gel, activated alumina and zeolites (molecular sieve) are used in heatless air dryer to diminish the content of water vapour in compressed air [14, 17].

Solid desiccants are porous materials that adsorb water through a variety of methods, including chemical adsorption onto the walls of pores, multilayer physical adsorption of water molecules, and capillary condensation into the holes. The crystalline structure of the surface area mainly attracts the moisture [14]. Several naturally occurring desiccants such as silica gel [2] which is non-toxic and easily available along with zeolites which are alkali aluminosilicate minerals that include alkaline earth metals. [17], thermally treated hydrides and oxides of aluminium known as activated alumina [17]. Synthetic zeolites also called as molecular sieves [3], activated carbon, polymer desiccants and composite desiccants [21] where the host desiccants are impregnated with other materials to enhance the adsorption and water retaining capacity are also used in dehumidification applications. Solid desiccants utilization in desiccant wheels [4, 7, 12, 13], heat pumps [18] and heat exchanger [19, 20] is also gaining momentum. Using artificial neural networks [9, 11] many solid desiccant systems have been analysed. Solid desiccants can be classified as follows (Fig. 1).

Solid desiccants provide the following advantages over liquid desiccants: (a) no leakage, (b) minute or negligible corrosion and environmental concerns and (c) minuscule maintenance [6]. Literature survey indicates the potential of research that can be done to investigate the use of activated alumina and molecular sieve 13× in heatless air dryer. For externally heated packed bed dryers, the majority of previous investigations were conducted using models or experimentation. To the best of the authors' knowledge, no previous study of solid desiccant dryers has included a complete investigation of a heatless air dryer using compressed air. Also, the effect of the dew point temperature has yet not been explored by most researchers in their work. This was the primary rationale for doing the current experimental study on the use of activated alumina and molecular sieve in Heatless Desiccant Air Dryer. A series of experiments were conducted in order to determine the dew point temperatures that could be achieved in the heatless air dryer that was constructed. Moreover, the influence of a heterogenous mixture of the two solid desiccants and its effect has been discussed in detail. The results reveal that changing the desiccant has a significant impact on system performance. The change in dew point temperature over a period of time has also been displayed.

Molecular sieve 13×. The molecular sieve is a crystalline aluminium silicate substance that can separate molecules of different proportions during sorption. [8]]. Dehumidification of air to very low dew points in the region of −40 °C to −60 °C is aided by molecular sieves. Molecular sieves are distinguished primarily on the basis of their pore size. Molecular sieve 13× having pore size of approximately 10 angstroms can be considered as the largest in the A type. Synchronized adsorption for bi-molecule and tri-molecule is observed, hence it is preferred basically for refinements of gases and liquids. It finds its applications in dehumidification and air conditioning and compressor air utilization [3, 8].

Activated alumina. Activated alumina is a group of aluminium hydrides and oxides formed by thermal dehydration or activation of aluminium trihydrate or gibbsite. The pore size of activated alumina is 1.5–6 nm, the surface area is 150–500

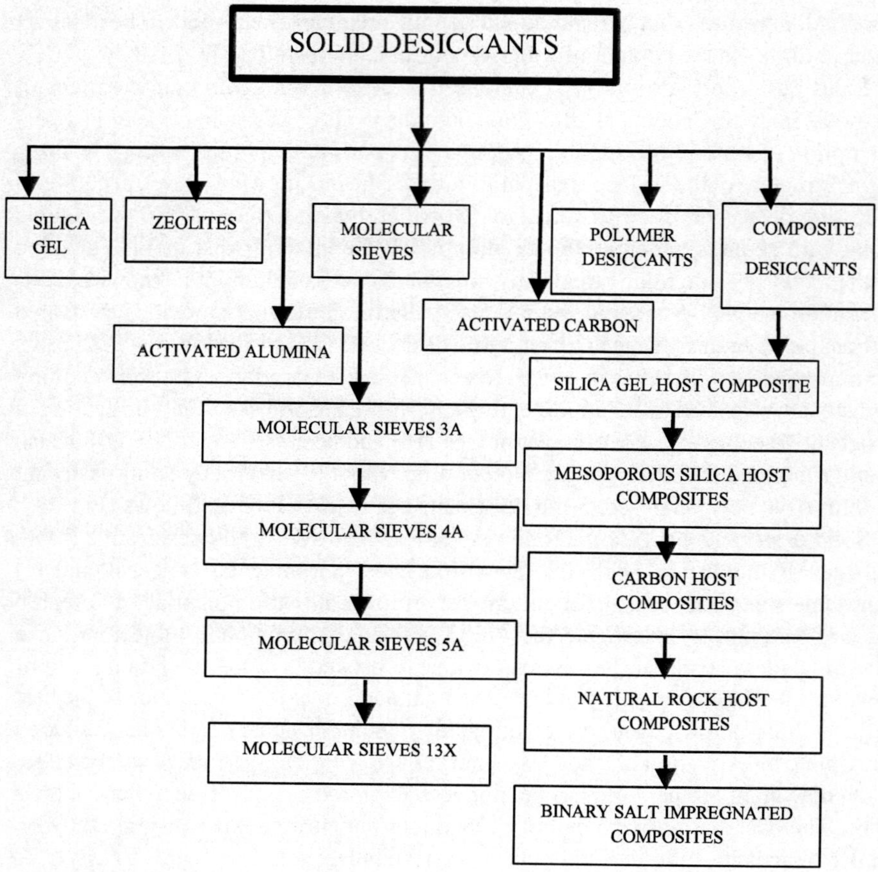

Fig. 1 Classification of solid desiccants [3, 8]

m^2/g, and the heat adsorption capability is 3000 kJ/kg. It is mostly employed in desiccant dehumidification and cooling applications owing to its excellent sorbent characteristics.

From Fig. 2, it can be observed that the surface pores of molecular sieve 13× are more uniformly distributed in comparison to activated alumina. Table 1, depicts the physical properties of activated alumina and molecular sieve 13× used in the experimentation.

2 System Description

A twin tower pressure swing heatless air dryer is designed to investigate the performance of activated alumina and molecular sieve 13×. Compressed air with moisture is passed through carbon filter, followed by the after filter and then tower A, where

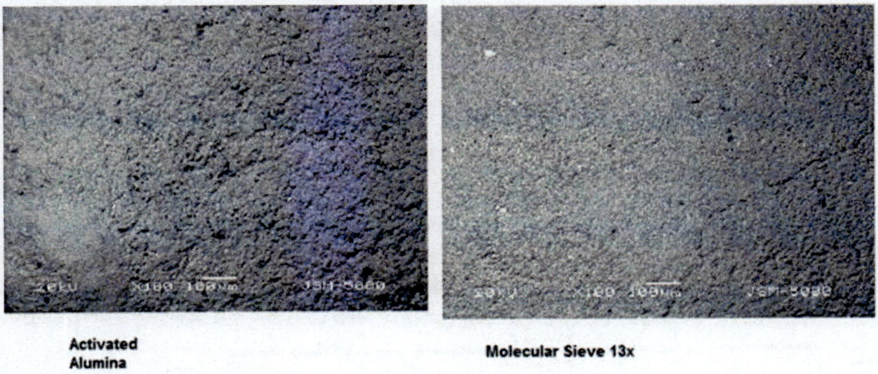

Fig. 2 SEM images of activated alumina and molecular sieve 13x

Table 1 Properties of activated alumina and molecular sieve 13x

Sr. No.	Properties	Activated alumina	Molecular sieve 13×
1	Bed crushing strength	95%	90%
2	Particle size	5–8 mm	3–5 mm
3	Tapped bulk density	800 kg/m^3	700 kg/m^3

the desiccant adsorbs moisture. The dryness of the compressed air is measured with the help of dew point metre. Fractional dry compressed air is passed through the tower-B, where desiccant which is in saturation state is regenerated. The cycle time for the switch between the towers is kept at 5 min. A timer is used along with a solenoid valve to change the airflow from tower-A to tower-B.

It is essential to estimate the moisture content of the compressed air for the design and operation of a heatless air dryer. Moisture content in the compressed air entering the dryer can be obtained by:

$$W = 593.335 \times \text{EXP}\,(0.05486 \times t_g) \times (-0.81452) \quad (1)$$

It is reasonable to assume that the desiccant bed adequately absorbs all of the moisture. Considering 8 h as the cycle length for the running of the dryer per day. The amount of moisture removed per cycle can be given as:

$$M = (\text{Q cW})/24 \quad (2)$$

Desiccant's water vapour capacity is generally measured as mass of water vapour adsorbed per unit mass of desiccant.

$$V = \frac{\text{Mass of water vapour adsorbed}}{\text{Mass of Desiccant}} \tag{3}$$

Accordingly, the volume is obtained and the towers are designed for the heatless air dryer for 10 CFM compressed air [5]. The following is a diagram of the experimental setup (Figs. 3 and 4).

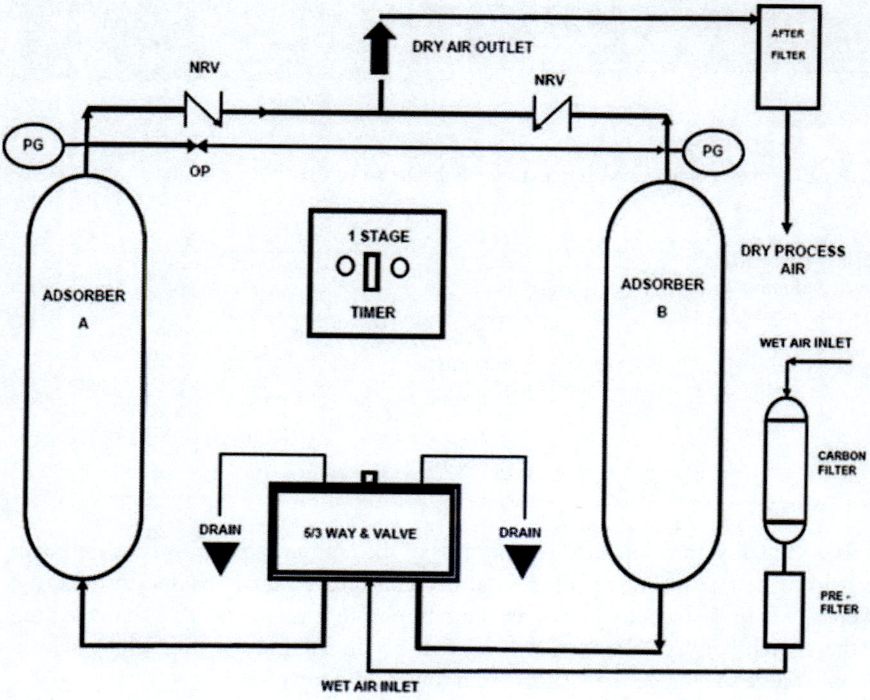

Fig. 3 Diagram of the experimental test rig

Fig. 4 Detailed view of the experimental set up. PG: pressure gauge. CT: control timer. HD: heatless air dryer. SV: solenoid valve. AF: after filter. CF: carbon filter. DPT: dew point transmitter. DL: data logger. T2: tower-B. T1: tower-A. PF: pre filter

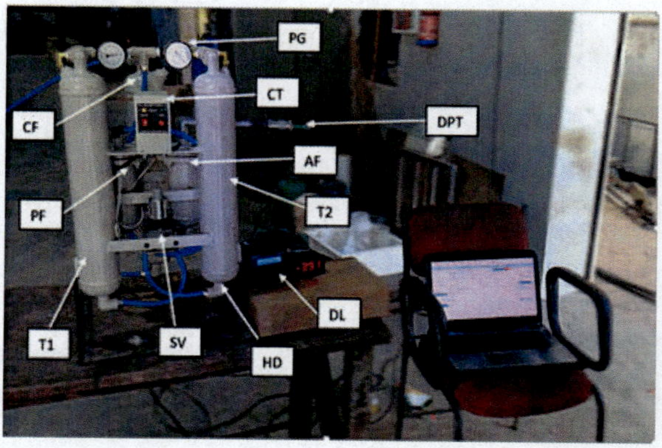

For this experiment heatless type desiccant air dryer (also known as pressure swing regenerative dryer), built for the capacity of 10 CFM is used. Compressor compresses the air, which is having a pressure of 7 bar. It is first supplied to the activated carbon filter in order to remove the oil content in the compressed air. As oil can be adsorbed by the solid desiccant, which can make the desiccant inactive. After passing through the activated carbon filter, the air passed through the pre filter which has the filtration efficiency of 5 μm. After the pre filter the air enters the desiccant tower-A or -B as per the operating condition.

The towers are completely packed with desiccant to ensure no desiccant particle is carried over in the compressed air stream, steel straps at the openings of the tower are installed. Desiccant material filled in the tower adsorbs the moisture of the compressed air and makes it dry. The dry air is now sent outside of the tower. Fraction of this dry air, referred as purge air, is bypassed via thin tube to the tower-B for the regeneration process. Remaining air, via non-return valve (also called check valve), is passed to the output line from where it goes to the after filter which removes dust particle if added by the desiccant (desiccant particles). Rendering clean, compressed, dry air.

The purge air is supplied to the tower-B from top. Bottom of the tower-B is kept open to the atmosphere, hence pressure in the tower-B is lower. The dry compressed purge air regenerates the desiccants. Now the cycle changes. Compressed air is passed through the tower-B whilst the tower-A gets regenerated. This cycle changes its direction after an interval of 5 min.

This cycle is controlled by a timer, which is attached with the solenoid valve, which controls the motion of the 5/2 direction control valve. This 5/2 control valve controls the direction of the compressed air (towards tower-A or tower-B) and opening of the tower-A or tower-B.

3 Instrumentation

The following instruments were utilized to measure the different variables such as temperature, pressure during the experiment.

(1) Dew point Metre (Accuracy: ±0.5 °C)
(2) Pressure gauge (Range: 25 kg/cm^2, Accuracy: ±0.25 kg/cm^2)
(3) Data logger

4 Experimentation Procedure

Case 1: Both towers fully filled with activated alumina. In this case both the towers are fully filled with the activated alumina as desiccant material. The total weight of the activated alumina, is 3.5 kg. The compressed air having 7 bar pressure and 45 °C temperature is passed at the rate of 10 CFM. When the dry air comes out from the dryer as an output, its pressure and temperature are 6.8 bar and 47 °C, respectively. The pressure dew points achieved, at every interval, are shown in Fig. 5.

Case 2: Both towers fully filled with molecular sieve 13×. In this case both the towers are fully filled with the molecular sieve as desiccant material. The total weight of the molecular sieve, is 2.9 kg. The compressed air having 7 bar pressure and 45 °C temperature is passed at the rate of 10 CFM. When the dry air comes out from the dryer as an output, its pressure and temperature are 6.8 bar and 48 °C, respectively. The pressure dew points achieved, at every interval, are shown in Fig. 6.

Case 3: Both towers are filled with the mixture of activated alumina and molecular sieve. In this case both the towers are fully filled with equal quantity of activated alumina and molecular sieve forming a heterogeneous mixture. The total weight of the mixture is 3.2 kg considering 1.75 kg of activated alumina and 1.45 kg of molecular sieve 13×. The compressed air having 7 bar pressure and 45 °C temperature is passed at the rate of 10 cfm. When the dry air comes out from the dryer as an output, its pressure and temperature are 6.8 bar and 50 °C, respectively. The pressure dew points achieved, at every interval, are shown in Fig. 7.

Temperatures along the bed length were recorded using three thermowells installed along the bed of tower A. The results cand be seen in Figs. 8, 9 and 10, respectively for all the three cases.

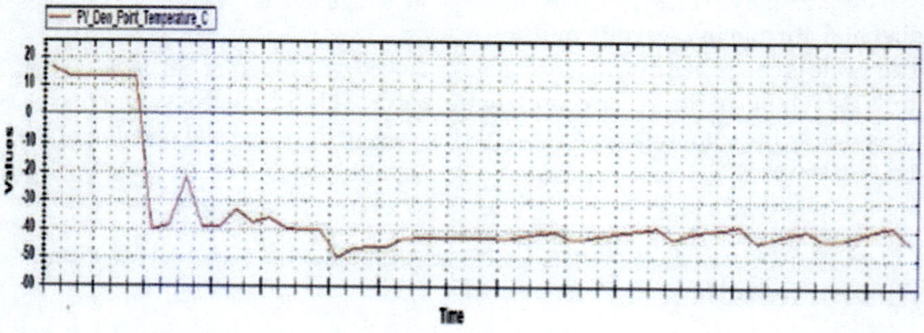

Fig. 5 Activated alumina dew point versus time plot

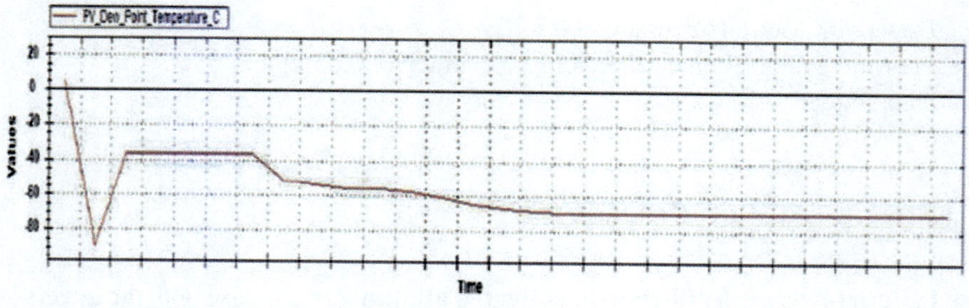

Fig. 6 Molecular sieve 13× dew point versus time plot

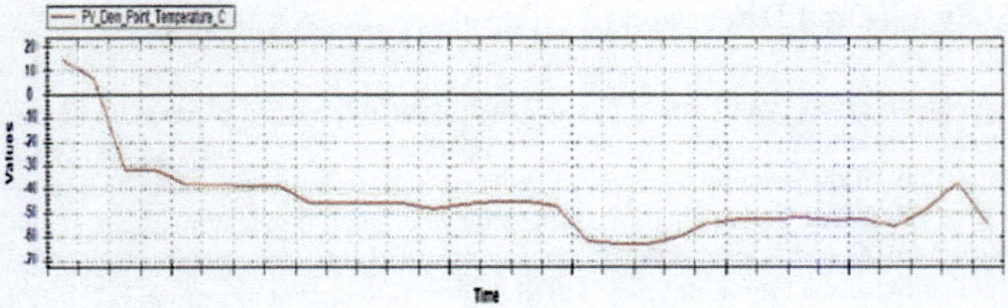

Fig. 7 Heterogenous mixture of activated alumina and molecular sieve 13x, dew point versus time plot

Fig. 8 Temperature variation in tower-A filled with molecular sieve 13× along the bed length

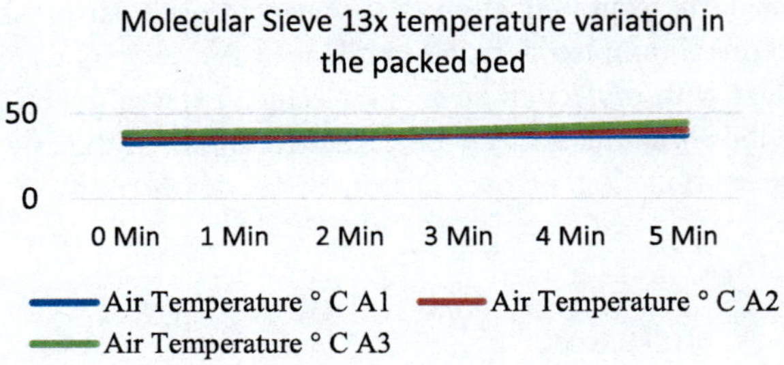

Fig. 9 Temperature variation in tower-A filled with activated alumina along the bed length

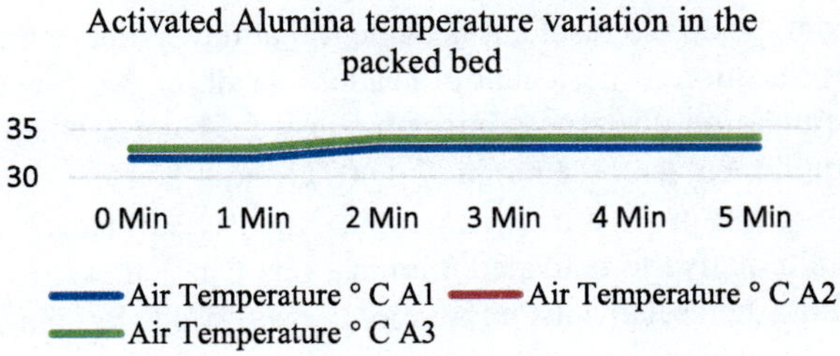

Fig. 10 Temperature variation in tower-A filled with mixture of molecular sieve 13× and activated alumina along the bed length

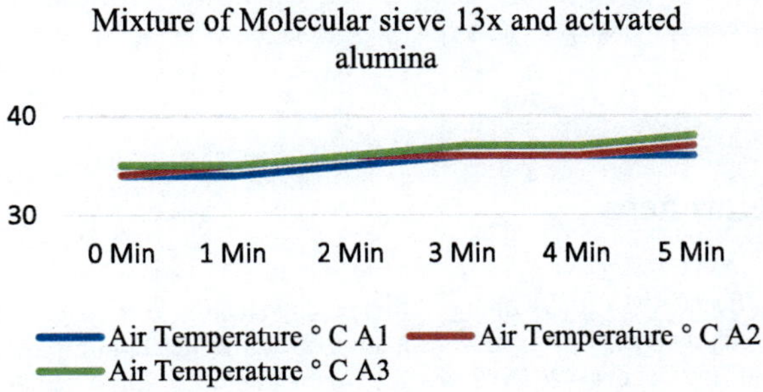

5 Results and Discussion

It can be observed from Figs. 5, 6 and 7 that when all physical parameters of the bed size, i.e. length and diameter along with parameters such as inlet pressure and temperature of the compressed air are kept same, the lowest dew point temperature up to −70 °C can be obtained by using molecular sieve 13×. Whilst dew point temperature of −40 °C can be obtained using activated alumina. However, when the combination of the two desiccants is used dew point temperatures around −50 °C can be obtained in the heatless air dryer.

It can be observed from Figs. 8, 9 and 10 that the temperature along the bed length is relatively uniform throughout. The increase in the temperature of the solid desiccant is an indication of the heat released during adsorption process. However, the maximum temperature gradient of $\Delta T = 7$ °C was observed when the bed was filled with molecular sieve 13×. Thus, externally cooling the packed bed was not found essential. Excess temperature can affect the rate of adsorption but owing to the short cycle time over heating of the desiccant was evaded.

6 Conclusion

From the above results it is evident that dew point temperature plays a vital role in the selection of desiccant in heatless air dryer. As dew point temperature indicates the measure of dryness in the process air. Hence, when the dew point temperature requirement is in the range of −40 °C and −50 °C instead of going for an oversized equipment or for a relatively more expensive desiccant, i.e. molecular sieve 13× as an alternative to activated alumina, one can achieve the said desired dew point by using a heterogeneous mixture of the activated alumina and molecular sieve 13×.

Acknowledgements We thank Dr D. B. Jani who provided insight and expertise that greatly assisted the research.

References

1. Botts AM (2019) Energy efficiency evaluation of blower heater non-purge compressed air dryers energy efficiency evaluation of blower heater non-purge compressed air dryers
2. Chen CH et al (2016) Silica gel/polymer composite desiccant wheel combined with heat pump for air-conditioning systems. Energy 94:87–99. https://doi.org/10.1016/j.energy.2015.10.139
3. Cui Q, et al (2005) Performance study of new adsorbent for solid desiccant cooling. Energy 30(2–4 SPEC. ISS.):273–279. doi:https://doi.org/10.1016/j.energy.2004.05.006
4. Dhar PL, Singh SK (2001) Studies on solid desiccant based hybrid air-conditioning systems. Appl Therm Eng 21(2):119–134. https://doi.org/10.1016/S1359-4311(00)00035-1
5. Gandhidasan P, Al-Farayedhi AA, Al-Mubarak AA (2001) Dehydration of natural gas using solid desiccants. Energy 26(9):855–868. https://doi.org/10.1016/S0360-5442(01)00034-2

6. Grossman G, Johannsen A (1981) Solar cooling and air conditioning. Prog Energy Combust Sci 7(3):185–228. https://doi.org/10.1016/0360-1285(81)90011-3
7. Higashi T et al (2018) Gravimetric method for sorption performance measurement of desiccant wheel and desiccant coated heat exchanger. Appl Therm Eng 144:639–646. https://doi.org/10.1016/j.applthermaleng.2018.05.079
8. Jani DB, Mishra M, Sahoo PK (2016) Performance prediction of solid desiccant—Vapor compression hybrid air-conditioning system using artificial neural network. Energy 103:618–629. https://doi.org/10.1016/j.energy.2016.03.014
9. Jani DB, Mishra M, Sahoo PK (2017) Application of artificial neural network for predicting performance of solid desiccant cooling systems—a review. Renew Sustain Energy Rev 80(May):352–366. https://doi.org/10.1016/j.rser.2017.05.169
10. Kannan VS, Arjunan TV, Vijayan S (2020) Experimental investigation of temperature swing adsorption system for air dehumidification. Heat and Mass Transfer/Waerme-und Stoffuebertragung 56(7):2093–2105. https://doi.org/10.1007/s00231-020-02841-w
11. Koronaki IP, Rogdakis E, Kakatsiou T (2012) Thermodynamic analysis of an open cycle solid desiccant cooling system using artificial neural network. Energy Convers Manage 60:152–160. https://doi.org/10.1016/j.enconman.2012.01.022
12. Panaras G, Mathioulakis E, Belessiotis V (2007) Achievable working range for solid all-desiccant air-conditioning systems under specific space comfort requirements. Energy Build 39(9):1055–1060. https://doi.org/10.1016/j.enbuild.2006.10.015
13. Panaras G, Mathioulakis E, Belessiotis V (2011) Solid desiccant air-conditioning systems—design parameters. Energy 36(5):2399–2406. https://doi.org/10.1016/j.energy.2011.01.022
14. Rambhad KS, Walke PV, Tidke DJ (2016) Solid desiccant dehumidification and regeneration methods—a review. Renew Sustain Energy Rev 59:73–83. https://doi.org/10.1016/j.rser.2015.12.264
15. Singh RP, Mishra VK, Das RK (2018) Desiccant materials for air conditioning applications—a review. IOP Conf Ser Mater Sci Eng 404(1). doi:https://doi.org/10.1088/1757-899X/404/1/012005
16. Sreenivasa CG et al (2013) A case study on mapping air dryer capabilities from agile manufacturing perspectives. Int J Serv Oper Manage 16(1):86–104. https://doi.org/10.1504/IJSOM.2013.055574
17. Sultan M et al (2015) An overview of solid desiccant dehumidification and air conditioning systems. Renew Sustain Energy Rev 46:16–29. https://doi.org/10.1016/j.rser.2015.02.038
18. Tu R, Hwang Y, Ma F (2017) Performance analysis of a new heat pump driven multi-stage fresh air handler using solid desiccant plates. Appl Therm Eng 117:553–567. https://doi.org/10.1016/j.applthermaleng.2017.02.005
19. Weixing Y et al (2008) Study of a new modified cross-cooled compact solid desiccant dehumidifier. Appl Therm Eng 28(17–18):2257–2266. https://doi.org/10.1016/j.applthermaleng.2008.01.006
20. Zhao Y et al (2016) A high performance desiccant dehumidification unit using solid desiccant coated heat exchanger with heat recovery. Energy Build 116:583–592. https://doi.org/10.1016/j.enbuild.2016.01.021
21. Zheng X, Ge TS, Wang RZ (2014) Recent progress on desiccant materials for solid desiccant cooling systems. Energy 74(1):280–294. https://doi.org/10.1016/j.energy.2014.07.027

A Three-Dimensional CFD Investigation of Nozzle Effect on the Vortex Tube Performance

Nitin Bagre, A. D. Parekh, and V. K. Patel

Nomenclature

L	Tube length, mm
D	Tube diameter, mm
VT	Vortex tube
3D	Three-dimensional
ΔT_{cold}	Difference of cold exit and inlet, K
ΔT_{hot}	Difference of hot exit and inlet, K
μ	Cold mass fraction

1 Introduction

A vortex tube is a mechanical device used for cooling and heating purposes. It is an eco-friendly device that operates on compressed air. The single injection of compressed air with high pressure generated the cool and hot air simultaneously. A French physicist Ranque invented the vortex tube in 1937. Further, Hilsch German scientist promoted the model in 1947. Therefore, it is called as the Ranque-Hilsch

N. Bagre (✉) · A. D. Parekh · V. K. Patel
Department of Mechanical Engineering, Sardar Vallabhbhai National Institute of Technology, Surat, India
e-mail: ds18me008@med.svnit.ac.in

A. D. Parekh
e-mail: adp@med.svnit.ac.in

V. K. Patel
e-mail: vkp@med.svnit.ac.in

J. Banerjee et al. (eds.), *Recent Advances in Fluid Dynamics*, Lecture Notes in Mechanical Engineering, https://doi.org/10.1007/978-981-19-3379-0_10

vortex tube. It is compact in size and used for various industrial applications. Experimental and CFD analysis by Rafiee and Sadeghiazad [1] of a convergent type vortex tube in which different throttle angles are selected to examine the performance. Results found the optimum angle of throttle valve for the highest magnitude of temperature separation 60°. Matveev and Jacob [2] a modified cyclonic type vortex chamber has been proposed and investigated. It is found that this modification in a vortex tube improves the performance up to 15% as compared to the original design. Li et al. [3] experimentally studied the internal flow mechanism of the vortex tube. In this work, probes and thermocouples are located on various positions to obtain the flow characteristics inside the tube. The study found that the reverse flow varies with the location and operating conditions inside the tube. Also, a low magnitude of static temperature found near the walls. Aghagoli and Sorin [4] concluded that for a specific cold mass fraction and pressure, the temperature separation can be improved by 10–78.9 °C for hot and 44.2–9.7 °C for cold outlet when CO_2 employed as a working fluid. Bianco et al. [5] a comparative analysis has been done to obtain the most feasible turbulence model to predict the temperature separation inside the double vortex tube arrangement. A limited impact was found by RSM-LRR and LES turbulence model in a double circuit vortex tube. Chen et al. [6] a CFD approach to calculated the temperature gradient in a vortex tube when H_2 gas used as an operating fluid. The results showed the existence of two different streamline structure with an inlet gas temperature fluctuating between 50 and 294 K. Kiran and Ashok [7] studied the effect of various parameters like cold orifice diameter, length and diameter ratio, and angles of throttle valve on the performance of vortex tube. According to this work, it is found that the loss of exergy can be decreased with an increase of orifice diameter of cold exit.

2 Objective

It has been observed from the past investigation that the design of vortex tube has great influences on the effectiveness of the vortex tube. The design parameters which affect the performance of the vortex tube are nozzle number, L/D ratio, vortex chamber, the material of vortex tube and cold orifice diameter. It is found in a study of Thakare and Parekh [8] when N_2 is used as a working fluid it delivered high-temperature separation compare to air and other gases. Attalla et al. [9] show that the optimum nozzle number can enhance the performance. Also, Kirmaci and Uluer [10] suggested that the amount of temperature separation of vortex tube improve as the nozzle number increases up to a certain limit. In the previous cases, the optimization of nozzle number has been done when air was used as a working fluid. So, it is decided to find an optimum number of nozzles for a vortex tube when nitrogen gas is employed as a working fluid. Also, the objective of the present work is to study the flow behaviour of nitrogen gas inside the tube and to predict the temperature magnitude at the outlets.

3 Governing Equations of CFD

Air is used as a working fluid considering compressible and Newtonian fluid. Which resulting the following equation and to solve the following equations, FLUENT 18.2 was used.

- Continuity Equation

$$\frac{\partial \rho}{\partial t} + \frac{\partial}{\partial x}(\rho u) + \frac{\partial}{\partial y}(\rho v) + \frac{\partial}{\partial z}(\rho w) = 0 \tag{1}$$

- Energy Equation

$$\frac{\partial \rho h_o}{\partial t} - \frac{\partial P}{\partial t} + \mathrm{div}(\rho h_o U) = div(\lambda grad \mathrm{T}_s) \tag{2}$$

- Momentum Equation

$$\frac{\partial}{\partial t}(\rho \vec{v}) + \nabla \cdot (\rho \vec{v}\vec{v}) = -\nabla p + \nabla \cdot (\overline{\overline{\tau}}) + \rho g + \vec{F} \tag{3}$$

4 Modelling and Boundary Conditions

To analyze nozzle's effects, various geometries have been created. In present work, three different geometries containing 1, 2, and 4 number of nozzles were carried out to analysis the performance and predict the temperature of the vortex tube with nitrogen (N_2) as a working fluid. The inlet pressure condition was kept 8 bar for all the cases, whereas the hot exit pressure varies and cold exit remains constant. The hot and cold exit diameter are 9 and 5 mm, whereas the length of the tube was 105 mm for all the cases. Figures 1, 2, and 3 show the generated vortex tube containing one, two, and four number of nozzles. Figure 4 represents the mesh generated for four number of nozzle of a vortex tube. Fluent 18.2 has been selected to solve the continuity, momentum, and energy equations with the help of standard k-ε turbulence.

5 Validation

It is found that the obtained result of present numerical simulation is in good agreement when equated with experimental and numerical outcomes of Dincer et al. [11] and Ouadha et al. [12] as shown in Fig. 2. It indicates that the simulation approach followed in this work is suitable. The remaining temperature difference is due to the boundary conditions.

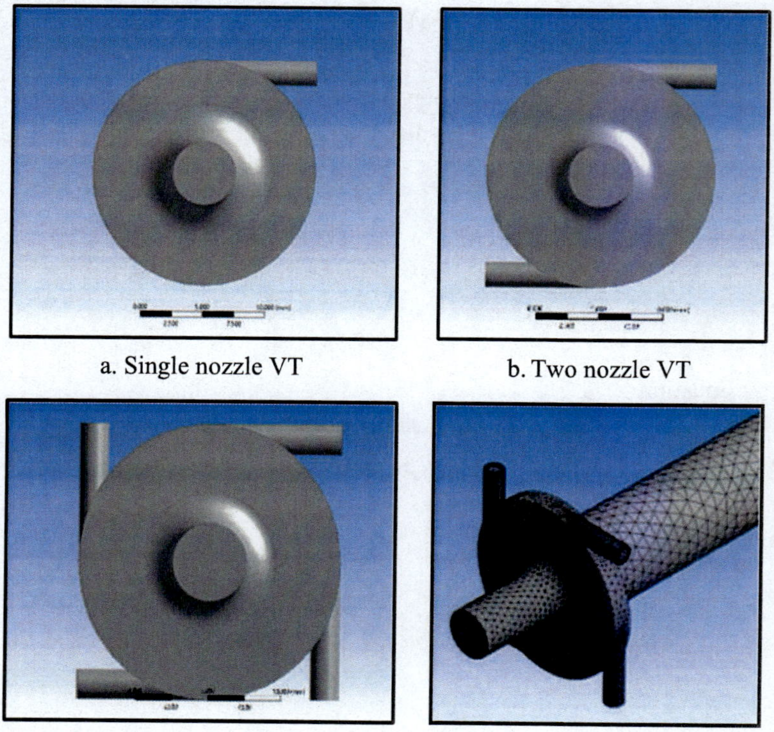

a. Single nozzle VT

b. Two nozzle VT

c. Four nozzle VT

d. Generated mesh for four nozzle VT

Fig. 1 Generated geometries and mesh

Also, the gas is considered to be ideal with no slip adiabatic conditions. The magnitude of temperature difference between hot and cold keeps on increasing as the cold mass fraction increases. The fluid particles move towards the peripheral wall increase the temperature separation. Also, the magnitude temperature separation dropped as it moves axially towards the end of the tube.

6 Result and Discussion

The study deals with the optimization of nozzle number for a vortex tube when nitrogen gas is utilised as an operating fluid. The vortex tube contained three various geometries in which the nozzle numbers are varied. The number of nozzles are $N = 1$, 2, and 4. The magnitude of temperature separation is obtained at 8 bar inlet pressure and cold mass fraction from 0.1 to 0.9. In Fig. 3, the deviation of streamline flows with respect to the nozzle number. It is found that the highest magnitude of velocity is found in two nozzle number, and in one and four nozzle number, the magnitude

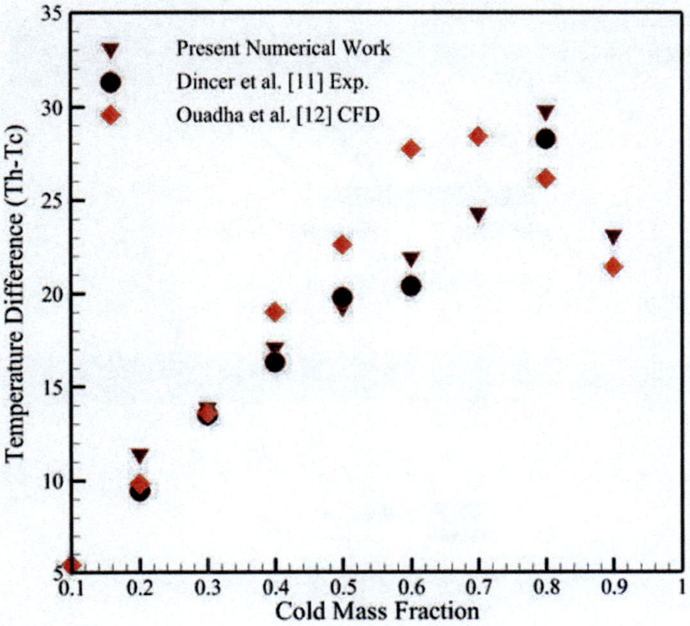

Fig. 2 Predicted temperature separation (hot and cold) values when pressure is 200 kPa

remains less. The prediction of cold temperature separation inside the vortex tube is shown in Fig. 5 for all three cases. The maximum cold temperature found for $N = 2$ as compared to the other geometry. According to Shamsoddini and Abolpour [13], the temperature gradient decreases due to increment of number of nozzles. The similar trend can be observed in this work as the nozzle number extended beyond $N = 2$, the magnitude of the temperature decreases.

The hot temperature separation keeps on increasing as the cold mass fraction increases shown in Fig. 6. The highest magnitude is found in single nozzle vortex tube as the nozzle number increased the hot temperature gradient decreases. It can be observed that vortex tube containing $N = 1$ and 2 achieved almost same temperature for nitrogen. The vortex tube having two nozzle succeed to reach the hot temperature separation 60 k at the cold mass fraction of 0.9. Moreover, in cold temperature separation, the temperature gradually decreases as the flow moves towards the high-cold mass fraction. The temperature separation contour shown in Fig. 4 which shows that the temperature gradient developed near the inlet wall and cold region. Also, it diminished as it moves towards the hot exit of the tube. The expansion of nitrogen clearly visible through this computational work in Fig. 4.

The temperature gradient is generated due to wall friction and the friction between the fluid adjacent layers. The sudden expansion of fluid inside the tube generated the pressure gradient and temperature gradient as the energy transfer takes place. The main advantage of 3D computational work is that the flow of nozzle can be visualize and can be investigated. Figure 7 shows the vector representation of flow inside the

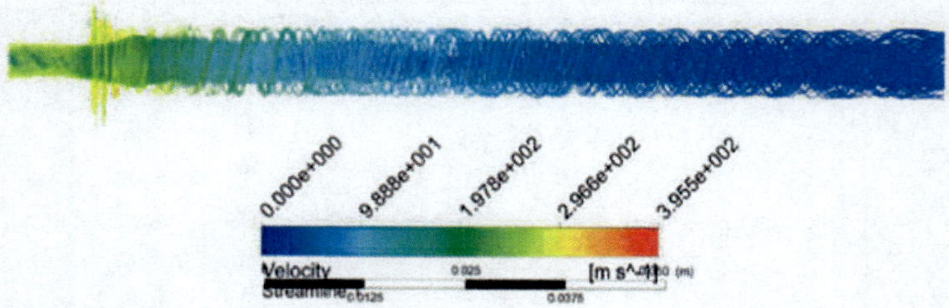

a. Single nozzle VT streamline

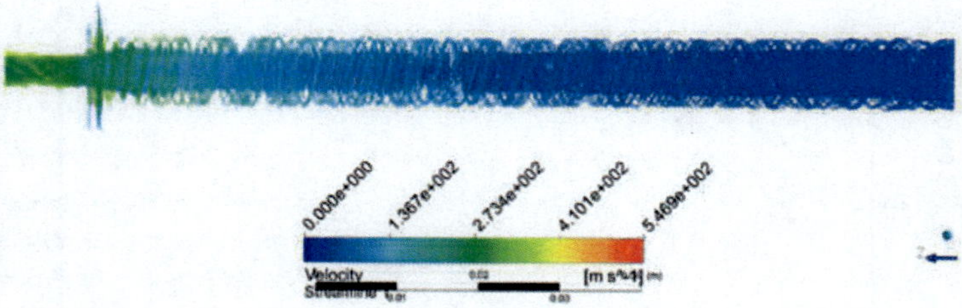

b. Two nozzle VT streamline

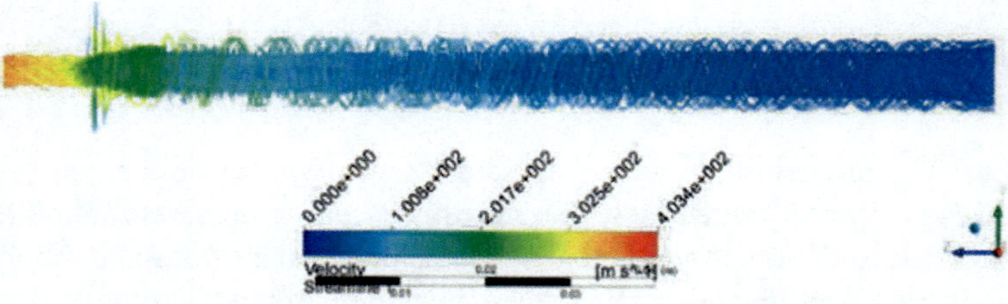

c. Four nozzle VT streamline

Fig. 3 Streamline flow for respective geometries

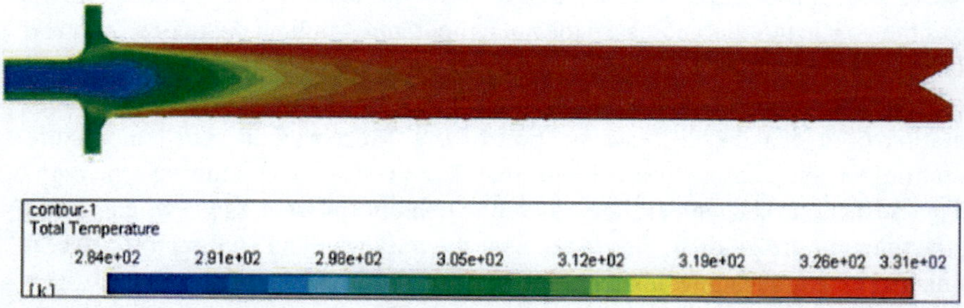

Fig. 4 Vortex tube temperature contour

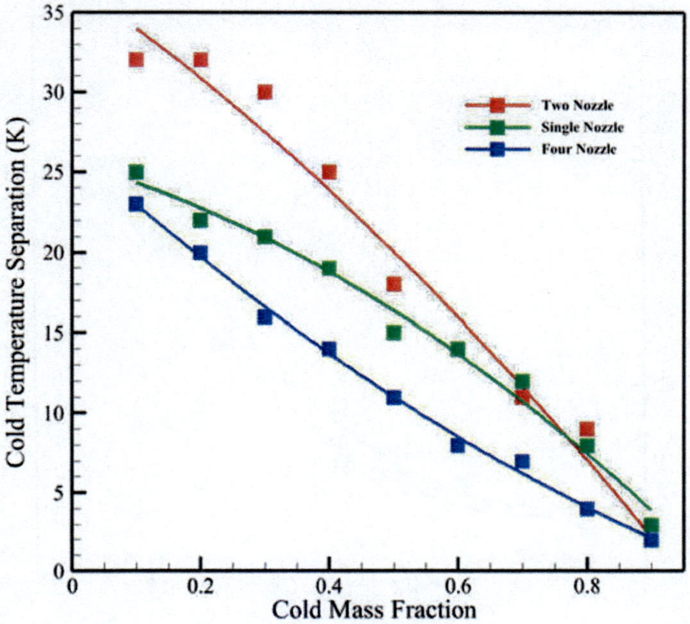

Fig. 5 Influence of cold temperature gradient along with cold mass fraction for three different computational domain

tube from inlet nozzle. It is found that as the flow admitted, it followed a specific path inside the tube. The fluid flow is relevant to the hot and cold exit. The streamline flow also justifies the flow separation of inside the tube.

The velocity varies regarding the radial distance as shown in Figs. 8 and 9. Here, the flow patter of axial velocity and tangential or swirl velocity has been studied. It is found that at the core of the tube, the magnitude of axial velocity is high. The representation of negative region indicating the fall of temperature values and the positive shows the rise of temperature. The negative region of axial velocity caused because of pressure gradient between flow field and cold stream of the tube. The tangential velocity controls the flow mechanism of vortex tube. The expansion of fluid works as the fluid admitted through high pressure, this will generate a high magnitude of tangential velocity. The highest magnitude of tangential velocity is 70 m/s. Figure 8 indicates the fall of velocity magnitude as it moves towards the end of the tube and increases as it moves towards the wall of the tube. It can be observed that a forced vortex is occupied in the most of the region, and free vortex is generated near the wall of the tube.

Figures 8 and 9 show the dominant behaviour of tangential velocity of fluid inside the vortex as equated to axial velocity. The radial velocity magnitude obtained to be negligible than the axial and tangential velocity of vortex tube. Therefore, it has not been described here. The total temperature distribution can be observed from Figs. 10, 11, and 12 predicted by standard k-ε turbulence model for all cases. It shows

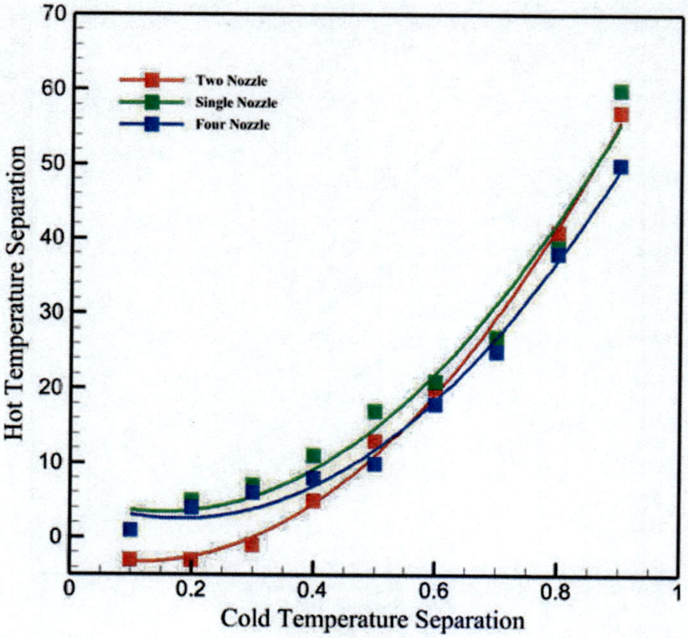

Fig. 6 Influence of hot temperature gradient along with cold mass fraction for three different computational domain

the total temperature deviation with radial distance. The position of total temperature is considered to be at the centre of the tube length $Z = 0.0525$ mm ($L = 105$ mm). The total temperature includes a part of kinetic energy and static temperature in it. It is mentioned that the tangential velocity has a great influence inside the tube due to the high-kinetic energy, so the total temperature can be correlate with the static pressure and tangential velocity in longitudinal path. This conclude that the total temperature separation is generated because of air expansion and pressure gradient in longitudinal direction and also due to the high-kinetic energy of working fluid. The fluid flow transfers from the centre to the end of the tube. These moments of flow generate the temperature and pressure gradient inside the tube. The fluid consists of kinetic energy, and this kinetic energy transforms into thermal energy. The graphical representation of total temperature shows a sharp drop of temperature near the wall. A sudden expansion of fluid clearly shown in Figs. 10, 11, and 12 which justify the flow separation occurred in the vortex tube for various number of nozzles. The reason behind the sharp drop of temperature near the wall is due to the shearing occurs between the layer of fluid and outer layers as the fluid moves in the radial direction. The variation of pressure can also be observed during this work. The pressure falls near the surfaces of the tube due to the impulsive expansion of nitrogen gas inside the tube. The fluid flow along the peripheral of the tube generated the compression of nitrogen causing the high temperature at cold and hot junction.

Fig. 7 Variation of flow inside the vortex chamber of vortex tube

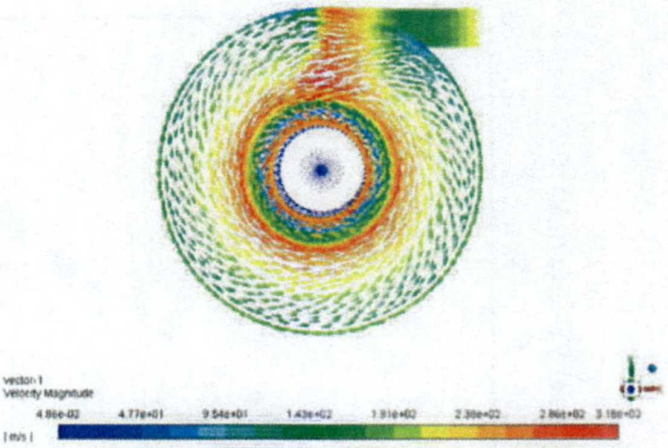

a. Vector representation of flow of single nozzle VT

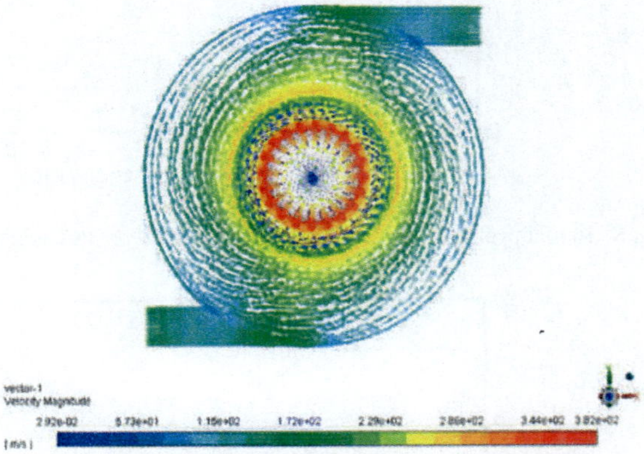

b. Vector representation of flow of two nozzle VT

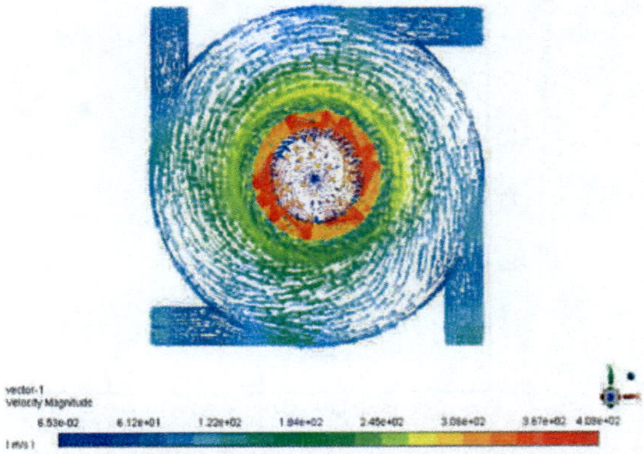

c. Vector representation of flow of four nozzle VT

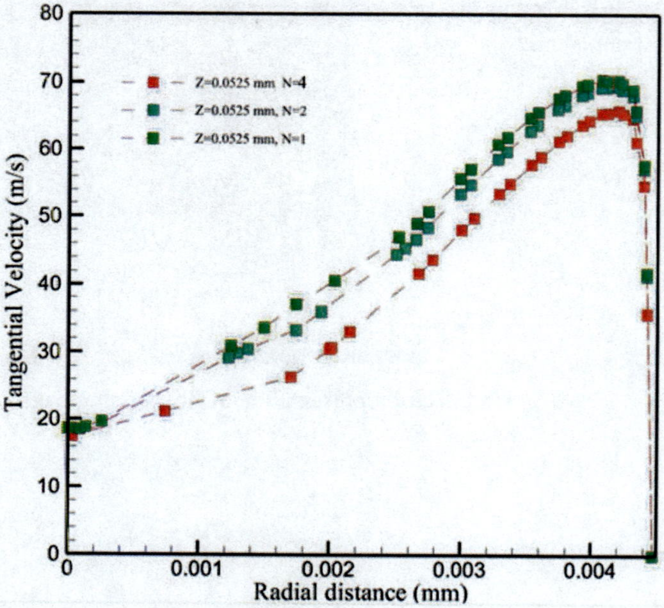

Fig. 8 Radial profile of tangential or swirl velocity for three different vortex tube ($N = 1$, 2 and 4)

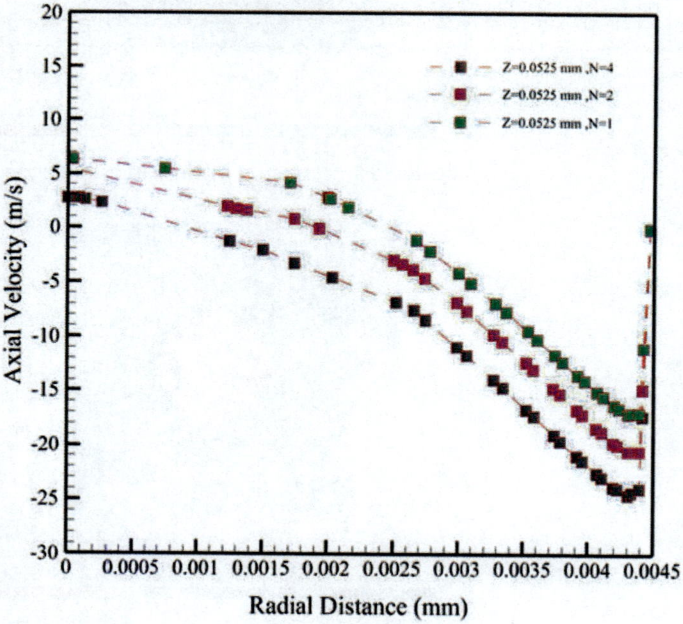

Fig. 9 Axial velocity distribution along with radial direction for three different vortex tube ($N =$ 1, 2, and 4)

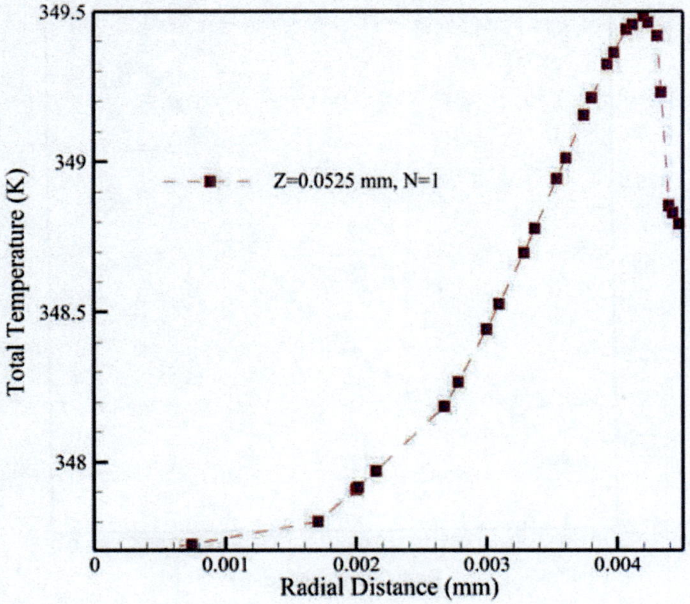

Fig. 10 Total temperature profile regarding radial distance ($N = 1$)

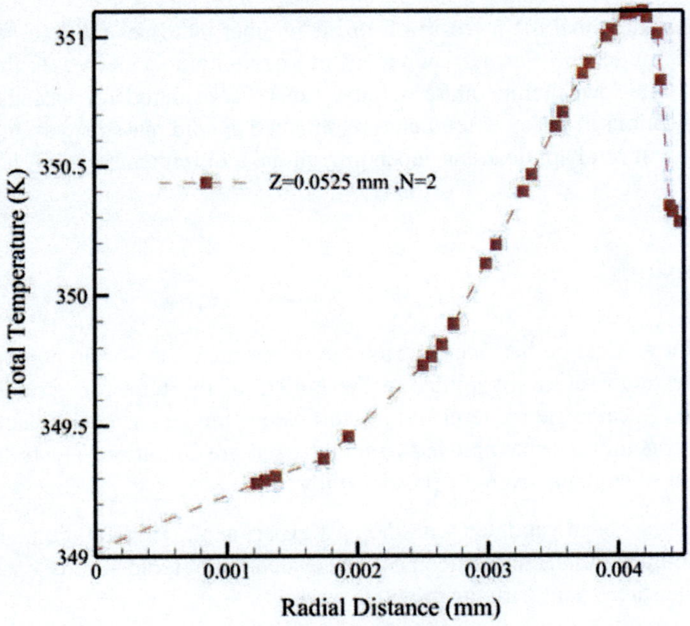

Fig. 11 Total temperature profile regarding radial distance ($N = 2$)

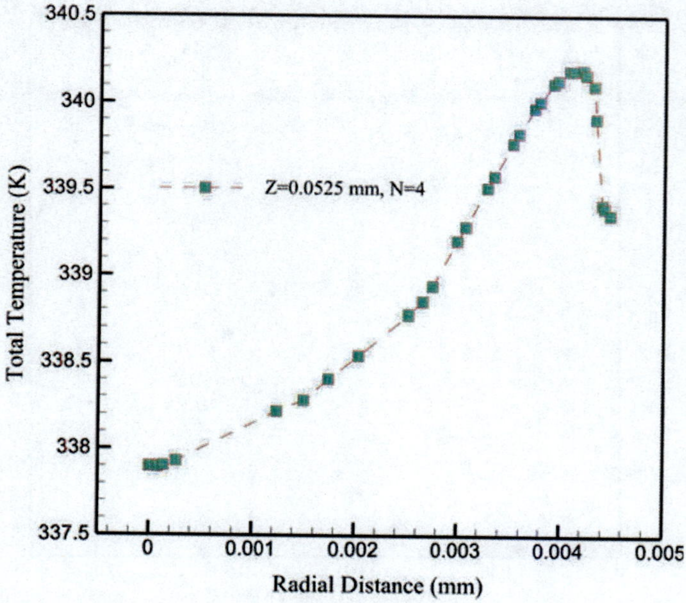

Fig. 12 Total temperature profile regarding radial distance ($N = 4$)

The results show that the increase of nozzle number increases the total temperature up to some extent. Further, increment of nozzle number decreases the total temperature. The lower temperature of fluid can be considered for various engineering applications. The lower temperature obtained at cold mass of working fluid is desirable for several applications depending on the requirements.

7 Conclusions

A numerical investigation has been carried out to optimize the nozzle number and to predict the temperature magnitude at the outlets of the tube for various cold mass fraction. To calculate the temperature, turbulence model has been selected. In this work, nitrogen has been employed with high-pressure condition. The following conclusion has been made from the present study:

1. The work has been validated with the past experimental investigation, and the results found a satisfactory outcomes. The evaluation of temperature separation shows the same trend with the previous work.
2. The investigation shows that as the number of nozzle increases, the magnitude of temperature decreases. The effectiveness of vortex tube can be increased by selecting an optimal nozzle number. The optimum number is $N = 2$ for present

work when nitrogen used as an operating fluid in a vortex tube. The vortex tube containing one and four nozzle reduces the magnitude of temperature separation.

3. The flow characteristics like axial velocity and swirl or tangential velocity have also been discussed for various radial distance of vortex tube.
4 It can be concluded that the maximum velocity is generated at the core of the vortex tube, whilst the intensity of flow is low at the peripheral of the tube.
5. The magnitude of total temperature is also found to be dropped at the wall, and it has high magnitude at the central part of the tube.

References

1. Rafiee SE, Sadeghiazad MM (2016) Experimental study and 3D CFD analysis on the optimization of throttle angle for a convergent vortex tube. J Mar Sci Appl 15(4):388–404. https://doi.org/10.1007/s11804-016-1387-1
2. Matveev KI, Leachman J (2019) Numerical investigation of vortex tubes with extended vortex chambers. Int J Refrig. https://doi.org/10.1016/j.ijrefrig.2019.08.030
3. Li N, Jiang G, Fu L, Tang L, Chen G (2019) Experimental study of the impacts of cold mass fraction on internal parameters of a vortex tube. Int J Refrig 104:151–160. https://doi.org/10.1016/j.ijrefrig.2019.05.002
4. Aghagoli A, Sorin M (2019) Thermodynamic performance of a CO_2 vortex tube based on 3D CFD flow analysis. Int J Refrig. https://doi.org/10.1016/j.ijrefrig.2019.08.022
5. Bianco V, Khait A, Noskov A, Alekhin V (2016) A comparison of the application of RSM and les turbulence models in the numerical simulation of thermal and flow patterns in a double-circuit Ranque-Hilsch vortex tube. Appl Therm Eng 106:1244–1256. https://doi.org/10.1016/j.applthermaleng.2016.06.095
6. Chen J, Zeng R, Zhang W, Qiu L, Zhang X (2018) Numerical analysis of energy separation in Ranque-Hilsch vortex tube with gaseous hydrogen using real gas model. Appl Therm Eng 140:287–294. https://doi.org/10.1016/j.applthermaleng.2018.05.017
7. Devade K, Pise A (2014) Effect of cold orifice diameter and geometry of hot end valves on performance of converging type Ranque Hilsch vortex tube. Energy Procedia 54:642–653. https://doi.org/10.1016/j.egypro.2014.07.306
8. Thakare HR, Parekh AD (2015) Computational analysis of energy separation in counter-flow vortex tube. Energy 85:62–77. https://doi.org/10.1016/j.energy.2015.03.058
9. Attalla M, Ahmed H, Ahmed MS, El–Wafa AA (2017) Experimental investigation of the effect of nozzle numbers on Ranque–Hilsch vortex tube performance. Exp Heat Transf 30(3):253–265, 2017. doi:https://doi.org/10.1080/08916152.2016.1233150
10. Kirmaci V, Uluer O (2009) An experimental investigation of the cold mass fraction, nozzle number, and inlet pressure effects on performance of counter flow vortex tube. J Heat Transfer 131(8):1–6. https://doi.org/10.1115/1.3111259
11. Dincer K, Baskaya S, Uysal BZ, Ucgul I (2009) Experimental investigation of the performance of a Ranque-Hilsch vortex tube with regard to a plug located at the hot outlet. Int J Refrig 32(1):87–94. https://doi.org/10.1016/j.ijrefrig.2008.06.002
12. Ouadha A, Baghdad M, Addad Y (2013) Effects of variable thermophysical properties on flow and energy separation in a vortex tube. Int J Refrig 36(8):2426–2437. https://doi.org/10.1016/j.ijrefrig.2013.07.018
13. Shamsoddini R, Abolpour B (2018) A geometric model for a vortex tube based on numerical analysis to reduce the effect of nozzle number. Int J Refrig 94:49–58. https://doi.org/10.1016/j.ijrefrig.2018.07.027

A New High-Resolution Flux-Corrected Algorithm for the 1-D Euler Equations of Gas Dynamics

R. M. Hemanth Kumar, M. Arun, V. Suryanarayanan, K. V. Nirmal Naathan, and Raushan Kumar

Nomenclature

$\vec{u}$	Conservative variable vector
$\vec{f}$	Flux vector
v	Velocity
e	Total energy
p	Pressure
u_i^n	Conservative variable at ith position and nth time step
Δt	Time step
Δx	Grid spacing

Greek Symbol

λ $\frac{\Delta t}{\Delta x}$

1 Introduction

Computational Fluid Dynamics (CFD) involves numerical solution of integral or partial differential equations that govern fluid motions. In gas dynamics, where the flow of fluid is inviscid and compressible, the Navier-Stokes equations reduce to the Euler equations. While solving the Euler equations numerically, capturing the

R. M. Hemanth Kumar · M. Arun · V. Suryanarayanan · K. V. Nirmal Naathan · R. Kumar (✉)
Department of Mechanical Engineering, Amrita School of Engineering, Amrita Vishwa Vidyapeetham, Coimbatore, India
e-mail: k_raushan@cb.amrita.edu

J. Banerjee et al. (eds.), *Recent Advances in Fluid Dynamics*, Lecture Notes in Mechanical Engineering, https://doi.org/10.1007/978-981-19-3379-0_11

discontinuities like shock and contact discontinuity accurately is a big challenge. Generally, CFD schemes perform well while capturing smooth solutions but fail to capture important flow features such as shock waves, contact discontinuities, corners of expansion waves and sonic points simultaneously. While capturing such flow features the solution shows anomalous behavior such as spurious oscillations, overshoots, undershoots, and smearing of the discontinuities and sharp corners over a range of cells.

Boris and Book [1] came up with the very early solution-sensitive finite-difference method popularly known as Flux-corrected Transport (FCT) scheme. They presented a two-step high-resolution flux-correction method to combine a first-order upwind method and Lax-Wendroff method. FCT improves the accuracy of shock and contact discontinuity computation. Zalesak [2] generalized the FCT of Boris and Book [1], where any two complementary schemes, a first-order and a second-order, can be combined. Thus, he paved the way for the formulation of an entire family of algorithms. Sod [3] compared the characteristics of various schemes that are used to solve nonlinear hyperbolic equations. The test problems used by him is the well-known shock tube problem. Laney [4] presents a modified shock tube problem in addition to the Sod test problem. This modified test problem involves expansive sonic points where many of the first-generation well-established algorithms like the Godunov [5] and Roe [6] schemes give nonphysical expansion shocks. Woodward and Colella [7] introduced the blast wave test. They also explain the three available approaches for treating discontinuities in the flow. They are artificial viscosity, blending of low- and high-order-accurate fluxes, and the use of nonlinear solutions. In 2012, Bernard Parent [8] proposed a new type of flux-limited scheme for system of conservation laws which has high-resolution and maintains positivity.

The objective of the present work is to develop a new numerical scheme that captures the discontinuities with improved accuracy and less computational effort. The reduced computational effort is achieved through reduced complexity of the algorithm when compared with well-known Riemann solvers [5, 6]. The development of the new scheme in this work is based on Zalesak's [2] generalized FCT and involves blending of Steger-Warming [9] and McCormack [10] schemes into one which behaves in such a way that it combines the best characteristics of the two parent schemes. Steger-Warming scheme [9] seemed to be a natural choice for the base first-accurate scheme as the flux-splitting is much simpler in comparison to Roe [6] type Riemann solvers. The second-order extension presented in Parent's paper [8] is a further evidence for its suitability. The performance of the scheme developed is evaluated using the Sod [3], Laney [4], and Woodward-Colella [7] test cases which are engineered to cover a range of solution characteristics, especially those that cause anomalies in the numerical solutions. This new algorithm can easily be extended to 2-D and 3-D cases using dimensional splitting that can be used for several practical applications [11–14].

2 Methodology

The Euler equations are

$$\frac{\partial \vec{u}}{\partial t} + \frac{\partial \vec{f}}{\partial x} = 0; \tag{1}$$

where $\vec{u} = \begin{bmatrix} \rho \\ \rho v \\ \rho e \end{bmatrix}$; and $\vec{f} = \begin{bmatrix} \rho v \\ \rho v^2 + p \\ (\rho e + p)v \end{bmatrix}$.

Steger-Warming Scheme The Steger-Warming scheme is given by:

$$\vec{u}_i^{n+1} = \vec{u}_i^n - \lambda\left(\vec{f}_{i+\frac{1}{2}}^n - \vec{f}_{i-\frac{1}{2}}^n\right); \tag{2}$$

where $\vec{f}_{i+\frac{1}{2}}^n = \vec{F}_i^+ + \vec{F}_{i+1}^-$

and $\vec{F}_i^+$ and $\vec{F}_{i+1}^-$ are split flux-vectors.

McCormack Scheme The McCormack scheme is given by:

$$\overline{\vec{u}}_i = \vec{u}_i^n - \lambda\left(\vec{f}\left(\vec{u}_{i+1}^n\right) - \vec{f}\left(\vec{u}_i^n\right)\right) \tag{3}$$

$$\vec{u}_i^{n+1} = \frac{1}{2}\left(\vec{u}_i^n + \overline{\vec{u}}_i\right) - \frac{\lambda}{2}\left(\vec{f}\left(\overline{\vec{u}}_i\right) - \vec{f}\left(\overline{\vec{u}}_{i-1}\right)\right) \tag{4}$$

The New Scheme The Zalesak's methodology results in a two-step predictor–corrector method given by Eqs. (5) and (6).

$$\text{Predictor}: \overline{\vec{u}}_i = \vec{u}_i^n - \lambda\left(\vec{f}_{i+\frac{1}{2}}^{(1)} - \vec{f}_{i-\frac{1}{2}}^{(1)}\right) \tag{5}$$

$$\text{Corrector}: \vec{u}_i^{n+1} = \vec{u}_i^n - \lambda\left(\vec{f}_{i+\frac{1}{2}}^c - \vec{f}_{i-\frac{1}{2}}^c\right) \tag{6}$$

where

$$\lambda \vec{f}_{i+\frac{1}{2}}^c = \text{minmod}\left(\overline{\vec{u}}_i - \overline{\vec{u}}_{i-1}, \lambda \vec{f}_{i+\frac{1}{2}}^{(2)} - \lambda \vec{f}_{i+\frac{1}{2}}^{(1)}, \overline{\vec{u}}_{i+2} - \overline{\vec{u}}_{i+1}\right) \tag{7}$$

In Eq. (5)–(7), $\vec{f}_{i+\frac{1}{2}}^{(1)}$ is the time averaged flux component of Steger-Warming scheme and $\vec{f}_{i+\frac{1}{2}}^{(2)}$ is the time averaged flux component of McCormack scheme.

3 Results and Discussion

The Riemann problem is one of the few problems that has been analytically solved for exact solution, thus the first two tests are the Sod's shock tube problems. The third test is the Woodward-Colella blast wave problem for which a reference solution [15] is available for comparison.

The results obtained using the new scheme is compared with McCormack scheme and Steger-Warming scheme. The results of the three schemes are plotted against the exact solution or the reference solution.

Test case 1

In test case 1, pressure, density, and velocity across fixed spatial domain (−10 to 10 m) at a time of $t = 0.01$ s is calculated. The initial condition for the Riemann problem is given as

$$w(x,0) = \begin{cases} w_L, & x < 0 \\ w_R, & x > 0 \end{cases}$$

where $w_L = \begin{bmatrix} \rho_L \\ u_L \\ p_L \end{bmatrix} = \begin{bmatrix} 1\text{kg/m}^3 \\ 0\text{m/s} \\ 100,000\text{N/m}^2 \end{bmatrix}$;

$$w_R = \begin{bmatrix} \rho_R \\ u_R \\ p_R \end{bmatrix} = \begin{bmatrix} 0.0125\text{kg/m}^3 \\ 0\text{m/s} \\ 10,000\text{N/m}^2 \end{bmatrix}.$$

A total of 101 grid points with the CFL number of 0.75 is used. This simple test includes all the important flow features like shock, contact discontinuity, expansion fan, etc. Figure 1 shows the geometry of the initial setup. It becomes a challenging test for a numerical scheme for the following reasons:

- Even at stopping time ($t = 0.01$ s) the shock and the contact discontinuity is separated only by 2.5 m.

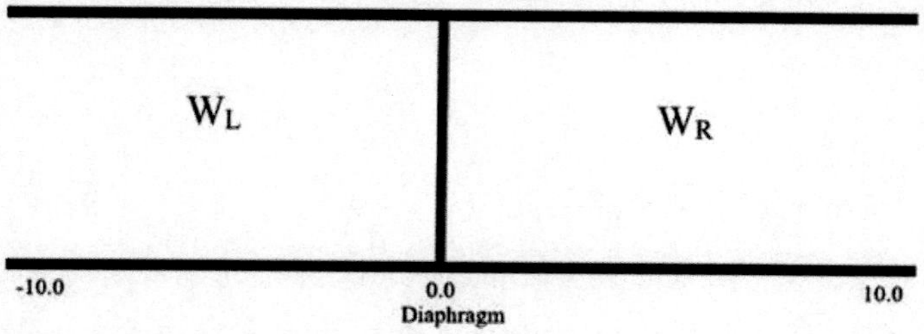

Fig. 1 Geometry of test case 1

- It is nearly impossible to capture shock and contact properly when they are within five or less grids in spatial domain.

Test case 2

In test case 2, pressure, density, and velocity across fixed spatial domain (−10 to 15 m) at a time of $t = 0.01$ s is calculated. The initial condition for the Riemann problem is given as

$$w(x,0) = \begin{cases} w_L, & x < 0 \\ w_R, & x > 0 \end{cases}$$

where $w_L = \begin{bmatrix} \rho_L \\ u_L \\ p_L \end{bmatrix} = \begin{bmatrix} 1\text{kg/m}^3 \\ 0\text{m/s} \\ 100000\text{N/m}^2 \end{bmatrix}$;

$$w_R = \begin{bmatrix} \rho_R \\ u_R \\ p_R \end{bmatrix} = \begin{bmatrix} 0.010\text{kg/m}^3 \\ 0\text{m/s} \\ 1000\text{N/m}^2 \end{bmatrix}.$$

In this test, 101 grid points with a CFL number of 0.9 is used. Figure 2 shows the geometry of the initial setup. This test is more stringent when compared to test case 1 for the following reasons:

- As in test case 1 the shock and contact are separated by a very small spatial distance. Thus, capturing the shock and contact discontinuity is harder.
- The expansion fan in this test case contains an expansion sonic fan at $x = 0$, which may cause formation of expansion shocks for various numerical methods.
- The shock contains a compressive sonic point.

In Test Case 1 the large pressure ratio is spread across in the long and smooth expansion fan, which makes, for several numerical schemes, easier to handle the large pressure ratio. Even some modern schemes find it hard to handle the large

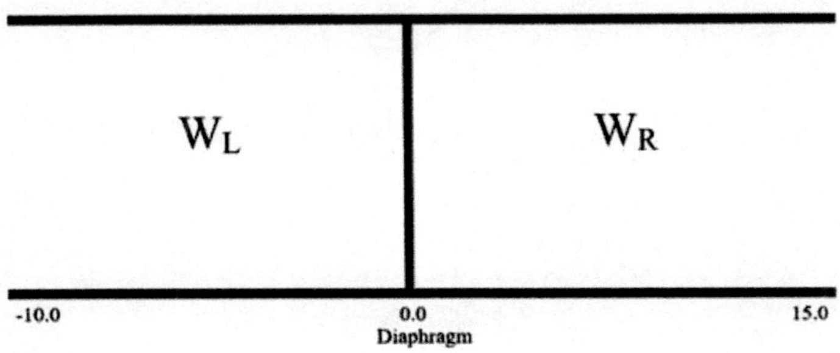

Fig. 2 Geometry of test case 2

pressure initially in the Test Case 2, although they tend to settle down after the initial large discontinuity settles down.

Test case 3

The initial conditions are two planar flow discontinuities with reflective boundary conditions at the left and right walls. Density is taken 1.0 unit everywhere and the equation of state for a perfect gas with adiabatic index of 1.4 is used. Pressure and geometry are as given in Fig. 3.

Figures 4, 5 and 6 are the results obtained for the test case 1 of the shock tube problem using the three schemes.

Figures 7, 8 and 9 are the results obtained for the test case 2 of the shock tube problem using the three schemes.

Figure 10 compares the results obtained using McCormack, Steger-Warming and the new scheme plotted against a reference solution [15].

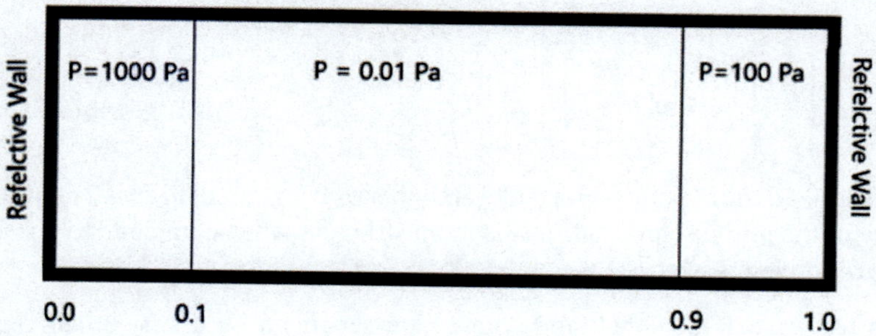

Fig. 3 Geometry and initial pressure distribution of test case 3

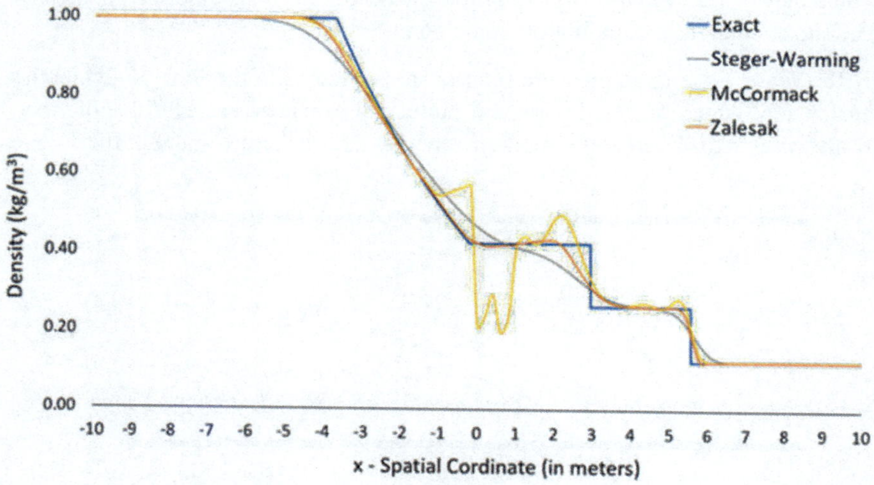

Fig. 4 Density profile (Grid = 101, CFL = 0.75)

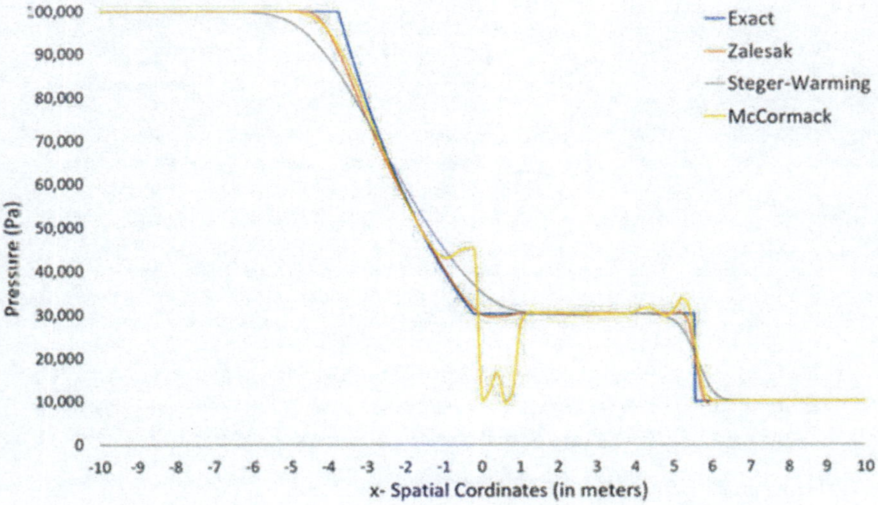

Fig. 5 Pressure profile (Grid = 101, CFL = 0.75)

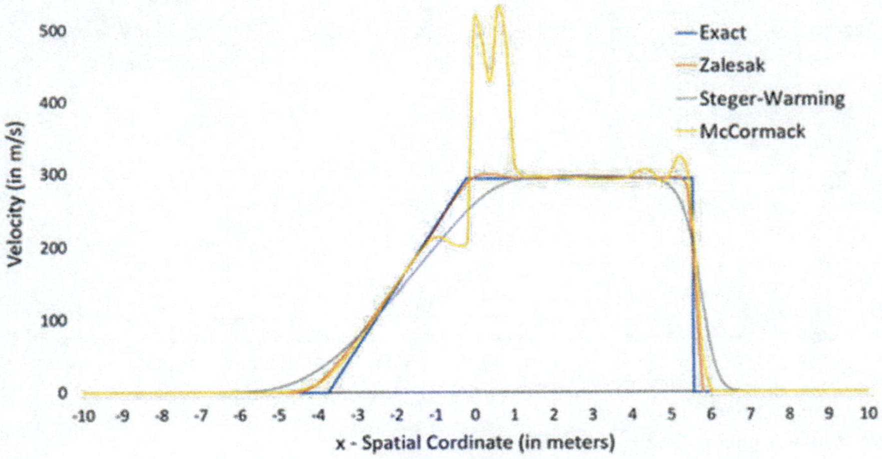

Fig. 6 Velocity profile (Grid = 101, CFL = 0.75)

It is evident that the new scheme captures the discontinuities and corners with better accuracy without any significant oscillations.

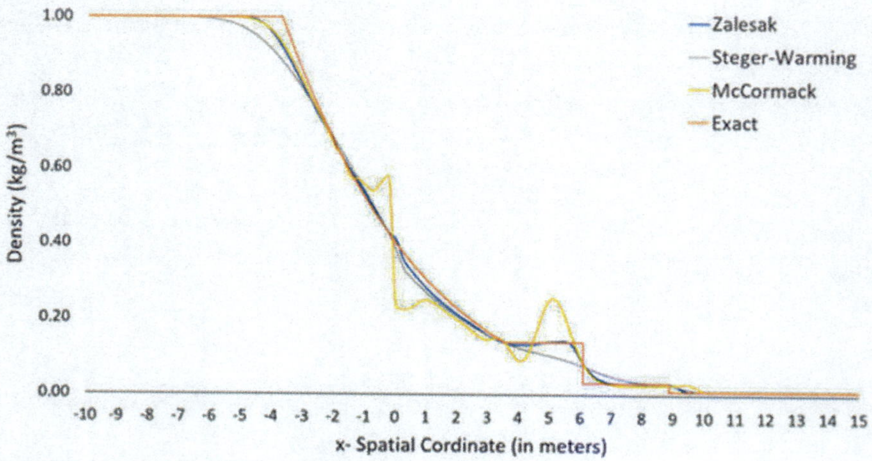

Fig. 7 Density profile (Grid = 101, CFL = 0.9)

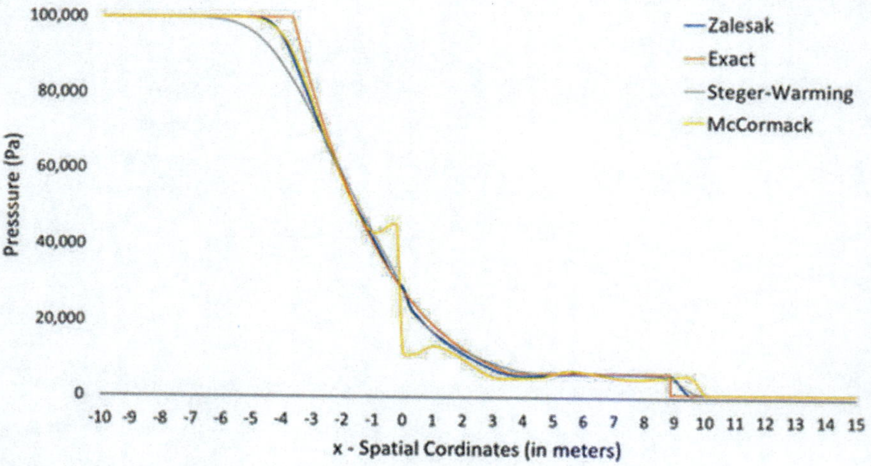

Fig. 8 Pressure profile (Grid = 101, CFL = 0.9)

4 Conclusions

In conclusion, the McCormack Scheme, being a second-order scheme cannot handle large gradients. Thus, resulting in large spurious oscillations around large gradients and sonic points. Steger-Warming on the other hand is a first-order accurate scheme, thus is relatively less accurate than the McCormack scheme. These two schemes make a perfect complementary pair to devise a flux-corrected scheme. The new scheme produces better results than either of the parent schemes, as it avoids the oscillations of McCormack in regions with steep gradients but has high accuracy of McCormack Scheme in smooth regions. The formulation simplicity of the scheme makes it easy

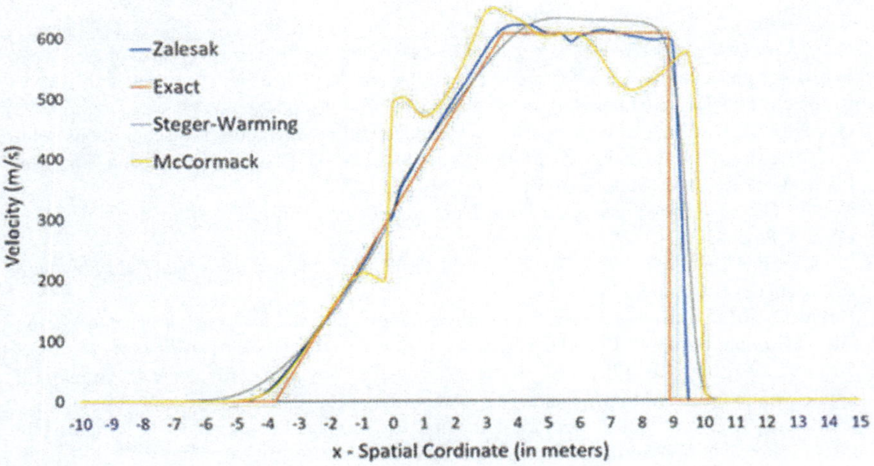

Fig. 9 Velocity profile (Grid = 101, CFL = 0.9)

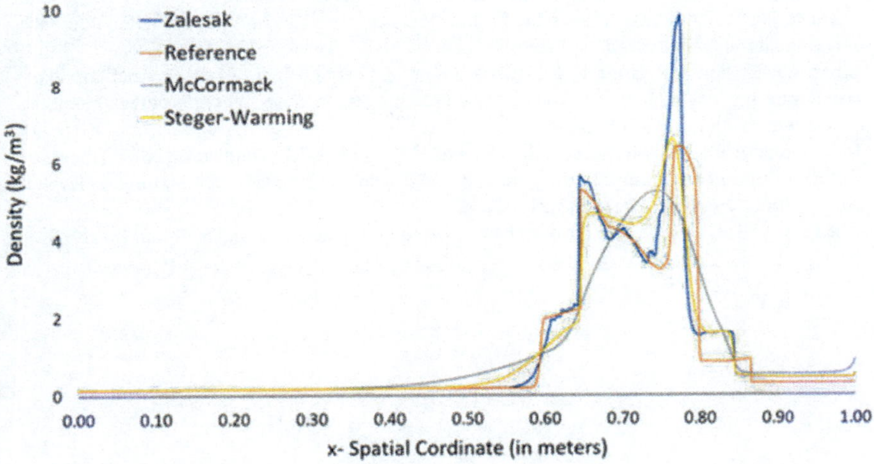

Fig. 10 Density profile (Grid = 1001, CFL = 0.2)

to implement and computationally less costly when compared to well-established sophisticated Riemann solvers without compromising on accuracy.

References

1. Boris JP, Book DL (1973) Flux-corrected transport. I. SHASTA, A fluid transport algorithm that works. J Comput Phys 11:38–69
2. Zalesak ST (1979) Fully multidimensional flux-corrected transport algorithms for fluids. J

Comput Phys 31:335–362
3. Gary AS (1978) A survey of several finite difference methods for systems of nonlinear hyperbolic conservation laws. J Comput Phys 27:1–31
4. Laney CB (1998) Computational gas dynamics
5. Godunov SK (1959) A difference scheme for numerical computation of discontinuous solutions of hydrodynamics equations. Math Sbornik 47:271–306. English translation in U.S. Joint Publications Research Service, JPRS 7226, 1969
6. Roe PL (1981) Approximate Riemann solvers, parameter vectors, and difference schemes. J Comput Phys 43:357–372
7. Woodward P, Colella P (1984) The numerical simulation of two-dimensional fluid flow with strong shocks. J Comput Phys V.54
8. Parent B (2012) Positivity-preserving high-resolution schemes for systems of conservation laws. J Comput Phys 231:173–189
9. Steger JL, Warming RF (1981) Flux vector splitting of the inviscid gasdynamic equations with application to finite-difference methods. J Comput Phys 263–293
10. MacCormack RW (1982) A numerical method for solving the equations of compressible viscous flow. Am Inst Aeronaut Astrophys J 20(9):1275–1281
11. Vishnu Chandar S, Sarath RS, Ajith Kumar R (2017) Aerodynamic characteristics of a square section cylinder : effect of corner arc. In: 2nd international conference for convergence in technology (I2CT) Siddhant College of Engineering, Pune
12. Chaitanya JS, Prasad A, Pradeep B, Sri Harsha PLN, Shali S, Nagaraja SR (2017) Vibrational characteristics of AGARD 445.6 wing in transonic flow. In: ICMAEM-2017, IOP conference series: materials science and engineering 225
13. Singh AP, Kishore VR, Yoon Y, Minaev S, Kumar S (2017) Effect of wall thermal boundary conditions on flame dynamics of CH4-air and H2-air mixtures in straight microtubes. Combust Sci Technol 189(1):150–168
14. Mohammed AN, Juhany KA, Sudarshan Kumar V, Kishore R, Mohammad A (2017) Effects of CO2/N2 dilution on laminar burning velocity of stoichiometric DME-air mixture at elevated temperatures. J Hazard Mater 333:215–221
15. Roe PL (1986) Characteristic-based schemes for Euler equations. Ann Rev Fluid Mech 18

Modelling of Subcooled Boiling in Corrugated Pipes

K. Madan, Kuldeep Singh, and A. Sathyabhama

Nomenclature

a_p	Amplitude, m
CP-1	Corrugated pipe-1
CP-2	Corrugated pipe-2
CP-3	Corrugated pipe-3
HTC	Heat transfer coefficient
D_P	Pipe diameter, m
FP	Flat pipe
g	Acceleration due to gravity, m^2/s
h_{avg}	Average HTC, W/m^{2}K
L	Length of the pipe, m
L_{CP}	Corrugated pipe length, m
L_e	Entrance length, m
δp	Pressure drop
r	Arbitrary radius along radial direction, m
RPI	Rensselaer Polytechnic Institute
$T_{subcooled}$	Subcooled temperature,
U_{in}	Inlet velocity, m/s
V_v	Velocity of vapour, m/s
V_w	Velocity of water, m/s
X	Arbitrary distance along the length of the pipe, m

K. Madan · A. Sathyabhama
Department of Mechanical Engineering, NITK Surathkal, Srinivasnagar, Mangaluru 575025, India

K. Singh (✉)
Gas Turbine and Transmissions Research Centre, University of Nottingham, Nottingham NG7 2TU, UK
e-mail: kuldeep.singh@nottingham.ac.uk

J. Banerjee et al. (eds.), *Recent Advances in Fluid Dynamics*, Lecture Notes in Mechanical Engineering, https://doi.org/10.1007/978-981-19-3379-0_12

Greek Symbols

μ	Dynamic viscosity of fluid, Pa-s
λ	Wavelength, m

1 Introduction

Subcooled flow boiling heat transfer takes place when the working fluid near the heated surface reaches the boiling temperature, when the subcooled condition is still prevalent in the major volume of the fluid. The subcooled flow boiling is very complicated phenomenon since it involves bubble nucleation, its growth and subsequent departure, its merger with the neighbouring bubble and the unequal dispersal of the bubbles in the region close to the surface. It is thus a painstaking task to precisely predict the subcooled boiling heat transfer but is essential for the design of the heat transfer device. Expensive experiments are required to evaluate the influence of design improvements on the heat transfer coefficient (HTC). Alternatively, numerical analyses can be used to either supplement or even to replace the experiments and are of high interest in industrial applications. For the simulation of boiling process, the interface tracking approaches developed for multi-phase flows are used by many researchers. Level-set method [1], volume-of-fluid method [2], colour function method, front-tracking method [3], phase-field method [4, 5] and Lattice-Boltzmann method [6] are the few interface capturing methods adopted in literature. The comprehensive macroscopic two-phase flow model currently in use is the two-fluid model. For subcooled boiling flows, the Eulerian–Eulerian method to describe the vapour and the liquid phases along with Rensselaer Polytechnic Institute (RPI) wall boiling model to explain subcooled boiling process has been verified to give moderately fair results by previous researchers [7]. Previous studies have employed RPI model with various combinations of closure models to estimate the key boiling parameters, namely the nucleation site density, bubble diameter at departure and frequency of bubble departure.

In the present work, subcooled flow boiling heat transfer in corrugated vertical pipe is numerically investigated. The main purpose of using corrugated tubes in heat transfer devices is to augment the heat transfer. The additional area available for the heat transfer due to the corrugations and the flow turbulence due to the ribs is responsible for the enhanced heat transfer in corrugated pipes. The percentage enhancement reported by different researchers is different. Experimental study of flow boiling heat transfer to R407C using tubes with corrugations carried by [8] showed heat transfer enhancement of 20%. The horizontal corrugated tubes of fixed corrugation depth and different corrugation pitch were used to increase the evaporation of R134A in the experimental investigation by [9]. The results showed that the increase of vapour quality, and two-phase equivalent Reynolds number increased the HTC in the corrugated tubes. Compared to the smooth tube, the maximum Nusselt number increased by 29% for the corrugated tube. The Nusselt number was not significantly affected by the corrugation pitch at low-average vapour quality. Heat transfer enhancement of about 40% was observed for evaporation of HCFC22 inside the tubes in commercially available enhanced tubes [10]. The experimental investigation by [11] on the heat transfer characteristics of saturated liquid nitrogen during the two-phase flow boiling in a horizontal corrugated tube made of stainless steel reported higher two-phase effective heat transfer compared to a normal tube of the same diameter. This is because the heat transfer area per unit length was higher in the corrugated tube.

From the literature reported above, it is apparent that the corrugated tubes transfer higher amount of heat as compared to the smooth tubes. Even though the two-phase flow boiling characteristics like the heat transfer and pressure drop in corrugated tubes are discussed in literature, the studies that focussed on the subcooled flow boiling regime are scarce. It is also evident from the literature that the tube geometry, operating conditions and working fluid influence the heat transfer enhancement in corrugated tubes, and hence, further investigation is necessary to appreciate the effects of these parameters. Further, the numerical investigations on the effect of corrugated pipe on subcooled flow boiling are not reported in literature. Consequently, this work deals with the numerical simulation of the flow boiling of water under subcooled condition to explore the possibility of enhanced heat transfer using corrugated tubes. Corresponding increase in pressure drop inside the enhanced tube too will be examined.

2 Problem Description

In the current study, subcooled boiling of water is investigated in a flat pipe and three different corrugated pipes. The investigated geometry is shown in Fig. 1.

The geometric parameters of the corrugated pipes are enumerated in Table 1. Subcooled water enters the pipe with 57.5 °C degree of subcooling at a mass flow

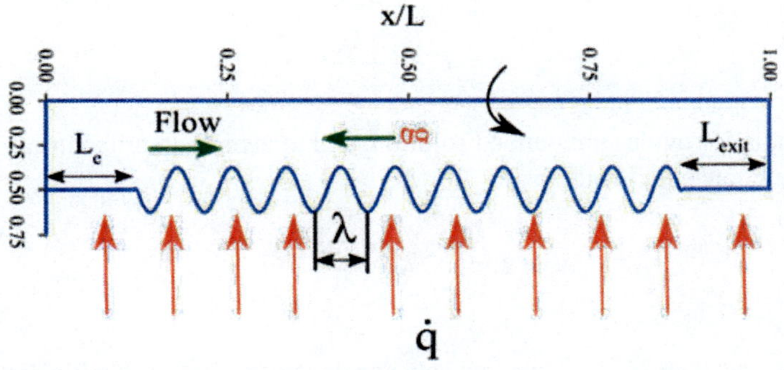

Fig. 1 Geometry of the computational domain

Table 1 Geometric parameters of corrugated pipes

L (m)	L_e (m)	L_{CP} (m)	D_P (m)	λ (m)	a_p (m)
2	0.25	1.5	0.0154	0.1, 0.125, 0.15	0.001925

rate of 900 kg/s m^2 and 45 bar pressure. Uniform wall heat fluxes of 500, 570 and 650 kW/m^2 were applied at the circumference of the tubes.

3 Mathematical Modelling

3.1 Wall Boiling Model

In the present study, wall boiling model developed by Kurul and Podowski [12] at RPI is used to model subcooled boiling. In this model, total wall heat flux ($\dot{q}_w$) is divided into convective heat flux ($\dot{q}_C$), quenching heat flux ($\dot{q}_Q$) and evaporative heat flux ($\dot{q}_E$) as expressed in Eq. (1).

$$\dot{q}_w = \dot{q}_C + \dot{q}_Q + \dot{q}_E \tag{1}$$

The convective heat flux ($\dot{q}_C$) accounts the heat transfer from the portion of the heated wall that is not covered by nucleating bubbles to the single-phase liquid. It is expressed in Eq. (2).

$$\dot{q}_C = h_C(T_w - T_l)(1 - A_b) \tag{2}$$

where h_C is single-phase HTC, T_w is wall temperature, T_l is liquid temperature and A_b is area covered by the nucleating bubbles. The area covered by nucleating bubbles is obtained from Eq. (3).

$$A_b = K\frac{N_w \pi D_w^2}{4} \tag{3}$$

This equation provides unbounded solution, and to avoid numerical instabilities, the area of influence is restricted as:

$$A_b = \min\left(1, K\frac{N_w \pi D_w^2}{4}\right) \tag{4}$$

where K is a constant, N_w is active nucleate site density and D_w is bubble departure diameter. The correlation proposed by Valle and Kenning [7] is used to determine the value of K which is given in Eq. (5).

$$K = 4.8e^{-\frac{Ja_{\text{sub}}}{80}} \tag{5}$$

where Ja_{sub} is subcooled Jacob number and expressed in Eq. (6).

$$Ja_{sub} = \frac{\rho_l C_{pl} \Delta T_{\text{sub}}}{\rho_v h_{fv}} \tag{6}$$

where ρ is density and subscripts l and v indicate liquid and vapour phase, C_{pl} is specific heat of liquid and ΔT_{sub} is degree of subcooling, i.e. $\Delta T_{\text{sub}} = T_{\text{sat}} - T_l$, h_{fv} is latent heat of vapourization. The nucleate site density in Eq. (4) is modelled by Eq. (7).

$$N_w = C^n (T_w - T_l)^n \tag{7}$$

where $C = 210$ and $n = 1.805$ as suggested by Lemmert and Chawla [8]. The diameter of departing bubble D_w is modelled by the correlation proposed by Tolubinsky and Konstanchuk [9]

$$D_w = \min\left(0.0014, 0.006\mathrm{e}^{-\frac{\Delta T_{\text{sub}}}{45}}\right) \tag{8}$$

The quenching heat flux ($\dot{q}_Q$) is a transient and cyclic averaged heat transfer associated with the liquid replacing the departing bubbles from the wall and it expressed in Eq. (9).

$$\dot{q}_Q = \frac{2k_l}{\sqrt{\pi \lambda_l T_{\text{periodic}}}} (T_w - T_l) \tag{9}$$

where k_l is thermal conductivity of liquid, λ_l is thermal diffusivity of liquid and T_{periodic} is periodic time.

The evaporative heat flux ($\dot{q}_E$) accounts the energy carried away by the departing bubbles as latent heat, and it is expressed in Eq. (10).

$$\dot{q}_E = V_d N_w \rho_v h_{fv} f \tag{10}$$

where V_d is volume of bubble based on the bubble departing diameter and f is bubble departure frequency.

The bubble departure frequency is modelled by the correlation proposed by Cole [10] as given in Eq. (11).

$$f = \frac{1}{T} = \sqrt{\frac{4g(\rho_l - \rho_v)}{3\rho_l D_w}} \tag{11}$$

3.2 Eulerian–Eulerian Model

In the present study, subcooled flow boiling is investigated numerically, and hence, the related governing equations: continuity, momentum and energy are solved by Eulerian–Eulerian approach. This approach solves the governing equations for each phase. The generalised form of governing equations is given in Eqs. (12)–(14).

$$\frac{\partial(\alpha_q \rho_q)}{\partial t} + \nabla.\left(\alpha_q \rho_q \vec{V_q}\right) = \sum\nolimits_{p=1}^{n} (\dot{m}_{pq} - \dot{m}_{qp}) \tag{12}$$

where α_q is qth phase vf, $\vec{V_q}$ is velocity magnitude, $\dot{m}_{pq}$ and $\dot{m}_{qp}$ mass transfer from qth phase to pth phase and vice-versa.

$$\begin{aligned}\frac{\partial(\alpha_q \rho_q \vec{V_q})}{\partial t} + \nabla.\left(\alpha_q \rho_q \vec{V_q}\vec{V_q}\right) &= -\alpha_q \nabla p + \nabla.\bar{\bar{\tau}}_q + \alpha_q \rho_q \vec{g}\,) \\ &+ \sum\nolimits_{p=1}^{n} \left(\vec{R}_{pq} + \dot{m}_{pq}\vec{V}_{pq} - \dot{m}_{qp}\vec{V}_{qp}\right) + (\vec{F}_q \\ &+ \vec{F}_{lift,q} + \vec{F}_{wl,q} + \vec{F}_{vm,q} + \vec{F}_{td,q}\end{aligned} \tag{13}$$

where $\bar{\bar{\tau}}_q$ is qth phase stress–strain tensor, $\vec{g}$ is acceleration due to gravity, $\vec{R}_{pq}$ is interaction force between phase, $\vec{V}_{pq}$ and $\vec{V}_{qp}$ are interphase velocities, $\vec{F}_q$ is external body force, $\vec{F}_{lift,q}$ is lift force, $\vec{F}_{wl,q}$ is wall lubrication force, $\vec{F}_{vm,q}$ is virtual mass force and is $\vec{F}_{td,q}$ turbulence dispersion force.

$$\begin{aligned}\frac{\partial(\alpha_q \rho_q h_q)}{\partial t} + \nabla.\left(\alpha_q \rho_q \vec{V_q} h_q\right) &= \alpha_q \frac{dp_q}{dt} + \bar{\bar{\tau}}_q.\vec{V}_q - \nabla.\vec{q}_q \\ &+ S_q + \sum\nolimits_{p=1}^{n} \left(Q_{pq} + \dot{m}_{pq} h_{pq} - \dot{m}_{qp} h_{qp}\right)\end{aligned} \tag{14}$$

where h_q is qth phase specific enthalpy, is source term, is intensity of heat exchange between the pth and qth phase.

3.3 Boundary Conditions

The computational domain investigated in the present study is two dimensional as shown in Fig. 1. The liquid flows upward in a vertical pipe of 2 m in length against gravity. The centre line of the pipe is assigned axisymmetric boundary conditions. At inlet, the mass flux boundary is applied; at outlet, the pressure boundary is applied, and the heat flux is applied to the pipe wall.

Table 2 Thermophysical properties considered in the present study

T (K)	ρ (kg/m^3)	μ (Pa-s)	C_p (j/kg K)	k (W/m k)
473.15	864.7	1.339e−04	4494	0.664
543.15	770.6	9.995 e−04	5067	0.5928

3.4 Fluid and Thermal Properties

Temperature dependence of the thermophysical properties of the liquid phase is considered in the present study, whereas the vapour phase properties are considered at the saturation temperature. Piecewise-linear polynomial variations of ρ, μ, C_p and k of fluid were considered [13] as given in Table 2.

3.5 Solution Procedure

The current simulation is carried out in ANSYS FLUENT 2021R1 [14] to solve the energy, momentum and continuity equation. The solution is considered as converged when residuals of fluid flow equation fall below 10^{-5} and that for energy below 10^{-8}. Phase-coupled SIMPLE [15] scheme is considered for velocity and pressure coupling.

3.6 Grid Sensitivity Study

For grid-independent analysis, three meshes were created for a corrugated pipe-1. The number of nodes was varied from 15×1000 for Mesh 1 to 30×2000 for Mesh2 and 40×2500 for Mesh3. The chosen conditions for this analysis were a mass flux of 900 kg/m^2-s, saturation pressure of 45 bar, and a heat flux of 570 kW/m^2. Figure 2 presents a typical grid used in the current study and the variation of non-dimensional wall temperature along the walls of pipe, considering all computational meshes. It is found that all cases resulted in almost identical solutions. The maximum deviation in the temperature obtained from Mesh1 and Mesh2 was less than 2% and that for Mesh2 and Mesh3 was less than 0.3%. Hence, Mesh2 is selected for further analysis.

3.7 Validation

The experimental results of Bartolomei et al. [16] and numerical results of Nemitalla et al. [7] for the test case of flat/smooth pipe are used for validation of the numerical model. Both operating conditions and the working parameter values are kept constant

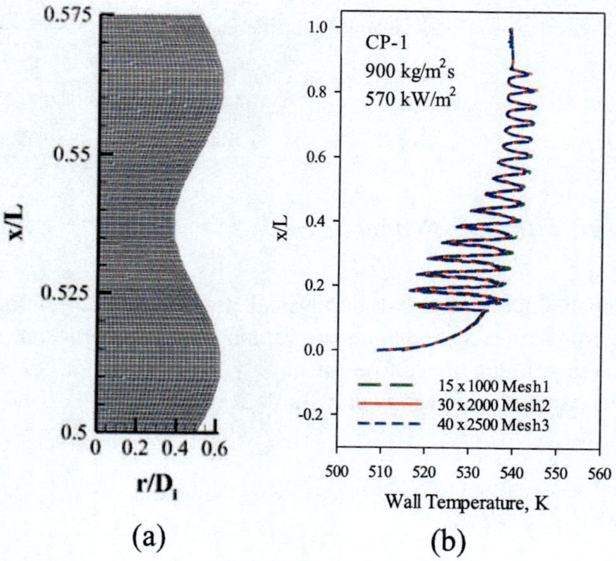

Fig. 2 **a** Typical grid used in the present study, **b** temperature variation along the length of the pipe for the investigated meshes

corresponding to the experimental conditions. Figure 3 shows comparisons between the experimental and numerical results of Bartolomei et al. [16] and Nemitalla et al. [7] along with the current model results. Both the wall temperature and the *vf* of vapour profiles against the non-dimensional axial length were used in the validation at an operating uniform heat flux of $q = 570$ kW/m^2, subcooled temperature of 57.5 °C. It can be observed that the present numerical results are in good agreement with both Nemitalla et al. [7] and Bartolomei et al. [16]. Hence, the present computational methodology can be utilised for parametric studies.

4 Results and Discussion

4.1 Effect of Corrugation Wavelength on Wall Temperature

The variation of wall superheat along the length of the pipe for all the four geometries considered in the present study is shown in Fig. 4. It is evident that the wall temperature fluctuates in corrugated pipes, higher on the crest and lower on the trough portion. This is because the fluid velocity in the crest region is significantly lower than that of the trough region (almost one order lower) as shown in Fig. 5a. This kind of temperature variation is due to a corrugated profile.

Figure 5b, c shows the velocity distribution of vapour phase along the length in a corrugated pipe and flat/smooth pipe. As the temperature rises above saturation

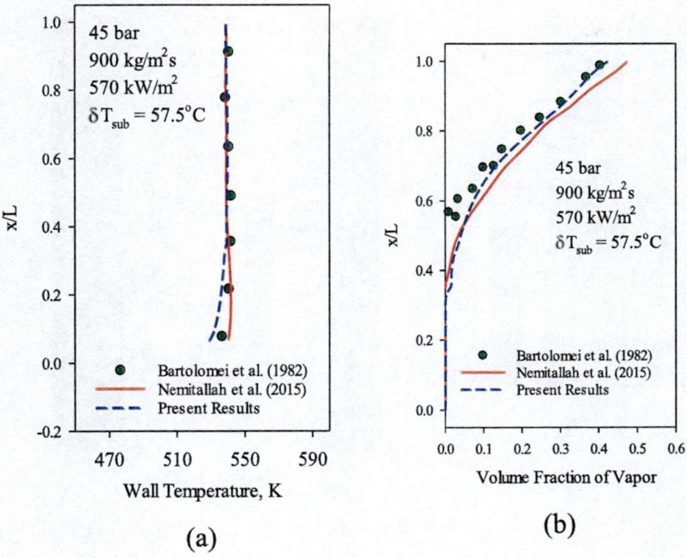

Fig. 3 Comparison of present results with the experimental data of Bartolomei et al. [16] and numerical results of Nemitalla et al. [7]: **a** non-dimensional length versus wall temperature and **b** non-dimensional length versus *vf* of vapour

Fig. 4 Variation of wall superheat along the length

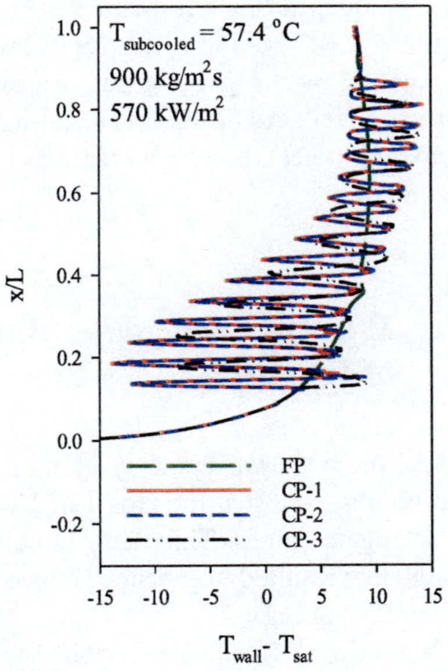

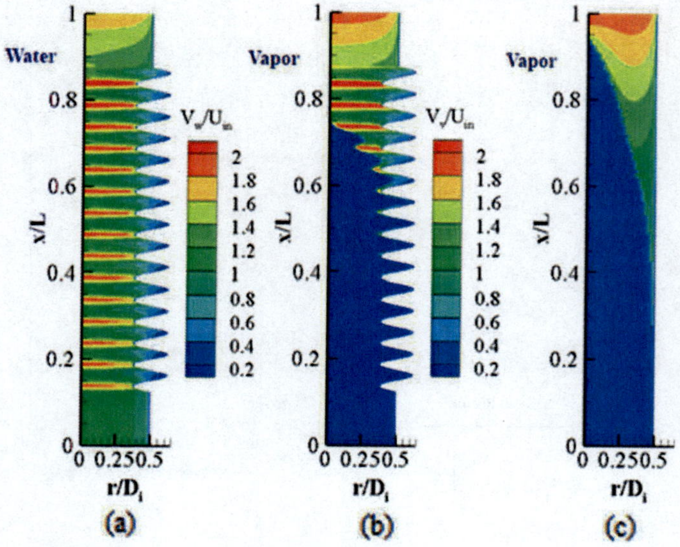

Fig. 5 Velocity contours of **a** water phase in corrugated pipe, **b** vapour phase in corrugated pipe, **c** vapour phase in flat/smooth pipe

temperature, nucleate formation is promoted in the region as shown in Fig. 4. But, bubble growth is suppressed due to lower temperature in the nearby area. Thus, two competing mechanisms of heat transfer are taking place. Suppression of nucleate formation reduces the boiling heat transfer. In the meantime, both convective and the evaporation heat transfer increase as corrugated profile promotes velocity variation and mixing.

4.2 *Effect of Corrugation Wavelength on Volume Fraction of Vapour*

Figure 6a, b shows vapour *vf* along the axial and radial directions, respectively. It can be observed that the Onset of Nucleate Boiling (ONB) is delayed in the case of corrugated pipes. This delay is because of the increased fluid velocity near the trough that resulted in enhanced convective heat transfer. This suppresses nucleation and delay boiling.

Variation of *vf* of vapour along the length of pipe confirms that more vapour is formed in the case of flat/smooth pipe as compared to the corrugated pipe as shown in Fig. 7. Vapour content at the outlet of flat pipe is around 7% higher as compared to corrugated pipes.

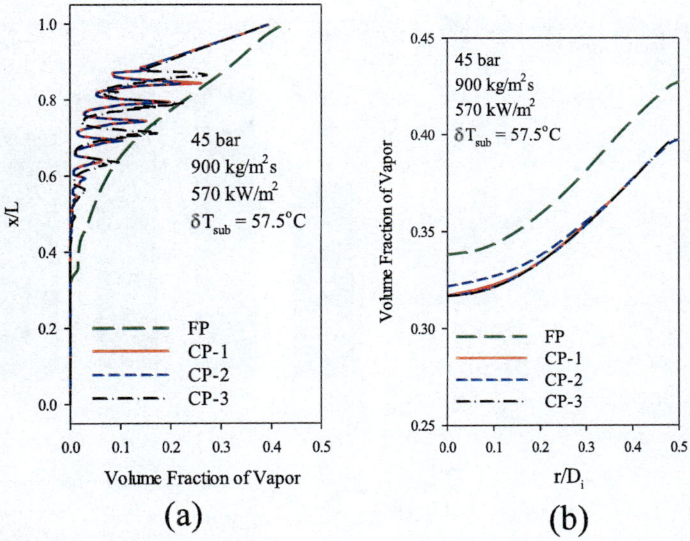

Fig. 6 Volume fraction (*vf*) of vapour **a** along the axial direction, **b** along the radial direction

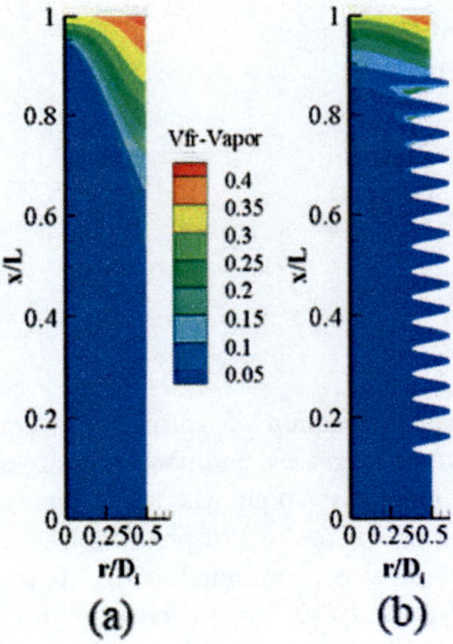

Fig. 7 Contours of *vf* of vapour in **a** flat/smooth pipe, **b** corrugated pipe

4.3 *Effect of Corrugation Wavelength on Heat Transfer and Pressure Drop*

The total length accesses the relative merit of the pipe profile averaged HTC. As represented in Fig. 8a, the HTC is higher for the corrugated pipes than the flat pipe. As discussed earlier, velocity magnitude near a trough is higher and increases convection

Fig. 8 **a** HTC versus geometry. **b** Pressure drop versus geometry

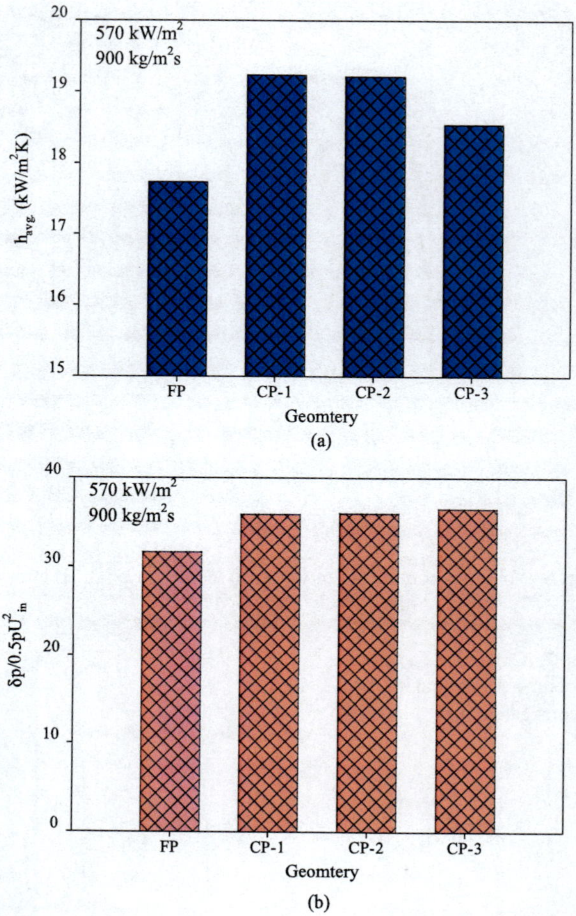

means nucleation site formation is suppressed. The contribution of convective heat transfer increases, and hence, improvement in the overall HTC is observed despite forming less vapour. Maximum improvement in the HTC is around 8.5% for CP-1. But, it is at the cost of pressure loss. The pressure drop in CP-1 is 13% more than that in FP as represented in Fig. 8b. Improvement in the HTC is almost identical for CP-1 and CP-2, but it decreases for CP-3. The wavelength is longest amongst the investigated cases for CP-3 that reduces mixing, and hence, improvement in HTC is not as much as it is observed for CP-1.

4.4 *Effect of Wall Heat Flux on Vapour Volume Fraction*

As the wall heat flux increases, the vapour *vf* increases, as shown in Fig. 9. Increased

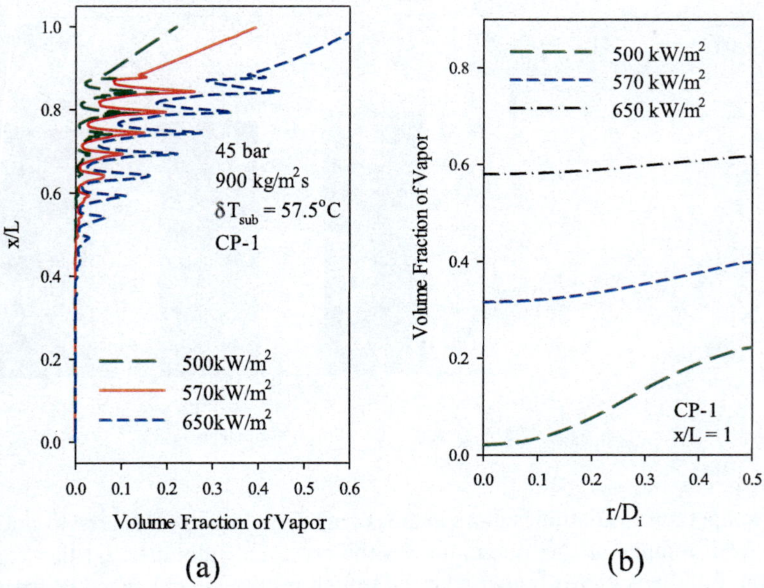

Fig. 9 Volume fraction (*vf*) of vapour **a** along the axial direction, **b** along a radial direction at different heat fluxes for CP-1

wall heat flux promotes nucleation, and hence, vapour *vf* increases. A *vf* of 20, 40 and 60% is observed at the outlet of the pipe when wall heat flux is varied from 500 to 650 kW/m^2 as shown in Fig. 9b.

4.5 *Effect of Wall Heat Flux on Heat Transfer*

The effect of wall heat flux on the averaged HTC is presented in Fig. 10. A marginal decrement (0.8%) in the averaged HTC is observed when heat flux is increased from 500 to 650 kW/m^2. As discussed earlier, the *vf* of vapour increases with increased wall heat flux. This increment in the vapour formation increases the contribution of quenching heat transfer and evaporative heat transfer but more vapour decreases the heat transfer because of the convection. Thus, overall averaged HTC decreases slightly with the increase in wall heat flux.

5 Conclusions

In this study, subcooled flow boiling heat transfer in corrugated vertical pipes is investigated numerically. Based on the presented results, it can be concluded that

Fig. 10 Variation of HTC with wall heat flux for CP-1

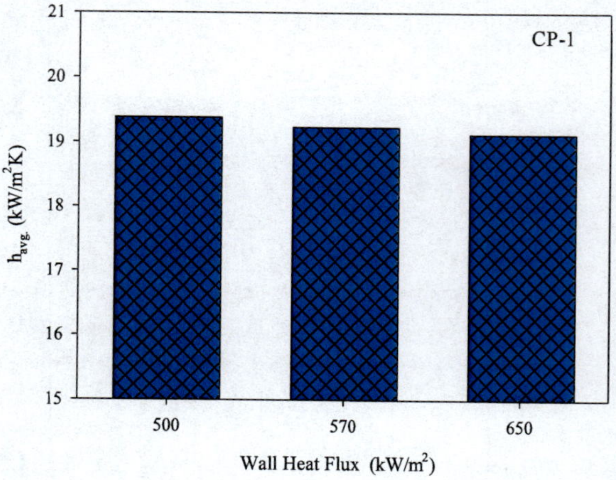

- Wall temperature variation is more in the corrugated pipes as compared to smooth pipe. Wall temperature is maximum on the crest and minimum on the trough portion. Fluid velocity is higher near the trough resulted in enhanced convective heat transfer on the trough portion. Fluid velocity is lower in the crest region, and nucleation formation is promoted in this region.
- The quantity of vapour formed is more in the case of the flat pipe as compared to the corrugated pipes.
- HTC is higher for the corrugated pipes as compared to the flat pipe. Heat transfer increases at the cost of pressure loss. Maximum improvement in HTC is around 8.5% for CP-1, and the pressure drop in CP-1 is 13% more than that in FP.

References

1. Gibou F, Chen L, Nguyen D, Banerjee S (2007) A level set based sharp interface method for the multiphase incompressible Navier-Stokes equations with phase change. J Comput Phys 222(2):536–555
2. Welch SWJ, Wilson J (2000) A volume of fluid based method for fluid flows with phase change. J Comput Phys 160(2):662–682
3. Tryggvason G, Esmaeeli A (2004) Computations of boiling flows. In: Proceedings ASME heat transfer engineering summer conference 2004, HT/FED 2004, 3(3):561–566
4. Sato Y, Niceno B (2017) Nucleate pool boiling simulations using the interface tracking method: boiling regime from discrete bubble to vapor mushroom region. Int J Heat Mass Transf 105:505–524
5. Szijártó R, Badillo A, Ničeno B, Prasser HM (2017) Condensation models for the water-steam interface and the volume of fluid method. Int J Multiph Flow 93:63–70
6. Dong B, Zhang Y, Zhou X, Chen C, Li W (2020) Numerical simulation of bubble dynamics in subcooled boiling along inclined structured surface. J Thermophys Heat Transf 1–12
7. Nemitallah MA, Habib MA, Ben Mansour R, El Nakla M (2015) Numerical predictions of flow boiling characteristics: current status, model setup and CFD modeling for different non-uniform heating profiles. Appl Therm Eng 75:451–460

8. Targanski W, Cieslinski JT (2007) Evaporation of R407C/oil mixtures inside corrugated and micro-fin tubes. Appl Therm Eng 27(13):2226–2232
9. Laohalertdecha S, Wongwises S (2011) An experimental study into the evaporation heat transfer and flow characteristics of R-134a refrigerant flowing through corrugated tubes. Int J Refrig 34(1):280–291
10. Thors P, Bogart J (1994) In-tube evaporation of HCFC-22 with enhanced tubes 1(4):365–377
11. Yang P, Zhang T, Hu L, Liu L, Liu Y (2021) Numerical investigation of the effect of mixing vanes on subcooled boiling in a 3 × 3 rod bundle channel with spacer grid. Energy 236:121454
12. Kurul N, Podowski M (1990) Multidimensional effects in forced convection subcooled boiling. In: Proceedings of the 9th international heat transfer conference, Jerusalem, pp 21–26
13. Fluent A, Guide T, Fluent A Tutorial : modeling nucleate boiling using ANSYS FLUENT introduction, pp 1–22
14. ANSYS Inc. (US) (2020) ANSYS fluent theory guide, Release2020R1
15. Patankar SV (1980) Numerical heat transfer and fluid flow, Hemisphere Publishing Corporation
16. Bartolomei G, Brantov V, Molochnikov Y, Kharitonov Y, Solodkii V, Batashova V, Mikhailov V (1982) An experimental investigation of true volumetric vapor content with subcooled boiling in tubes. Appl Therm Eng 29(3):132–135

Experimental Studies on Thermoelectric Generator Used in IC Engines

D. Srinu and G. Ganesh Kumar

1 Introduction

Huge number of empirical and semi-empirical studies on heat recovery was discussed in the literature. Samir et al. have discussed about the performance, design, and assembly of thermoelectric generator (TEG) [1]. The motto was to improve the energy conversion efficiency by studying on thermoelectric generator plates. It also gives us the clear definition of Seebeck effect. A hollow circular shaft graded by piezoelectric material with constant angular was discussed by Ghorbanpour et al. [2]

Li-Ling Liao and Ming-Ji Dai [3] described on CAD analysis of a TEG with experimental design comprising twelve pairs of thermoelectric pillars. Further, they described about performance by means of slope method and edged reuse of energy. Utilization of thermoelectric modules for applications which are at high temperature has been described by Ziabari and Suhir [4]. They worked on finite element analysis on TEM to decrease shearing and thermal stresses. They have further compared analytical with finite element analysis.

A study on deposition of electrical conductive coatings on aluminum matrix has been discussed by Urena and Utrilla [5] which comprises 75% of SiC particles to be used for electronics packaging.

A rectangular micro-channel method has been discussed by Sandip et al. [6] for the case of a counter-flow heat exchanger. Two methods were used; first is on one-dimensional model while the other on CFD model. A comparative study was performed.

Novel concepts to generate metal matrix composites by injection of silicon carbide particles into molten aluminum have been discussed by Eslmian et al. [7]. An experimental procedure for this process was discussed in detail in their study. Bensaid and Brignone [8] have presented about thermoelectric performance.

D. Srinu (✉) · G. Ganesh Kumar
Kakatiya Institute of Technology and Science, Warangal, India
e-mail: srinudaravath@gmail.com

J. Banerjee et al. (eds.), *Recent Advances in Fluid Dynamics*, Lecture Notes in Mechanical Engineering, https://doi.org/10.1007/978-981-19-3379-0_13

2 Experimental Setup

To generate high quality of electrical energy without kinetic energy a thermoelectric generator was designed and fabricated for the purpose. Performance of thermoelectric converters will depend on material parameters. The design parameters such as length of the leg, heat transfer conditions play a prominent role on performance of a TEG. The main motto of this work is to develop a fabricated model of a TEG for generation of power. A model was designed in CATIA, and analysis was performed in ANSYS.

Figure 1 shows a simulated model of the TEG in CATIA. This model has been further fabricated as shown in Fig. 2. An appropriate model of TEG was designed to generate maximum output voltage. Number of trails have been performed and analyzed in ANSYS for different shapes such as cube, cuboid, cylinder, and triangular prism that can be easily fabricated. It is observed that a triangular prism is an optimum design for fabrication and produce high power with lesser losses in thermoelectric generator (TEG). This TEG comprises a counterflow heat exchanger. Six TEMs are placed in two lines with three in each line. A cooling room was fabricated on the other side. Figure 3 shows the assembly of the same. Conduction mode of heat transfer to modules is taking place from copper plate that is exposed to exhaust gases on hot side, while convection heat transfer is taking place through aluminum plate on the cold side of the module. There is loss occurs due to thermal resistance existing among the hot copper plate and cold aluminum plate. Hence, care has been taken to maintain good surface finish by special polishing material. Baffles were utilized to increase the heat transfer rate in the TEG. Here, steel has been used to fabricate the body as well as baffles.

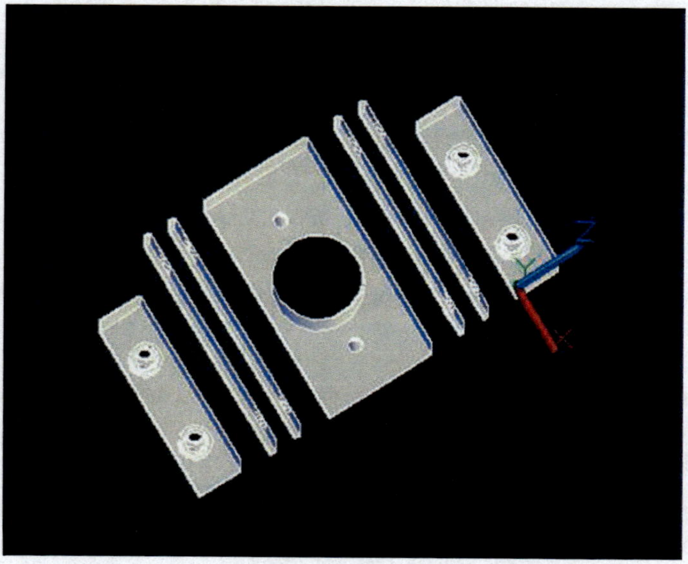

Fig. 1 Simulated model of thermoelectric generator

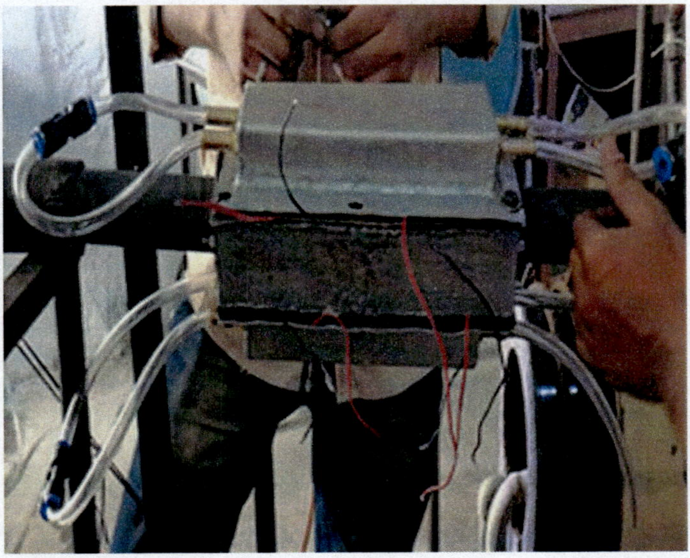

Fig. 2 Fabricated model of a thermoelectric generator

Fig. 3 Location of modules used in TEG

Fig. 4 CI engine with TEG

Modules are placed in between the hot copper plate and cold aluminum plate. The exhaust gases that are emitted through pipe channel were directed into the TEG chamber to interact with copper plates arranged on the three sides of the assembly. 5-mm-thickness plates were used to make aluminum plate, and 3 mm was chosen for copper plates.

Figure 4 displays the thermoelectric generator assembled to 4-S 1-Cylinder, CI engine with brake drum dynamometer that was chosen to test the TEG. A continuous flow of water into the cooling chamber allows the aluminum plate to maintain the lower as well as constant temperature on cooling side of the module. Exhaust gases flow through the lower portion of a heat exchanger as hot gases come in contact with the plate, and the heat is transferred to modules via aluminum plate.

A load test was performed before as well as after the assembling of TEG to ensure that there will not be any effect on engine performance. This test was conducted from 0 to 12-kg load.

If the exhaust gases were struck, then there may be back pressure due to exhaust gases that lead to fall in engine performance. An appropriate model was selected to ensure that there is no back pressure in the chamber. It is observed that the more the temperature difference, higher is the electrical power generated. It is also observed that high power is generated at 12 kg. Also, large temperature difference is maintained to generate the maximum power out of it. Hence, an appropriate design of a heat exchanger has become a criterion to increase the rate heat transfer. Large number of trails was performed on hexagonal, rectangular, and triangular-shaped heat exchangers to analyze the power generation. It is found that the triangular prism is best suited.

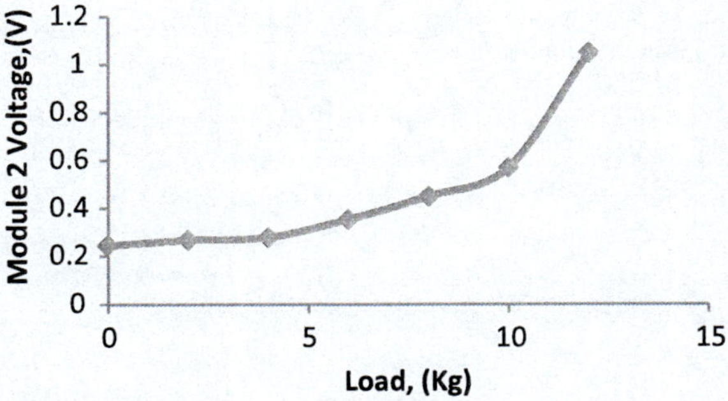

Fig. 5 Variation of voltage for a given load for module 2

The power generated is being utilized to operate the electrical devices used in vehicle as well as to recharge the battery in automobiles. To study the performance, the parameters measured are engine speed, TEG output, TEG coolant inlet and exit, and surface temperatures for different loads and speeds with constant torque. Exhaust gas analyzer is utilized to check for the exhaust gas toxicity. TEG was place in downstream of the exhaust headers next to catalytic converter to produce higher exhaust gas temperature as it decreases the efficiency if the generator is located upstream of the catalytic converter. 0–1200 °C K-type thermocouples with digital measuring unit were used to measure exhaust gas temperature on hot and cold side of TEM. Back pressure of the exhaust gas was measured using a U-tube mercury manometer.

3 Results and Discussion

Large number of experimental studies is conducted to check the effect of performance of an engine due to assembly of TEG. As an illustration, a study was done to discuss about the effect of load on voltage is done. It can be seen that as load increases voltage in thermoelectric generator increases (Fig. 5).

Figure 6 shows variation of voltage in module 2 versus load of an engine. It shows as the load increases voltage in module 2 increases and reaches maximum.

4 Conclusion

Performance of TEG modules utilized in CI engines for generation of an electrical energy was studied in detail. Further, their performance was discussed to generate the power from exhaust gases released from engines.

Fig. 6 Variation of voltage for given load for module 3

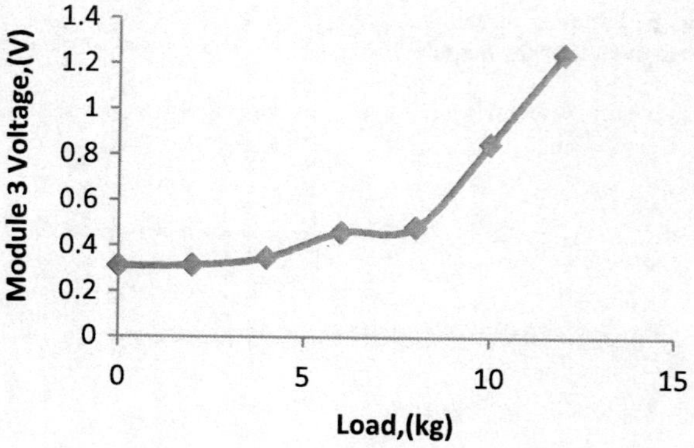

References

1. Ghorbanpour Arani A, Kolahchi R, Mosallaie Barzoki AA (2010) Effect of material in—homogeneity on electro-thermo-mechanical behaviors of functionally graded piezoelectric rotating shaft. J Appl Math Model 35. ISSN:2771-2789
2. Ziabari A, Suhir E, Shakouri A (2013) Minimizing thermally induced interfacial shearing stress in a thermoelectric module with low fractional area coverage. J Microelectron 45. ISSN: 547-553
3. Bensaid S, Brignone M, Ziggioi A (2011) High efficiency thermo-electric power generator. Int J Hydrogen Energy 37. ISSN: 1385-1398
4. Ziabari A, Suhir E, Shakouri A (2013) Minimizing thermally induced interfacial shearing stress in a thermoelectric module with low fractional area coverage. J Microelectron 45. ISSN: 547-553
5. Urena A, Utrilla MV (2007) Electroless multilayer coatings on aluminium-silicon carbide composites for electronics packaging. J Eur Soc 27. ISSN: 3983-3986
6. Eslamian M, Rak J, Ashgriz N (2007) Preparation of aluminium/silicon carbide metal matrix composites using centrifugal automization. J Powder Technol 184. ISSN: 11-20
7. Saha SK, Baelmans M (2014) A design method for rectangular microchannel counter flow heat exchangers. Int J Heat Mass Transf 74. ISSN: 1-12
8. Liao L-L, Dai M-J, Liu C-K, Chiang L-N (2013) Thermo-electric finite element analysis and characteristic of thermoelectric generator with inter metallic compound. Int J Microelectron Eng 120. ISSN: 194-199

The Influence of Baffle Cut and Baffle Distance on a Shell-Side Fluid Flow of a Shell-and-Tube Heat Exchanger

Juluru Pavanu Sai and B. Nageswara Rao

1 Introduction

A heat exchanger is a device which is used to exchange the heat from one fluid to other. The types of heat exchangers are shell-and-tube heat exchanger, plate and shell heat exchanger, plate fine heat exchanger, plate-type heat exchanger, and finned tube heat exchangers are used in a variety of industries such as chemical plants, refrigeration, oil refining, power generation, and so on [1–3] for applications such as heating, cooling, condensation, and evaporation. From the several types of heat exchangers, shell-and-tube heat exchangers (STHXs) are the most widely employed in industrial applications. In general, shell-and-tube heat exchangers occupy more than half of all heat exchangers due to its durable construction, ease of maintenance, and advancement potential [4, 5]. The quantity of heat transfer that occurs using the smallest heat transfer area with the best heat transfer coefficient determines the heat exchanger's efficiency. It is difficult to determine the optimum design parameters. Due to the complicated geometry and fluid characteristics of heat exchanger [6–8]. Various techniques have been developed to reducing energy consumption and increase heat transfer performance. Several researchers have published their findings in design of various types of baffles aimed at decreasing pressure drop or improving heat transmission. The following are some examples: helical baffles with non-continuous, baffle with flower structure, Wang et al. [9], middle-overlapped helical baffles, Zhang et al. [10], helical baffle with continuous, Peng et al. [11], and ladder-type fold baffles, Jain et al. [12]. Tubular Exchanger Manufacturers Association (TEMA) publishes design guidelines and standards for heat exchangers on a regular basis. TEMA recommended correlation-based analytical methodologies are used for modeling of shell-and-tube

J. P. Sai · B. Nageswara Rao (✉)
Department of Mechanical Engineering, Vignan's Foundation for Science, Technology & Research, Vadlamudi, Guntur, Andhra Pradesh 522213, India
e-mail: Director_ies@vignan.ac.in

J. Banerjee et al. (eds.), *Recent Advances in Fluid Dynamics*, Lecture Notes in Mechanical Engineering, https://doi.org/10.1007/978-981-19-3379-0_14

heat exchangers [13]. Due to the accumulation of industry experience and operational data, as well as the improvement of equipment, these approaches have been continuously enhanced since their inception. To achieve an optimal heat exchanger, the sizing and rating of the heat exchanger are based on correlation methodologies. A design professional can understand the impact of a given variable on the heat exchanger's performance. Although setting the tube-side characteristics is rather simple, getting the exact combination for the shell side is quite difficult. Because of the various leakage channels and bypass streams inside the various flow zones, the shell-side design is quite difficult. Leakages and streams are largely depending on the shell design and size. Computational fluid dynamics (CFD) is a well-known technology in the industry for visualizing flow and temperature fields on the shell side [14–16]. Furthermore, it can make assessing flaws easier by pointing the designer in the appropriate direction.

In shell-and-tube heat exchangers, only, a few researchers have used a full-length model to do the numerical analysis of heat and fluid flow, according to the literature. Furthermore, no significant attempts have been made to explore the effect of baffle angle on heat exchanger performance.

The present study examines the thermo-hydraulic performance of a shell-and-tube heat exchanger with segmental baffles with different baffle cuts using CFD and analytical methods for three different baffles cuts 25, 30, and 35%. The performance of proposed baffle cuts of shell and tube heat exchanger were examined by both CFD and analytical methods and proven good.

2 Computational Analysis

In the present study, we are taken the baffles that are placed along the shell at a different orientations alternatively with one baffle cutting face looking up and another one baffle cut face looking downward, and this way orientation repeats; it creates the flow path for fluids across the tube bundles [17, 18]. The shell-and-tube heat exchanger model consists of different parameters; one of the parameters to be optimized is baffle spacing and baffle cut. The three different baffles spacing are 120, 86, and 67 mm along with baffle cut 25, 30, and 35% with a variable mass flow rate in the shell side. Baffles are used to create support to the tubes for structural strength and to reduce the tremble as well as to redirect the flow through the bundle to increase the heat transfer coefficient [19]. The modeling is done in Solid edge ST9, and meshing is done in ANSYS 19.2, respectively, as show in Figs. 1, 2, and 3.

The fluids flow between the shell, and the tube heat exchanger has been simulated in a 3D model by solving the relevant governing equations, which are shown below.

Conservation of mass: $\nabla \, . \, (\rho V) = 0$

$$\text{x} - \text{Momentum} : \nabla.(\rho u \tilde{V}) = -\frac{\partial \text{p}}{\partial \text{x}} + \frac{\partial \tau xx}{\partial \text{x}} + \frac{\partial \tau yx}{\partial \text{y}} + \frac{\partial \tau zx}{\partial \text{z}} \tag{1}$$

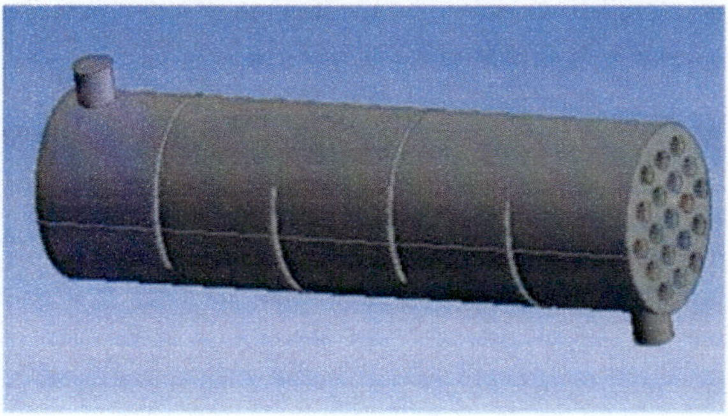

Fig. 1 3D solid edge modeling of STHE

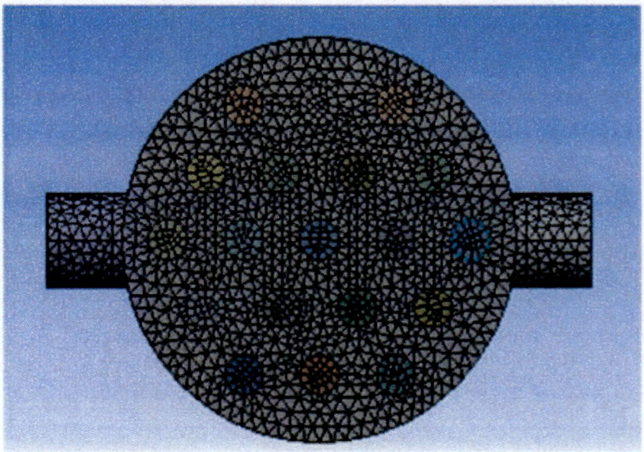

Fig. 2 Side mesh view of STHE

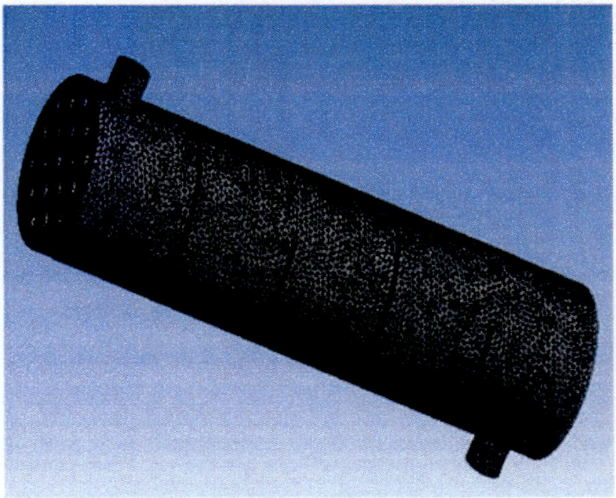

Fig. 3 3D mesh view of STHE

$$\text{y}-\text{Momentum}:\nabla.(\rho\upsilon\tilde{V})=-\frac{\partial \text{p}}{\partial \text{y}}+\frac{\partial\tau xy}{\partial \text{x}}+\frac{\partial\tau yy}{\partial \text{y}}+\frac{\partial\tau zy}{\partial \text{z}}+\rho\text{g} \tag{2}$$

$$\text{z}-\text{Momentum}:\nabla.(\rho w\tilde{V})=\frac{\partial \text{p}}{\partial \text{z}}+\frac{\partial\tau xy}{\partial \text{x}}+\frac{\partial\tau yy}{\partial \text{y}}+\frac{\partial\tau zy}{\partial \text{z}} \tag{3}$$

$$\text{Energy}:\ \nabla.\left(\rho e\tilde{V}\right)=-P\nabla.\tilde{V}+\nabla.(K\nabla T)+q+\phi \tag{4}$$

The corresponding equation for calculating the (ϕ) as shown in Eq. (5)

$$\phi=\mu[2[\left(\frac{\partial \text{u}}{\partial \text{x}}\right)^2+\left(\frac{\partial \upsilon}{\partial \text{y}}\right)^2+\left(\frac{\partial \text{w}}{\partial \text{z}}\right)^2]+\left(\frac{\partial \text{u}}{\partial \text{y}}+\frac{\partial \upsilon}{\partial \text{x}}\right)^2+\left(\frac{\partial \text{u}}{\partial \text{z}}+\frac{\partial \text{w}}{\partial \text{x}}\right)^2$$
$$+\left(\frac{\partial \upsilon}{\partial \text{x}}+\frac{\partial \text{w}}{\partial \text{y}}\right)^2]+\lambda(\nabla.\tilde{V})2 \tag{5}$$

The shell and tube heat exchanger was designed by considering design parameters and boundary conditions. The dimension related to the shell-and-tube heat exchanger is shown in Table 1. The three dimensional model of shell and tube heat exchanger was developed using sold edge ST9 and created the meshing while imported into ANSYS19.2.

The boundary limits related to the shell-side working fluid water with entry temperature 30 °C and the corresponding equilibrium temperature related to the tube walls 80 °C are considering for this study. The leakage of steam takes between the baffles and shell; tubes and baffles are negligible and are thus ignored. The realizable K-ε turbulence model is used. The heat flux is zero due to perfectly insulated.

Table 1 indicates the dimensions of segmental baffle shell-and-tube heat exchanger (SBSTHX). The dimensions of shell, tube, baffle, nozzle, tube sheet, length, and baffle thickness are selected from different tables provided in TEMA STANDARDS.

Table 1 Shell-and-tube heat exchanger geometrical dimensions

S. No.	Design parameters	
1	Shell diameter (D_s)	130 mm
2	Number of baffles	4, 6, 8
3	Tube pitch (P_t)	25 mm
4	Tube outer diameter (d_o)	12.7 mm
5	Effective length (L)	500 mm
6	Baffle cut (B_c)	25, 30 and 35%
7	Number of tubes	19
8	Throttle valve	4 numbers for each (hot and cold flows inlet and outlet)
9	Mass flow rate (m_s)	0.5, 0.6 and 0.7 kg/s

Table 2 Grid independence of STHXs

Grid	Elements	Nodes	Cold outlet temperature (K)
Coarse	247,598	281,613	307.231
Medium	308,648	358,093	309.851
Fine	353,648	378,493	310,059
Very fine	416,570	432,721	310.072

2.1 *Generation of Grid*

The three-dimensional model is to be discrete into the 17 million tetrahedral elements by using ANSYS 19.2 to capture both temperature and velocity boundary layers. The entire model is discretized using accurate tetrahedral mesh elements that reduce computational effort. The tetrahedral mesh along the wall surface provides for more precise capture of the boundary layer gradient. Grid independence is used to ensure that the numerical results are accurate. A few grid dependency tests were run to check that the optimum computational mesh was achieved. With segmental baffles and a grid system that showed the least convergent time and solution accuracy, three sets of grids were computed for STHE. Table 2 shows the findings of the grid independence investigation.

Above grid, system elements were taken. Figures 1, 2, and 3 show the three-dimensional solid edge ST9 modeling of STHE, two-dimensional view, and three-dimensional mesh view of STHE. ANSYS 19.2 software is used to solve the numerical analysis.

3 Validation

The CFD study findings are found for varied shell-side mass flow rates of 0.5, 0.6, and 0.7 kg/s. The model analysis was done with the combustion of fluid flow rate, baffle spacing, and baffle cut of 0.5 kg/s, 86 mm, and 25% of these results is validated using data from Ozeden and Tair [15]. The output temperature of the shell side is found to match the literature results, with a variation of less than 1%.

4 Results and Discussions

Table 3 shows the results of the analysis for the abovementioned design parameters and boundary conditions. Figure 4a represents the velocity streamline, whereas Fig. 4b represents the temperature distribution, and Fig. 5a represents the three-

Table 3 Different parameters obtained by CFD results

CASE	Result	CFD analysis			
	Baffle spacing (m)	Mass flow rate (kg/s)	Shell-side		Total heat transfer rate (W)
			Heat transfer coefficient (W/m^2 K)	Pressure drop (Pa)	
Case 1 BC = 25%	0.12	0.5	1173.82	63.33	3218.56
	0.086	0.6	1376.06	181.81	3757.32
	0.067	0.7	1561.91	414.92	4124.12
Case 2 BC = 30%	0.12	0.5	1020.34	64.89	3143.56
	0.086	0.6	1225.63	198.53	3676.25
	0.067	0.7	1468.308	401.02	4060.74
Case 3 BC = 35%	0.12	0.5	908.232	54.32	2902.13
	0.086	0.6	1110.167	152.64	3293.95
	0.067	0.7	1296.57	332.85	3699.65

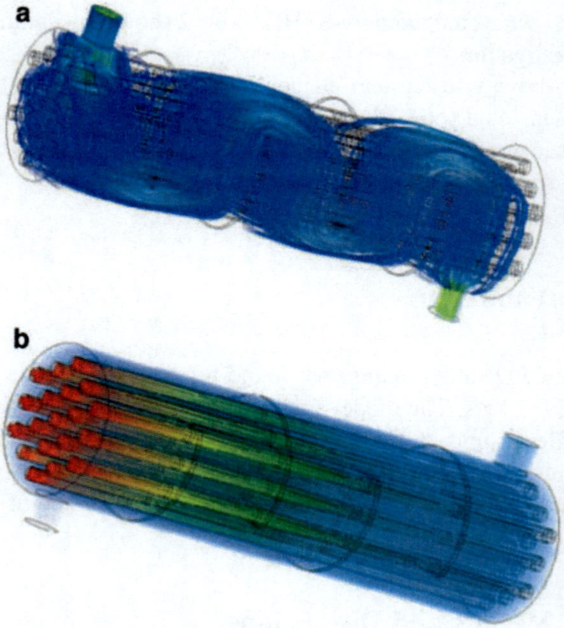

Fig. 4 **a** 3D-velocity streamline. **b** 3D-temperature distribution

dimensional pressure distribution; Fig. 5b represents streamline flow in tube layout. It was found that fluid passing near the baffles caused turbulence eddies. They resulted in a larger shell-side pressure drop, and the distance between the baffle and heat transfer coefficient was shown to be directly proportional to the increased total heat

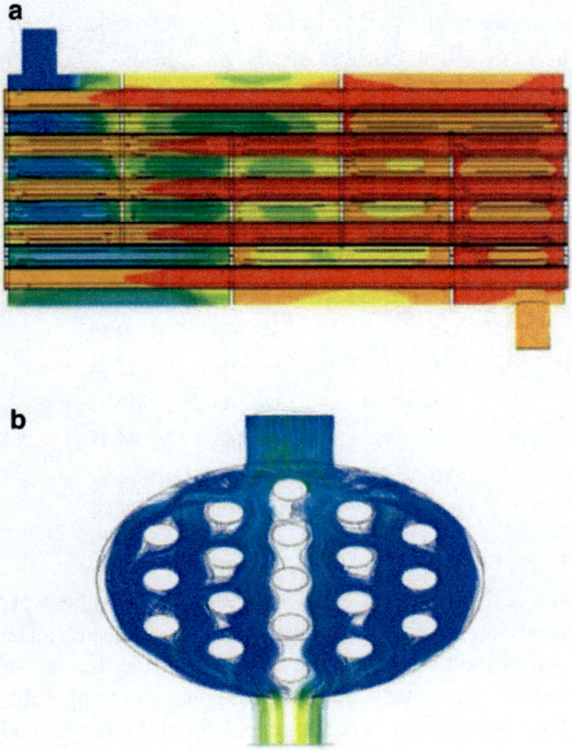

Fig. 5 **a** Pressure distribution. **b** Temperature distribution

transfer rate. Table 3 shows the significant ranges of pressure drop and heat transmission for the baffle cut. Table 4 shows the outcomes of the Bell–Delaware contest. The simulation values for the 35% baffle cut results are closer to the analytical values. The percentage error between CFD and Bell–Delaware results are presented in Table 4.

5 Conclusions

CFD simulations results showed that the heat transfer coefficient (W/m^2 K), pressure drop (Pa), and total heat transfer rate (W) increased with the decreasing the baffle distance. Coming to the baffle point, 25% baffle cut at 0.067 baffle space provides a higher total heat transfer rate, heat transfer coefficient, and pressure drop comparing with another baffle cut and baffle spacing. The Delaware method utilizes empirical correlations for the heat transfer coefficient and friction factor in flow perpendicular to banks of tubes; these are referred to as ideal tube bank corrections; the simulation values for the 35% baffle cut results are closer to the Bell–Delaware method. CFD

Table 4 Different parameters results obtained by Bell–Delaware method

CASE	Results	Bell–Delaware method (Analytical)			
	Baffle spacing (m)	Mass flow rate (kg/s)	Shell-side		Total heat transfer rate (W)
			Heat transfer coefficient (W/m^2 K)	Pressure drop (Pa)	
Case 1 BC = 25%	0.12	0.5	1006.68	51.33	3018.02
	0.086	0.6	1218.706	152.81	3456.6
	0.067	0.7	1411.91	341.95	3838.7
Case 2 BC = 30%	0.12	0.5	943.034	52.89	2942.03
	0.086	0.6	1145.896	158.74	3374.29
	0.067	0.7	1328.308	357.92	3752.22
Case 3 BC = 35%	0.12	0.5	842.462	44.53	2794.91
	0.086	0.6	1020.167	132.86	3212.06
	0.067	0.7	1186.57	297.85	3579.21

simulations are used to illustrate the fluid flow structures of heat exchangers for various combinations of structural and hydrodynamic factors such as baffle spacing, baffle cut, and mass flow rate. To resolve the fluid flow inside the shell-and-tube heat exchanger, this simulation visualizes the streamline path and temperature fields. The recirculate of the fluid is collected in some areas behind the baffle, and the counter flow windows are underutilized. By reducing the baffle spacing or alternatively, increasing the number of baffles, this problem is avoided, and the heat exchanger's heat transfer is improved.

Table 5 In different cases percentage of error

Baffle spacing (m)	Mass flow rate (kg/s)	Results		Percentage of error (%)		
				Case 1	Case 2	Case 3
0.12	0.5	Shell side	heat transfer coefficient	14.2	7.6	7.2
			Pressure drop	18.9	18.5	18.0
		Total heat transfer rate		6.2	6.4	3.7
0.086	0.6	Shell side	Heat transfer coefficient	19.4	7.6	8.1
			Pressure drop	16	20	13
		Total heat transfer rate		8	8.2	2.5
0.067	0.7	Shell side	Heat transfer coefficient	9.6	9.5	8.5
			Pressure drop	17.6	10.7	10.5
		Total heat transfer rate		6.9	7.6	3.2

References

1. Kern DQ (1950) Process heat transfer. McGraw-Hill, New York
2. Bell KJ, Shah RK (2000) CRC handbook of thermal engineering. CRC Press, Florida
3. Sai JP, Nageswara Rao B (2020) Efficiently and economic optimization of shell and tube heat exchanger using bacteria foraging algorithm. SN Appl Sci 2:13. https://doi.org/10.1007/s42452-019-1798-0
4. Wang Q, Chen G, Chen Q, Zeng M (2010) Review of shell-and-tube heat exchangers with helical baffles. Heat Transfer Eng 31(10):836–853
5. You Y, Fan A, Huang S, Liu W (2012) Numerical modelling and experimental validation of heat transfer and flow resistance on the shell side of a shell and-tube heat exchanger with flower baffles. Int J Heat Mass Transf 55:7561–7569
6 Ulson RK, Sinnot JF, Richardson (1996) Chemical engineering design, Butterworth Heinemann, Boston MA
7. Caputo AC, Pelagagge PM, Salini P (2008) Heat exchanger design based on economic optimisation. Appl Therm Eng 28(10):1151–1159
8. Peters MS, Timmerhaus KD (1991) Plant design and economics for chemical engineers. McGraw-Hill, New York
9. Wang Y et al (2011) Experimental investigation of shell-and-tube heat exchanger with a new type of baffles. Heat Mass Transf 47(7):833–839
10. Zhang J-F et al (2009) Experimental performance comparison of shell-side heat transfer for shell-and-tube heat exchangers with middle-overlapped helical baffles and segmental baffles. Chem Eng Sci 64(8):1643–1653
11. Peng B et al (2007) An experimental study of shell-and-tube heat exchangers with continuous helical baffles. J Heat Transf 129(10):1425–1431
12. Jian W et al (2015) Numerical investigation on baffle configuration improvement of the heat exchanger with helical baffles. Energy Convers Manage 89:438–448
13. Gaddis D (ed) (2007) Standards of the tubular exchanger manufacturers association, 9th edn. TEMA Inc., Tarrytown (NY)
14. Sunden B (2007) Computational heat transfer in heat exchangers. Heat Transf Eng Heat Transf Eng 28(11):895–897
15. Ozden E, Tari I (2010) Shell side CFD analysis of a small shell-and-tube heat exchanger. Energy Convers Manage 51:1004–1014
16. Bhutta MMA, Hayat N, Bashir MH, Khan AR, Ahmad KN, Khan S (2012) CFD applications in various heat exchangers design: a review. Appl Thermal Eng 32:1–12
17. Hadidi A, Nazari A (2013) Design and economic optimization of shell-and-tube heat exchangers using biogeography-based (BBO) algorithm. Appl Therm Eng 51(1–2):1263–1272
18. Şencan Şahin A, Kılıç B, Kılıç U (2011) Design and economic optimization of shell and tube heat exchangers using artificial bee colony (ABC) algorithm. Energy Conv Manage 52(11):3356–3362
19. Taher FN, Movassag SZ, Razmi K, Azar RT (2012) Baffle space impact on the performance of helical baffle shell and tube heat exchangers. Appl Therm Eng 44:143–149

Experimental Investigation on Droplet Regimes and Droplet Impact on Horizontal Tube Array

Kandukuri Prudviraj, Sandip Deshmukh, and Katiresan Supradeepan

1 Introduction

The heat exchanger is a system used in both cooling and heating applications, which allows thermal energy to be transferred between two or more fluids at distinct temperatures. Falling film heat exchangers with horizontal tube array intently used in many industrial applications such as refrigeration [1], desalination [2], food processing industries, chemical engineering, and petroleum refineries. Figure 1 portrays the association of droplet flow behavior in a wide range of industrial applications [3]. These devices have various advantages over flooded evaporators, including a reduced refrigerant charge, less pressure drop, as well as functioning over modest changes in temperature [4]. In the case of falling film heat exchangers, for a given Reynolds number (Re), the gravity-driven test liquid passes downhill from the previous tube to its next subsequent tube. The liquid film characteristics and inter-flow patterns between the tubes affect the heat and mass transfer. It is essential to understand the different intermediate phases in the droplet flow regime and inter-tube droplet flow characteristics under various parametric conditions such as Re is critical. The explicit details of droplet flow regimes were meticulously examined using high-speed digital camera and an image processing approach.

Hu et al. [5] conducted experimental studies to examine the inter-tube falling film flow patterns and formulated mathematical equations based on Reynolds number (Re) and Galileo number (Ga). Chen et al. [6] analyzed flow pattern transformations between the tubes using the numerical and experimental methods for different flow rates. Mohamed et al. [7] investigated the impact of tube rotation on fluid flow regimes. It is observed that the transitions begin at low Re for the rotating tube

K. Prudviraj (✉) · S. Deshmukh · K. Supradeepan
Department of Mechanical Engineering, Hyderabad Campus, Birla Institute of Technology & Science, Pilani, Hyderabad, India
e-mail: p20180448@hyderabad.bits-pilani.ac.in

J. Banerjee et al. (eds.), *Recent Advances in Fluid Dynamics*, Lecture Notes in Mechanical Engineering, https://doi.org/10.1007/978-981-19-3379-0_15

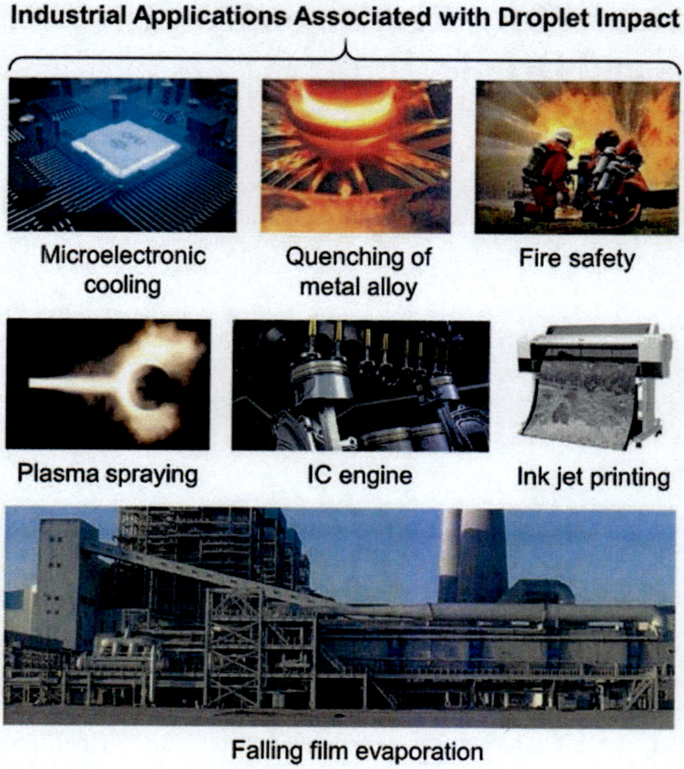

Fig. 1 Industrial applications of droplet flow regime [3]

compared to stationary. The inter-tube flow mode patterns between the flat tubes were experimentally investigated by Wang et al. [8].

Roques et al. [9] and [10] examined the flow regime transitions for the fin tubes based on the fin density and inter-tube distance. [11] performed experiments to examine the flow patterns by varying feeder height and orifice spacing.

The study of Bustamante et al. [12] revealed the effect of distributor type on falling liquid hydrodynamics. The maldistribution of a liquid film for different Re was investigated using box and tube-based distributors. Qu et al. [13] studied flow modes using six liquid distributors of different configurations.

Yung et al. [14] investigated various phases of droplet flow, including droplet detachment and neckbreaking. The authors formulated an equation for calculating droplet diameter. Killion et al. [15] conducted experiments to study the droplet flow characteristics over the tubes. The flow parameters are quantified using image processing with the aid of region of interest (ROI) method. Zhang et al. [16] investigated experimentally to examine the different regimes in the droplet flow, and surface area and volume of a droplet were measured using MATLAB image processing. Li et al. [17] and Wang et al. [18] experimentally studied the droplet flow characteristics and impact modes in a static liquid film with the aid of a high-speed photographic device. The outcomes are measured by the image analysis technique.

1.1 Limitations in the Existing Literature

According to the abovementioned literature study, some of the experimental and numerical research works have been performed to understand the primary and intermediate flow regimes. In the open literature, majority of the studies have focused on several factors such as flow mode transitions, based on the geometry of the distributor, test tube diameter, type of fluid, and fluid properties. The substantial photographic evidence of different droplet flow regimes and droplet impact mechanisms between the tubes is not yet explored in the existing literature to the best knowledge of authors. From the experimental findings, a detailed insight of different phases in the droplet flow regime is far from ample. To some degree, this impedes the proper knowledge of droplet flow behavior and formulation of model equations for the different droplet impacting modes. To assess the impact of droplet flow behavior on design specifications of falling film condensers and evaporators, it is essential to study the inter-tube droplet flow characteristics for different Re.

1.2 Focus of the Present Research

The goal of this work is to use a high-speed photographic camera to investigate the different phases in the droplet flow regime. The flow parameters were evaluated for different Re with the aid of image processing methodology. A series of experimental investigations were carried out to conceptualize the distinct phases in droplet flow regime for Re ranging from 41 to 250 for the same tube-to-tube spacing of 10 mm, using Phantom V440L high-speed digital camera. Furthermore, the study included comprehensive description of images acquired and edge detection employing the *Sobel* image analysis approach. The pivotal information on variation in droplet flow regimes for different Re were elucidated.

1.3 Experimental Approach

The experimental setup is depicted schematically in Fig. 2. The test rig's main components are inlet water circulation unit, a concentric tube liquid distributor, and copper tubes. A storage unit, centrifugal pump, bypass flow line, flow meter, and regulating valves comprise the test liquid flow system. The test liquid is pumped from the storage unit to the circulation unit. Water is utilized as a test liquid in the experiments. Water

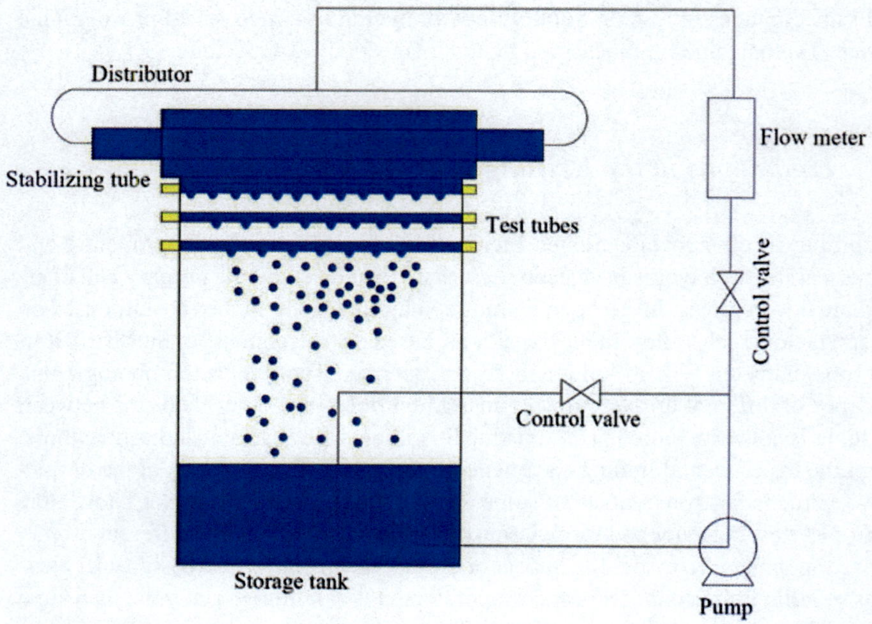

Fig. 2 Schematic of the horizontal tube test setup

flow to the feeder through the circulation unit is controlled by manually operated ball valves. A rotameter is utilized to measure the flow rate in the pipeline.

Figure 3 shows the schematic view of concentric tube liquid sprayer. The sprayer is configured to spread liquid uniformly across the tube walls. The liquid sprayer is made up of an inner tube with dimensions of 75 mm and 500 mm, the diameter and length of the outer pipe are 110 mm and 450 mm, respectively. A continuous channel of 2 mm width and 400 mm size is furnished at the top side of the inner pipe. The outer pipe contains 2-mm-diameter holes with a center distance of 4 mm at the underside and a span of 400 mm. Initially, water fills the inner tube from both the endpoints and departs via a continuous slot provided at the top side. Afterward, the liquid falls from the bottom side of the outer pipe through a succession of tiny holes.

The test setup comprises three copper tubes of the same length 500 mm and inner and outer diameters of 25.4 mm and 31 mm, respectively. For all test tubes, the investigative length for experimenting is 400 mm. The very first tube closest to the sprayer is known as a stabilizing tube, while the lower tubes are known as test tubes. The space between the liquid sprayer and stabilizing tube is close to 2 mm, to attenuate the impact of liquid film from underside the sprayer holes. The key role of the stabilizing tube is to help in the uniform dispersion of droplets.

Figure 4 illustrates the image acquisition unit utilized in this article. The investigative portion is evenly lighted by a 100 W LED and a diffuser sheet. Phantom V440L a high-speed camera is operated to capture and record the flow regimes between the tubes with a resolution of 1920 × 1080 pixels and 1500 frames per sec.

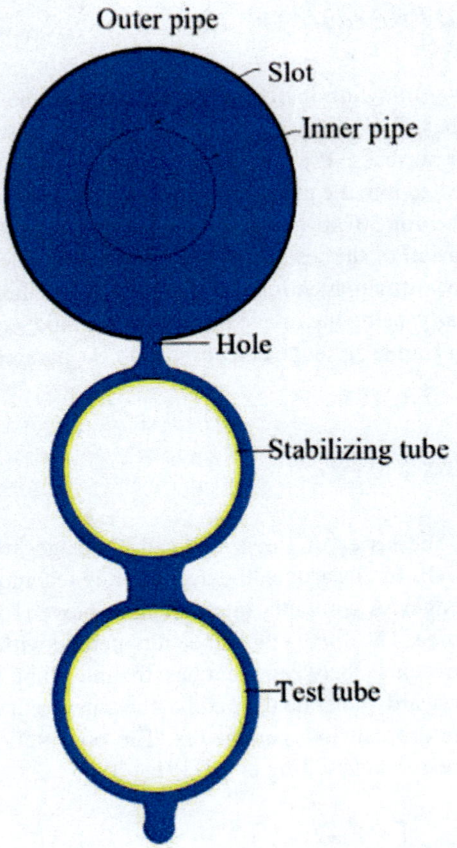

Fig. 3 Schematic view of liquid distributor

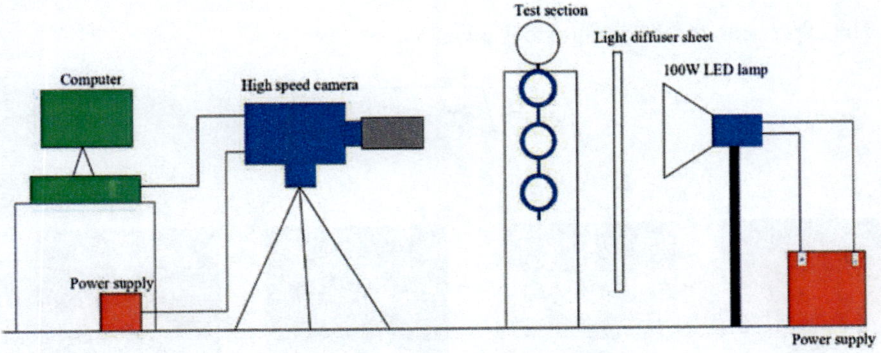

Fig. 4 Image acquisition setup

1.4 Experimental Procedure

The study employs the following methodology to examine the falling liquid flow pattern between the tubes. To begin, use acetone to wipe the tube surfaces to eliminate dust from the exterior surfaces. Prior to the beginning of the test, all tubes were precisely leveled and fitted into the tube holding units at the prescribed tube spacing. The water is distributed for 30–40 min to guarantee that the tubes are thoroughly wetted. The center portion of the test setup is opted for the video capture to avoid "edge effect." It is noteworthy to mention that there are two or three falling instances in the test site. Eventually, using the image processing method, convert the captured videos to a sequence of frames for edge detection and flow parameter quantification.

1.5 Image Analysis

Figure 5 demonstrates the series of steps involved in image processing for edge detection using MATLAB. In this article, the backlighting technique is employed to detect high-definition edges. A computer interface is employed to operate the high-speed photographic device. The *Sobel* edge detection method with gradient function is adopted. When no flow exists between the tubes, the inter-tube distance is used to calibrate the scale. Afterward, generate the pixel data values to the actual dimension in order to estimate the droplet flow parameters. The convolution kernels for the *Sobel* function are expressed below, Jing et al. [19],

$$\mathrm{Gx} = \begin{bmatrix} -1 & -2 & -1 \\ 0 & 0 & 0 \\ +1 & +2 & +1 \end{bmatrix} \mathrm{Gy} = \begin{bmatrix} -1 & 0 & +1 \\ -2 & 0 & +2 \\ -1 & 0 & +1 \end{bmatrix} \tag{1}$$

The expression for the amplitude is given by,

$$|G| = \sqrt{Gx^2 + Gy^2} \tag{2}$$

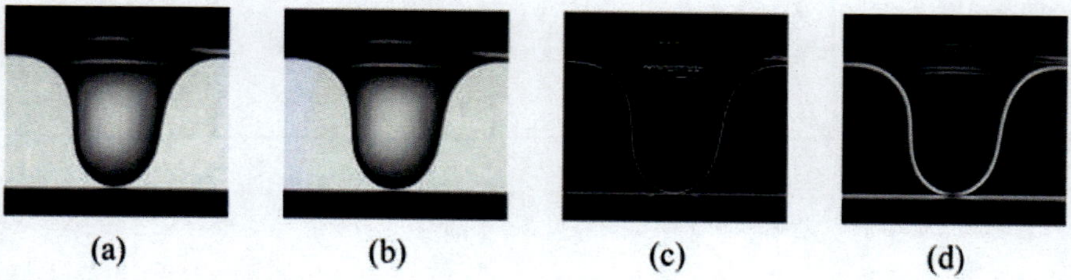

Fig. 5 Image analysis **a** original, **b** gray scale, **c** Sobel image, **d** gradient image

2 Model Verification

The Re can be expressed as follows,

$$\mathrm{Re} = \frac{4\Gamma}{\mu_L} \tag{3}$$

where Γ is liquid sprinkle density for one side (kg/m s) and μ_L is the dynamic viscosity of a liquid. The test liquid properties are presented in Table 1 at room temperature.

The flow-regulating valves are used to control the liquid supply to the distributor, and the liquid sprinkle density can be measured using flow meter readings. The valves are gradually opened to achieve the desired flow pattern between the tubes. Figure 6 illustrates the comparison of present results for the different flow regimes with the values obtained from the existing literature by Hu et al. [5]. The transitional Re fits a pattern compatible with published literature for various flow regimes, and mean absolute error is 8.37.

Table 1 Working fluid properties

Test liquid	Surface tension (N/m)	Viscosity (Pa-s)	Density (kg/m^3)
Water	0.0712	0.001	998.7

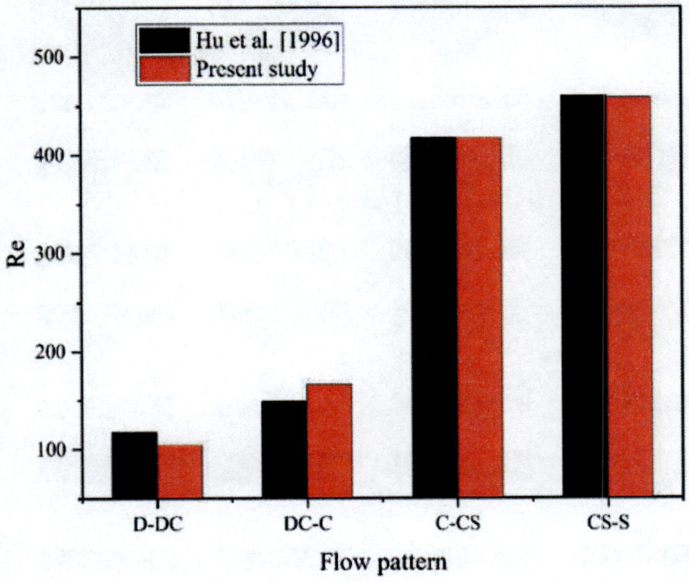

Fig. 6 Comparison of flow regimes

3 Results and Discussion

The objective of the present study experimentally visualizes the different phases in the droplet flow and impacting modes for different Re. A high-speed photographic unit and the *Sobel* image analysis method are used to detect the droplet boundaries. The droplet flow parameters were measured using MATLAB image analysis. Since the falling film hydrodynamics occur abruptly, the flow mechanism between the tubes is obscured to the human eye. The detailed description of different stages in the droplet flow pattern and variation in impacting mode with the Re can be demonstrated in this segment using photographic pieces of evidence. The inter-tube flow regimes can be divided into the following main modes such as "Droplet flow (D)," "Column flow (C)," "Sheet flow (S)," and intermediate flow regimes are "Droplet-Column flow (DC)" and "Column-Sheet flow (CS)" by Hu et al. [5]. Likewise, the distinct phases in droplet flow and impact modes were presented from the experimental outcomes.

The Re is found to have a significant influence on droplet flow regimes for a given inter-tube distance. The visualization of flow modes between tubes is especially challenging and not easy to ascertain for lower tube-to-tube distance. When the inter-tube distance is 10 mm, the transformation of flow regime from droplet to the column pattern involves more contemplation and difficult to find out.

Figures 7, 8, 9, and 10 represent the various droplet flow regime phases, which primarily include droplet development, neck formation, droplet impacting, neck break and retraction, and columnar droplet development.

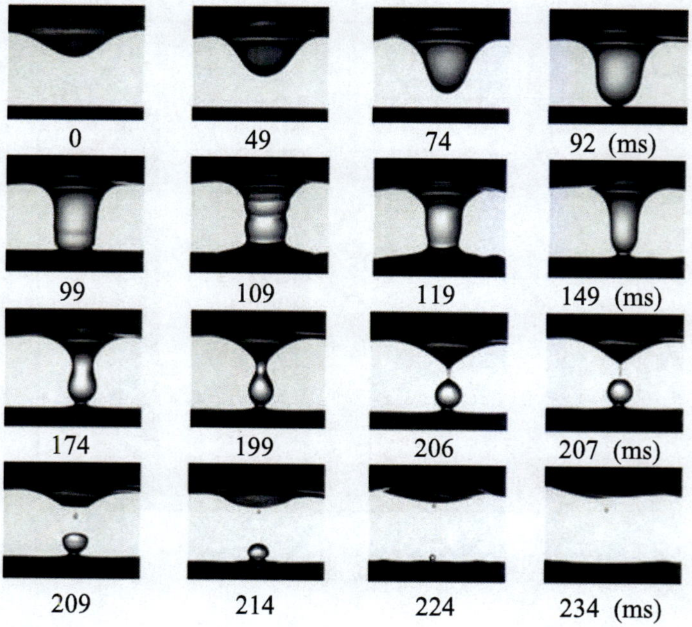

Fig. 7 Droplet flow regime, Re = 41

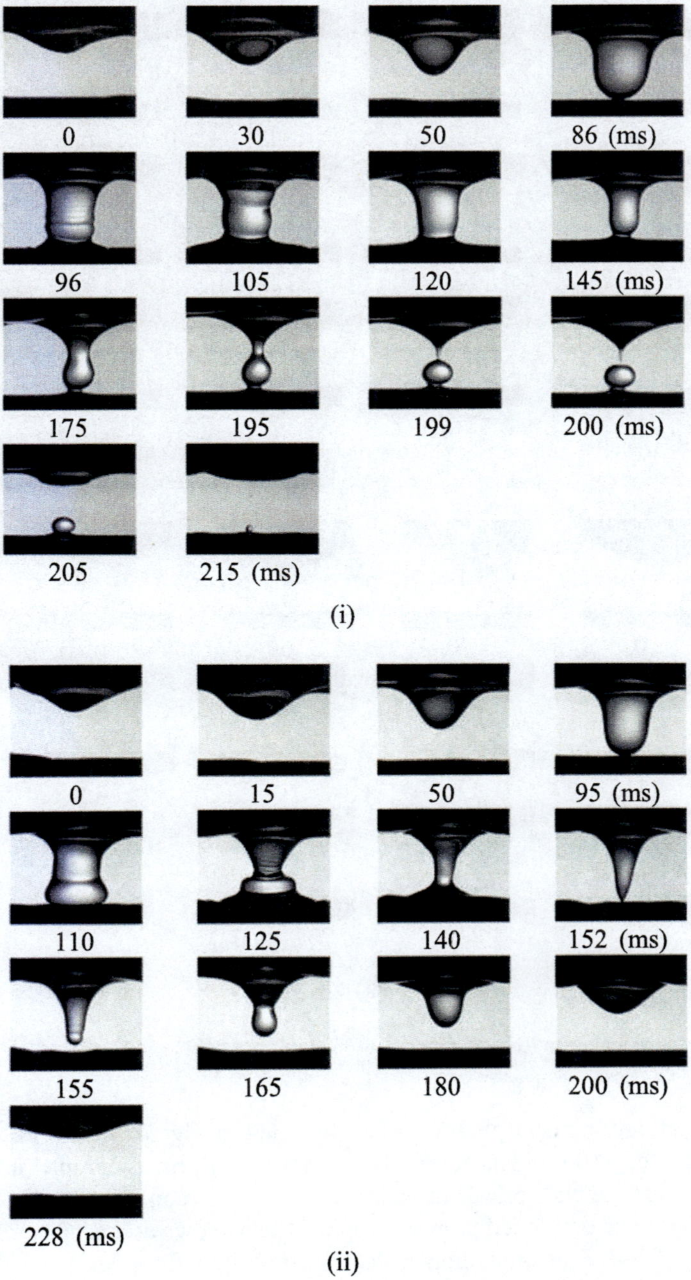

Fig. 8 Droplet flow regime, Re = 83. **a** Mode-1, **b** mode-2

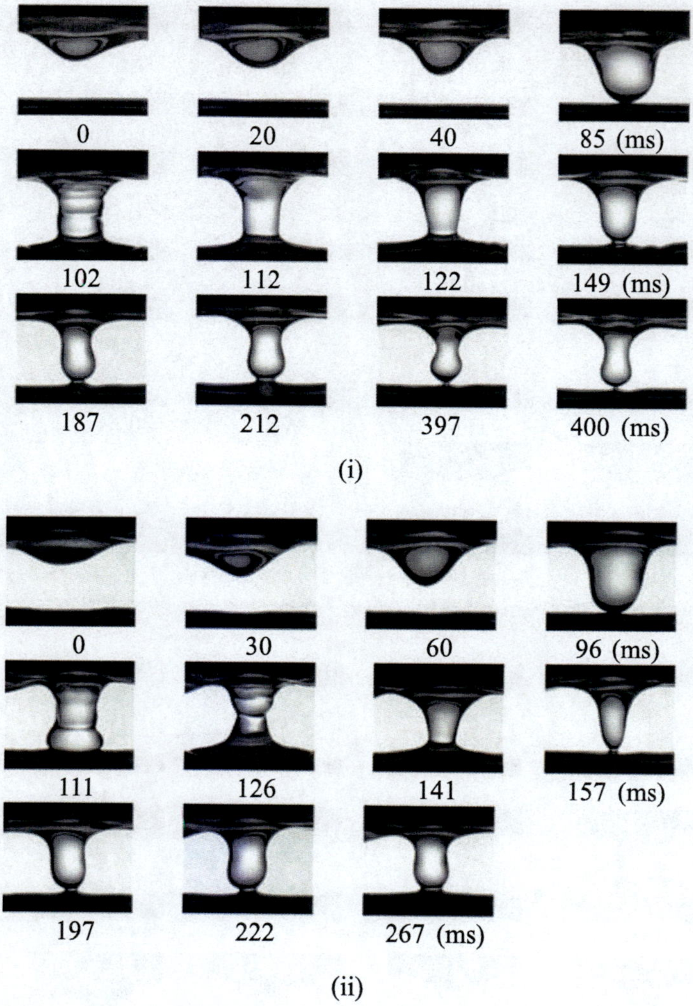

Fig. 9 Droplet flow regime, **a** Re = 125, **b** Re = 166

The droplet development phase can be illustrated in Fig. 7 at 0–92 ms, when the tube distance is 10 mm and Re is 41. The developed primary droplet interacts with the tube surface and spreads in both the axial and circumferential directions. In this sequence, the developed primary droplet is also known as a pendant droplet, also mentioned this behavior, Killion et al. [15]. At this point, the developed droplet has a hemispherical shape at the droplet head. The formation of a droplet neck is observed during the process of droplet spreading from 99 to 119 ms in Fig. 7. The contact angle increases at this moment, and the development of a secondary droplet is visible in Fig. 5 at 149 ms. Afterward, the secondary droplet neck weakens due to inadequate liquid supply to overcome gravitational force. As a result, the neck

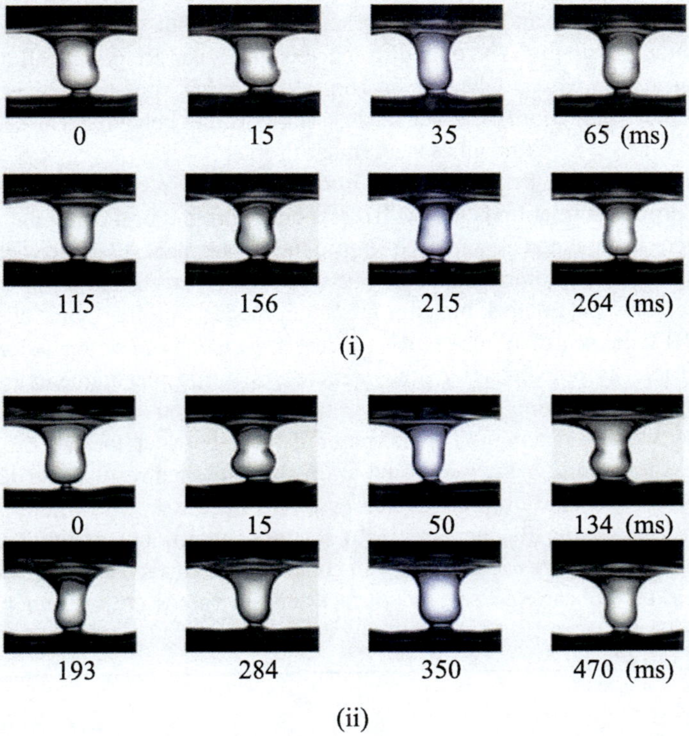

Fig. 10 Droplet flow regime, **a** Re = 208, **b** Re = 250

becomes thinner and thinner, eventually breaking, and forming a pendulous tiny droplet. This is referred to as a detached spherical droplet phase. The remaining liquid film is then retracted by the upper tube, as shown in Fig. 7 at 207–234 ms. This phase is known as neck break and retraction.

For Re = 83, two distinct droplet phases were observed. The droplet flow regime in the first phase is similar to the previous scenario when Re = 41, except that the retraction time is shorter in this segment. At this Re, two observable modes of the droplet impact, neck formation and retraction stages, have been noticed. At this point, the primary droplet makes contact with the tube's surface and begins to spread across it. The development of the neck can be seen in Fig. 8 at 110–140 ms, with the shape of the neck being a V with a connected oval droplet at the head. The sharp interface at the droplet head is identified at 152 ms in Fig. 8, and the developed neck becomes thinner and begins to break. There is no secondary droplet formation in this impacting mode. Finally, as shown in Fig. 8, at 155–228 ms, the upper tube withdraws the remaining pendulous liquid in the form of a droplet.

Droplet flow hydrodynamics for Re values of 125 and 166 can be seen in Fig. 9. The droplet flow regime follows the same trend for Re values 125 and 166. At this point, the formation of a primary droplet, a neck, and a second droplet is observed.

There are no neck break and retraction phases in this case due to an adequate supply of liquid to the distributor. The only difference is that when Re is 125, all dropping sites are inactive, whereas when Re is 166, all dropping sites are active. On the other hand, individual droplets, at active sites, show similar behavior. The secondary droplet that forms is also known as a columnar droplet.

Figure 10 depicts the behavior of the droplet flow pattern between the tubes. In this phase, droplet formation began with a secondary droplet or a columnar droplet. Because there is an adequate supply of liquid in this segment, there are active and continuous columnar droplets connected between the tubes around all of the sites. In this stage, there are no neck break or retraction modes. Another point of concern from Fig. 10 sequence of frames is the modest movement and deformation of the columnar droplet as Re increases. As the Re increases, the droplet diameter increases, which can be depicted from Fig. 11.

The departure site spacing (λ) is the center distance between the two droplets or columns, as shown in Fig. 12. According to the experimental results, the departure site spacing between the droplets decreases as Re increases. The liquid sprinkle density increases as the Re increases, and so does the amount of fluid gathered beneath the tubes, leading to an increase in cohesive and surface tension forces. As Re increases, the cohesive forces and surface tension forces are overwhelmed by

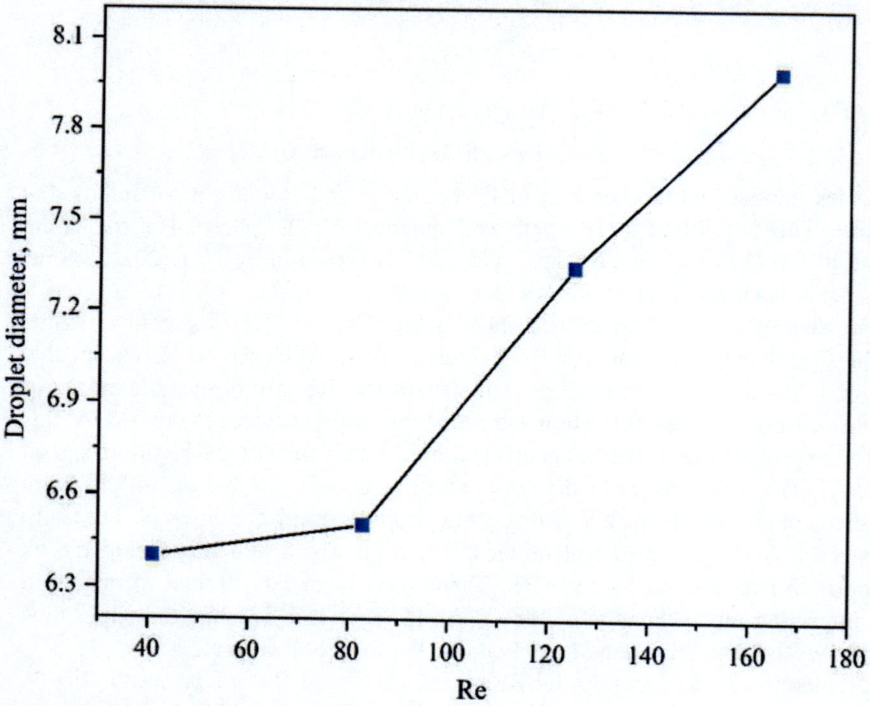

Fig. 11 Variation in primary droplet diameter

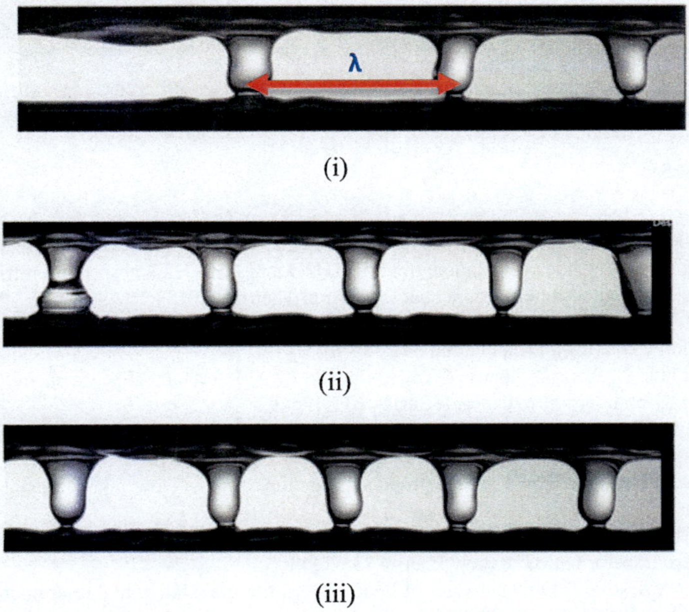

Fig. 12 Departure site spacing, **a** Re = 166, **b** Re = 208, **c** Re = 250

the gravitational force. The λ value for the Re 166, 208, and 250 is approximately 23.84 mm, 18.16 mm, and 14.52 mm, respectively. As a result, all of them dropping points in the flow field are active. The wetting ratio across the tube surface is essential to heat transfer. The development of dry spots on the tube surface has an adverse influence on the operation; on the other hand, the data on departure site spacing for different Re are useful to designers in preventing this problem. Furthermore, a large departure site spacing between the droplets may lead to the formation of dry regions over the tube.

4 Conclusions

The present investigation provided a detailed description of droplet flow regimes and impact modes for Re differing from 41 to 250, with an inter-tube distance of 10 mm. A high-speed camera was used to capture the flow mechanism using a backlighting approach. The *Sobel* image analysis method is used to detect the boundaries, and flow parameters were measured by image analysis. The visual evidence revealed that the distinct phases in the droplet flow pattern can be mainly classified as follows: primary droplet development, secondary droplet formation, columnar droplet, neck break, and retraction. The droplet impact modes were identified to differ as the

Reynolds number increased. As the Re increases, the size of the droplet increases, and departure site spacing decreases.

References

1. Narváez-Romo B, Chhay M, Zavaleta-Aguilar EW, Simões-Moreira JR (2017) A critical review of heat and mass transfer correlations for LiBr-H2O and NH3-H2O absorption refrigeration machines using falling liquid film technology. Appl Therm Eng 123:1079–1095
2. Mabrouk AN, Fath HE (2015) Technoeconomic study of a novel integrated thermal MSF–MED desalination technology. Desalination 371:115–125
3. Wang J, Liang G, Wang T, Zheng Y, Yu H, Shen S (2021) Interfacial phenomena in impact of droplet array on liquid film. Colloids Surf A 615:126292
4. Bustamante JG, Garimella S (2014) Dominant flow mechanisms in falling-film and droplet-mode evaporation over horizontal rectangular tube banks. Int J Refrig 43:80–89
5. Hu X, Jacobi AM (1996) The inter tube falling film part 1—flow characteristics. Mode transitions and hysteresis. ASME J Heat Transfer 118:616–625
6. Chen J, Zhang R, Niu R (2015) Numerical simulation of horizontal tube bundle falling film flow pattern transformation. Renew Energy 73:62–68
7. Mohamed AMI (2007) Flow behavior of liquid falling film on a horizontal rotating tube. Exp Thermal Fluid Sci 31(4):325–332
8. Wang X, Hrnjak PS, Elbel S, Jacobi AM, He M (2012) Flow modes and mode transitions for falling films on flat tubes. J Heat Transfer 134(2)
9. Chen J, Zhang J, Ma Z (2019) Falling film mode transitions on horizontal enhanced tubes with two-dimensional integral fins: effect of tube spacing and fin structures. Exp Thermal Fluid Sci 101:241–250
10. Roques JF, Thome JR (2003) Falling film transitions between droplet, column, and sheet flow modes on a vertical array of horizontal 19 FPI and 40 FPI low-finned tubes. Heat Transfer Eng 24(6):40–45
11. Wang X, He M, Lv K, Fan H, Jacobi AM (2013) Effects of liquid supply method on falling-film mode transitions on horizontal tubes. Heat Transfer Eng 34(7):562–579
12. Bustamante JG, Garimella S (2019) Experimental assessment of flow distributors for falling-films over horizontal tube banks. Int J Refrig 101:24–33
13. Qu Z, Ma Z, Chen J, Zhang J (2019) Falling film flow mode transitions on an array of horizontal tubes under nonuniform liquid distribution conditions. Exp Thermal Fluid Sci 109:109901
14. Yung D, Lorenz JJ, Ganic EN (1980) Vapor/liquid interaction and entrainment in falling film evaporators
15. Killion JD, Garimella S (2004) Pendant droplet motion for absorption on horizontal tube banks. Int J Heat Mass Transf 47(19–20):4403–4414
16. Zhang H, Yin D, You S, Zheng W, Li B, Zhang X (2019) Numerical and experimental investigation on the heat and mass transfer of falling film and droplet regimes in horizontal tubes LiBr-H2O absorber. Appl Therm Eng 146:752–767
17. Li J, Zhang H, Liu Q (2019) Characteristics of secondary droplets produced by a single drop impacting on a static liquid film. Int J Multiph Flow 119:42–55
18. Wang J, Fan Z, Wang D, Lu S, Zhang Y (2021) Coulomb split evolution behavior in different growth stages of droplets. Colloids Surf A 611:125847
19. Jing M, Du Y (2020) Flank angle measurement based on improved Sobel operator. Manuf Lett 25:44–49

Flow Control Using MVG in Shock Wave/Boundary Layer Interaction

N. Nishantt and Nikhil A. Baraiya

Nomenclature

MVG	Micro vortex generator
H	Height of MVG in mm
SL	Separation line
Re_θ	Reynolds number based on momentum thickness
ILES	Implicit large eddy simulation
RANS	Reynolds-averaged Navier Stokes
c/h	Chord to height ratio
d	Diameter in mm
r	Radius in mm
u	Velocity in m/s

Greek Symbols

α	Skew angle in degrees
β	Pitch angle in degrees
δ	Boundary layer thickness in mm
θ_s	Half-span angle of MVG in degrees
θ_r	Ramp angle of MVG in degrees

N. Nishantt (✉) · N. A. Baraiya
Sardar Vallabhbhai National Institute of Technology, Surat, India
e-mail: nishanttn99@gmail.com

N. A. Baraiya
e-mail: nikhil@med.svnit.ac.in

J. Banerjee et al. (eds.), *Recent Advances in Fluid Dynamics*, Lecture Notes in Mechanical Engineering, https://doi.org/10.1007/978-981-19-3379-0_16

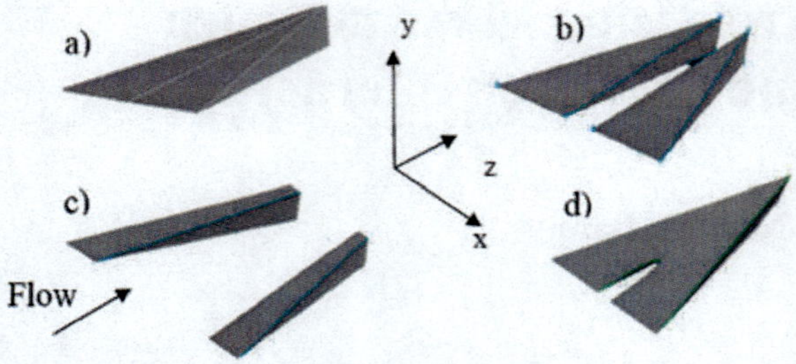

Fig. 1 GAMBIT model of **a** micro ramp, **b** split micro ramp, **c** ramped vane, and **d** slotted micro ramp

1 Introduction

Beginning from the twentieth century, flow separation became the limelight after a concept called "boundary layer theory" was introduced by Ludwig Prandtl in 1904. This showed, there lies a thin region over a solid surface where the effects of friction are experienced only very near an object moving through a fluid [2]. After World War II, research and development of supersonic flights began and it further added complications in name of "shock-induced flow separation." Chang et al. [8] say the shock wave imposes a huge pressure rise such that the subsonic part of the boundary layer close to surface gets decelerated and further increases to an extent where the advancement of flow is non-viable and the flow reverses, which thickens the boundary layer.

This thickening reduces the total pressure recovery and aerodynamic efficiency of inlets resulting to higher drag. Several active and passive control techniques were introduced and researched worldwide such as boundary layer suction and blowing by porous cavities [23], micro vortex generators shown in Fig. 1, micro bumps, synthetic jets [13, 26], a combination of MVG and bleed [24] ,and injection through MVG [25]. Since active techniques added energy to the flow by using separate set of control components which increased weight and occupied more area, passive techniques were preferred, as it employed a strategical design of geometry based on the application. In recent years, it has been applied successfully both in the internal and external flow fields [12]. This paper reviews and summarizes the recent advances in using MVG for control of shock-induced flow separation, and the physics involved behind these micro mechanical systems.

2 MVG—Flow Topology

Employing MVG alters the near wall flow by producing sets of vortices that entrain high momentum mean free flow to low momentum region. Babinsky et al. [5] investigated the potential of micro ramps, height (h) varying from 30 to 90% of boundary layer thickness (δ) in Mach number (M) = 2.5 flow, which suggested five pair vortex tube system emanating from the ramp surface as shown in Fig. 2. A counter rotating pair of primary vortices originating from the slant edges of ramp surface, secondary vortices originating from the plate surface, and a leading edge corner separation is giving rise to horse shoe vortex system.

Li et al. [15] numerically verified the above experiment by implicit large eddy simulation (ILES) and fifth order bandwidth optimized weighted essential non-oscillatory (WENO) scheme at momentum thickness-based Reynolds number Re_θ = 5760. Body-fitted grid strategy was implemented, and the study was conducted for grid numbers (span-wise × normal × stream-wise) =128 × 192 × 1600. The code is validated for same as of [16] by study of inviscid, M = 4 flow around a half cylinder, and correctness of the code was confirmed. He discussed about surface separation topology and proposed a framework of a separation system which features six different separation lines from over and around the micro ramp shown in Figs. 3 and

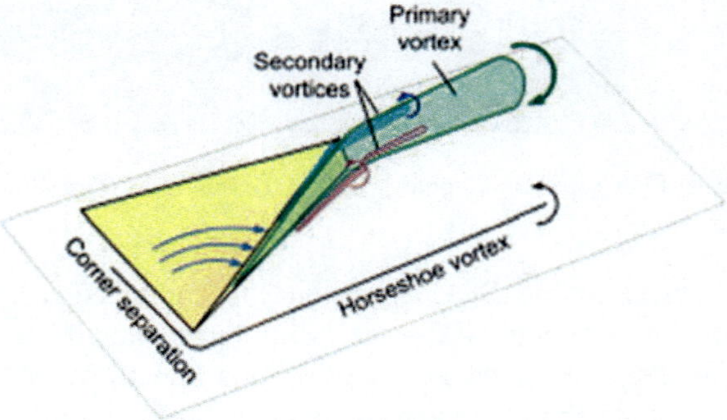

Fig. 2 Half view of vortex pair emanating from micro ramp [5]

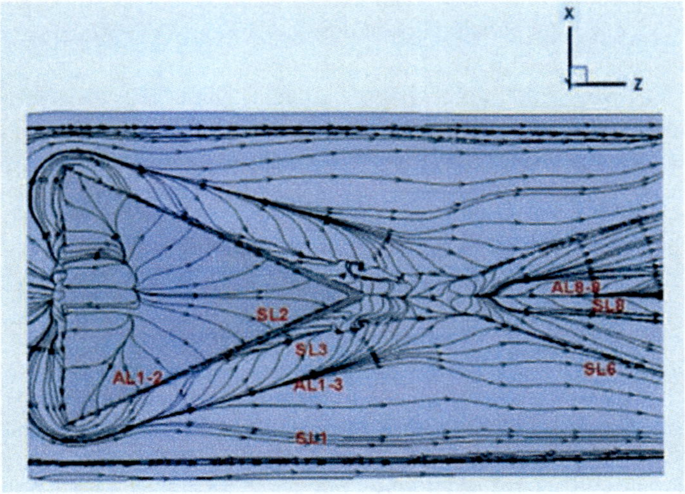

Fig. 3 Framework of a separation system [15]

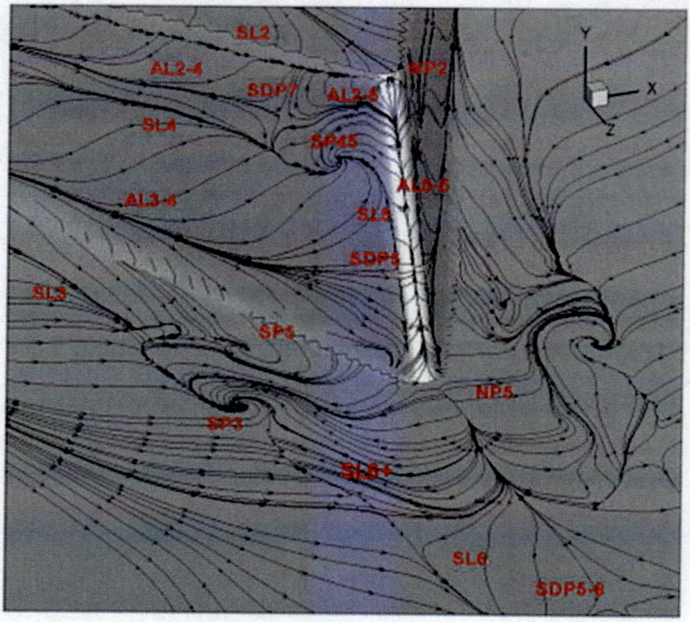

Fig. 4 Flow separation topology behind the foot of MVG trailing edge [15]

4. Separation line (SL1) corresponds to leading edge separation giving rise to horse shoe vortex system, SL2—separation line behind the formation of primary vortices, SL3—first pair of secondary separation line on the plate surface beside the MVG, SL4—first pair of secondary separation line on the side face of the MVG, SL5—first secondary separation line lies along the trailing edge of MVG, SL6—second pair of secondary separation line where valid experimental support is still needed, and SL8—second secondary separation line at downstream of MVG. Many small scale separations between SL3–SL6 are marked as SL6+.

Between SL2, SL4, spiral point SP45, and nodal point NP2, there are elusive separation from saddle point SDP7. The main distinct find of this numerical investigation was pair of spiral points behind the MVG which was experimentally confirmed [17] at university of Texas, Arlington. In between each separation line, there lies an attachment line AL.

Li and Liu [16] used ILES and simulated $M = 2.5$ flow over a micro ramp vortex generator for $Re_{\theta} = 1440$. This study revealed the existence of many inflection points generating vortex rollers in the shear layers causing Kelvin–Helmholtz (K-H) instability as shown in Fig. 5.

The inducement of rotational velocity at the undisturbed points of sine wave caused the K-H instability to raise, and vortex rings were observed to break in turbulent fashion as stated by Friedlander and Lipton-Lifschitz [11]. These K-H instabilities produced ring vortices found to break shock reflection, such that less

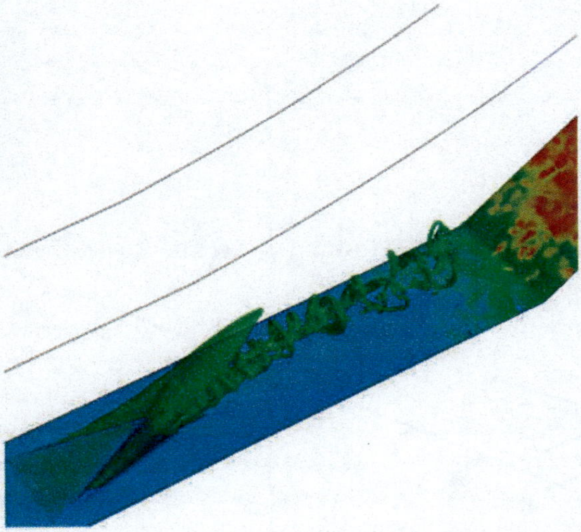

Fig. 5 Iso-surface of pressure showing generation vortex rings by MVG due to K-H instability [16]

strengthened weak oblique shocks are formed. Yan et al. [29] experimentally verified Li and Liu [16] work and confirmed that the momentum deficit caused K-H instability, which produced chains of vortex rings. This momentum deficit is caused due to the entrainment nature of vortices behind the MVG. Babinsky et al. [5] (Flow parameters already discussed above) measured velocity at three different locations downstream of the MVG and observed a high momentum region near the wall (gray region) and momentum deficit region behind the center of MVG represented in dark region in Fig. 6. The vortex core (inside the dark region) was moving away from the wall as the flow proceeded downstream because of mutual upwash produced by the counter rotating primary vortex pair, and vorticity was found to decay at far downstream.

Ashill et al. [3] conducted experiment using Laser Doppler Anemometer over $h = 0.5\delta$ forward, backward, counter rotating vanes, and single vane type of MVGs in DERA boundary layer tunnel. The circulation decay along stream-wise distance was compared for the four types of MVGs shown in Fig. 7.

The closed proximity of counter rotating pair of vortices from joint vane and forward wedge caused mutual interaction, and strength is reduced. The vortices were near the wall in case of backwards wedge but had more decay at far downstream. The spaced vanes had much better circulation than joined vanes and experienced less drag. It can also be seen that the circulation of counter rotating vanes is almost double than the forward wedge. To add with that, CFD predictions underestimated vortex strength by 20%, over predicted the size of the vortices and the vortex path also deviated after 10h distance from the trailing edge of the MVG. Zhang et al. [31, 32] proposed dissymmetric ramp (half portion of a micro ramp) of $h = 0.7\delta$ in $M = 2.5$ flow studied numerically by RANS simulation. Vorticity in dissymmetric ramp was

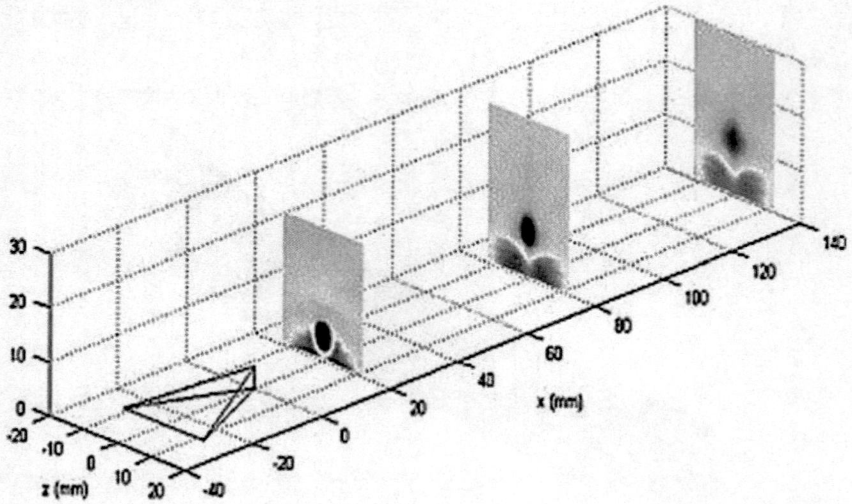

Fig. 6 Stream-wise flow field development [5]

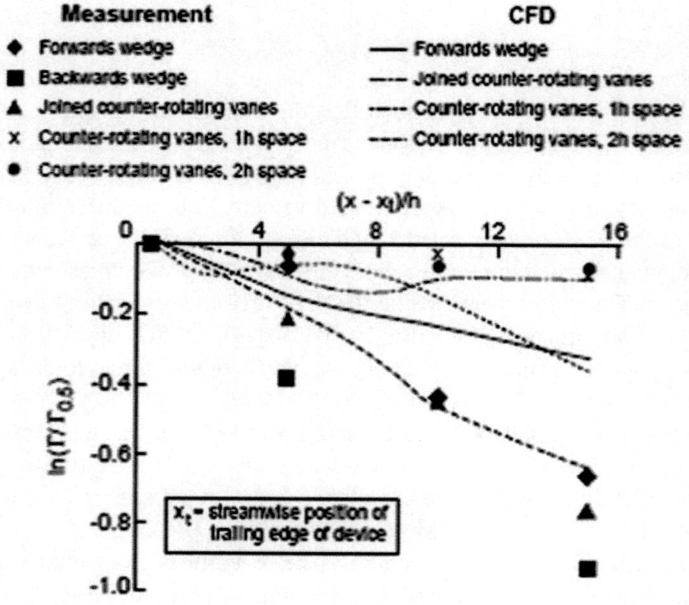

Fig. 7 Stream-wise vorticity decay between various types of MVGs [3]

stronger and not counter rotating as of conventional micro ramp. The stream-wise core height comparison between micro ramp and dissymmetric ramp is shown in Fig. 8. The vortex core raises at constant speed when compared to the one produced by normal conventional micro ramp, so the vortex stays longer in boundary layer and better flow control is obtained.

The effect of MVG's geometry on wake characteristics was studied by [22]. The flow conditions were $M = 2.0$, $\mathrm{Re}_\theta = 2.4 \times 10^4 \pm 0.1 \times 10^4$, and $\delta = 6$ mm, and a constant Anderson micro ramp [1] height of 4 mm is chosen and effect of change in half-span angle (θ_s) with constant ramp angle (θ_r) and vice versa on vortex core distance and strength. Tangential component (u sinθ_s) is added over the length of the edge, by which a larger radius of vortex is obtained as radius is proportional to length of the edge C, shown in Fig. 9.

A brief summary on the observations is as follows, when ramp angle is increased for constant half-span angle, the width of geometry and circulation decreased and the distance between the vortex cores increased. Increasing half-span angle for constant ramp angle resulted in increasing width of geometry and circulation. It has also been

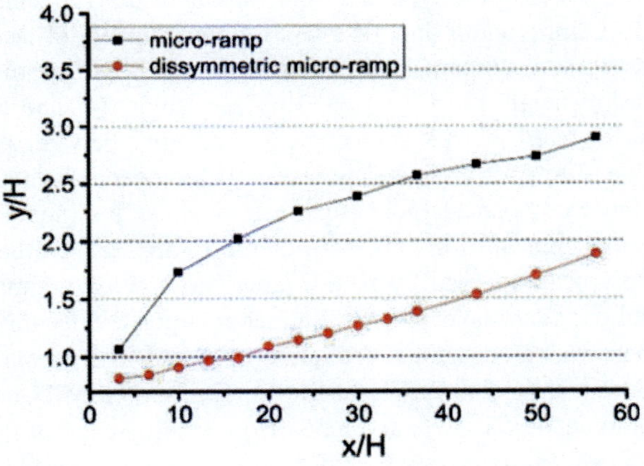

Fig. 8 Comparison of stream-wise vortex core height development downstream between micro ramp and dissymmetric ramp [31, 32]

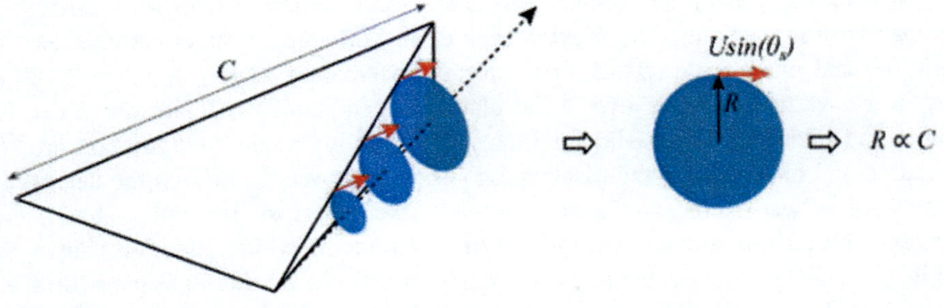

Fig. 9 Addition of tangential momentum at various edge-wise locations [22]

found that the half-span angle affects the location of wake while ramp angle does not have an effect since the wall normal location of wake is undisturbed by the captured momentum. Lower half-span angles and higher ramp angles added more momentum and produced stronger vortices.

3 MVG—Shock Wave/Boundary Layer Interaction

The emergence of ramp type compression engines raised many uncertainties in the name of shock wave/boundary layer interaction and the flow separation due to that was a trepidation. Beginning from 1940s, there were many flow control techniques being suggested and one of the promising method that alleviated separation and reduced drag penalty with decent pressure recovery was micro vortex generators. In this section, recent advances and development of flow control using MVG in shock wave/boundary layer interaction and its performance comparison between the various types, span-wise, and stream-wise placement effect will be discussed briefly. Zhang et al. [31, 32] were working on hypersonic inlets, where the 12° shock generator model was kept in a finite width duct of a wind tunnel capable of Mach 3.5 flow. Three cases were studied, without control (baseline case), with traditional micro ramps [1] of chord to height ratio (c/h) = 11.6, span angle 24° and highly swept micro ramps (HSMR) of c/h = 18.8, span angle 6°, and angle between plate and line joining trailing edge being 53°. The baseline case had a huge separation, whereas with two traditional micro ramp placed 15.42δ upstream of shock impingement location, flow separation was suppressed downstream of the ramp, but in other span-wise positions (especially near side wall) was still large and had a flow unstart, caused by vortex side wall boundary layer interaction that enlarged the separation than the baseline case. There was a substantial reduction observed in separation when the location of MVG was shifted to 2.07δ. In third case, nine HSMRs having shock impingement location at 1.35δ, 3.7δ, and 5.55δ from leading edge of the ramp was compared. The wall shear stress lines for third case is shown in Fig. 10. Initially at 1.35δ, leading edge flow separation occurs, while impinging at 3.7δ and 5.55δ, the leading edge separation is under control and the separation bubbles is broken into small regions and the low momentum flow from the side wall region also reduced compared to traditional micro ramps. This is due to the ability of HSMR to restrict the span-wise movement of the flow. Verma et al. [28] studied the effect of skew angle (α) and pitch angle (β) of air jet vortex generators (AJVG) on $M = 2.18$ flow over a vertical cylinder over a flat plate. Sixteen AJVGs of diameter (d) = 0.6mm and inter-jet spacing 10d for four sets of varying angle configuration are compared and observed, all the cases had good control over the flow as the height of the λ—foot was decreasing (indicating the weakening of the separation shock). The separation and bow shock strength for the given jet injection pressure is shown in Fig. 11. AJVG having skew and pitch angle of 90° placed like in vertices of a triangle (MJ4 in Fig. 11), reduced the separation shock strength around 55%, but the bow shock strength reduced for very low jet injection pressure when compared to

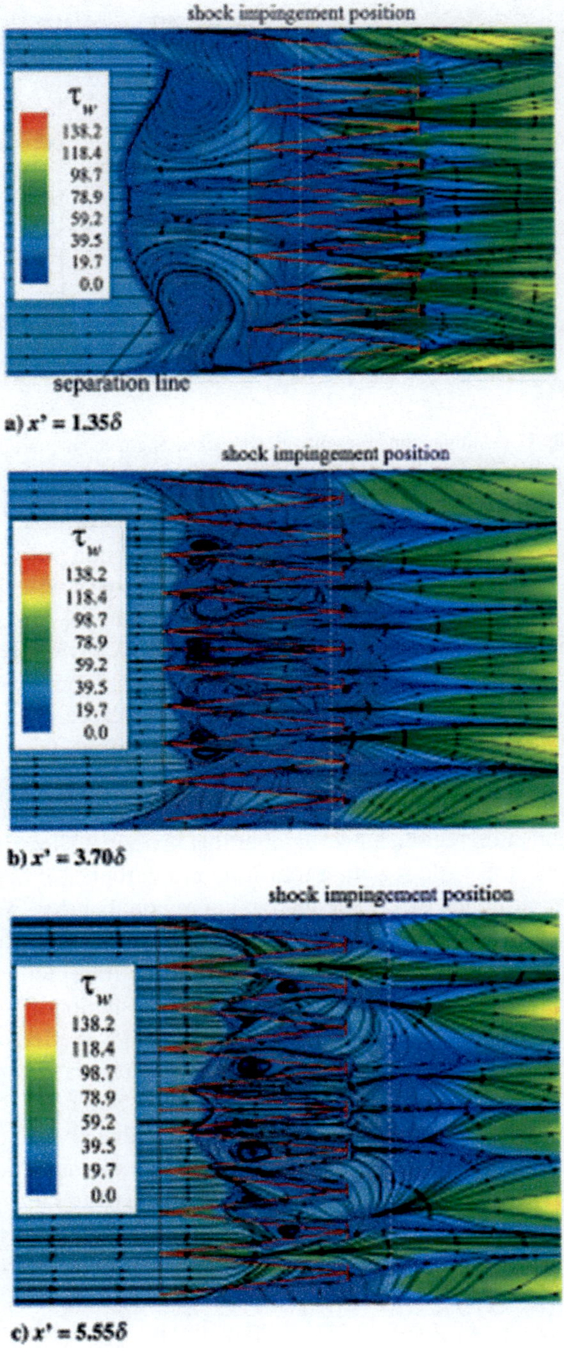

Fig. 10 Numerical results of wall shear stress lines of HSMR for shock impingement location **a** 1.35δ, **b** 3.7δ, and **c** 5.55δ [31, 32]

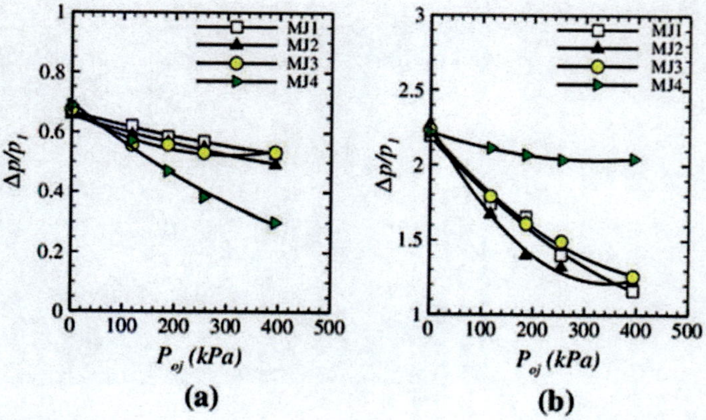

Fig. 11 Comparison of **a** separation shock strength and **b** bow shock strength for different injection pressure for four configurations [28]

other configurations with increasing jet injection pressure. This is due to the jet to jet interaction laterally to which a 45° pitching and skewing is suggested for flow control at higher injection pressure. Estruch-Samper et al. [10] studied the effect of MVG in controlling shock induced high speed laminar layer separation over hypersonic blunt cylinder. An 8° flare region was placed at 212 mm from the cylinder nose, and two MVG shapes namely diamond and square placed 40 mm before the flare region.

The flow control was measured in terms of heat transfer, initially without control before the flare region, the heat transfer reduced stream-wise indicating the onset of flow separation. Diamond MVG also showed the same characteristics before the interaction region, however, after that there was shoot in heat transfer caused by the MVG wake (flow getting transition into turbulent flow). Square MVG also showed similar trends before the interaction region, but there was separation well ahead of the MVG, where strong separation shock was seen compared to diamond case. As the MVG height increased from 0.17δ, the transition region was shifting ahead of the flare region close the laminar layer separation region. A range of 0.29δ–0.58δ was found suitable, and MVG height of 0.3δ was effective in reducing induced drag, total pressure loss, interference effects, and unsteadiness. Yan et al. [30] numerically studied the flow over micro ramp at $M = 2.5$ using ILES. The MVG model is same as that of studied by Babinsky et al. [5] with a change of 70° to trailing edge angle for the ease of grid generation. To reduce the numerical errors, body-fitted grids were implemented. Similar grid number and code validation of [16] were employed. λ_2 method [9] was used to investigate the coherent structures which revealed ring-shaped chain of vortices (K-H instability produced vortices) perpendicular to the plate surface. The lower part of the vortex hits the shock and decelerates the speed, while the upper part impinges upon the shock and exists until they are convected out far downstream. This vortex/shock interaction reduced the separation length of 8.2–8.4h at span-wise domain to 6–6.5h and the incompressible momentum thickness increased from 0.46δ to 0.49δ, incompressible shape factor improved from 1.5 to

1.39, which indicates a transfer of high momentum from mean flow into the low momentum region by the action of vortices.

Sharma et al. [21] conducted numerical studies using REACTMB on slotted/novel vortex generator (SLVG) of $h = 2, 3, 4$ mm for three different slot radius $r = 0.3$ h, 0.4 h, and 0.6 h in $M = 2.5$ flow over 7° wedge corner. A high stream-wise velocity was added by slot along the center plane, and found that the jet effect increases with increasing MVG height. An 84 and 52% reduction in separated region was observed for SLVG (r = 0.6h) trailing edge kept 8.0 and 28.5 mm upstream of shock impingement location, respectively, whereas normal micro ramp was giving 53.1 and 49% reduction when kept 8 and 28.5 mm ahead of the shock impingement location, respectively. The presence of slot also slows down the rate of vortex lift off and performs much better when kept near the shock impingement location, when compared to normal conventional micro ramps. Verma and Manisankar [26] say all control devices seen to introduce span-wise variation in the form of corrugations in the separation line. Five types of MVG namely Ashill et al. [4], Anderson et al. [1], Split-Anderson [19], trapezoidal, and ramped vane kept 10δ upstream of the separation location, having constant $h = 1$ mm and $\theta_s = 24°$ except Ashill type having $\theta_s = 14°$ as studied in $M = 2.05$ flow. Other than trapezoidal all the other control devices showed less compression shock angle compared to flow over Anderson MVG, indicating reduction in shock/boundary layer interaction. Height modification of 1.75 mm and smaller inter-vane spacing of 0.85δ applied to the ramped vane, further improved the results by shifting the separation downstream. This is because of the span-wise spacing that reduces the mutual upwash between counter rotating vortices and makes them to sustain longer downstream offering better control. Lee et al. [14] did RANS computing for flow in external compression supersonic inlet and used ILES for vortex concepts, studied the effects of span-wise, height-wise variations applied to ramped vane kept in $M = 1.3$ flow. Out of four cases, the ramped vane having 50% more height with 2.5h trailing edge spacing and 8.36h leading edge spacing eliminated the centerline separation. Further, [18] conducted numerical studies on six different MVGs comprising basic ramp (BR), BR with 45° inclined edge (BR45), blunt ramp (BLR), split ramp (SR), ramped vanes (RV), and thick vanes(TV) of $h = 0.7\delta$ in $M = 4$ flow over an 20° inclined fin. The stream-wise velocity contour at δ and 8δ suggested, the vortices produced by BR, BR45 and blunt ramp are almost similar, BR45 had lesser magnitude, whereas for BLR the vortices had more space between them, but all had same rapid decay downstream. The SR also had spacing between the vortices but the intra vane spacing reduced the interaction between the vortices. The RV also produced analogous vortex structure of SR, but boundary thickening between the vortex pair was large. The TV also produced similar vortices like SR but their vorticity magnitude was higher and flow entrainment existed over large span-wise extent resulting in better boundary layer thinning and vorticity decay rate. So the devices with intra vane spacing are better in controlling the separation and thick vanes hold the superiority. The wall shear stress comparison is shown in Fig. 12.

Vaisakh and Muruganandam [25] suggested jet injection from atmosphere through micro ramps kept in blow down wind tunnel capable of $M = 1.48$ flow. Anderson micro ramp [1] was used with a hole of diameter 2 mm drilled at the trailing edge that

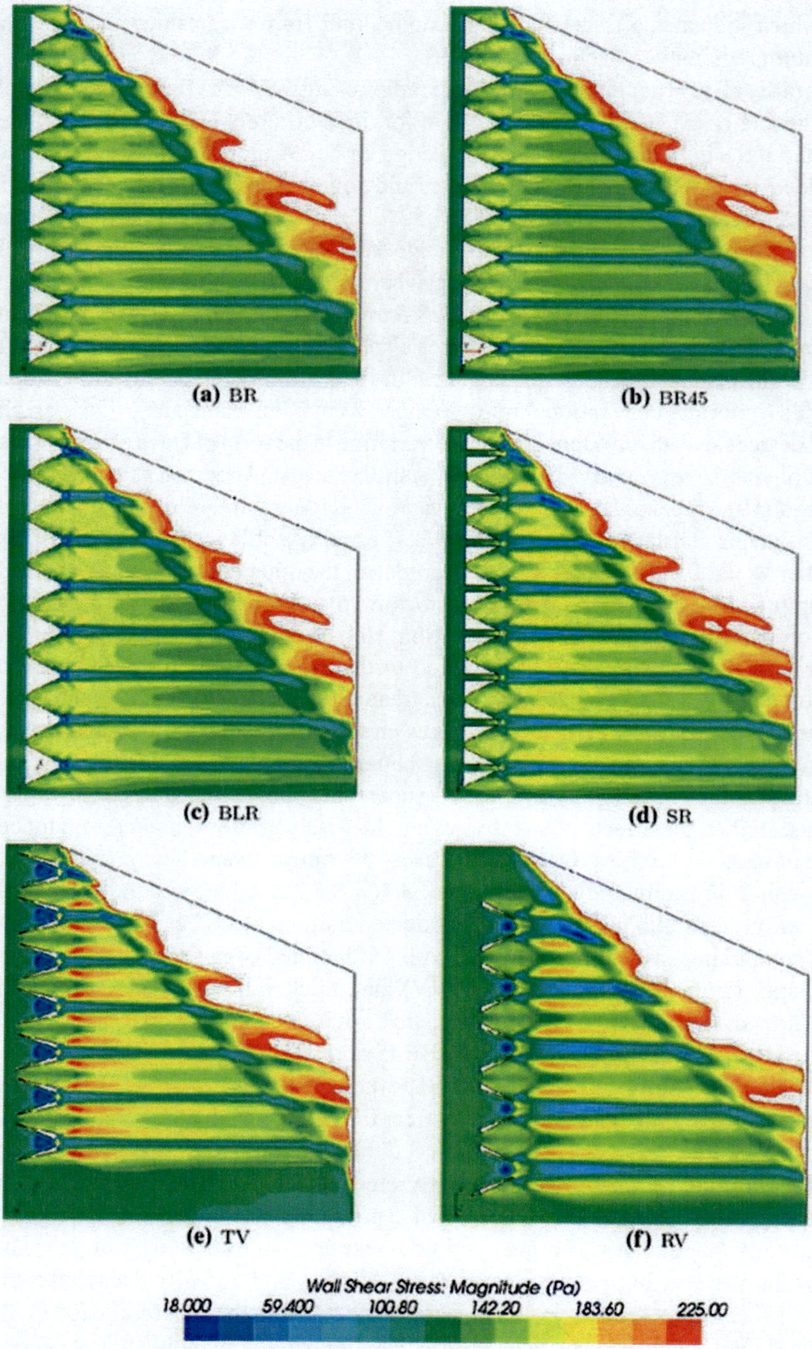

Fig. 12 The shear stress distribution comparison between various MVG [18]

injects free stream jet. Three cases were discussed, a baseline case (without control) having a centerline separation, micro ramp case that reduced separation and MVG with jet injection case where the separation region was broken into small separation bubbles with more corner separation than micro ramp case. This enhanced corner separation is due to the coupling between the corner and center line separation shown by (Burton and Babinsky [7]. Injection technique also reduced shock oscillations and found to be an advantage to high speed intake flows. Verma and Manisankar [27] studied the effect of height and stream-wise location of rectangular and ramped vane type MVG having 1 h distance at trailing edge in $M = 2.05$ flow. Ramped vane having $h = 1.7, 2.72, 3.4$ mm had their separation location moved downstream and showed wedge type, spade type corrugations in their separation line pattern. Similarly, for rectangular vane the separation line is shifted downstream and the corrugations were well formed spade distributed uniformly span-wise. Rectangular vanes having $h = 0.5\delta$ provided a 38% reduction in separation whereas ramped vanes with $h = 0.5, 0.8,$ and 1.0δ showing 18%, 32%, and 32%, respectively. Rectangular vane placed at 5δ upstream of the separation line was a preferred choice of control as the drag associated with it was less compared to other configurations. Ramaswamy and Schreyer [20] studied the effect of AJVG over a 24° compression corner facing $M = 2.52$ flow which produced $\delta = 10.4$ mm. Twenty three circular orifices having a jet diameter of 0.1δ, $\alpha = 90°$, and $\beta = 45°$ were studied for three different jet spacing 0.38δ, 0.76δ, and 2.4δ. Without control, the separation length observed was around 5.1δ, where the flow separated 2.6δ upstream of the ramp and reattached 1.11δ downstream of the ramp. For 0.38 case, the separation shock upstream of the jet combine to form a single bow shock wave. In 0.76 case, the shocks showed a wrinkled pattern and for 2.4 case, each jet has individual bow shock reducing the jet shock interaction. 0.76 case offered good control and at center line stream-wise plane showed 52% reduction in separation area and a maximum of 73% in between the jets. The jet penetration was felt around 0.6–0.7δ, and it didn't affect the flow far downstream which will help in overcoming drag penalties and increased efficiency. Bagheri et al. [6] numerically investigated the effects of MVG in low aspect ratio ducts. A 20° ramp corner of height 3.024 mm is kept in $M = 2.05$ flow generating $\delta = 5.4$ mm which completely submerges the compression corner. Four MVGs namely Ladybug body, NACA 4412 airfoil body, outer body of NACA 4412 airfoil, and inner body of NACA 4412 airfoil each of height 10 micro meter is placed at the separation initiation location that was obtained without control. All the control cases showed increase in the angle of the lambda foot, particularly inner body type MVG had highest angle of 95°, which gave highest shear stress over the bottom surface. This was mainly because of the more uniform distribution of velocity and its ability to reduce flow discontinuities intensely which allows the flow to overcome the adverse pressure gradient without much energy loss compared to other cases.

4 Conclusion

In this paper, we discussed MVG as a flow control device in shock wave/boundary layer interaction and its flow topology gave an insight on suppressing or eliminating the shock induced flow separation utilizing vortex pairs. Following are some of the points that concludes this study:

(1) MVGs are devices that entrain high momentum flow to the low momentum region by the action of vortex system, which gives rise to ring like vortices that has the capability to break the shock reflection.
(2) Thick vanes produce stronger vortices, but ramped vanes are preferred over thick vanes because of its structural rigidity and their similar ability to sustain their circulation for far downstream distance. The vorticity decay rate along downstream distance for dissymmetric ramps is better than normal conventional ramps.
(3) Decreasing the half-span angles and increasing the ramp angles adds more momentum to the flow and increases the strength of the vortices. Highly swept micro ramps, due to their ability to suppress span-wise flow movement is a good option to use near the wall where vortex/boundary layer interaction that enlarges separation.
(4) Slotted vortex generators work efficiently when placed near the separation location, whereas AJVG with 45° pitching and skew angle reduces and delays the flow separation with less drag penalty. MVG with jet injection is also another promising method that can be employed to eliminate center line separation that favors the performance of high speed flow intakes.
(5) MVGs increase the lambda foot angle causing the shock waves to lose its strength, thereby reducing the energy loss associated with flow separation.

Acknowledgements I would like to thank my colleagues Vijender Singh and Abhishek Bhupendra Gade for their constant support.

References

1. Anderson BH, Tinapple J, Surber L (2006) Optimal control of shock wave turbulent boundary layer interactions using micro-array actuation. In: Collection of technical papers—3rd AIAA flow control conference, vol 2, pp 880–893. https://doi.org/10.2514/6.2006-3197
2. Anderson JD (2005) Ludwig Prandtl's boundary layer. Phys Today 58(12):42–48. https://doi.org/10.1063/1.2169443
3. Ashill PR, Fulker JL, Hackett KC (2001) Research at DERA on sub boundary layer vortex generators (SBVGs). In: 39th aerospace sciences meeting and exhibit. https://doi.org/10.2514/6.2001-887
4. Ashill PR, Fulker JL, Hackett KC (2002) Studies of flows induced by sub boundary layer vortex generators (SBVGs). In: 40th AIAA aerospace sciences meeting and exhibit. https://doi.org/10.2514/6.2002-968

5. Babinsky H, Li Y, Ford CWP (2009) Microramp control of supersonic oblique shock-wave/boundary-layer interactions. AIAA J 47(3):668–675. https://doi.org/10.2514/1.38022
6. Bagheri H et al (2021) Effects of micro-vortex generators on shock wave structure in a low aspect ratio duct, numerical investigation. Acta Astronaut 178:616–624. https://doi.org/10.1016/j.actaastro.2020.08.012
7. Burton DMF, Babinsky H (2012) Corner separation effects for normal shock wave/turbulent boundary layer interactions in rectangular channels. J Fluid Mech 707:287–306. https://doi.org/10.1017/JFM.2012.279
8. Chang PK et al (1970) Separation of flow
9. Dong Y, Yan Y, Liu C (2016) New visualization method for vortex structure in turbulence by lambda2 and vortex filaments. Appl Math Model 40(1):500–509. https://doi.org/10.1016/J.APM.2015.04.059
10. Estruch-Samper D et al (2014) Micro vortex generator control of axisymmetric high-speed laminar boundary layer separation. Shock Waves 25(5):521–533. https://doi.org/10.1007/S00193-014-0514-7
11. Friedlander S, Lipton-Lifschitz A (2003) Localized instabilities in fluids. In: Handbook of mathematical fluid dynamics, vol 2, pp 289–354. https://doi.org/10.1016/S1874-5792(03)80010-1
12 Huang W et al (2020) Recent advances in the shock wave/boundary layer interaction and its control in internal and external flows. Acta Astronautica. Elsevier Ltd. https://doi.org/10.1016/j.actaastro.2020.05.001
13. Jenkins LN, Althoff Gorton SA, Anders SG (2002) Flow control device evaluation for an internal flow with an adverse pressure gradient. In: 40th AIAA aerospace sciences meeting and exhibit. https://doi.org/10.2514/6.2002-266
14. Lee S, Loth E, Babinsky H (2011) Normal shock boundary layer control with various vortex generator geometries. Comput Fluids 49(1):233–246. https://doi.org/10.1016/J.COMPFLUID.2011.06.003
15. Li Q et al (2011) Numerical and experimental studies on the separation topology of the MVG controlled flow at M = 2.5. https://doi.org/10.2514/6.2011-72
16. Li Q, Liu C (2011) Implicit LES for supersonic microramp vortex generator: new discoveries and new mechanisms. Model Simul Eng 2011. https://doi.org/10.1155/2011/934982
17. Lu FK, Pierce AJ, Shih Y (2010) Experimental study of near wake of micro vortex generators in supersonic flow. In: 40th AIAA fluid dynamics conference, vol 1. https://doi.org/10.2514/6.2010-4623
18. Martis RR, Misra A (2017) Separation attenuation in swept shock wave–boundary-layer interactions using different microvortex generator geometries. Shock Waves 27(5):747–760. https://doi.org/10.1007/S00193-016-0690-8
19. Nolan WR, Babinsky H (2012) Comparison of micro-vortex generators in supersonic flows. In: 6th AIAA flow control conference. https://doi.org/10.2514/6.2012-2812
20. Ramaswamy DP, Schreyer A-M (2020) Control of shock-induced separation of a turbulent boundary layer using air-jet vortex generators 59(3):927–939. https://doi.org/10.2514/1.J059674
21. Sharma P, Varma D, Ghosh S (2016) Novel vortex generator for mitigation of shock-induced flow separation 32(5):1264–1274. https://doi.org/10.2514/1.B35962
22. Tambe S, Schrijer F, van Oudheusden B (2021) Relation between geometry and wake characteristics of a supersonic microramp. AIAA J 1–13. https://doi.org/10.2514/1.j059868
23. Tilton N, Cortelezzi L (2015) Stability of boundary layers over porous walls with suction 53(10):2856–2868. https://doi.org/10.2514/1.J053716
24. Titchener N, Babinsky H (2013) Shock wave/boundary-layer interaction control using a combination of vortex generators and bleed 51(5):1221–1233. https://doi.org/10.2514/1.J052079
25. Vaisakh S, Muruganandam TM (2017) Control of boundary layer separation in supersonic flow using injection through microramps. In: 30th international symposium on shock waves, vol 2, pp 1183–1188. https://doi.org/10.1007/978-3-319-44866-4

26. Verma SB, Manisankar C (2017) Assessment of various low-profile mechanical vortex generators in controlling a shock-induced separation 55(7):2228–2240. https://doi.org/10.2514/1.J055446
27. Verma SB, Manisankar C (2018) Control of incident shock-induced separation using vane-type vortex-generating devices 56(4):1600–1615. https://doi.org/10.2514/1.J056460
28. Verma SB, Manisankar C, Akshara P (2014) Control of shock-wave boundary layer interaction using steady micro-jets. Shock Waves 25(5):535–543. https://doi.org/10.1007/S00193-014-0508-5
29. Yan Y et al (2012) Numerical discovery and experimental confirmation of vortex ring generation by microramp vortex generator. Appl Math Model 36(11):5700–5708. https://doi.org/10.1016/J.APM.2012.01.015
30. Yan Y et al (2016) Numerical study of micro-ramp vortex generator for supersonic ramp flow control at Mach 2.5. Shock Waves 27(1):79–96. https://doi.org/10.1007/S00193-016-0633-4
31. Zhang B et al (2015) An improved micro-vortex generator in supersonic flows. Aerosp Sci Technol 47:210–215. https://doi.org/10.1016/j.ast.2015.09.029
32. Zhang Y et al (2015) Control of shock/boundary-layer interaction for hypersonic inlets by highly swept microramps 31(1):133–143. https://doi.org/10.2514/1.B35299

A Review on Design and Optimisation of Axial Fan

Vijender Singh and Nikhil A. Baraiya

Nomenclature

A	Axial spacing
C_{FR}	Chord length of front rotor
$N_{FR, RR}$	RPM of front and rear rotor
dBA	Decibel annoyance level
σ	Specific speed

Greek Symbols

δ	Specific diameter
Q	Flow rate m^3/s
D	Tip diameter
n	Rotational speed RPM
$\eta_{t-s,i}$	Total to static ideal efficiency
ψ	Pressure coefficient
φ	Flow coefficient

V. Singh (✉) · N. A. Baraiya
Sardar Vallabhai National Institute of Technology, Surat, India
e-mail: vijender23062000@gmail.com

N. A. Baraiya
e-mail: nikhil@med.svnit.com

J. Banerjee et al. (eds.), *Recent Advances in Fluid Dynamics*, Lecture Notes in Mechanical Engineering, https://doi.org/10.1007/978-981-19-3379-0_17

1 Introduction

Axial blowing fans are pretty commonly used in industries for ventilation as well as cooling purpose. Diverse applications of the axial fans present a problem of required optimisation according to the purpose. These requirements could vary from high pressure ratio, higher mass flow rate for cooling, low acoustic level to swirl free flow. Intensive research has been conducted to fulfil the required parameters without compromising the performance or efficiency of the system. Performance map has been calculated to provide the best efficiency points near the desired operation point.

2 Literature Review

The following sections will discuss briefly about the various efforts that have been made in achieving these aspects.

2.1 Contra-Rotating Fan

Shigemitsu et al. [30] did the study on operation of contra-rotating fan at partial flow rates to reduce the total energy required for cooler fans. Using aerofoil blades and operating at 60% of design flow rates, the performance curve was obtained for both contra rotating and radial fan alone. Max efficiency for contra rotating and radial rotor was 44.4 and 34.8% at design flow rates. Slightly reduced but stable pressure rise was observed at partial flow rates with back flow on tips of both rotors and flow inclined towards hub. Due to off design conditions, swirl was not completely eliminated at outlet giving pressure losses.

In the research conducted by Nouri et al. [25] have used an already developed code for axial fan [24] with some modifications for contra rotation at Reynolds number above 10^5 to study the impact of speed ratio (N_{RR}/N_{FR}) and axial spacing ($A = S/C_{FR}$). The concept here is to keep the first rotor same as reference with modification in second rotor to obtain the swirl free flow. The tests were carried out for both fans separately and together to precisely get the performance curves for individual as well as combined. Rotating alone at 2000 rpm the front rotor had a flow rate of ~3650m^3/h with efficiency of 46.2%. The rear rotor showed a much lower flow rate while operating alone at 1800 rpm due to the alterations in geometry. The combination had an increase of up to 20% in efficiency at 65% at design flow rate (3600 m^3/h). Keeping N_{FR} constant at 2000 rpm, with rear rotor speed varying, it was observed that high efficiency around 66% at speed ratios ϵ (0.8, 1.2) with peak 66.5% at ratio 1.05, i.e. 2100 rpm. It was also shown that relative spacing had a really low impact on performance when operating at design conditions, whereas high variation

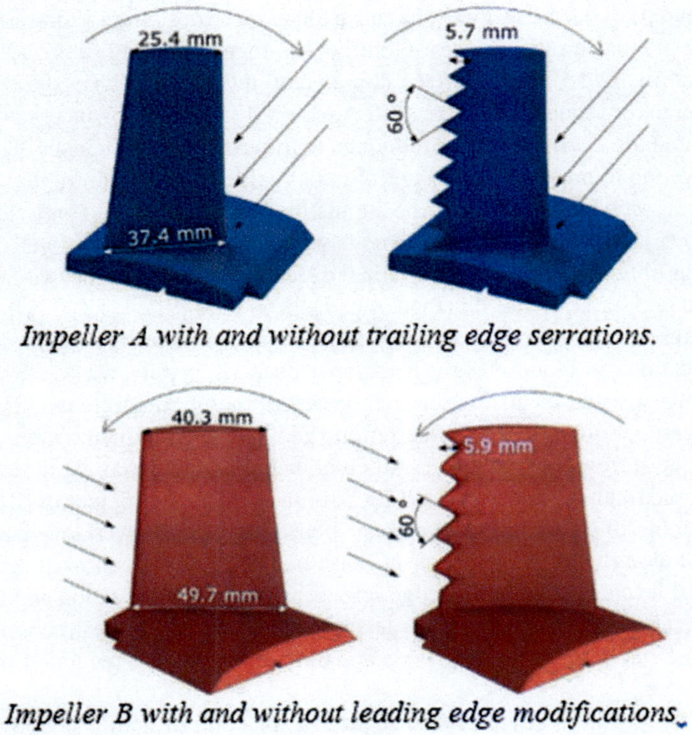

Fig. 1 Blade modification on contra rotating fan [11]

in spacing leads to deterioration effects. They concluded that the resulting contra-rotating fan (CFR) could be operated at three desired combinations, fixed flow rate, constant efficiency and constant pressure rise with high performance efficiency in all cases (Fig. 1).

The effect of inflow distortion and turbulence was studied by Mistry and Pradeep [22] for both hub strong distortion and tip strong distortion using a low porous screen for flow structure measurement and performance mapping. They concluded that increasing the speed of the second rotor improves the performance even in distorted flow. The effects of presence of screen revealed performance decrease in hub complex flow and increase in case of tip complex flow.

Further insights were taken into the flow distortion in the contra-rotating fans by Manas and Pradeep [16, 23] as they studied the effect of radially distorted flow and found similar results that showed the dependence of performance and stall on speed of second impeller. Increasing the rpm of second impeller stopped the pre-swirl on first rotor and damped out the stalling frequencies. It was found that near stall frequencies are closely related to the blade pass frequencies and during full stall, a continuous band of lower and higher frequencies coexisted.

The research conducted by Friebe has been focussed on designing a contra-rotating fan using a popular turbo simulator CFturbo and then validating the results using experimental methods. The swirl free flow is required in many areas, where swirl may counteract or reduce the desired effect required from the flow and can induce unprecedented losses which lead to reduction in overall system efficiency. The swirl can be converted to pressure rise by use of outlet guide vanes or contra rotating fan the latter being studied here. Doing the simulations and carrying out particle image velocimetry (PIC) studies, it was observed that maximum efficiency as well as swirl free flow was obtained when there was small difference in speed of the two fans. But the required pressure rise was obtained at high speed ratio [9].

Further research was carried out in a span of year to investigate the acoustics and modifications in blade design. Rotational beam algorithm was developed to study the intensity and localising the sound sources on the rotating fans. Blades of both fans were modified by use of serrations and sinusoidal structures on trailing and leading edge, respectively Fig. 1. All combinations of these configuration were tested, and maximum efficiency of 57% was found at flow coefficient of 0.28 with pressure number of 0.4. Third octave band acoustic images showed localisation of noise source along with phenomenon of noise masking. Results showed reduction of about 1 dB levels, where trailing edge serrations gave noise reduction and leading edge modification gave favourable noise source alterations. Further modifications can be carried out by inclusion of sweep and dihedral for higher reduction in noise [11].

Dong et al. [5] did study on noise reduction for contra-rotating fans by using micro-perforated panels around the casing and observed that acoustics benefits were acquired when the panels were placed near first rotor. The maximum pressure loss observed was less than 10 Pa when panels were placed around second rotor. Noise levels were reduced by 3.5 dBA levels near the design point. Using enlarged cavity, above panels of front rotor damped all noise level above 35 dBA. The noise near the stall could not be reduced by the given modifications.

2.2 *Blade Skew and Sweep*

The paper by Wright and Simmons [35] takes a look at the research done up to the date on low speed axial fans mainly used in heat exchanger applications to reduce the noise emission by use of blade sweep. Performance and weighted sound pressure level (SPL) test were carried out for three sets of prototypes with blade sweep of 0°, 40°, and 50°, respectively with hub-to-tip ration of 0.55 and results were compared with computational results. The aerodynamic model given by [31] has been used to predict the performance and matched quite nearly with results. The efficiency for all the three fans was nearly constant at about ~42% for unswept and 40% for swept blades. Sweeping of blades producing a significant less noise level around 10 dB in broadband and 7 dB in tonal component while giving a higher performance margin around the design point. They concluded that the noise produced was mainly because of turbulence ingestion due to disturbances at inlet condition and separation happening at the blade surface.

The paper by Beiler [3] presented the results and comparison of numerical and theoretical results with experimental validation for performance and acoustics characteristics of axial fans with skewed blades. Both directions of skewness were investigated along with variation in sweep angle along the blade length with negative sweep at hub and positive sweep at tip. Numerical simulations were carried out with k-ε model with directionally sensitive upwind discretisation with physically based advection term. Flow was assumed to be incompressible with fourth order pressure–velocity linkage. Grid with 10^4 and 10^5 nodes was used along with assumption of negligible tip clearance for moderately loaded blades. Discrepancy was observed with low grid sizes, while the highly dense mesh gave values approaching theoretical results. Theoretical analysis at design point of $\varphi = 0.18$ showed efficiencies of order of 95% in ideal scenario. Whereas numerical and experiments ruled out the value as about 57% for forward sweep and 54% for backward. It was observed that flow separation occurred at lower flow rates in forward and at higher flow rates in backward, leading to higher noise by the latter fan. Swept-forward blades improved the fan performance with more uniform outlet flow having lower dynamic losses with relatively less noise emissions.

The authors [26] studied and compared the aerodynamic and acoustics performance of radial, forward and backward skewed blades for a low pressure axial fan using generic algorithm for optimisation with constant fan skew angle of 8.3°. They obtained noise reduction of about 4.3 dBA with extension in stall margin up to 6% for forward skewed impeller compared with radial impeller. But the efficiency loss of 3.53% occurred along with 5.63% loss in total pressure rise. Radial blade gave the highest pressure rise with high noise.

Within the next year, they developed a genetic algorithm coupled with back propagation method and artificial neuron network to perform the optimisation task by using 3D Reynolds-averaged Navier–Stokes (RANS) model. The main optimisation parameter taken in account was the blade stacking line which gave optimum skew angle of 6.1°. Three dimensional, incompressible, viscous flow was simulated with Spalart–Allmaras one-equation turbulence model. The experimental results indicated an increase in total pressure efficiency by 1.27% and total pressure rise by 3.56% at design point of $\varphi = 0.245$ with decrease of 6.5 dBA SPL levels. Also, the acoustic noise shifted from tonal to broadband region for most part of performance map. Detailed flow field was studied for this blade and showed decrease in aerodynamic losses at hub and shroud with little increase near mid span with significant reduce in overall losses leading to reduced blade tip loading resulting in high pressure rise with lower noise [36].

A similar study was carried out for all three types of blade skewness to understand the flow field better. Experimental procedure was carried out using hot wire anemometer. Skew angle of 8.3° was used for both skewed fans. The results obtained showed the variation happening in pressure rise and loss along the span combined with variation in radial and axial flow velocity at outlet of the three types of fans. It was observed that the stable operation ranges for forward, backward, and radial were

25.56%, 12.19%, and 19.87%, respectively around design point of $\varphi = 0.25$. The pressure rise was highest for radial > forward > backward skewed blades. Losses were highest for backward skewed with lowest in forward. Flow fields for radial blade inherited high roughness and wakes producing higher noise. Forward blade gave more uniform flow field, whereas backward blades flow was mostly stalled due to low flow rates. In mid span of blade, backward blades showed reduced losses and blade loading because of body forces [14].

Optimisation for axial fan with varying pressure rise along blade span was carried out by Pascu et al. [27]. They developed a model to generate the blade shape according to operating conditions and verified it with cascade testing. Varying blade outlet angle along with blade solidity according to required pressure rise, three models were developed and simulated numerically. The results displayed about 7–28% increase in static efficiency, 2–9% increase in total efficiency and 4–17% increase in polytropic efficiency for the blade designed for operating at higher flow rates. The computed design gave almost similar static pressure rise but gave about 11% less torque compared to reference fan with around 10% rise in total to static efficiency. The parametric method used to develop the blade shape optimisation technique produced significant results showing high efficiency gain in non-free vortex design blades.

Hurault et al. [10] carried out the research on effect of sweep on velocity and turbulence distribution downstream of fans using Reynolds stress model. Constant temperature anemometry was used to accurately determine the velocity distribution downstream. It was seen that the forward sweep tends to decrease the swirl or radial component of flow, whereas backward sweep increases it. Turbulent kinetic energy production is lower for radial than the sweep blades near the tip. While for the swept blades, it tends to increase downstream causing higher turbulence at end.

Bamberger and Carolus [1, 20] did the study of performance and acoustics for fans with highly swept blades along with effect of change in sweep angle of leading and trailing edge on acoustics. Sweep angles used at hub, mean, and tip are $-55°$, $+3°$, and $+55°$, respectively. The sweep angles were kept the same as reference for the stacking line with constant chord length along blade span only leading and trailing edge was optimised for design condition. Increase in efficiency near the design point was observed along with decrease in acoustic noise. But since too many modification factors were included at once it is not feasible to mark the role of each parameter in the results obtained. Flow contours using numerical tools showed highly attached flow near trailing edge along with higher pressure rise at blade leading edge.

Lotfi et al. [15]used a numerical method involving genetic algorithm process to optimise the blade using Navier–Stokes solver taking into account the three dimensional flow domain. Global optimisation was carried out using the algorithm for problem domain containing fan of 508 mm outer diameter with design speed of 1500 rpm equipped with 27 blades with fixed tip clearance of 1.2% of blade height. After a long process of 200 h calculation, the final converged design showed increase of 1.1% in efficiency, 0.45% in pressure coefficient, with 0.3% increase in mass flow rate due to optimisation of blade profile. The results showed much better attached flow with more pressure rise near hub and leading edge which delays separation at trailing edge with increase in overall efficiency at off design condition. Inclusion of sweep and lean leads to further efficiency increase with decrease in acoustic noise. It was found that backward sweep with lean in rotation direction leads to decrease in performance, whereas forward sweep gave respectable results.

Zhang et al. [38] studied the effect of blade sweep along with optimising the blade profile for a geared fan for turbofan engine. The goal was to reduce the rpm and acoustics while increasing the efficiency. Resulting blade showed ultra-high load with large camber reducing overall flow losses. For the same mass flow rate and pressure ratio, efficiency of 96% was maintained after reducing rpm from 3700 to 2240. Forward sweep showed reduction in relative Mach number thus reducing the shock losses. With increase in blade design complexity, stator design difficulty increases.

Work done by Engineering and Africa (2018) [7] takes into account the performance evaluation and comparison of radial and forward swept fan along with acoustic characteristics for both the fans. Two similar fans of 630 mm diameter with 8 blades (one radial fan and other forward swept with maximum sweep of 11.51°) were tested in a ducted facility with ISO 5801 [32] standards at constant 1440 rpm. It was found that radial fan gave higher pressure rise during throttling but gave less efficiency (54%) compared to forward swept blade (57%). Noise levels for radial fan were 10 dBA levels which were higher than forward swept fan. The spectrogram taken over total period of 6 s showed the tonal noise component at 192 Hz for both fans due to blade pass frequency, whereas broadband noise was much higher for radial compared to forward swept. It was concluded that noise level reduction of forward swept blade is due to reduction in loudness. The annoyance levels were found to be a function of fan operating point with forward sweep being less sensitive to changes in flow rate.

Ye et al. [11, 37] did the work on variable pitch axial fan with forward skewed blades with outlet guide vanes. Fan with tip diameter for 1500 mm with 0.6 hub-to-tip ratio with 14 rotor blades and 15 guide vane blades running at 1200 rpm was designed using ANSYS Gambit for both radial and skewed blades. Experimental validation was carried out to validate the results obtained by Fluent using RANS 3D model. The blade skew angle was varied between 1° and 8° to see the variation in performance and efficiency. It was observed that at design flow rate, pressure rise increased up to angle of 4° with maximum gain of up to 3%. Efficiency curve gave maximum value at 3°, but the stable high values were observed at angle 1° to 2° for varying flow rates. For varying blade pitch angle of 29°, 32°, and 35°, variations were observed for varying skew angles. At 32° angle, best pressure rise of 2.99% was observed at best efficiency point, whereas 35° angle showed better performance at higher flow rates. The flow structure investigation showed increase in flow velocity in middle and lower parts of blade which increases the flow rate and delays the separation at blade root giving larger stall margin. The tip loading is decreased leading to lower tip leakage losses.

Zhang et al. [39] did the study of three-dimensional numerical investigation of abnormal blade combinations, i.e. the blade setting angle is set apart from the reference blade to see the blade stress changes and stall conditions. Blade angles of +9°, −9°, ±9°, ∓9° were used to create the geometry for the 24 blades on each rotor of the two stage axial fan. Gambit is used for discretisation of zones and meshing. It was seen that number of stall cells was independent of the blade angle combinations with higher pressure rise with positive deflection and vice versa. For blade stress, the stress is higher in non-stall conditions with higher deformation but lower in stalled cases. Also the deformation depended both on aerodynamic and centrifugal loading, whereas stress changed with centrifugal loads only. Similar vibrations modes were observed with highest deformation at tip.

2.3 Blade Stacking

The paper by Lee et al. [13] takes into account the effect of blade geometry modification including stacking line, blade lean, maximum thickness and its location to improve the total efficiency of blade. Using a gradient-based algorithm for determining optimum value, the CFD analysis was done by using RANS model coupled with response surface approximation [29, 33]. The reference blades had values of blade camber as zero with blade lean of $-19°$ with optimised values of 0.001 m with blade lean of $-16.54°$. The efficiency of the optimised design was increased up to 87.4% from reference of 85.9%. But the blade loading for the optimised blade was found to be more than the reference and flow streamline were more attached and symmetrical near the mid span and hub. Flow field showed more variation with change in blade lean than the blade thickness.

Another important work done in field of modifying the blade stacking is by Masi et al. [18]. A complete study was carried out to improve the performance of a high-pressure rotor only axial fan without loss in total efficiency by doing forward sweep of blade stacking line for controlled vortex blades. CFD simulations as well as validation was carried out. Grid independence study showed that fan total pressure coefficient was less sensitive to grid change compared to total efficiency. Using two layer k-ε, realisable model gave the most efficient results at lower grid size reducing computation time [20, 34]. Performing the experiment and simulations, various modifications regarding the blade stack, sweep, hub-to-tip ratio, vortex field, and tip clearance were studied to develop an algorithm for generating most efficient blade stacking line. The results showed that sweep alone cannot achieve appreciable performance or efficiency gain. Similar angles for sweep and stacking surface must be used. Using 3D aerofoil stacking in cascade analysis, the complex flow computation times could be reduced significantly. Using this model, they were able to obtain 10.56% gain in total to static efficiency and 36.04% gain in pressure coefficient at flow rate coefficient of $\varphi = 0.716$.

Similar design method was applied to low pressure fan with high hub-to-tip ratio to study the impact of forward sweep on performance and to give a simple theory to justify this impact. Swept wing theory given by Busemann was used to calculate sweep angle and then modified using blade stacking method with controlled vortex design previously described. Peak efficiency value is increased by 11.5% and pressure coefficient by 46.7% at flow coefficient $\varphi = 0.716$. It was concluded that the performance gain is due to high positive circulation gradient along forward swept

CVD blade, and the effect is pretty much better than forward swept only blades. It was also observed that inclusion of forward sweep left efficiency unchanged with about 2% increase in pressure rise. The stall margin of fan got extended by up to 10% range. 3D stacking gives higher performance at flow rates above design point but effects worsen near the stall point at low flow rates [18, 19].

The same design method was applied to rotor only tube axial fan with hub-to-tip ratio of 0.44 to improve the overall efficiency. Cylindrical surface method for blade stacking is used and analysed here. Three blade prototypes have optimised and tested. One from reference fan, second with optimised blade stacking and third with stacking plus sweep optimised blade. Tests were performed in low Reynolds number range of 20,000–90,000. The results verified the theories given in [18, 21] for 3D stacked and forward sweep blades. At low blade loading, 3D stacked blade had higher pressure rise with lower efficiency with advancement of blade stall. The optimisation of reference blade generated a 5% increase in total efficiency from base. It was concluded that further studies are required in optimisation of blade stacking angle on curved conical surfaces [18, 20].

Another work by the same authors [17] in Fan conference 2018 that focussed on studying the effectiveness of cylindrical surface blade stacking in small tube axial fan with inclusion of sweep. The numerical method was used to design the unswept blade with 3D blade stacking and later the design was modified to include sweep (forward sweep) along with inclusion of extra sweep of 6° at blade tip for 315 mm rotor. Experimental and numerical investigation was carried out on 560 mm diameter rotor due to availability of prototype. Forward swept blade showed increased pressure rise with extra stall margin. The tip modified blade showed progressive increase in both efficiency and pressure rise for operating range from peak pressure to stall. Tip swept blade obtained about 10% increase in pressure rise with 2% increase in efficiency compared to peak results of forward sweep blade. The results achieved expected results, and the suggestion was made to use the optimisation strategy for high Reynolds number conditions to extend the validation of results.

Krömer et al. [12] did the aero-acoustic study for controlled vortex fan with different blade stacking. The sweep angle and dihedral angles were varied to generate four sets of rotor with backward (−50°), forward (+50°), forward–backward (+50° to −50°) and backward–forward (−50° to +50°) blades. Performance curves indicated that aerodynamic performance depended mainly on type of sweep in outer part of blade. Acoustics measurements showed presence of subharmonic humps which were caused due to flow blockage in tip clearance flow. Sound sources were present in outer part of blade for all types with minor difference in location. Combining the different sweep types improved the blade stress condition due to shifting of centre of gravity.

2.4 Cordier Diagram Optimisation

Epple et al. [8] took the task for deriving proper theoretical relation for obtaining the results similar to the original Cordier diagram [6]. Another aspect was to do parameterisation of the variables so as to include the effects of blade shape and angles in the empirical results obtained by Cordier which does not take into these factors. The Cordier diagram is between specific speed and specific diameter that gives a relation between flow rate, rpm, pressure and diameter. The formula for both parameter is given below.

$$\sigma = 2\sqrt{\pi}n\frac{\sqrt{Q}}{(2\frac{\Delta p_{\text{opt}}}{\rho})^{\frac{3}{2}}} \tag{1}$$

$$\delta = \frac{\sqrt{\pi}}{2}D^4\sqrt[4]{\frac{2\Delta p_{opt}/\rho}{Q^2}} \tag{2}$$

The work done by Epple et al. [8] led to derivation of an analytical relation between the three basic parameters of turbo machinery, flow coefficient, pressure coefficient and total to static ideal efficiency ($\eta_{t-s,i}$) and was verified to be valid for both axial as well as radial machines.

$$\eta_{t-s,i} = 1 - \frac{1}{2\psi}(\varphi^2 + \psi^2) \tag{3}$$

Figure 2 shows the overlapped graphs of original Cordier diagram along with derived validation and results seem to fully capture the extent of variation in best efficiency points similar to original Cordier diagram.

The work done by Bamberger and Carolus [2, 23] focusses on identifying the upper limit of efficiency for axial rotor only, axial fan with guide vanes, centrifugal rotor only, centrifugal with volute fans based on empirical loss models—a theoretical efficiency limit and a more realistic approach using CFD simulation. The band of high efficiency in Cordier diagram (grey band Fig. 3) has been used for finding the peak efficiency points for each type of fan. Using the theoretical model that involves only the inevitable losses, i.e. internal friction, shock losses and exit losses, defines a limit that cannot be surpassed in real life. The optimisation problem is solved with the conjugate-gradient method [23] with CFD simulations performed in OpenFOAM with steady state RANS and SST model at a Reynolds number of 10^6. Theoretical efficiency showed up to 100% efficiency in case of guide vanes and volute and around 90% in rotor only case due to exit loss. Whereas CFD showed achievable results with peak value of total to static efficiency of 68, 75, 77, 86 for the four sets of fans, respectively. It was observed that efficiency decreased with decrease in size as exit losses got more severe and friction losses played a major role in radial fans compared to axial.

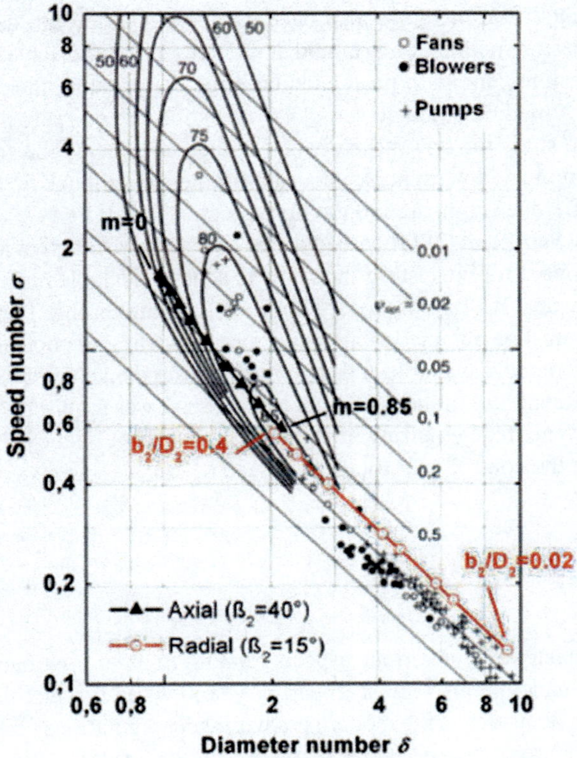

Fig. 2 Cordier diagram [6] with theoretical curves [8]

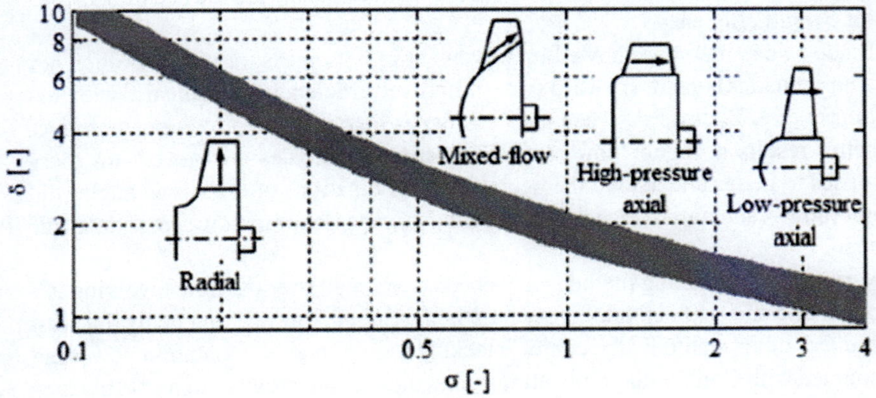

Fig. 3 Cordier diagram with high-efficiency grey band [2, 23]

Some other modifications have also been carried out in the fan design. One of them is the tip design modifications aimed at reducing noise. Bizjan et al. [4] tested two blade tip designs, one with prolonged baseline design and other with curved-back tip design. A method based on velocity distribution and pressure measurements was developed to study the flow structure and the results showed the coincidence of best efficiency point and lowest noise emission for the curved-back tip design which showed large difference compared to the first design.

Work done by Saul et al. [28] focussed on the effects of variation in Mach number and Reynolds number for fan test rigs and how to achieve scalable results by varying the ambient pressure. It was observed that to study compressible flows, at least a Mach no. of 0.4 must be obtained in all rigs. Pressure scaling method showed better results than other models available at time for estimating the value of best efficiency point at varying Reynolds number. A higher pressure rise was required for simulating flow at high Mach no. thus requiring high power input. This effect can be controlled by implementing the control over rotational speed of fan.

3 Conclusion

Much work has been done in various aspects relating to axial flow fans with focus limited to performance improvement in earlier years, while it is getting shifted to flow structure and acoustics in the modern era with strict regulations coming up time to time. Following are some aspects we learned from this study:

- Contra-rotating fans are useful in scenario with requirement of higher flow rates with uniform distribution and speed ratio and axial spacing are the deciding factors for overall efficiency.
- Blade sweep methods showed improvement in both pressure rise and efficiency with acoustics benefit. Forward sweep improved efficiency and acoustics with loss in pressure. Backward sweep showed higher losses at low flow rates but gives better results at higher flow rates. Flow structure studies are needed for these blades to better understand the operation. Combinations of the sweep angles did not show much improvements while giving more variance in flow across the blade span.
- Various blade stacking methods have been developed over the time involving use of genetic and artificial intelligence methods. Cylindrical stacking being the most studied, other methods like conical stacking require high computation power but coupled with Cordier diagram optimisation, these could provide highest efficiency possible.
- It can be concluded that performance should be considered primary and acoustics parameters should be inspected when no further improvements can be made in the efficiency factor for the fan.

References

1. Bamberger K, Carolus T (2012) Optimization of axial fans with highly swept blades with respect to losses and noise reduction. Noise Control Eng J 60(6):716–725
2. Bamberger K, Carolus T (2020) Efficiency limits of fans. Proc Inst Mech Eng Part A J Power Energy 234(5):739–748
3. Beiler MG (1999) Computation and measurement of the flow in axial flow fans with skewed blades. J Turbomach 121(1):59–66
4. Bizjan B et al (2016) Energy dissipation in the blade tip region of an axial fan. J Sound Vib 382(July):63–72
5. Dong B et al (2021) Noise attenuation and performance study of a small-sized contra-rotating fan with microperforated casing treatments. Mech Syst Signal Process 147(6):107086
6. Eck B (1973) FANS—design and operation of centrifugal, axial-flow and cross-flow fans, pp 1–611
7. Engineering M, Africa S (2018) Comparison of sound quality metrics for axial flow fans with straight and forward swept blades. In: Proceedings of Fan 2018—international conference on fan noise, aerodynamics, applications and systems, pp 1–11
8. Epple P, Durst F, Delgado A (2011) A theoretical derivation of the Cordier diagram for turbomachines. Proc Inst Mech Eng C J Mech Eng Sci 225(2):354–368
9. Friebe C et al (2018) Design and investigation of a multistage axial contra-rotating fan. In: Proceedings of Fan 2018—international conference on fan noise, aerodynamics, applications and systems, pp 1–12
10. Hurault J et al (2010) Experimental and numerical study of the sweep effect on three-dimensional flow downstream of axial flow fans. Flow Meas Instrum 21(2):155–165
11. Krause R et al (2019) Investigations on noise sources on a contra-rotating axial fan with different modifications. In: E3S web of conferences, 111
12. Krömer F, Müller J, Becker S (2018) Investigation of aeroacoustic properties of low-pressure axial fans with different blade stacking. AIAA J 56(4):1507–1518
13. Lee KS, Kim KY, Samad A (2008) Design optimization of low-speed axial flow fan blade with three-dimensional RANS analysis. J Mech Sci Technol 22(10):1864–1869
14. Li Y, Ouyang H, Du Z (2007) Experimental research on aerodynamic performance and exit flow field of low pressure axial flow fan with circumferential skewed blades. J Hydrodyn 19(5):579–586
15. Lotfi O et al (2016) GT2006-90659, pp 1–11
16. Manas MP, Pradeep AM (2020) Stall inception in a contra-rotating fan under radially distorted inflows. Aerosp Sci Technol 105:105909
17. Masi M et al (2018) Effectiveness of blade forward sweep in a small industrial tube-axial fan. In: Proceedings of fan 2018—international conference on fan noise, aerodynamics, applications and systems, pp 1–13
18. Masi M, Castegnaro S, Lazzaretto A (2016) Forward sweep to improve the efficiency of rotor-only tube-axial fans with controlled vortex design blades. Proc Inst Mech Eng Part A J Power Energy 230(5):512–520
19. Masi M, Lazzaretto A (2016) GT2015-43029 (2006), pp 1–13
20. Masi M, Lazzaretto A (no date) CFD models for the analysis of rotor-only industrial axial-flow fans (April 2012), pp 18–20
21. Masi M, Piva M, Lazzaretto A (2016) GT2014-27176, pp 1–16
22. Mistry C, Pradeep AM (2014) Experimental investigation of a high aspect ratio, low speed contra-rotating fan stage with complex in flow distortion. Propul Power Res 3(2):68–81
23. Nelles O (2020) Nonlinear system identification
24. Noguera R et al (1993) Design and analysis of axial flow pumps, ASME-PUBLICATIONS-FED, 154
25. Nouri H et al (2012) Design and experimental validation of a ducted counter-rotating axial-flow fans system. J Fluids Eng Trans ASME 134(10):1–6

26. Ouyang H et al (2006) Experimental study on aerodynamic and aero-acoustic performance of low pressure axial flow fan with circumferential skewed blades. Hangkong Dongli Xuebao/J Aerosp Power 21(4)
27. Pascu M et al (2009) Parmeterization of the total pressure distribution along a low-pressure axial fan blade according to the design requirements. In: ASME international mechanical engineering congress and exposition, proceedings, pp 1343–1352
28. Saul S, Matyschok B, Pelz PF (2018) Fan model test at varying ambient pressure: efficient product validation at full scale Reynolds and Mach number. In: Proceedings of fan 2018—international conference on fan noise, aerodynamics, applications and systems
29. Seo S-J, Choi S-M, Kim K-Y (2006) Design of an axial flow fan with shape optimization. Trans Korean Soc Mech Eng B 30(7):603–611
30. Shigemitsu T et al (2010) Performance and flow condition of contra-rotating small-sized axial fan at partial flow rate. Int J Fluid Mach Syst 3(4):271–278
31. Smith LH, Yeh H (1963) Sweep and dihedral effects in axial-flow turbomachinery. J Basic Eng 85(3)
32. Standard I (2007) International Standard ISO using standardized airways
33. Toc T (2002) Gt-2002-30445 aerodynamic design optimization, pp 15–17
34. Wilcox DC (1994) Turbulence modeling for CFD, 1st edn, Aiaa
35. Wright T, Simmons WE (1990) Blade sweep for low-speed axial fans (1989). Proc ASME Turbo Expo 1:151–158
36. Yang L, Hua O, Zhao-Hui D (2007) Optimization design and experimental study of low-pressure axial fan with forward-skewed blades. Int J Rotating Mach 2007:1–10
37. Ye X et al (2019) Prediction of performance of a variable-pitch axial fan with forward-skewed blades. Energies 12(12)
38. Zhang J et al (2017) Aerodynamic design of an ultra-low rotating speed geared fan. Aerosp Sci Technol 63:73–81
39. Zhang L et al (2019) Study on static and dynamic characteristics of an axial fan with abnormal blade under rotating stall conditions. Energy 170:305–325

Numerical Simulation of Flow Past Elliptic Cylinder Using Smoothed Particle Hydrodynamics

Justin Antony and Ranjith Maniyeri

Nomenclature

a	Major axis of the ellipse
b	Minor axis of the ellipse
c	Speed of sound (10 × U)
D	Diameter of ellipse along major axis
$f(\boldsymbol{x}_i)$	Field variable of ith particle at position vector **x**
h	Smoothing length
i, j	Indices for denoting ith and jth particle
N	Number of neighbouring particles in the support domain
$\boldsymbol{u}$	Velocity vector
U	Inlet velocity
$\boldsymbol{x}$	Position vector
Δx	Initial particle spacing
W_{ij}	Smoothing kernel function
$\nabla_i W_{ij}$	Derivative of the smoothing kernel function

Greek Symbols

α, β	Indices denoting components of field variables

J. Antony · R. Maniyeri (✉)
Biophysics Laboratory, Department of Mechanical Engineering, National Institute of Technology Karnataka, Surathkal, Mangalore, Karnataka 575025, India
e-mail: mranji1@nitk.edu.in

J. Banerjee et al. (eds.), *Recent Advances in Fluid Dynamics*, Lecture Notes in Mechanical Engineering, https://doi.org/10.1007/978-981-19-3379-0_18

1 Introduction

The flow past bluff bodies of various cross-section have been an interesting and challenging problems in the field of computational fluid dynamics (CFD) to understand the flow characteristics and to find correlations between parameters like Re and St. Many researchers extensively investigated flow past bluff bodies both experimentally [13, 14] and numerically [8]. However, most of the studies are devoted to circular and square cylinders, only a limited number of studies are conducted on flow over elliptic cylinders. Elliptic cylinders have smaller wake region and drag coefficients, which make them best models for many practical applications such as heat exchangers, blades, airfoils, etc. [10]. The orientation of flow with respect to the major and minor axes also influences the flow pattern. Unsymmetrical vortices are formed in case of inclined elliptic cylinders, and it results in a negative lift when the first vortex is shed away from the surface 1.

Smoothed particle hydrodynamic (SPH) is a meshless particle method, which is originally invented 3 for solving astrophysical problems. Later, its application extended into many areas including CFD. The meshless nature and the ability of nodes to act as material component are the two features that makes SPH attractive. SPH can handle complex problems involving complex geometries much easily compared to the conventional grid-based methods. In this work, a numerical model has been developed using an improved version of smoothed particle hydrodynamics (SPH) known as decoupled finite particle method (DFPM) 15 to solve the problem of flow past elliptic cylinder of axis ratio (b/a) 0.5.

2 Methodology

The DFPM proposed by Zhang and Liu 15 is an improved version of SPH in which the field variables and its derivatives at a node are evaluated from the neighbouring nodes using a smoothing kernel function as in Eq. (1) and (2).

$$< f(\boldsymbol{x}_i) >= \frac{\sum_{j=1}^{N} \frac{M_j}{\rho_j} f(\boldsymbol{x}_j) W_{ij}}{\sum_{j=1}^{N} \frac{M_j}{\rho_j} W_{ij}} \quad (1)$$

$$< \nabla \cdot f(\boldsymbol{x}_i) >= \frac{\sum_{j=1}^{N} \frac{M_j}{\rho_j} f(\boldsymbol{x}_j) \cdot \nabla_i W_{ij}}{\sum_{j=1}^{N} \frac{M_j}{\rho_j} (\boldsymbol{x}_j - \boldsymbol{x}_i) \cdot \nabla_i W_{ij}} \quad (2)$$

A novel discrete scheme proposed by Huang et al. [4] is used to evaluate the Laplacian of field function as in Eq. (3).

$$< \nabla^2 f(\boldsymbol{x}_i) >= 2\frac{\sum_{j=1}^{N} \frac{M_j}{\rho_j} \left(f(\boldsymbol{x}_j) - f(\boldsymbol{x}_i)\right) W_{ij}}{\chi \psi} \quad (3)$$

where $\chi = 15/7\pi h^2$ and $\psi = 31h^4\pi/210$. From the above approximations, the SPH formulations of continuity and momentum equations can be written as given in Eqs. (4) and (5), respectively.

$$\frac{D\rho_i}{Dt} = -\rho_i \frac{\sum_{j=1}^{N} \frac{M_j}{\rho_j}\left(\boldsymbol{u}_j^{\beta} - \boldsymbol{u}_i^{\beta}\right) \cdot \frac{\partial W_{ij}}{\partial \boldsymbol{x}_i^{\beta}}}{\sum_{j=1}^{N} \frac{M_j}{\rho_j}(\boldsymbol{x}_j - \boldsymbol{x}_i) \cdot \nabla_i W_{ij}} \tag{4}$$

$$\frac{D\boldsymbol{u}_i^{\alpha}}{Dt} = -\frac{1}{\rho_i} \frac{\sum_{j=1}^{N} \frac{M_j}{\rho_j}(p_j - p_i)\frac{\partial W_{ij}}{\partial \boldsymbol{x}_i^{\alpha}}}{\sum_{j=1}^{N} \frac{M_j}{\rho_j}(\boldsymbol{x}_j - \boldsymbol{x}_i) \cdot \nabla_i W_{ij}} + 2\frac{\mu_i}{\rho_i} \frac{\sum_{j=1}^{N} \frac{M_j}{\rho_j}\left(\boldsymbol{u}_j^{\alpha} - \boldsymbol{u}_i^{\alpha}\right) W_{ij}}{\chi\psi} + \frac{F_i}{M_i} \tag{5}$$

The above SPH formulations are used to discretize the governing equations at each node of the problem domain, and the resulting set of equations are solved explicitly with simple forward marching in time.

2.1 Artificial Equation of State

There are two approaches to implement incompressibility condition in SPH. First, using an artificial equation of state [11] in which pressure is a function of density. In this approach, any variations in density will result in a corresponding change in pressure that limits the variations in density variations to less than 3%. This approach is known as weakly compressible SPH (WCSPH). Second one known as incompressible SPH (ISPH) [11] uses a Poisson equation, which is solved implicitly to obtain the pressure fields. In this work, we adopted the WCSPH method due to its simplicity in modelling and lower computational costs. The artificial equation of state used in this paper is given in Eq. (6).

$$p_i = c^2(\rho_i - \rho_0) \tag{6}$$

2.2 Artificial Particle Displacement

The distribution of particles in the domain has a greater influence in the solution accuracy of SPH. Non-uniform particle distributions sometimes result in voids or clumps of particles that result in sudden termination of simulations. In order to avoid such errors, it is necessary to maintain uniform particle distribution. An artificial particle displacement algorithm proposed by Shadloo et al. [11] has been adopted, here in which all particles are given an artificial displacement at the end of each time step that shifts the particles to areas of less concentration and thus maintaining uniform distribution of particles.

2.3 *Density Diffusive Term*

The WCSPH method is known to have spurious oscillations in density and thus in pressure field due to the use of artificial equation of state. In order to avoid such spurious oscillation, a diffusive term suggested by [9] has been incorporated in the continuity equation in present work.

3 Problem Description

A stationary elliptic cylinder with axis ratio of 0.5 is considered for the simulations within a square domain of size 16D × 16D. A fluid with density $\rho = 1000$ kg/m^3 and viscosity $\mu = 1.0 \times 10^{-3}$ is considered to flow past the elliptic cylinder. The flow characteristics for different Re are analysed by varying the inlet velocity of the fluid. The schematic diagram of the domain is given in Fig. 1.

Fig.1 Schematic diagram

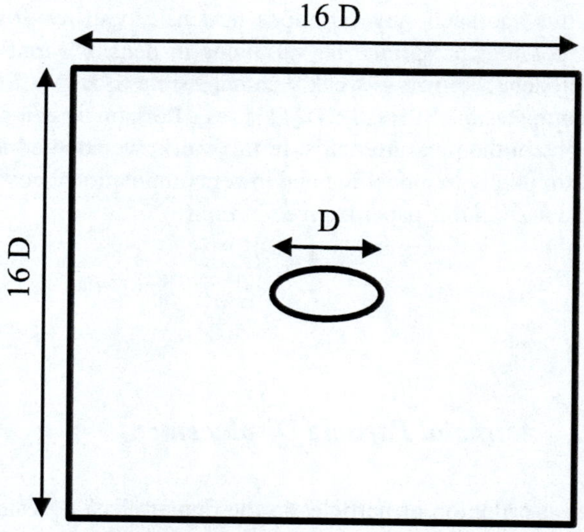

3.1 Numerical Simulation

The problem domain is discretized into 160 × 160 uniformly distributed particles (nodes) in a Cartesian coordinate system with an initial particle spacing of D/10. The particles that fall within the ellipse are removed to generate the fluid particles. A lower resolution of ten is used in all simulations due to limitations in the computational resources. A FORTRAN code is developed to execute the simulations.

3.2 Boundary Condition

The boundary conditions are applied using virtual particles as in Marrone et al. [9]. Virtual particles are the particles distributed in the boundary regions with same initial spacing as fluid particles. The properties of field variables at virtual particles are assigned to impart the respective boundary conditions. Each virtual particle is linked to a ghost node in the fluid domain, which is the mirror image of the virtual particle with respect to the boundary. The properties evaluated at these ghost nodes using SPH approximations are used to evaluate the properties of virtual particles. For no-slip boundary, the velocity components of the virtual particle are the negative value of that at ghost node while the density remains the same. For free-slip boundary, the velocity and density of virtual particles are same as that of ghost nodes. The inlet and outlet boundary conditions are applied as in Tafuni et al. [12]. In the present work, no-slip boundary conditions are applied to the cylinder walls and free-slip boundary condition is applied to both top and bottom walls of the channel. The inlet and outlet boundaries are applied on left and right boundaries, respectively. A uniform velocity profile is given at the inlet, which is varied to obtain different Re. The initial positions of fluid and virtual particles are shown in Fig. 2.

3.3 Time Step Size

The time step size determined using the CFL condition suggested by Korzani et al. [5], given in Eq. (7)

$$\Delta t \leq 0.25 \min\left(\frac{h}{c}\right) \tag{7}$$

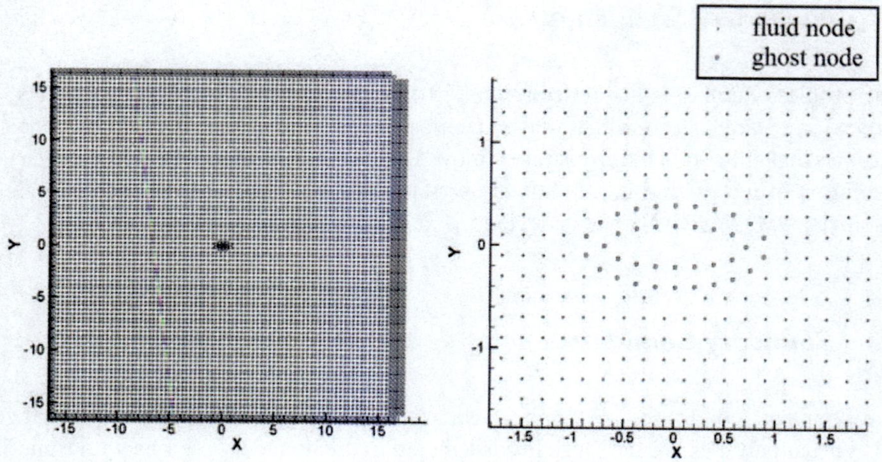

Fig. 2 Initial position of particles (left) and zoomed view of elliptic cylinder (right)

3.4 *Other Simulation Parameters*

In all simulations, the cubic spline kernel [6] is used with a smoothing length of 1.2 times the initial particle spacing. The speed of sound is chosen as ten times the upstream velocity. The constants in the artificial particle displacement and density diffusion term are taken to be 0.4 and 1.0, respectively. The simulations are run in personal laptop (DELL INSPIRON 3420, i3-2370M CPU, and 4 GB RAM) and hence, a low resolution ($D/\Delta x$) of ten is considered in all simulations.

4 Results and Discussion

Figure 3 shows the comparison of streamlines of present work with that of Maniyeri 7 for Re $=$ 10. It is observed that at Re $=$ 10 no wakes are formed behind the elliptic cylinder. The fluid flow remains attached to the cylinder for the whole time of simulation. Figures 4 and 5 show the streamlines for Re $=$ 30 and Re $=$ 40, respectively. Here, we can observe that symmetric vortices are formed behind the cylinder. The comparison with the result of Maniyeri 7 shows that the wake regions are in good agreement. However, a slight variation can be observed in the length of wake, which is due to insufficient data points near the cylinder to capture the vortices accurately. Simulation with higher resolution will increase the accuracy of the results. Similarly, Fig. 6 shows the streamlines for Re $=$ 60. Here also, symmetric vortices are formed which are in good agreement with the results of Maniyeri 7. The vortices at Re $=$ 30 and 40 show slight variations from the benchmark results. This is because number of data points (particles) is very less in the vicinity of the cylinder walls to capture the vortices. Adaptive particle refining proposed by Chiron et al. [2] can be

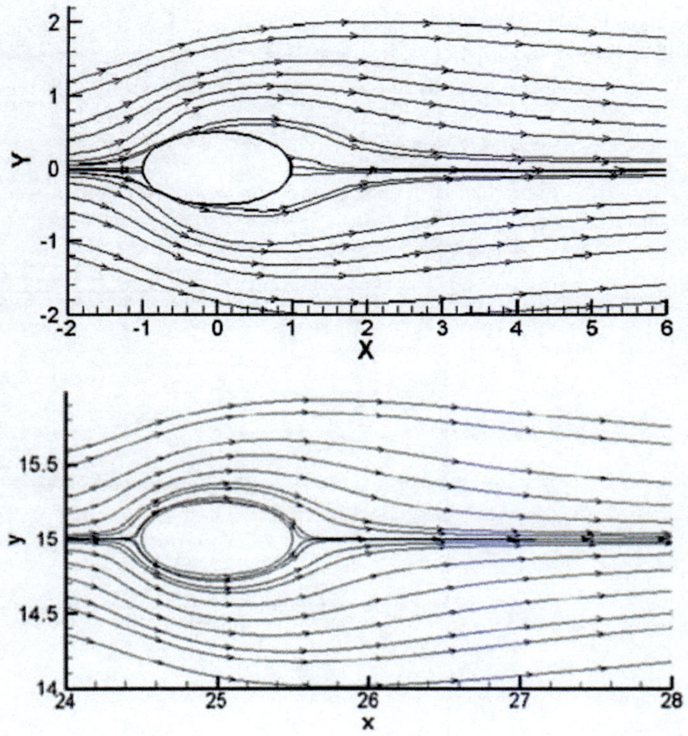

Fig. 3 Comparison of streamlines of present study (top) with Maniyeri 7 (bottom) for Re = 10

adopted to improve the results further refining particles near cylinder without large changes in computational time. The results for Re = 60 show good agreement with the benchmark results even at low resolutions. In case of flow past circular cylinder, the vortex shedding begins at a Re of 47 [8]. However, in case of ellipse, the vortex shedding has not started even at Re = 60. This shows that elliptic cylinder has low flow resistance compared to the circular cylinder, which is evident from small wake region behind the elliptic cylinder.

5 Conclusions

The flow past an elliptic cylinder with axis ratio 0.5 has been simulated for different Re ranging from 10 to 60 using a numerical model developed based on modified smoothed particle hydrodynamics. The wake regions obtained for different Re are in good agreement with the existing results in the literature. The results obtained even with the low resolutions show good agreement qualitatively with the existing results. The wake regions formed behind the elliptic cylinder is smaller compared to that of circular cylinder of same diameter, and no vortex shedding has observed till

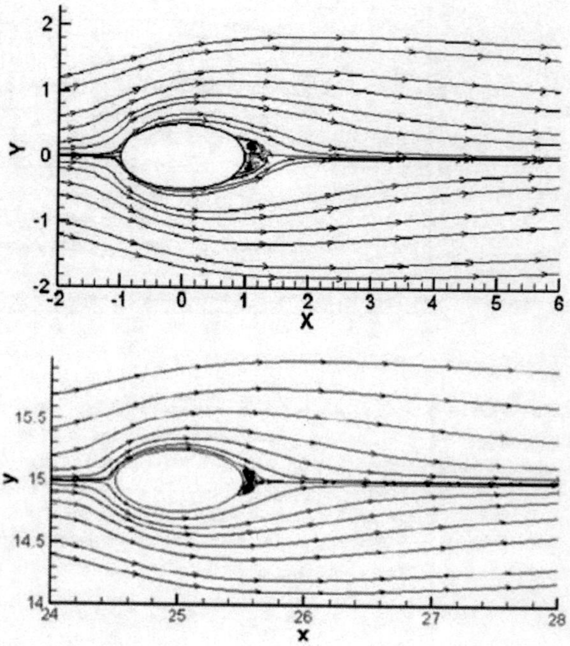

Fig. 4 Comparison of streamlines of present study (top) with Maniyeri 7 (bottom) for Re = 30

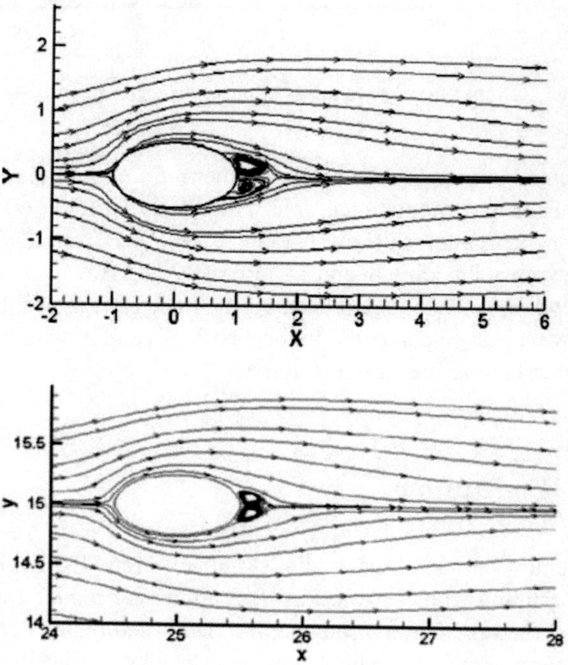

Fig. 5 Comparison of streamlines of present study (top) with Maniyeri 7 (bottom) for Re = 40

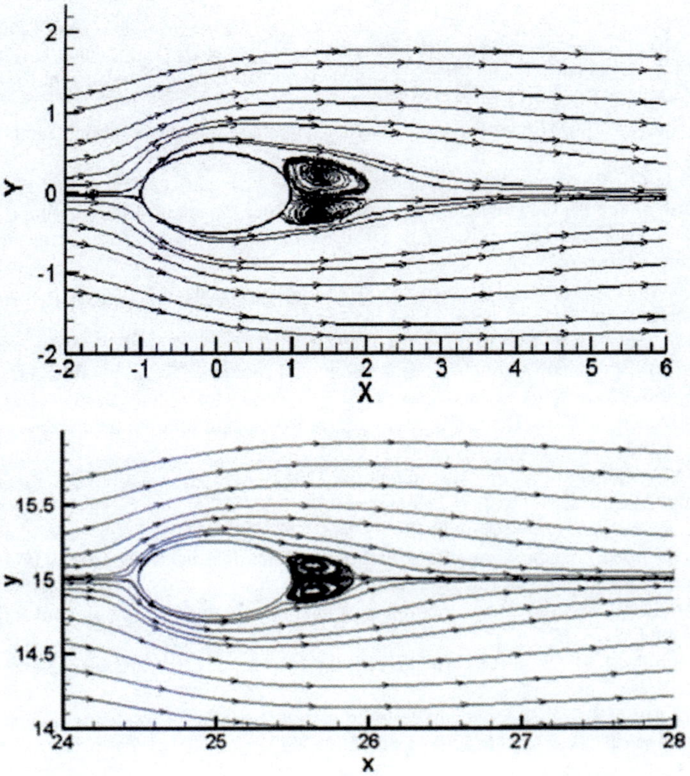

Fig. 6 Comparison of streamlines of present study (top) with Maniyeri 7 (bottom) for Re = 60

Re = 60. The developed model can be further improvised by incorporating adaptive particle refining technique in which the particles are refined near the cylinder to get more accurate results without compromising computational cost. We believe that this model can be further extended to simulate complex problems involving movable and deformable objects.

References

1. Badra HM, Dennis SCR, Kocabiyikc S (2001) Numerical simulation of the unsteady flow over an elliptic cylinder at different orientations. Int J Numer Meth Fluids 37:905–931
2. Chiron L, Oger G, Leffe MD, Touzé DL (2017) Analysis and improvements of adaptive particle refinement (APR) through CPU time, accuracy and robustness considerations. J Comp Phys 354:552–575
3. Gingold RA, Monaghan JJ (1977) Smoothed particle hydrodynamics: theory and application to non-spherical stars. Mon Not R Astron Soc 181(3):375–389
4. Huang C, Lei JM, Liu MB, Peng XY (2016) An improved KGF-SPH with a novel discrete scheme of Laplacian operator for viscous incompressible fluid flows. Int J Numer Meth Fluids

81:377–396
5. Korzani MG, Galindo-Torres SA, Scheuermann A, Williams DJ (2017) Parametric study on smoothed particle hydrodynamics for accurate determination of drag coefficient for a circular cylinder. Water Sci Eng 10(2):143–153
6. Liu GR, Liu MB (2003) Smoothed particle hydrodynamics: a meshfree particle method, World scientific
7. Maniyeri R (2019) Numerical simulation of viscous flow past elliptic cylinder. In: Chandrasekhar U, Yang LJ, Gowthaman S (eds) Innovative design, analysis and development practices in aerospace and automotive engineering (I-DAD 2018). Lecture notes in mechanical engineering. Springer
8. Maniyeri R (2014) Numerical study of flow over a cylinder using an immersed boundary finite volume method. Int J Eng Res 3(4):213–216
9. Marrone S, Colagrossi A, Antuono M, Colicchio G, Graziani G (2013) An accurate SPH modeling of viscous flows around bodies at low and moderate Reynolds numbers. J Comp Phys 245:456–475
10. Perumal DA, Kumar GVS, Dass AK (2012) Lattice Boltzmann simulation of viscous flow past elliptical cylinder. CFD Lett 4(3)
11. Shadloo MS, Zainali A, Yildiz M, Suleman A (2012) A robust weakly compressible SPH method and its comparison with an incompressible SPH. Int J Numer Meth Eng 89:939–956
12. Tafuni A, Dominguez JM, Vacondio R, Crespo AJC (2018) A versatile algorithm for the treatment of open boundary conditions in smoothed particle hydrodynamics GPU models. Comput Methods Appl Mech Eng 342:604–624
13. Tritton DJ (1959) Experiments on the flow past a circular cylinder at low Reynolds numbers. J Fluid Mech 6:547–567
14. Zdravkovich MM (1997) Flow around circular cylinders, vol 1. Fundamentals, Oxford University Press, New York
15. Zhang ZL, Liu MB (2018) A decoupled finite particle method for modeling incompressible flows with free surfaces. Appl Math Model 60:606–633

Visualisation Studies on Bubbles Formation and Propagation in Pool Boiling

V. S. Vaishak, R. Soundararajan, P. Sidharth Shivakumar, M. Christopher Jacob, and T. J. S. Jothi

1 Introduction

The mechanism of bubble formation in nucleate pool boiling heat transfer is one of the most extensively researched fields closely related to the efficient design of modern heat transfer systems. The various parameters associated with the dynamics of bubble formation of the local boiling phenomenon have been studied for a very long time. Nucleation results in the generation of embryonic vapour bubbles at cavities present over the heater surface immersed inside the fluid. The bubbles absorb heat from the heater surface and grow in size. The bubble departs from the heater surface after attaining a specific size and moves to the fluid's top surface. The bubbles move along their path to reach the top surface of the fluid. The mechanism of bubble formation during nucleate pool boiling is believed to be directly associated with the heat transfer process. Therefore, studies have been developed to establish various correlation models for estimation of the bubble dynamics parameters during nucleate pool boiling heat transfer.

Bubble dynamics refers to the whole process of formation of a bubble starting from the generation of an embryonic bubble over the heater surface to its departure towards the fluid's top surface. Some of the various bubble parameters of importance include bubble departure frequency, departure diameter, bubble velocity, contact angle, etc. Many prediction models have been established for predicting the bubble departure diameter over numerous applications. The Fritz model [1] has been recognised as one of the most efficient models for calculating the departure diameter of bubbles

V. S. Vaishak · R. Soundararajan (✉) · P. Sidharth Shivakumar · M. Christopher Jacob · T. J. S. Jothi
Department of Mechanical Engineering, National Institute of Technology Calicut, Kattangal, Kerala, India
e-mail: srajan21@iitk.ac.in

T. J. S. Jothi
e-mail: tjsjothi@nitc.ac.in

J. Banerjee et al. (eds.), *Recent Advances in Fluid Dynamics*, Lecture Notes in Mechanical Engineering, https://doi.org/10.1007/978-981-19-3379-0_19

for pure liquids and mixtures. The Fritz model considers the effect of various forces acting on the bubble surface such as buoyancy force, adhesive forces, and surface tension force to estimate the bubble departure (d_B) diameter during pool boiling.

$$d_B = 0.0208 * \theta^2 \sqrt[2]{\frac{\sigma}{g(\rho_l - \rho_v)}} \quad (1)$$

where $\theta = 35°$ for mixture and 45° for water.

Lee et al. [2] performed experiments on subcooled boiling flows to study the various local flow parameters for a two-phase flow such as vapour velocity, void fraction, and liquid velocity. It was observed that the velocity of the vapour bubbles increased with a decrease in inlet subcooling. The bubble boundary layer, which appeared to increase with the heat flux, was found to decrease with the inlet subcooling and mass flux. The size of the bubbles formed was observed to affect the boundary layer. These results were verified using CFD simulation. Goel et al. [3] carried out experiments on subcooled nucleate pool boiling for studying the bubble departure characteristics. The bubble departure frequency and departure diameter were determined by image analysis. Their extensive experiments concluded that the departure diameter increased with the heater size, wall superheat and inclination angle. The departure frequency was found to have a positive correlation with the inclination angle, wall superheat, and degree of subcooling. Further studies enabled them to notice that the surface roughness and liquid subcooling had a damping effect on the departure diameter, but the departure frequency was not affected by the surface roughness. However, as the size of the heater increased, the departure frequency appeared to decrease. Welle [4] conducted experiments in vertical air–water flows at atmospheric conditions and measured the local values of bubble velocity, void fraction, and bubble diameter by implementing the resistivity probe technique. Even though the author obtained reliable values for local void fraction, he suggested that the resistivity probe's accuracy should be improved for measuring higher velocities and getting reliable bubble size distributions. This can be improved by decreasing the distance between the probe tips and increasing the sampling frequency. Thiagarajan et al. [5] experimented on nucleate pool boiling heat transfer over microporous structure in plain copper surfaces. Using high-speed visualisation, they obtained bubble parameters such as bubble departure frequency, bubble departure diameter, and active nucleation site density. They concluded that on all surfaces, the active nucleation site density and the departure frequency increases with an increase in temperature, whereas the bubble diameter was not affected by the rise in temperature.

Mohanty et al. [6] emphasised the need of establishing a generalised correlation for nucleate pool boiling by using different parameters associated with the dynamics of bubble formation. The surface tension force was determined as the principal force is responsible for the adherence of bubbles over the heater surface. The departure of bubbles from the heater surface after its growth was mainly attributed to the buoyancy force acting on the bubble surface. They measured the active nucleation site density using their self-devised correlation which can be applied only in the case of low to moderate heat flux. The author also observed that a faster-growing vapour bubble attained a bigger diameter at its moment of departure. Michaie et al. [7] noticed an increase in the bubble size and detachment frequency, accompanied by a drop in pressure when the bubble changes its shape in the allowable pressure range of the test matrix. The bubble, initially, at atmospheric pressure changes its near-spherical form to that of an oblate spherical form as the pressure is reduced to the lowest value in the test matrix. This transformation in shape is achieved via a vapour column or a mushroom shape as a result of the rapid growth of subsequent bubbles. They also discovered that as the saturation pressure drops, the bubble detachment becomes difficult due to surface tension force. This coupled together with low-pressure condition results in the bubble attaining a longer development period with maximum volume compared to higher pressure conditions.

Zang et al. [8] used an electric discharge to develop a bubble inside a low-pressure tank for improving the effect of buoyancy force. The tests were conducted for generating a bubble in an infinite field positioned above a rigid boundary but below a free surface. The bubble appeared to collapse and the mode of collapse varied depending on the dominant force acting over the bubble surface. Baz-Rodriguez et al. [9] developed a correlation for predicting the terminal rise velocity of isolated single bubbles. By analysing the motion of a rising bubble in stagnant liquid, they formulated an equation for the terminal velocity. Moreover, by using experimental data trends, the effect of spiral trajectories was estimated and included in the formulation. The formulation was based on the concept of mechanical energy balance and the force balance from boundary layer theory. In their work, Chu et al. [10] used images captured from a high-speed camera to conduct an experimental study to estimate the bubble dynamic parameters such as bubble nucleation frequency and departure diameter for subcooled flow boiling conditions in a vertical annulus. The bubble nucleation frequency and departure diameter were found to depend on active nucleation site density over the heater surface. These characteristics varied with the subcooling, heat flux, and mass flux. The bubble departure diameter data was complied with Unal's model [11]. However, at higher pressure conditions, enough data for the study of bubble dynamics was not available. This paper presents a study of the visualisation of bubble growth and propagation in nucleate pool boiling. The bubble formation over the heated surface of the boiling vessel was visualised from two different orientation of the camera. The image analysis was carried out in ImageJ software. From the data obtained from the software, the results were obtained through plots and studied.

2 Experimental Setup

The study is carried out in two different vessels to capture the bubble formation phenomenon using visualisation. A steel vessel of diameter 80 mm is used to boil the water to determine the diameter of the bubble. The top view of the boiling vessel is captured using a 720p 30fps camera. The camera is carefully adjusted and placed on rigid support to prevent any vibration or disturbance. The top view is used for the visualisation of bubble formation over the heater surface. This is used to study the different bubble dynamics parameters such as the active nucleation site density and bubble diameter. Proper illumination is provided by using front lighting to capture the growth of bubbles.

A transparent vessel is selected to observe both the velocity and the bubbles' path to reach the free surface during pool boiling. An 800 ml capacity borosilicate glass vessel of diameter 120 mm is used to study the bubble velocity. The transparent glass vessel's side view is captured using the camera to determine the bubble departure velocity. A mercury thermometer is used for measuring the temperature of water.

2.1 Experimental Methodology

ImageJ software was used to process the bubble formation videos to determine the following bubble parameters: bubble diameter, bubble velocity, and bubble nucleation density. The whole experiment was captured at a recording speed of 30 frames per second, and the image size was set to 1280 × 720 pixels. A suitable calibration scale was selected to convert the pixels to mm. To determine the bubble parameters, the image frames were extracted from the video and converted into 8-bit per pixel image format with greyscale intensities. The 8-bit format was used in which the matrix takes values from 0 to 255, where 0 indicates white while 255 indicates black. A threshold value was set according to the image intensity. The matrix entry would be one if the intensity was above the threshold limit, whereas zero was below the limit. After the application of the threshold operator, the watershed operator is used for segmentation. The watershed operator is used to separate the bubble particles that are in contact with each other. The bubbles in the image appear as black dots or circles after the threshold and watershed operation's simultaneous usage. The image was then analysed by using the "Analyse particle" tool in the software. The tool was used to detect the black circles to determine the bubble radius distribution. The bubble diameter was then manually evaluated from the Image processing software, ImageJ.

The average bubble diameter was determined from the total bubble pixel area of all the bubbles in the image. The measurement uncertainty in estimating bubble diameter is associated with the accuracy of the calibration scale used. The standard uncertainty in bubble diameter measurement is expected to be in the range of 5% of the bubble diameter. The top view of the bubble formation was analysed (Fig. 2) to obtain the average bubble diameter and number of active nucleation sites. The nucleation site density (n) was obtained directly from the software. The average bubble diameter was determined by using the total area of the bubbles.

$$D = 2\sqrt[2]{\frac{A}{n\pi}} \tag{2}$$

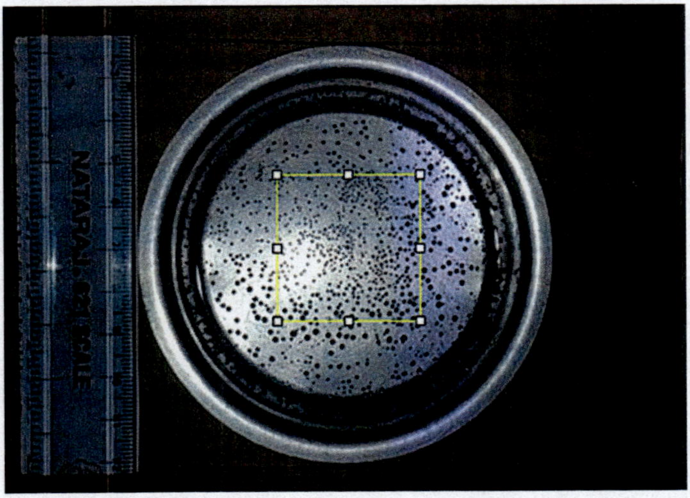

Fig. 1 Region of interest for image analysis (top view)

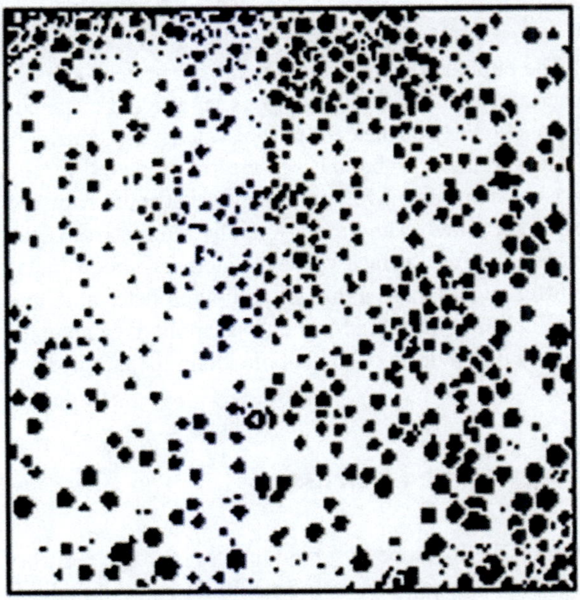

Fig. 2 Image obtained after thresholding and watershed operation at a time of 20 s

The bubble velocity was determined manually after applying a series of operators for image processing. The shadow operator was applied to the images to produce a shadow effect of the bubble moving through the fluid to reach the free surface. The images were sharpened using the Sharpen operator to increase the contrast of the image frames. The binary operator Subtract was then applied to adjust the pixel setting to further increase the bubble images' clarity. The average velocity of departure of the bubbles at different instances was obtained by analysing two adjacent image frames, one with the bubble on the heater surface and the other at the instant of departure of the bubble from the surface. The coordinates of the centre of the bubble at the two frames were identified to determine the distance moved by the bubble between two consecutive image frames. The Euclidean distance formula was applied to obtain the distance using the centre of the bubble's coordinates.

$$s = \sqrt[2]{(x_2 - x_1)^2 - (y_2 - y_1)^2} \tag{3}$$

Five successive bubbles from six different sites on the heater surface are identified and monitored to measure the bubble velocity. The bubbles are tracked in the image frames from the initiation phase till the bubble collapses. The position of the bubbles is measured every few image frames, over which a significant change in velocity is detected. The images were recorded at a recording speed of 30 fps. The time between two consecutive frames was found out to be 0.033 s. The velocity is then determined by using the distance-time formula.

$$\text{Speed} = \frac{\text{Distance}}{\text{Time}} \tag{4}$$

The bubble velocity is assumed to have been subjected to measurement bias and uncertainty arising from failure to capture the instance of departure of the bubbles. The uncertainty of the calibration scale is also considered. The standard uncertainty in measurement of bubble velocity is predicted to be in the range of 20% of the measurement. The average velocity distribution of the bubble along its path was also determined. The velocity of the bubble from the moment of departure till it reached the free surface was evaluated. The coordinates of the centre of the bubble at different instances along its path were determined and tabulated in an excel file. The Euclidean distance formula was applied to calculate the distance, and hence, the velocity distribution of the bubble along its path is determined.

3 Results and Discussion

From Fig. 3, it can be observed that the average radius of the bubbles formed on the heater surface increases over time. Initially, the bubbles were generated on the heater surface, which absorbed heat from them. The heat absorbed allowed the bubbles to grow on their nucleation site in the bubble's early growth stage. The bubble gets detached from the heater surface as it grows beyond the bubble departure diameter. The path followed by the bubble to reach the top surface is not the same for all the bubbles. The centre of the bubble was shifted from the nucleation site as it grew on the heater surface. Hence the path followed by the bubble was random for each bubble. It depends on the contact angle made by the bubbles. Over time, as the heater

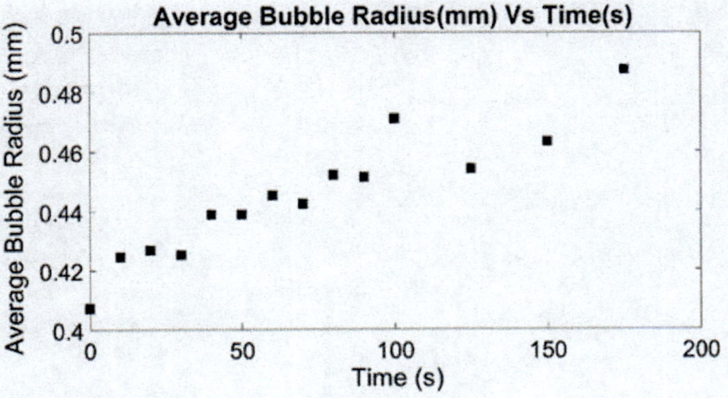

Fig. 3 Average bubble radius (mm) versus time

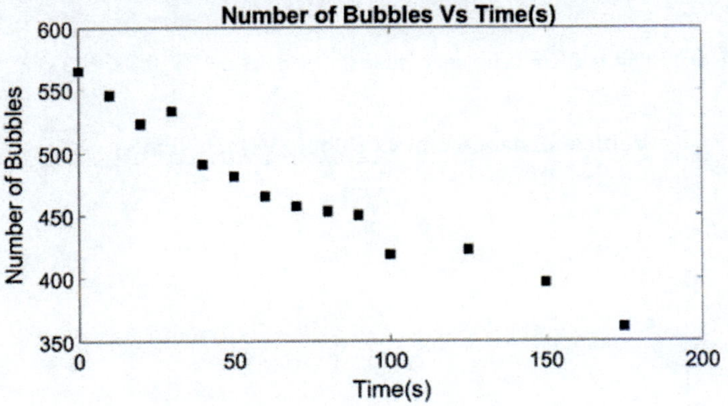

Fig. 4 Number of bubbles versus time (s)

surface temperature increases, the bubbles on the heater surface absorbs more heat and grows in size. Hence, the variation of the average bubble diameter across the nucleation sites was mainly attributed to the microstructure of the bubble nucleation cavities and the heater surface temperature [10].

Figure 4 shows the plot between the average number of bubbles versus time. The Average number of bubbles on the heater surface denotes the active bubble nucleation sites. From the plot, it was observed that the number of bubbles formed on the heater surface decreases over time. This was in contrast to the average diameter of the bubbles, which increased over time. As the temperature increases, the bubbles formed on the heater surface at adjacent nucleation sites merge to form a larger bubble. The larger bubble is then detached from the surface to reach the top surface. Hence, the decrease in the average number of bubbles formed was mainly attributed to the temperature effect [12]. Figure 5 (Scale 1 cm = 39.36 pixels) shows the motion of a single bubble and the container's vertical distance and reaching the surface.

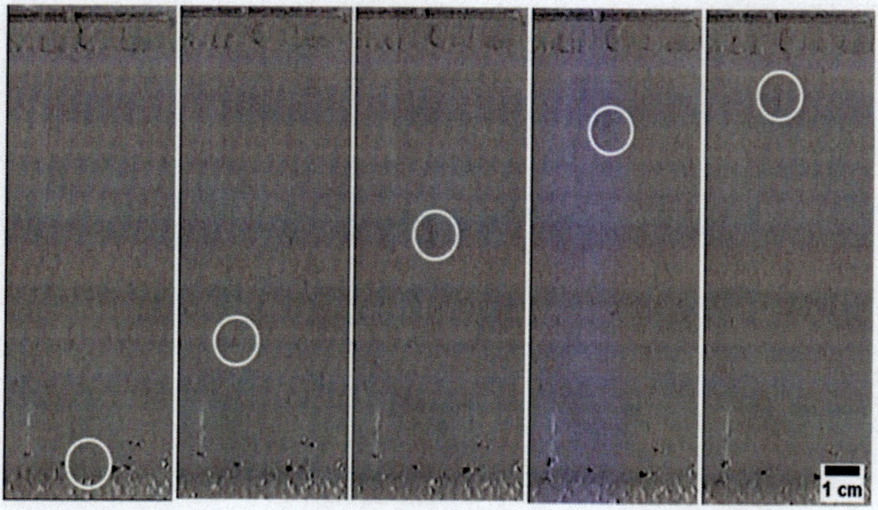

Fig. 5 Bubble travelling at different instant of time ($t = 0$ s, 0.1 s, 0.2 s, 0.3 s, 0.33 s)

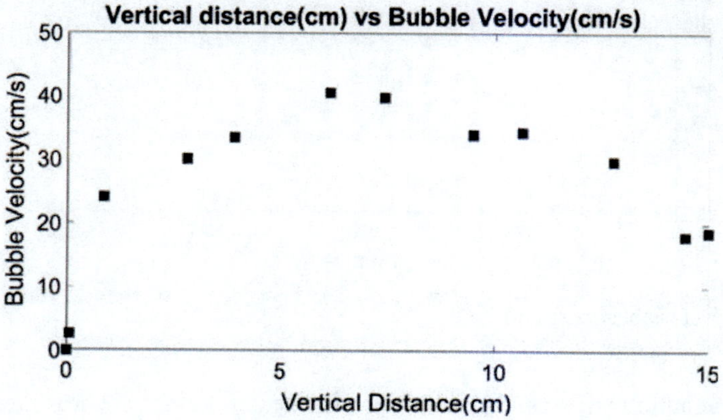

Fig. 6 The variation of bubble velocity with the vertical distance

Figure 6 shows the plot between the average bubble velocity with the vertical distance. The average velocity of bubble departure was determined to be 2.8 cm/s. The bubbles' average time to reach the top surface was also calculated and was found out to be 0.45 s. The average bubble velocity initially increases as the bubble leaves the heater surface and travels along its path to reach the free surface. The bubble velocity increases till it reaches a maximum somewhere along its path, and then, the velocity decreases [9]. The velocity and shape of the bubbles along the path depend on the inertial force and surface tension acting on the bubbles.

4 Conclusion

A visualisation study has been done on the bubble formation and their propagation in a pool boiling container. It was evident from the study that the average radius of the bubbles increases over time. The average number of bubbles from the heater surface decreased over time. This indicated that the number of nucleation sites decreases over time. This decrease in the average number of bubbles is due to the temperature effect. The average velocity of bubble departure initially increases with the vertical distance, and after attaining a maximum velocity, the average velocity decreases along its path.

References

1. Fritz (1935) Berechnung des maximal volumens von Dampfblasen. Phys Z 36:379–384
2. Lee TH, Park GC, Lee DJ (2002) Local flow characteristics of subcooled boiling flow of water in a vertical concentric annulus. Int J Multiphase Flow 28:1351–1368
3. Goel P, Nayak AK, Kulkarni PP, Joshi JB (2016) Experimental study on bubble departure characteristics in subcooled nucleate pool boiling. Int J Multiphase Flow 89:163–176
4. Van der Welle R (1985) Void fraction, bubble velocity and bubble size in two-phase flow. Int J Multiphase Flow 11:317–345
5. Thiagarajan SJ, Yang R, King C, Narumanchi S (2015) Int J Heat Mass Transf 89:1297–1315
6. Mohanty RL, Das MK (2017) A critical review on bubble dynamics parameters influencing boiling heat transfer. Renew Sustain Energy Rev 78:466–449
7. Michaie S, Rullière R, Bonjour J (2017) Experimental study of bubble dynamics of isolated bubbles in water pool boiling at subatmospheric pressures. Exp Thermal Fluid Sci 87:117–128
8. Zhang AM, Cui P, Cui J, Wang, Qian (2015) Experimental study on bubble dynamics subject to buoyancy. J Fluid Mech 776:137–160
9. Baz-Rodríguez S, Aguilar-Corona A, Soria A (2012) Rising Velocity for single bubbles in pure liquids. Revista Mexicana de Ingeniería Química 11:269–278
10. Chu I-C, No HC, Song C-H (2011) Bubble lift-off diameter and nucleation frequency in vertical subcooled boiling flow. Int J Nuclear Sci Technol 48(6):936–949
11. Unal HC (1976) Maximum bubble diameter, maximum bubble growth time and bubble-growth rate during the subcooled nucleate flow boiling of water up to 17.7 MN/m2. Int J Heat Mass Transf 19:643–649
12. Coulibaly A, Bib J, Christopher DM (2019) Experimental investigation of bubble coalescence heat transfer during nucleate pool boiling. Exp Thermal Fluid Sci 104:67–75

Predictive Analysis of Air-Cooled Condenser by Considering Fouling Using Machine Learning Algorithm

N. D. Shikalgar, Prabhakar R. Gujari, S. N. Sapali, and Vyankatesh D. Chavan

1 Introduction

Air-cooled condensers (ACC) is used widely for various applications (example: refrigeration, thermal power plant and cold storage). The ACC can be a substitute of the water-cooled condenser in power generation purpose when water resources are scarce [1]. The ACC for power generation purposes contains many fans to condense the refrigerants past the finned tubes. The fans can be located anywhere near the finned tubes to pass the ambient air with a certain velocity. The ACC has a greater impact on efficiency due to fouling occurring at heat exchanger surfaces. The performance of ACC depends mainly on environmental temperature, so the performance has the least impact during summer. There are several methods to monitor the performance of water-cooled heat exchangers that depends on the conditions like vacuum available in the condenser, also the temperature of inlet and outlet. So for an optimal solution, there are several algorithms to calculate these values and predict the condenser performance, [2] and we can obtain the heat transfer coefficient and heat transfer rate of these condensers [3]. But these values are not applied to ACC as mass flow rate and temperature are not known. Some important parameters to calculate the performance of ACC are the ambient temperature of the surrounding and mass flow rate over ACC. Some of the causes of reduction in performance of ACC are: (1) fouling around finned tubes and (2) some environmental conditions like ambient temperature, the air velocity and also some internal factors like a load of the motor fan. For proper analysing of condenser, investigation of the relationship between the condenser variables (parameters) [4].

Air-cooled condensers can be analysed in three ways. In the first approach, there is an analytical study of the empirical correlations between heat transfer and various

N. D. Shikalgar · P. R. Gujari (✉) · S. N. Sapali · V. D. Chavan
Department of Mechanical Engineering, College of Engineering Pune, Pune, India
e-mail: pgujari0@gmail.com

J. Banerjee et al. (eds.), *Recent Advances in Fluid Dynamics*, Lecture Notes in Mechanical Engineering, https://doi.org/10.1007/978-981-19-3379-0_20

variables. But the empirical relations will not indicate the performance due to degradation which might take place during operation [5]. In another approach is by computational fluid dynamics (CFD) where there is the numerical study (simulation) on the fouled condenser, but as the fouling keeps on changing with time, this approach is practically difficult to execute [6–8]. In the last approach, there has to be a study on the data, which can be utilized in some mathematical algorithm (ML) or artificial intelligence (AI) [9]. In this approach, much emphasis is given on the past operation or data to get the relation between input variables and desired target (output). These approaches do some regression analysis on data. But there is the requirement of the past data which is one of the main criteria. So, either one has to individually perform the experimentation or can use the experience of some user. In industry, their cab timely inspection of these datasets. The data is sent to the training model and some deviation or error can be found out.

This project work is used in industry in monitoring the plant to estimate the performance based on algorithms provided by machine learning such as multi-linear or neural networks (NN) [10]. The main idea is to get the variables that can be monitored for the performance of ACC.

2 Mathematical Modelling

The data points in these projects have been worked under two algorithms.

Hypothesis

$$h(x) = \theta_0 + (\theta_1 * x) \tag{1}$$

Where θ_0 and θ_1 are parameters, h is the hypothetical function and x is the input variable.

Choose θ_0 and θ_1 such that $h(x)$ is very close to the y (actual value) training example (x, y).

Cost function:

$$J(\theta_0, \theta_1) = \frac{1}{2 * m} * \sum_{i=1}^{m} \left(h\left(x^i\right) - \left(y^i\right)\right)^2 \tag{2}$$

$$\theta_0 = \theta_0 - \frac{\partial}{\partial \theta_0} J(\theta_0, \theta_1) \tag{3}$$

$$\theta_1 = \theta_1 - \frac{\partial}{\partial \theta_1} J(\theta_0, \theta_1) \tag{4}$$

Algorithm 1. Hypothesis for multivariable and multiparameters.

$$h(x) = \theta_0 * x_0 + \theta_1 * x_1 + \theta_2 * x_2 + \theta_3 * x_3 \quad (5)$$

Algorithm 2. Hypothesis for single variable quadratic polynomial

$$h(x) = \theta_0 + \theta_1 * x + \theta_2 * x^2 + \theta_3 * x^3 \quad (6)$$

$$P = 0.889 + 0.1911 * V - 0.033 * V^2 - 0.1109 * V^3 - 0.1280 * V^4 \quad (7)$$

$$P = 0.9144 + 0.1834 * V - 0.06048 * V^2 - 0.1483 * V^3 - 0.1738 * V^4 \quad (8)$$

Where P and V refers to compressor power and velocity.

Equations 7 and 8 have been developed by regression method with the help of Python software for the new and old condenser. These equations have an error of up to 4.16% and 6% for new and old condensers.

Equations 7 and 8 are a non-dimensional form and are formed by scaling of variables.

$$RE = 0.8347 + 0.265 * V + 0.041 * V^2 - 0.05862 * V^3 - 0.10113 * V^4 \quad (9)$$

$$RE = 0.674971 + 0.2576 * V + 0.0853 * V^2 + 0.0040 * V^3 - 0.0338 * V^4 \quad (10)$$

where RE refers to refrigerating effect.

Equations 9 and 10 have been developed by regression method with the help of Python software for a new and old condenser. These equations have an error of up to 2 and 1% for new and old condensers.

Note: Equations 9 and 10 are a non-dimensional form and are formed by scaling of variables.

$$\mathrm{COP} = 0.70350 + 0.28203 * V + 0.08831 * V^2 - 0.01142 * V^3 - 0.062511 * V^4 \quad (11)$$

$$\mathrm{COP} = 0.55845 + 0.244208 * V + 0.11282 * V^2 + 0.0488 * V^3 + 0.01701 * V^4 \quad (12)$$

Equations 11 and 12 have been developed by regression method with the help of Python software for the new and old condenser. These equations have an error of up to 2% for both the condensers.

Note: Equations 11 and 12 are a non-dimensional form and are formed by scaling of variables.

$$m = 0.87801 + 0.214833 * V - 0.007141 * V^2 - 0.0879 * V^3 - 0.11166 * V^4 \quad (13)$$

$$m = 0.90566 + 0.181126 * V - 0.064811 * V^2 - 0.15549 * V^3 - 0.18315 * V^4 \quad (14)$$

where m refers to a mass flow rate of refrigerant.

Equations 13 and 14 have been developed by regression method with the help of Python software for the new and old condenser. These equations have an error of up to 6% for the new condenser and 9% for the old one.

2.1 *Multi-Variable Linear Polynomial Regression*

$$P = 0.75405 * r + 0.11346 * V + 0.86592 * Q - 0.5459 * m + 1.65502 * RE - 1.9890 * COP \quad (15)$$

$$P = 0.04532 * r - 0.2369 * V - 1.053065 * Q + 0.1884 * m + 0.5039 * RE - 0.7291 * COP \quad (16)$$

Equations 15 and 16 state the new and old condensers and have the relative error of 0%, which proves that these algorithms can predict the more relevant data of the variables than the single variable quadratic polynomial of a higher power.

3 Methodology

As a part of the research paper, a detailed analysis of the variables that affect the performance of the vapour compression refrigeration system (VCRS) due to fouling has been performed using an experimental setup. Also, by some machine learning algorithms, some correlations have been made to study the variables dependency on each other. Multiple condensers are to be used in the project to experimentally test each of them, with R410a as a refrigerant and installing a heater in the cooling chamber with a cooling coil to estimate the actual cooling effect with heating effect at constant temperature to estimate the actual cooling effect. The two condensers used is selected from the Pune location, the first condenser is around1 a year old and the second one is around 15 years.

Figure 1 shows a schematic diagram of setup, where the test condenser will be there in the environmental conditions and the heater and the cooling coil will be there inside the cooling chamber of 1TR. There will be a motor fan attached to the condenser so that the effect of varying speed can be a criterion to judge the performance affected by fouling. After collecting all the data points, the data can be used to develop correlations in different forms of algorithms.

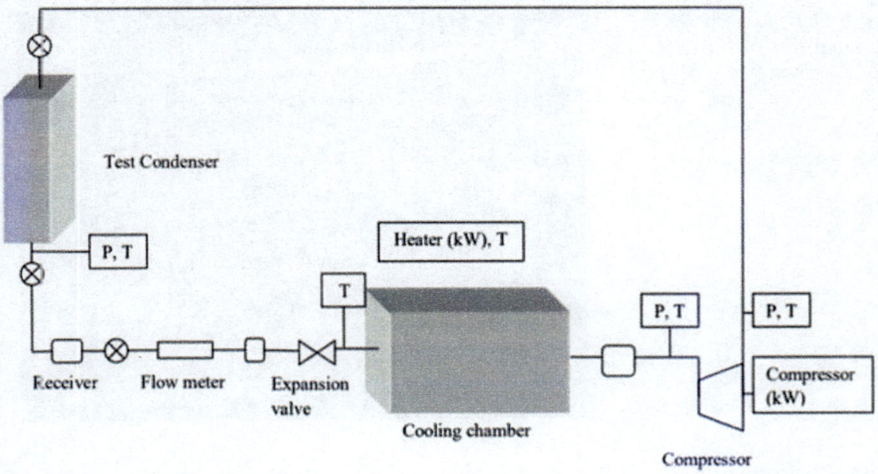

Fig. 1 Schematic diagram of the experimental setup

4 Experimentation

The experimental setup prepared is shown in Fig. 2. The system comprises air-conditioning components (compressor, condenser, cooling chamber, evaporator, regulator and expansion device). There is a control panel consisting of (suction and discharge pressure gauge, main supply ON/OFF switch, compressor ON/OFF switch, dimmer, temperature indicators and heater energy metre, compressor energy metre and timer, switch to reset a timer, condenser fan regulator and refrigerant flow

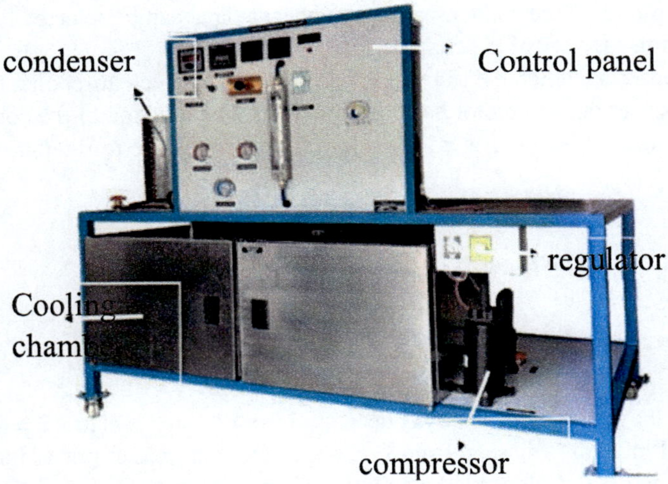

Fig. 2 Multiple condenser testing setup

Fig. 3 Old condenser for test setup

metre). The refrigerant temperature was measured using a K-type thermocouple, and a total of 8 thermocouples was attached to the setup.

Firstly, ensure that the door of the calorimeter chamber is closed properly so that there is no heat and mass exchange with the surrounding. The entire chamber needs to fill with atmospheric air before the start of the experimental readings. Then after, switch on the main supply of the control panel followed by the compressor switch. The RPM of the condenser fan has to be adjusted with a regulator, for the current process the rpm is kept on decreasing after alternate 30 min. To get the required temperature in the chamber, some current for the heater needs to be set by dimmer. Then wait for the steady-state to be achieved inside the calorimeter chamber. Then start taking the readings of temperature at eight locations. Once all the data values are noted for the setup, then change it with the old fouled condenser shown in Fig. 3. Note that the old condenser is selected in such a way that the condenser capacity and surface area of both of the condensers are equivalent. That can be selected based on the refrigeration capacity of the chamber.

The condenser on which the test was carried out is the forced air-cooled type for which a condenser fan and motor has been provided. The function of the condenser is to convert high-pressure refrigerant vapour into high-pressure refrigerant liquid.

5 Results and Discussion

5.1 Compressor Power Versus Time

As discussed above that, the velocity is decreased from $V1 = 3$ m/s to $V2 = 1.5$ m/s in every small interval. It is seen from Fig. 4 that, the compressor power increases up to 31.64% in case of old fouled condenser while it increases by 17.57% for the new condenser.

Fig. 4 Experimental results for compressor power versus time

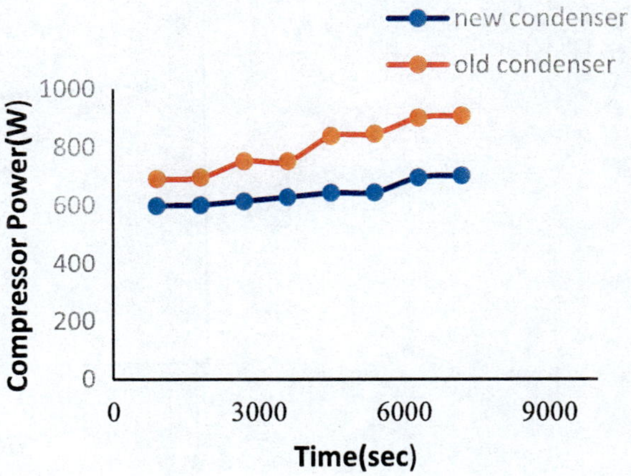

Fig. 5 Experimental results for refrigerating effect versus time

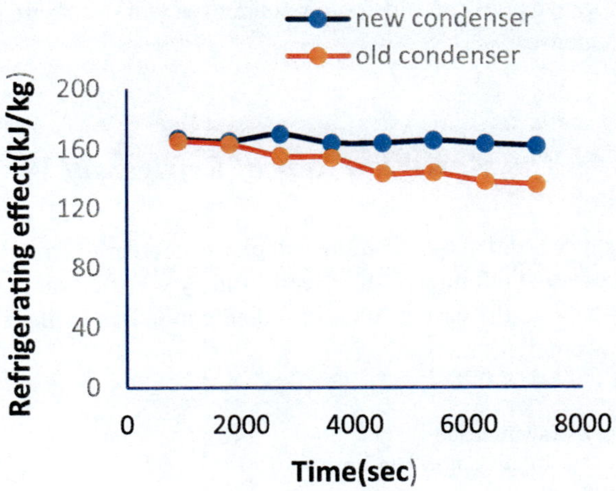

5.2 *Refrigerating Effect Versus Time*

As discussed above that, the velocity (V) is decreased from V1 = 3 m/s to V2 = 1.5 m/s in every small interval. It is seen from Fig. 5 that, the refrigerating effect decreases up to 22.22% in the case of the old fouled condenser while it decreases by 1.35% for the new condenser.

5.3 *Coefficient of Performance Versus Time*

As discussed above that, the velocity is decreased from V1 = 3 m/s to V2 = 1.5 m/s in every small interval. It is seen from Fig. 6 that, the COP decreases up to 39.45%

Fig. 6 Experimental results for COP versus time

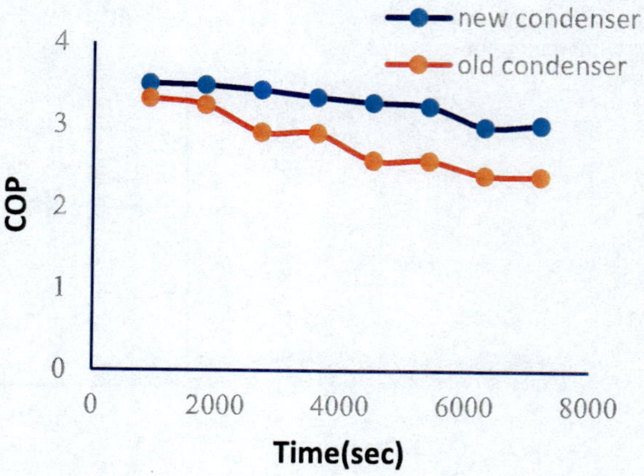

in the case of the old fouled condenser while it decreases by 16.66% for the new condenser.

5.4 *The Mass Flow Rate of Refrigerant Versus Time*

As discussed above that, the velocity is decreased from V1 = 3 m/s to V2 = 1.5 m/s in every small interval It is seen from Fig. 7 that, the mass flow rate increases up to 44.57% in the case of the old fouled condenser while it increases by 8.43% for the new condenser.

Fig. 7 Experimental readings for mass flow rate versus time

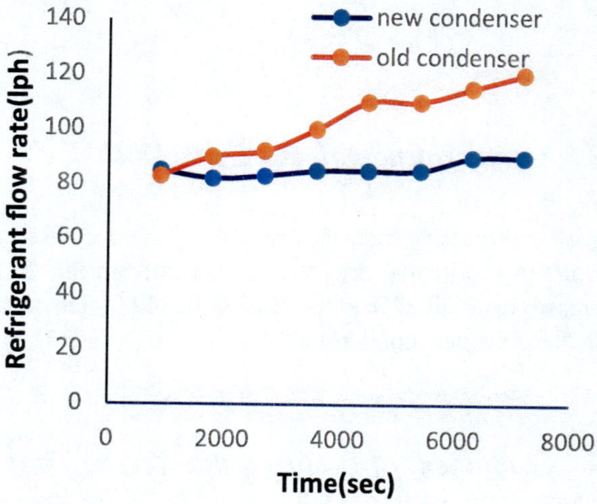

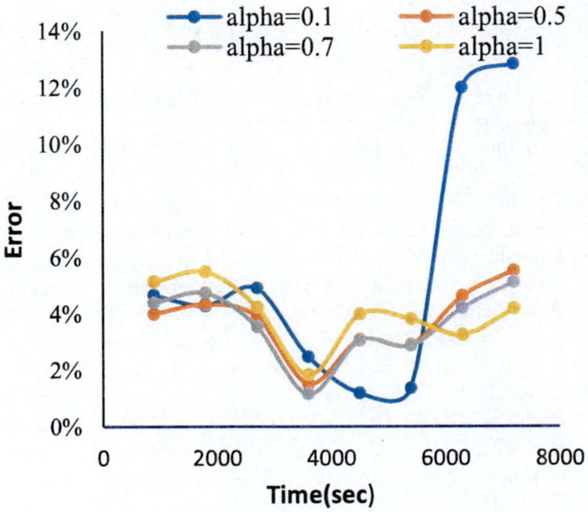

Fig. 8 Different value of gradient selected for solving cost function

5.5 *Effect of Various Gradient Descents on the Solution of the Compressor Power.*

When the gradient descent $\alpha = 1$ is taken in the cost function to calculate the value of coefficients that resulted in the minimum error (4.16%) for the larger time frame. So $\alpha = 1$ is taken as the best gradient descent for the analysis in the present paper (Fig. 8).

6 Conclusion

In this study, the performance of a new and old condenser is experimentally investigated and validated with help of a machine learning algorithm. There is degradation in the thermal performance of a VCR system is observed during the study. The major facts obtained during the study are discussed as follows:

1. A multivariable linear algorithm is found to have less error (<1%) with the ability to access huge data. When it comes to single variable quadratic polynomials, the regression technique shows around 5% error in the analytical and experimental values.
2. As the fouling increases pressure ratio, refrigerating effect has less role to play in compressor power, while air velocity and the mass flow rate have an important role to play to keep the heater capacity or cooling chamber requirements.

References

1. Kehlhofer R, Rukes B, Hannemann F, Stirnimann F (2009) Combined-cycle gas & steam turbine power plants, 3rd ed. PennWell Corp
2. Brummel HG, LeMieux DH, Voigt M, Zombo PJ (2005) Online monitoring of gas turbine power plants, Siemens Power Generation
3. He WF, Dai YP, Wang JF, Li MQ, Ma QZ (2013) Performance prediction of an air-cooled steam condenser using UDF method. Appl Therm Eng 50:1339–1350
4. Hotchkiss PJ, Meyer CJ, Backström TWV (2006) Numerical investigation into the effect of cross-flow on the performance of axial flow fans in forced draught air-cooled heat exchangers. Appl Therm Eng 26:200–208
5. Liu P, Duan H, Zhao W (2009) Numerical investigation of hot air recirculation of air-cooled condensers at a large power plant. Appl Therm Eng 29:1927–1934
6. Yang LJ, Du XZ, Yang YP (2012) Wind effect on the thermo-flow performances and its decay characteristics for air-cooled condensers in a power plant. Int J Therm Sci 53:175–187
7. Ertunc HM, Hosoz M (2006) Artificial neural ne1twork analysis of a refrigeration system with an evaporative condenser. Appl Therm Eng 26:627–635
8. Ertunc HM, Hosoz M (2008) Comparative analysis of an evaporative condenser using artificial neural network and adaptive neuro-fuzzy inference system. Int J Refrig 31:1426–1436
9. Ghettini S, Sorce A, Sacile R (2020) Data-driven air-cooled condenser performance assessment: model and input variable selection comparison EDP sciences 2020197:10003
10. Li X et al (2018) A data-driven model for the air-cooling condenser of thermal power plants based on data reconciliation and support vector regression. Appl Therm Eng 129:1496–1507

Analysis of Losses in Centrifugal Pump with Two Different Outlet Diameter of Impeller

A. Hari Krishna, Maitrik Shah, and Beena D. Baloni

Nomenclature

a	Channel width
b	Blade width
C_d	Dissipation coefficient
C_{fr}	Frictional coefficient
Ch	Head coefficient
Cq	Flow coefficient
D	Diameter
D_h	Hydraulic diameter
g	Gravitational acceleration
H	Head
h	Head loss
K_s	Shock factor
K_f	Friction factor
L_b	Length of the blade
Q	Flowrate
s	Seal length
t	Blade thickness
U	Circumferential velocity
V_p	Volute through flow velocity
V_d	Volute circulatory velocity
V_f	Flow velocity
W	Relative velocity

A. Hari Krishna (✉) · M. Shah · B. D. Baloni
Department of Mechanical Engineering, SVNIT, Surat, India
e-mail: amballakrishna@gmail.com

J. Banerjee et al. (eds.), *Recent Advances in Fluid Dynamics*, Lecture Notes in Mechanical Engineering, https://doi.org/10.1007/978-981-19-3379-0_21

Subscripts

1	Impeller inlet
1q	Impeller inlet throat
2	Impeller outlet
3	Volute inlet
4	Volute outlet
av	Average
imp.	Impeller
vol.	Volute
sh	Shock
s	Seal
d	Diffusion
f	Friction
BEP	Best efficiency point

Greek Symbols

β	Blade angle
η	Efficiency
σ	Slip factor
α_v	Volute angle

1 Introduction

In the early seventeenth century, many industries have chosen positive displacement pumps for various applications. However, due to its complex configurations in design, limited applications, and flow capacity, recent investigations have been shifted towards centrifugal pumps. Centrifugal pumps are simple in design, and low maintenance is required. Gradually, it became an integral part of many industries compared to other turbomachines. As centrifugal pumps are having vast applications all over the world such as water supply and circulation, irrigation, and transfer of chemicals, fluids, etc.; they are the machines in the field of the turbomachines, accounted for most of the electrical power is consumed. With the use of high-grade energy like electric power, any machine has to give the best out of it and operate with high efficiency. But in the centrifugal pumps, owing to its construction and its components, many losses are incurred in the energy and lead to reduced output. Even though they are less efficient compared to positive displacement type pumps, many people are studying for years to improve their efficiency because of its low maintenance and design simplicity. This made authors to learn more about the centrifugal pumps and the reasons behind most of the energy loss. To carry out this study, it is necessary to know about the different losses in the centrifugal pump and their effect on the pump's performance. Even though a lot of research works are done to improve the centrifugal pump performance. Still many unsolved issues, concerns with increasing the efficiency of these pumps that need to be worked on.

Some of the most important studies are based on the changing the pump shape, particularly the impeller and diffuser. Since an impeller is a dynamic part that energizes the fluid; its size and geometry play an important part in the performance of a pump. Any change in impeller shape or dimensions affects the impeller's inlet and outlet velocity triangles which results in significant change in pump performance. Patel [1] studied about the effect of impeller blade exit angle on the pump performance by Gulich mathematical modeling. Three pumps of different specific speeds are considered for the study and validated the predicted result with the manufacturer's data. The predicted results are in good agreement, and moreover by increasing the exit blade angle, there is an increment in head and efficiency of the pump for all the rated speeds. Many researches have chosen various approaches to predict the losses and effect of the impeller geometry on the pump performance. Omar [2] have developed a theoretical procedure to estimate the pump performance and predicted the pump with two different impellers at varying speeds by using a prediction program. They compared predicted data with the experimental data. They concluded that there is a good agreement between calculated and experimental data with differences of up to 15% between both the results. Fosile [3] developed a pump design program in MATLAB and opted four different methods to estimate head flow characteristics for a pump operating at 1480 rpm. He concluded that the Gulich loss calculation model [4] is giving good results and having valid agreement with the experimental results with variation of up to 8%. EL-Naggar [5] have performed one dimensional flow analysis in centrifugal pump. He manipulated Euler and energy loss equation into non-dimensional terms to find out the pump performance characteristics at different discharge coefficients. The predicted result and experimental result are compared, and he concluded that there is deviation in head coefficient due to slip factor variation and volumetric leakage. Elshestawy [6] have performed numerical study of slip factor and factors affecting pump performance. He concluded that increase in blade number will increase the slip factor but also reduction in flow area and it leads to blockage of the flow. Kumar et al. [7] have aimed to estimate the optimum set of loss models for centrifugal pumps. They made an attempt to investigate the pump performance characteristics by considering various impeller geometries and also concluded that there will be direct influence on the pump performance by varying geometrical and hydraulic parameters. John [8] given optimistic design of centrifugal pump and he differentiated the shock losses into individual losses at each section of pump. He explained the recirculation losses and concluded that, at off-design conditions, the flow does not follow the blade wall and it results in various flow patterns.

Win et al. [9] have predicted the performance of centrifugal pump. They considered certain loss models like shock losses, friction losses, volute losses, leakage losses, and recirculation losses to estimate the actual head for a pump operating at 1450 rpm. They observed that the major losses are shock losses at the impeller inlet occurred due to mismatch of fluid and blade angles and also found that the shock losses are present for almost all the flow conditions of the pump.

In the present work, impellers having different diameters, i.e., 174 and 160 mm are examined experimentally and theoretically to obtain the pump performance characteristics. Traditional method and Gulich loss calculation method are selected to obtain the pump performance characteristics, and these pump performance characteristics are represented in terms of non-dimensional coefficients by using mathematical modeling in C program. Later on, the validation of calculated result with the experimental data is done.

2 Loss Calculation Methods

2.1 *Traditional Method Shock and Friction Factors*

This method is used by researchers [10, 11] for finding out the hydraulic losses with considering friction losses and vortex dissipation losses (shock losses) [10, 11]. And these are calculated by using the following equations

$$H = \sigma H_{\text{theo.}} - k_{\text{s}}(Q - Q_{\text{BEP}})^2 - k_{\text{f}}Q^2 \tag{1}$$

Here, K_{s} and K_{f} are obtained from the experimental result, and these are different for each pump.

2.2 *Gulich Loss Calculation Method*

In this method, hydraulic losses are mainly divided into impeller losses and volute losses in a pump. The losses in the impeller are divided into shock losses at inlet, friction loss, and diffusion losses in the impeller. The volute loss consists of volute through flow velocity loss, friction loss, and diffusion losses in volute casing.

2.2.1 Hydraulic Losses in Impeller

Impeller hydraulic loss head is given as

$$h_{\text{imp.}} = h_{\text{imp\_ s}} + h_{\text{imp\_ f}} + h_{\text{imp\_ d}} \tag{2}$$

Impeller inlet shock loss is given as

$$h_{\text{imp\_ s}} = C_{\text{sh}} \frac{\left(w_1 - w_{1\text{q}}\right)^2}{2g} \tag{3}$$

Impeller friction loss is given as

$$h_{\text{imp\_ f}} = 4C_{\text{fr,imp}} \frac{l_{\text{b}}}{D_{\text{h}}} \frac{w_{\text{av}}^2}{2g} \tag{4}$$

Impeller diffusion loss is given as

$$h_{\text{imp\_ d}} = 0.25 \frac{w_1^2}{2g} \tag{5}$$

2.2.2 Hydraulic Losses in Volute

Volute hydraulic loss head is given as

$$h_{\text{vol.}} = c_{\text{fv}} \frac{v_{3\text{p}}^2}{2g} + C_{\text{d}} \frac{V_{3\text{d}}^2}{2g} + \frac{v_{3\text{f}}^2}{2g} \tag{6}$$

2.2.3 Theoretical Head

It is given by correlating Newton's second law and the conservation of momentum. Theoretical head for the centrifugal pump is given as

$$H_{\text{theo.}} = \frac{u_2}{g}(u_2 - v_{\text{f2}} \cot \beta_2) \tag{7}$$

2.2.4 Actual Head

Actual head is obtained by subtracting the hydraulic losses from the theoretical head and it is given as

$$H_{\text{act}} = \sigma H_{\text{theo.}} - h_{\text{imp.}} - h_{\text{vol.}} \tag{8}$$

3 Experimental Setup

Figure 1 shows the experimental setup used to estimate the head flow characteristics. The tests are carried out for the pump having an impeller and volute casing with dimensions listed in Table 1.

Fig. 1 Experimental setup

Table 1 Dimensions of centrifugal pump

Overall pump dimensions			
Rotational speed		2978 rpm	
Suction pipe diameter		152.4 mm	
Discharge pipe diameter		101.6 mm	
Impeller dimensions			
D_1	100 mm	b_1	20 mm
D_2	174 mm	b_2	20 mm
β_1	24°	t	5 mm
β_2	19°	D_h	36 mm
a_1	52 mm	D_s	50 mm
a_2	90 mm	L_s	70 mm
z	6	L_b	65 mm
Volute dimensions			
b_3	38 mm	D_4	90 mm
D_3	189 mm	α_v	62°

3.1 Uncertainty Analysis

The uncertainty analysis is carried out using the GUM method and estimated uncertainties in the head, flow, power, and efficiency as 0.53, 0.83, 0.57, and 1.33% and estimated repeatability and reproducibility in the head, flow, power, and efficiency as 0.74, 1.63, 2.56, and 0.82%. The estimated uncertainties are within the standard limits as per IS13538 [12].

4 Results and Discussion

Traditional method of calculating losses required two input parameters, i.e., K_s and K_f. These two parameters are different for each pump. The obtained K_s and K_f for different diameter impeller are listed in Table 2.

Figures 2 and 3 show the comparison of experimental analysis head flow curve obtained by traditional approach for both the impellers. From the figures, it can be observed that the experimental and calculated curves are not matching, except the

Table 2 Obtained data from experiments

174 mm impeller diameter	160 mm impeller diameter
$K_s = 25{,}117.868$	$K_s = 28{,}427.83$
$K_f = 17{,}882.6068$	$K_f = 17{,}618.5021$
HQ = 0 = 42.67 m	HQ = 0 = 31.52 m
HQ = QBEP = 30.39 m	HQ = QBEP = 21.83 m

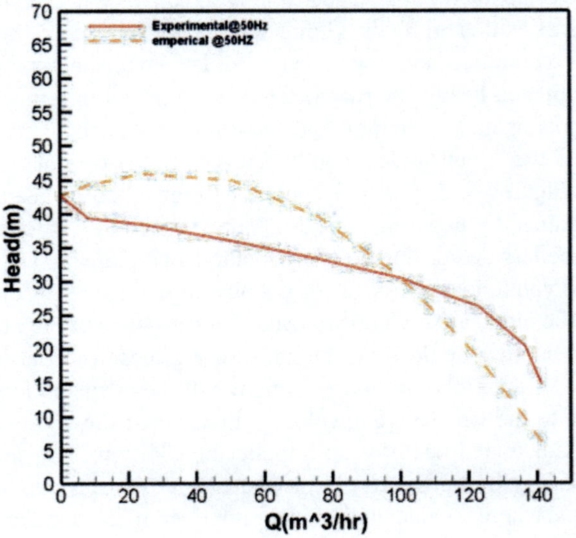

Fig. 2 Comparison of Head flow characteristics between experimental and traditional approach for the impeller of 174mm diameter

Fig. 3 Comparison of Head flow characteristics between experimental and traditional approach for the impeller of 160mm diameter

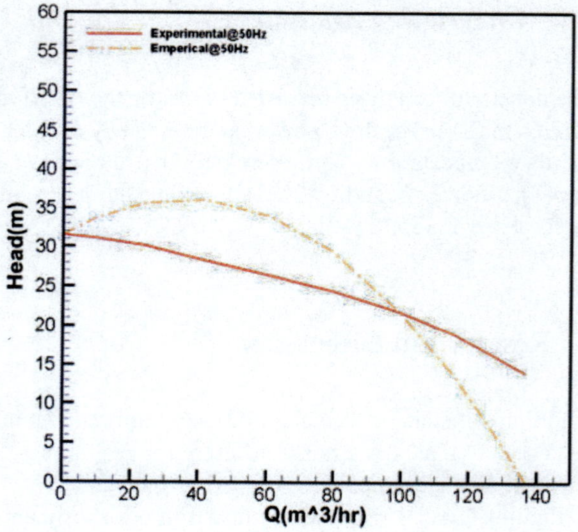

points at zero flow and flow at best efficiency point condition. The calculated data plot is deviating knowingly when it tried up to match with the measured plots.

From the graphs, one can predict that shock losses are more at $Q = 0$, and frictional losses are more at higher flow rates. These leads to unstable pump performance characteristics. So, this method is not suitable as there is a requirement of stable performance characteristics for any pump. Even though this method is not able to give good closer results to the measured values, it will be helpful to find the order of the head magnitude.

Gulich loss calculation approach is used to calculate various losses incurred in the centrifugal pump. The actual head can be predicted by subtracting the hydraulic losses estimated by the Gulich approach from Euler's head with slip consideration. The computed head developed is validated with the experimental measured head. The computed head flow parameters are expressed in terms of non-dimensional coefficients using C program, and these non-dimensional coefficients are also validated with the dimensionless coefficients acquired experimentally.

Figures 4, 5, 6, and 7 show the comparison of head flow characteristics curve obtained by experimental results and Gulich loss calculation methods for both the impellers. According to Gulich method, hydraulic losses are categorized into impeller and volute losses. For the lower flow regimes, volute losses (indicated by blue line in the plots) are high due to frictional resistance offered by the volute channel and it decelerates the flow. For higher flow regimes, impeller losses (indicated by yellow line in the plots) are more, as there will be diffusion losses and entry shock losses due to sudden change in velocity direction as shown in Figs. 4 and 6.

Calculated head for an impeller of 174 mm diameter is having a difference of 0.55% at best efficiency point and 14% at shut-off condition with the experimental head, whereas calculated head for impeller of 160 mm diameter is having a difference

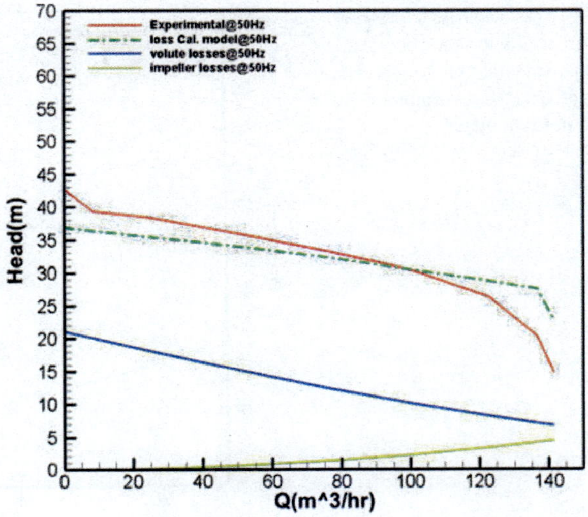

Fig. 4 Comparison of Head flow characteristics between experimental and gulich loss approach for the impeller of 174mm diameter

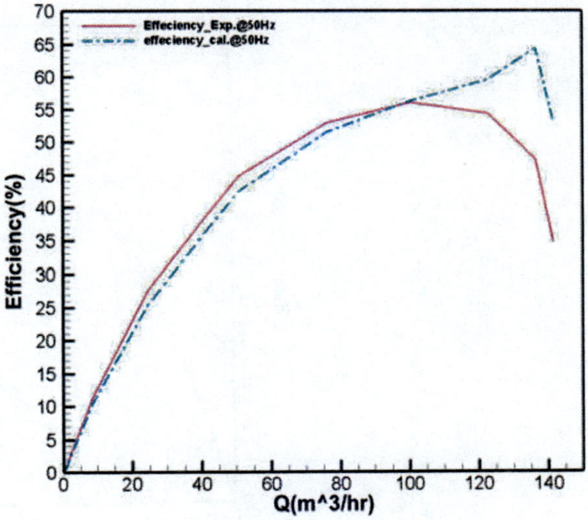

Fig. 5 Comparison of efficiencies for the impeller of 174 mm diameter

of 0.47% at best efficiency point and 4% at shut-off condition with the experimental head.

In similar manner, calculated efficiency for an impeller of 174 mm diameter is having a difference of 0.625% at best efficiency point with the experimental result, and calculated efficiency for an impeller of 160 mm diameter is having a difference of 0.549% at best efficiency point with the experimental result.

From the figures, it is evident that the calculated head and efficiency by Gulich loss equations are in reasonable agreement with the experimental results for both the impellers. It is clear that there is an over prediction of results for higher flow

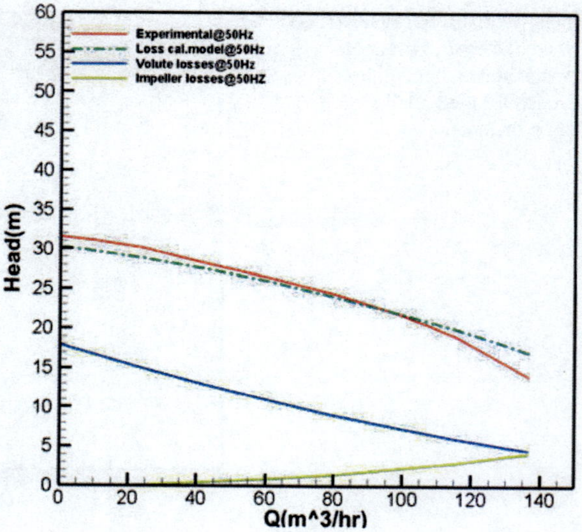

Fig. 6 Comparison of Head flow characteristics between experimental and gulich loss approach for the impeller of 160mm diameter

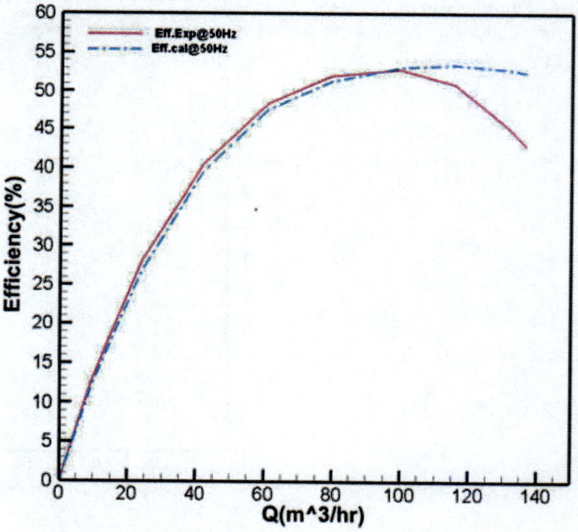

Fig. 7 Comparison of efficiencies for the impeller of 160 mm diameter

regimes than the best efficiency point by Gulich loss calculation method. The over prediction of the results happens because of assumptions like no swirl at inlet and not able to capture accurate results mathematically. Comparatively, 160 mm diameter impeller having lesser variation in the result than the 174 mm diameter impeller and the same effect has shown for the efficiencies. For the trimmed impeller, the head flow curve is shifted from the actual plot as shown in Fig. 6, and it is due to the lower circumferential speeds, lower flow velocities, and less power requirement. If all the pump input parameters are known, it seems likely to give good prediction of

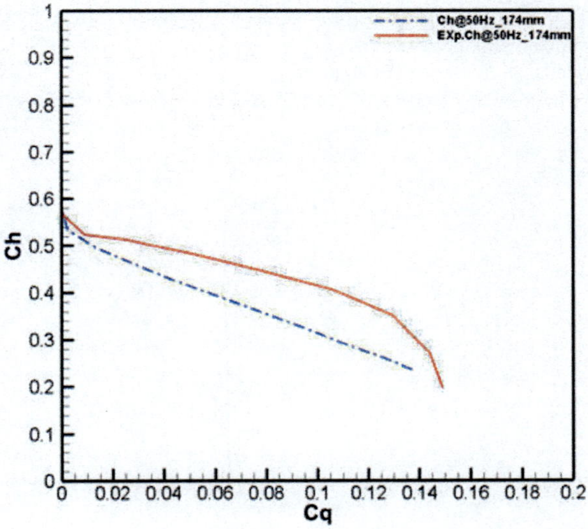

Fig. 8 Comparison of head flow coefficients for the impeller of 174 mm diameter

head flow characteristics with this method, and the pump characteristics are steady in nature even they are overpredicted at higher flow regime.

Gulich equations are modified by using mathematical modeling in order to estimate dimensionless head flow characteristics using C program. Non-dimensional results or numbers can be used to allow equations and data to be applied universally across the centrifugal pump projects as long as certain conditions are satisfied.

Figure 8 shows the comparison of experimental dimensionless head coefficient and predicted head coefficient by using C program. The maximum deviation of predicted head coefficient with experimental head coefficient is in the order of 10^{-2}, and the difference in the result at best efficiency point is 0.08.

Figure 9 shows the comparison of experiment and predicted efficiency by using C program. From the figures, it is evident that the predicted dimensionless head coefficient and efficiencies are in good agreement with the experimental measured data and the variation of calculate efficiency is 6.3% with the experimental efficiency at best efficiency point.

Figures 10 and 11 show the comparison of experimental obtained result and predicted result using C program for an impeller of 160 mm diameter.

The maximum deviation of predicted head coefficient with experimental head coefficient is in the order of 10^{-2}, the difference in the result at BEP is 0.056, and the variation of calculate efficiency is 6.2% with the experimental efficiency at BEP.

For low-flow rate coefficients, impeller head loss coefficients are low, and volute head losses are high and it will become vice versa as the flow increases, i.e., impeller volumetric head loss coefficient is high at low Cq and decreases till the flow reaches maximum. At the lower flow rate coefficients, there will be inlet circulation of the flow, this will add power to the impeller but the fluid does not get energized from the impeller, and it leads to an increase of shaft head coefficient. And it decreases the efficiency at lower flow rates.

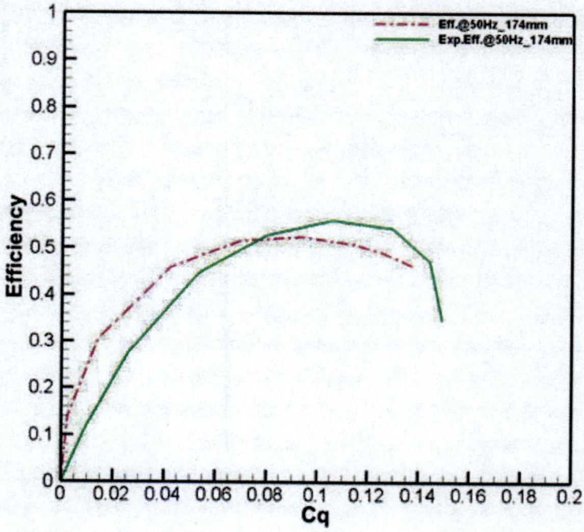

Fig. 9 Comparison of predicted efficiencies for the impeller of 174 mm diameter

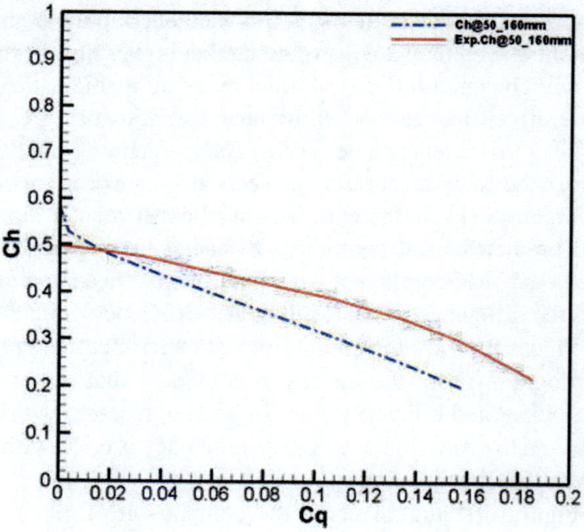

Fig. 10 Comparison of head flow coefficients for the impeller of 160 mm diameter

The present methodology using the C program to find out the head coefficient and efficiency is over predicting the efficiency value for the low-flow rates. The variation of head coefficient and efficiency is due to assumption of constant slip factor for the experimental head coefficient calculation. In the predicted program, slip factor is varied for each iteration, and this leads to some inaccuracy in the results.

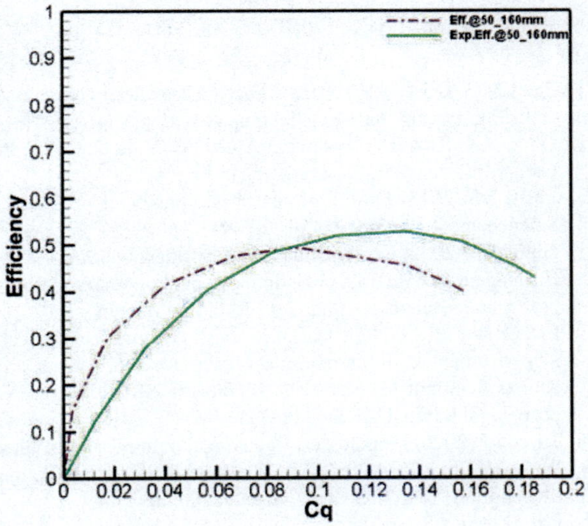

Fig. 11 Comparison of predicted efficiencies for the impeller of 160 mm diameter

5 Conclusions

In this study, hydraulic losses in the pump have been estimated using the Gulich loss calculation approach in order to calculate the actual head developed. Head and flow characteristics are represented in terms of dimensionless coefficients by mathematical modeling of Gulich loss equations using C program, and the predicted results are validated with the experimental results. The characteristics are difficult to estimate but the empirical method and loss calculation model with approximation have been tested and compared to the measured values.

From the present study, the following conclusions can be drawn

- Gulich loss model is giving acceptable steady head flow characteristics and valid agreement of these results are obtained with average variation of 14 and 6.22% for the 174 and 160 mm diameter impellers due to selected loss equations.
- Traditional method of calculating loss models will be helpful to identify the order of the head magnitude, but it is not reliable for predicting the head flow characteristics.
- The approximation with no swirl condition and the other friction factors may lead to overprediction of the results for the higher flow regime.
- All the dimensionless parameters are in reasonable agreement with small deviations due to variable slip factor and volumetric losses. And at lower flow coefficients, there will be inlet circulation of flow patterns, leads to decrease in efficiency of the pump.

References

1. Patel MG, Doshi AV (2013) Effect of impeller blade exit angle on the performance of the centrifugal pump. Int J Eng Technol Adv Eng 3(1)
2. Omar AK, Khaldi A, Ladoubhi A (2016) Prediction of centrifugal pump performance using energy loss analysis. Austr J Mechan Eng
3. Fosile SS (2013) Design of centrifugal pump for produced water. Norwegian University of Science and Technology
4. Gulich JF (2008) Centrifugal pumps. Springer, Berlin
5. EL-Naggar MA (2013) A one dimensional flow analysis for the prediction of centrifugal pump performance characteristics. Int J Rotating Machin
6. Elshestawy (2012) Numerical study of slip factor in centrifugal pump and study factors affecting its performance. In: International conference on mechanical egineering. Yangzhon, China
7. Kumar S, Singh M, Kumar S, Mohapatra SK (2017) An Optimum set of loss models for centrifugal pump. IJARSE J 06(12)
8 Tuzson J (2000) Flow losses. Centrifugal pump design. Wiley, New York, pp 189–199
9. Win HH, Htike TT, Myo MAA (2019) Performance prediction of centrifugal pump. IRE J 3(2)
10. Stepanoff AJ (1957) Centrifugal and axial flow pumps, 2nd edn. Wiley
11. Lazarkiewicz S, Troskolanski AT (1965) Impeller pumps. Pergamon Press
12. Modia A, Shah M, Baloni BD (2021) Uncertainty analysis of surface pump as per IS13538

Design Optimization of Splitter, Venturi Valve, and Charlotte Valve Using CFD

Sudarshan B. Ghotekar, Ashish Kinge, Ajay Ballewar, Amit Belvekar, and Yogesh Bhalerao

1 Introduction

The ongoing COVID-19 pandemic had caused surge in number of patients, that too in terms of waves of infections causing strain on healthcare system which has not adapted to deal for treating multiple patients in the limited supply of medical equipment such as ventilators for ARDS.

In the recent months of the pandemic, the world has seen slowdown in the manufacturing and industrial sector which made it hard for manufacturing the parts conventionally with mold building, machining, and injection molding for plastic products. While 3D printing has emerged as the revolutionary technology for prototyping, which can be also optimized for the mass production of the plastic printable products. Many private institutions carried out their independent projects involving the concepts of optimizing the use of ventilators in the period of shortage and urgency.

Ventilator as a whole setup is a costly equipment to manufacture as well as to operate, which reduces the supply for those in market because of economic reasons, while many researchers have been developing and improving the technologies which involves making of custom parts and using them directly. Parts like splitter, Venturi valve, and Charlotte valve have been developed and are getting registered for legal medical use.

Researchers focusing on the increasing the capacity of ventilator like Paladino [1] had conducted an experiment of ventilating four on a single full featured ventilator for 12 h, and a similar article by Chen [2] describes ventilating two simulation lungs using *Y* splitter, also demonstrating the change in compliance of one lung has no effect on other. Lewith [3] had explained concepts and mechanism of respiration like

S. B. Ghotekar (✉) · A. Kinge · A. Ballewar · A. Belvekar
School of Mechanical and Civil Engineering, MIT Academy of Engineering, Alandi, Pune, India
e-mail: sudarshanghotekar@gmail.com

Y. Bhalerao
Mechanical Engineering and Design, University of East Anglia, Norwich, UK

J. Banerjee et al. (eds.), *Recent Advances in Fluid Dynamics*, Lecture Notes in Mechanical Engineering, https://doi.org/10.1007/978-981-19-3379-0_22

tidal volume, compliance, peak inspiratory pressure (PIP), positive end expiratory pressure (PEEP). Zuckerberg [4] in an experiment demonstrated on an artificial lung aimed to provide inspiratory pressure with the help of low cost and effective custom designed device helps provide the response of the ventilator data for the study in this paper.

Researchers have also studied the application of Venturi air flow sensor to be used for human pulmonary ventilation measurement. Titheradge [5] has given a simple design and ease of maintenance; the Venturi method provides a potentially inexpensive, reliable alternative to common methods that are prone to damage and relatively expensive to purchase and maintain. Researchers from university of Leeds, Leeds Teaching Hospital NHS Trust (LTHT) and Bradford Teaching Hospitals NHS Foundation Trust have been working to repurpose existing ventilation systems. This Venturi valve works by having one end connected to the oxygen supply by a patient's bed and the other end connected to the tubing that goes to the person's oxygen mask.

Then, there is the Charlotte valve designed by an Italian team of engineers for converting a snorkeling mask into a mask for noninvasive type ventilator applications. They have printed it and also claim to have tested it on a patient. The valve is comparatively new to the world and is yet to be certified as medically safe but the idea to convert the snorkeling mask into ventilator mask is very much serviceable and potentially an aid for this emergency. That is why it was also included in the study.

In the present study, CFD analysis of different components of the ventilator system has been carried out aimed to determine the flow characteristics, later the designs had been modified to minimize the turbulent losses in the system. CFD analysis of the available design has been done which is available from many 3D printing institutions working on providing the modification idea for the ventilator due to the current pandemic. The splitter is used to split the flow from the ventilator to multiple patients. The Charlotte valve geometry was simulated for transient conditions. There were scopes for improvement in the valve geometry and have been successfully made and demonstrated in this paper.

2 Methodology

Computational fluid dynamics is an effective method for solving complex fluid problems with iterative solver, which can perform steady as well as transient simulations. The analysis is done with the help of ANSYS Fluent 19.1.

Table 1 Transient data of inspiration

		One person	Two person
	L/min	kg/s	kg/s
Time	Flow	m flow	m flow
0	0	0.000000	0.000000
0.1	75	0.001531	0.003063
0.2	64	0.001307	0.002613
0.3	49	0.001000	0.002001
0.4	35	0.000715	0.001429
0.5	22	0.000449	0.000898
0.6	14	0.000286	0.000572
0.7	9	0.000184	0.000368
0.8	6	0.000123	0.000245
0.9	6	0.000123	0.000245
1	5	0.000102	0.000204

2.1 *Splitter*

2.1.1 Design

The splitter has many designs; the model 1 is a planar splitter with inner and outer diameter 15 mm and 22 mm, respectively. The model 2 has 22 mm attachment. Model 1 and model 2 outlet can be fitted directly in ventilator piping, while the inlet geometry of model 2 is required to be attached with a flexible plastic coupling.

2.1.2 Boundary Conditions and Analysis Settings

The data had been obtained for the pressure and mass flow rate for the ventilator generated capacity, which is obtained for PIP, PEEP applied by ventilator Paladino [1], Chen [2], Zuckerberg [4]. The data is extracted from the ventilator generated graphs and the inspiratory time is set to as 1 s. The compliance condition of 20 mL/cm H_2O with PIP as 20 cm H_2O and PEEP as 5 cm H_2O with surface roughness of 0.4 mm. Table 1 shows the transient flow conditions for two-way splitter. The CFD model used here if *k* omega SST which is ideal for flow across closed geometry having boundary interaction.

2.1.3 Modified Geometry

Based on the analysis of streamlines, the geometry is modified to get streamlined with the flow, to make flow less turbulent. For model 1, the edge where the flow is getting divide is made longer and blunter at corner, to divide the flow effectively reducing the backflows across adjacent cells. Modified geometry can be seen in Table 2. For the model 2, instead of sharp triangular corners as considered in original design, a globular geometry has been tried, also the connecting edges have been highly filleted to ensure smooth flow. The globular geometry allows the flow to circulate freely without stagnating near the corners.

Table 2 Ventilator splitter CFD results

Model	Streamlines	Turbulent kinetic energy contour
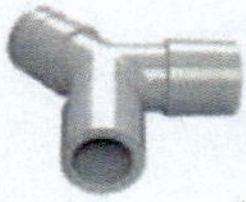 Model 1	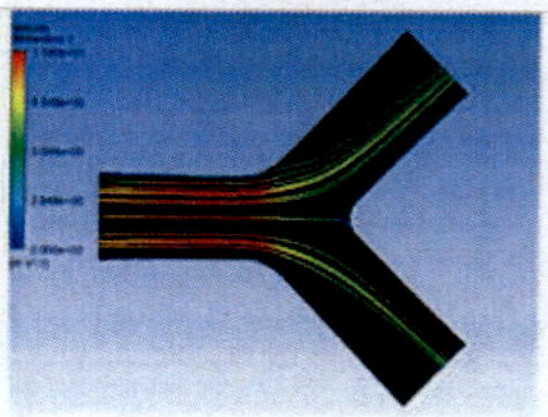 **Fig. 1 a** Model 1 (original)	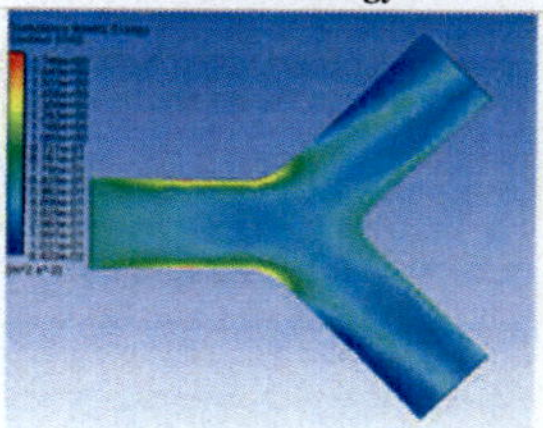 **Fig. 1 b** Model 1 (original)
 Model 1 Modified	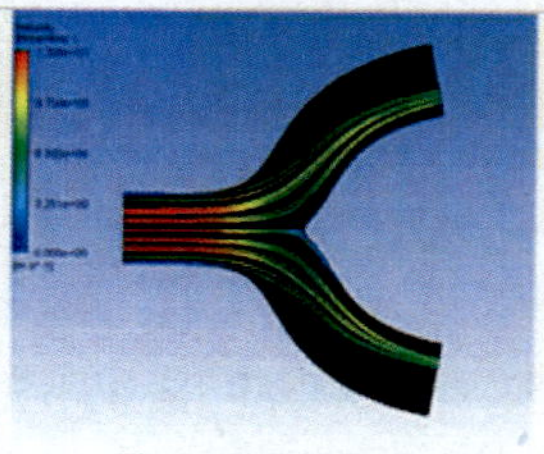 **Fig. 1 c** Model 1 (modified)	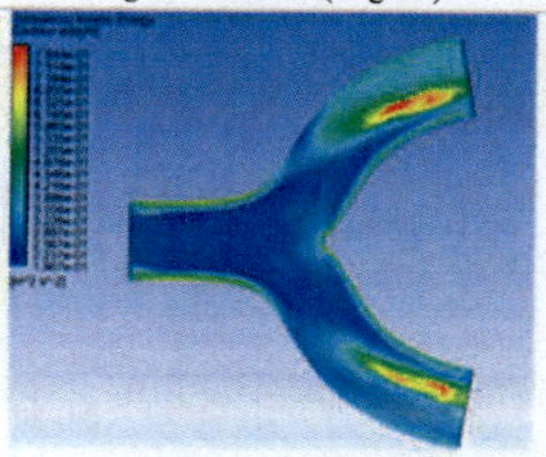 **Fig. 1 d** Model 1 (modified)
 Model 2	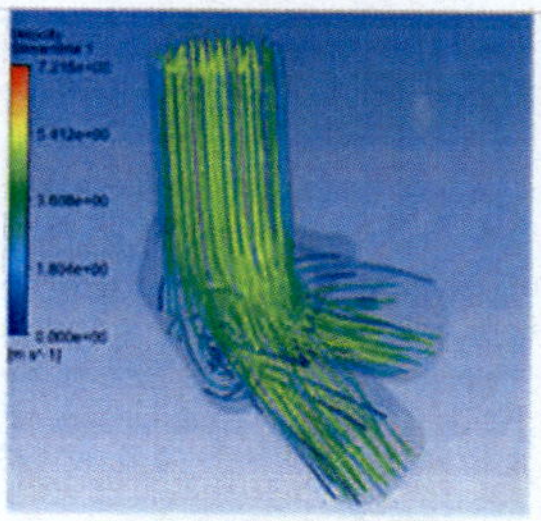 **Fig. 1 e** Model 2 (original)	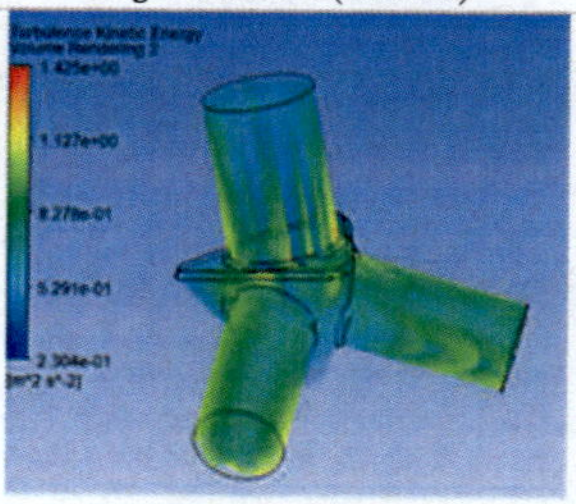 **Fig. 1 f** Model 2 (original)
 Model 2 modified	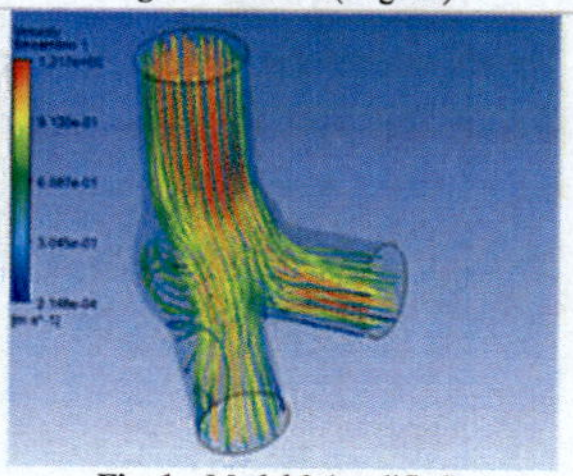 **Fig. 1 g** Model 2 (modified)	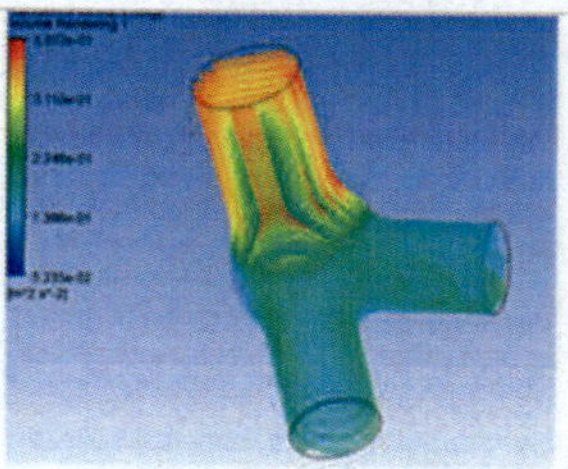 **Fig. 1 h** Model 2 (modified)

2.2 *Venturi Valve*

2.2.1 Design

There are three main types of Venturi valve designs available depending on the trajectory of the nozzle. The first type is conventional one with cone like structure having one end connected to the oxygen supply and other end connected to the corrugated tube. Near the inlet, there are ports to allow atmospheric air to enter in order to dilute the 100% oxygen supply so that patient does not get neat oxygen. Other types developed include a traditional Venturi like structure having converging and diverging sections. This type has two oxygen inlets, one from the side and one in the pathway. Atmospheric inlet is around the main inlet. The same model has another variant having only one oxygen inlet from the side and atmospheric inlet in the pathway. Italian engineers have devised an innovative design like the first type with an additional bleed valve in the design.

2.2.2 Boundary Conditions

The Venturi valve has been simulated for obtaining the suction at the inlet which is open to atmosphere and the specific quantity of oxygen has been supplied through the small tube and nozzle in the throat section of the Venturi valve geometry. The air inlet and outlet were kept as open to atmosphere without and fluid velocity. The oxygen inlet is given from the small tube, delivering 2 and 4 lpm of oxygen, for which, the models have been simulated differently. The surface roughness tolerance is kept as 0.4 mm for approximately reflecting 3D printed surface property.

2.3 *Charlotte Valve*

The Charlotte valve geometry obtained from the Isinnova team was first analyzed for noninvasive same transient conditions of a normal breathing mass as that used for the splitter. It showed turbulence due to two sharp 90° turns as shown in Fig. 4a then using CATIA the design is modified to make the flow streamlined.

3 Results

Splitter In the results, the streamline and turbulent kinetic energy of fluid has been analyzed which helps to determine the location of possible eddies formation in the flow. The effect of the direction change on the flow can be seen in comparison with both original and updated designs.

Table 3 Venturi valve CFD results for 2 lpm

Model	Streamlines	Volume fraction and velocity vectors
 Venturi valve 1	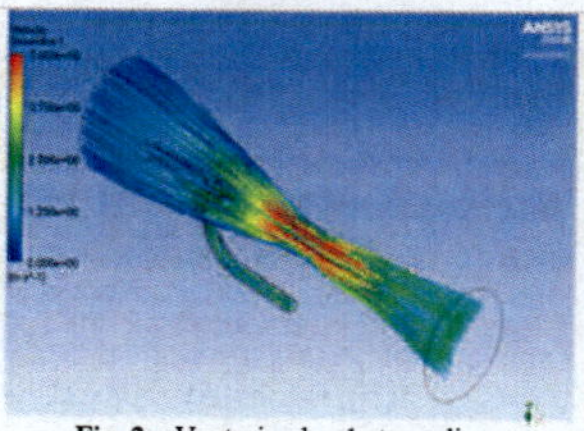 **Fig. 2 a** Venturi valve 1 streamlines	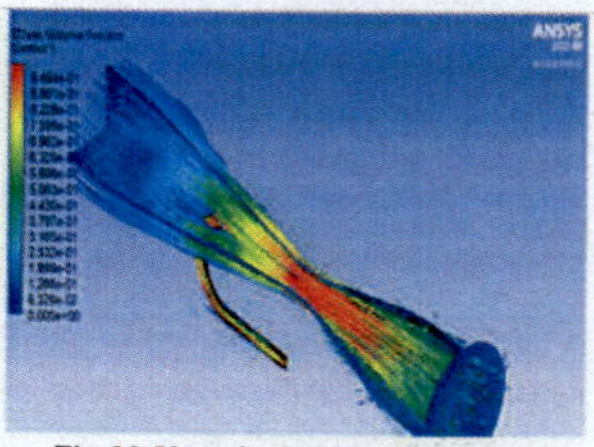 **Fig. 2 b** Venturi valve 1 volume fraction
 Venturi valve 2	 **Fig. 2 c** Venturi valve 2 streamlines	 **Fig. 2 d** Venturi valve 2 volume fraction
 Venturi valve 3	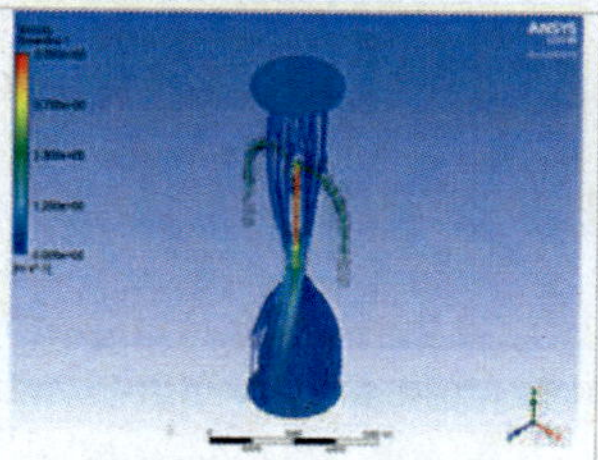 **Fig. 2 e** Venturi valve 3 streamlines	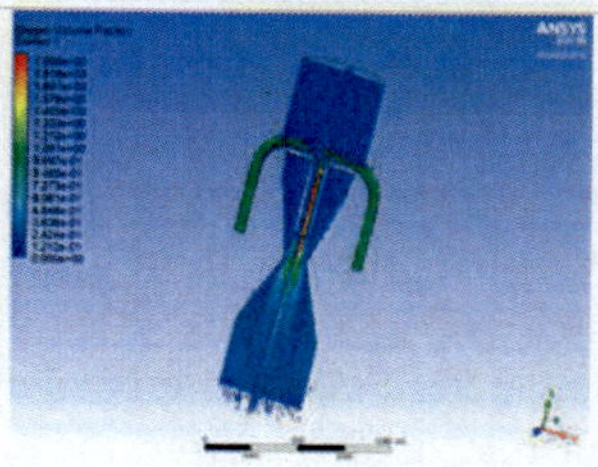 **Fig. 2 f** Venturi valve 3 volume fraction

Venturi valve The CFD results have been visualized in Tables 3 and 4 showing streamlines and volume fraction. Table 5 summarizes the results obtained for original and modified design while changing oxygen flow rates.

3.1 *Charlotte Valve*

The Charlotte valve design obtained from Isinnova engineers as shown in Table 6 (model 1) has some imperfections which led to these iterations. The part had some inward facing dents on the inside; the particular reason for this circumstance is merging of two curve surfaces. In the second model, i.e., Table 6 (model 2) those dents have been taken care of along with making the flow streamlined. The streamlined flow is desirable so that moist air does not condense in the valve. Hence, as shown in Fig. 4g, the valve is flowing full and the outlet velocity is also higher than others along with least turbulence.

Table 4 Venturi valve CFD results for 4 lpm

Model	Streamline flow	Volume fraction
 Venturi valve 1	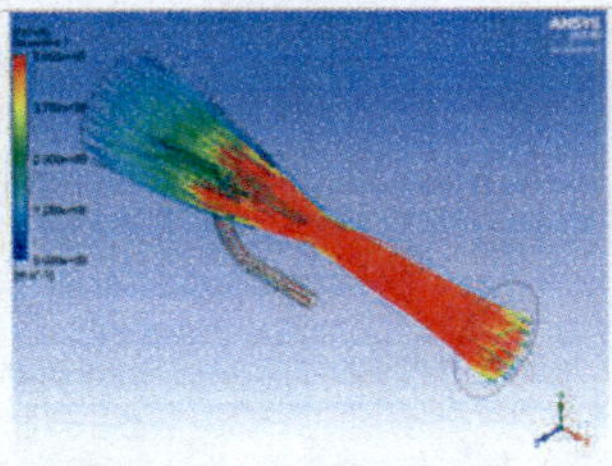 **Fig. 3 a** Venturi valve 1 streamlines	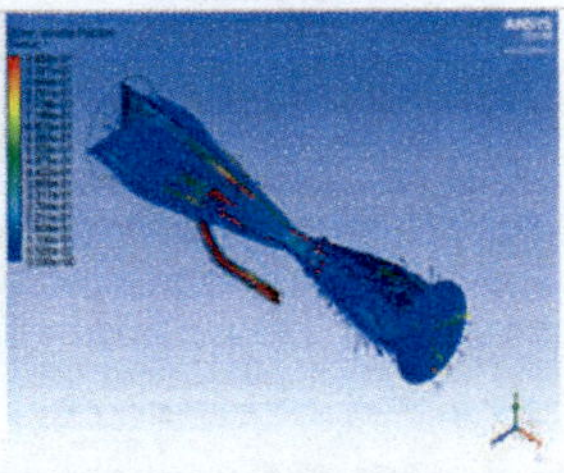 **Fig. 3 b** Venturi valve 1 volume fraction
 Venturi valve 2	 **Fig. 3 c** Venturi valve 2 streamlines	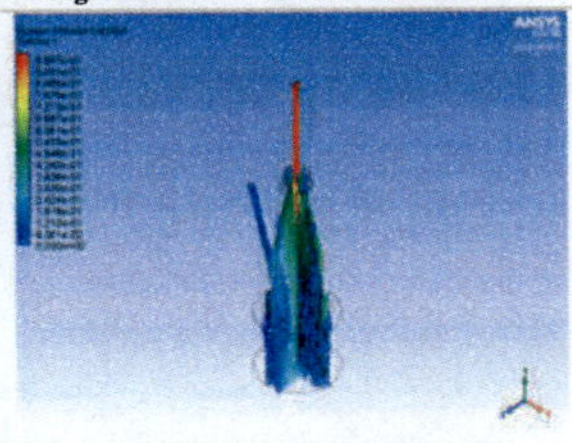 **Fig. 3 d** Venturi valve 2 volume fraction

Table 5 Venturi valve results of oxygen percentage and outlet mass flow

Model	2 lpm		4 lpm	
	Oxygen %	Mass flow outlet (lpm)	Oxygen %	Mass flow outlet (lpm)
Venturi valve 1	24	55	23	56
Venturi valve 2	40	3.58	36	7.21
Venturi valve 3	33	16	–	–

4 Conclusion

Finally, we have come up with a 3D printable model of all the three valves which can replace the already existing valves without any difficulty and are more efficient. They also can easily be modified for custom requirements.

For the splitter, two models have been modified and optimized to their respective better versions. Turbulence zone formed between the high-velocity streamlined flow and the low-velocity boundary layer flow near the outer wall has been dealt with using curvature implementation. For model 2, the turbulent kinetic energy has been found to be less and flow to be more organized than model 1. The acceptance of globular geometry instead of pyramidal has minimized the turbulence further.

The Venturi valves in operation have particular combination of flow rate (lpm) and Fi02 (in %). The output flow rate depends on the diameter of the nozzle, suction pressure at vena contracta and length of the pipe. By varying the above parameters, we are able to obtain different combinations of flow rate and Fi02. During the analysis,

Table 6 Charlotte valve CFD results

Sr. No.	Model	Streamlines	Velocity contour at exit
1	 Model 1	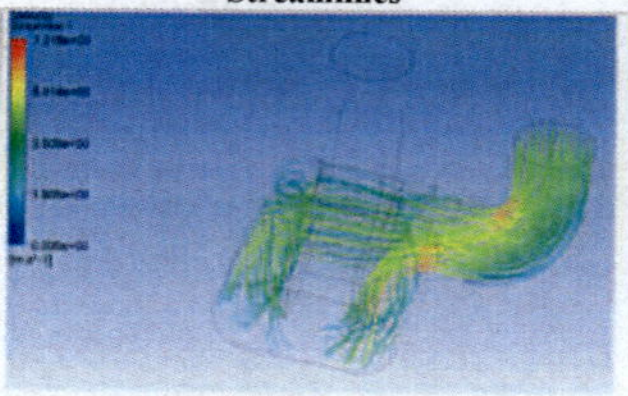 **Fig. 4 a** Model 1 streamlines	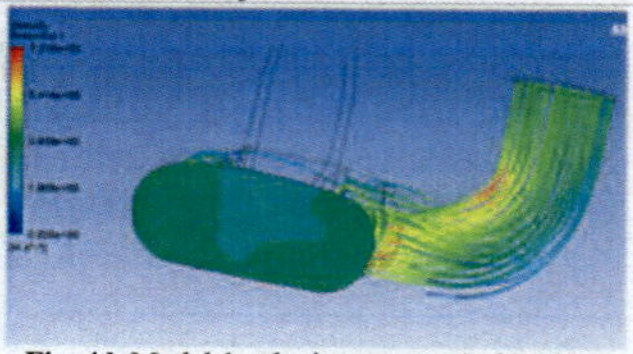 **Fig. 4 b** Model 1 velocity contour (exit) with streamlines from PEEP connector
2	 Model 2	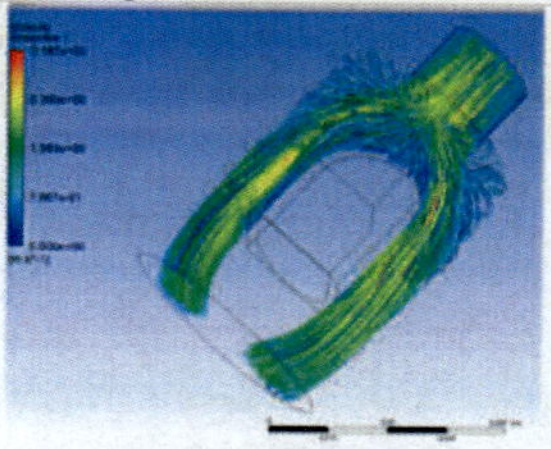 **Fig. 4 c** Model 2 streamlines	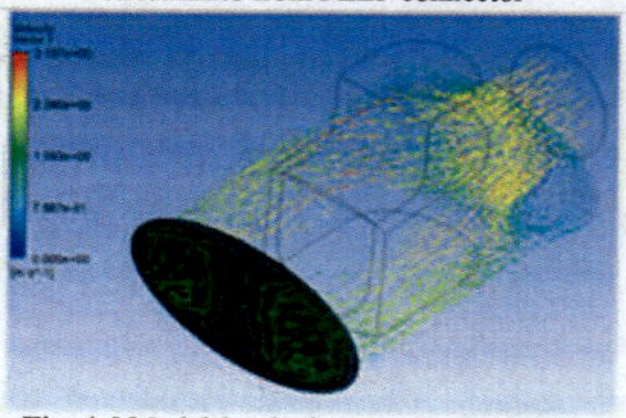 **Fig. 4 d** Model 2 velocity contour (exit) with vectors from inlet
3	 **Model 3**	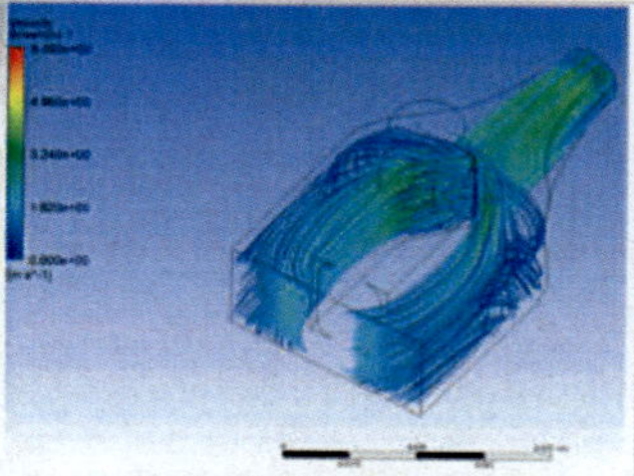 **Fig. 4 e** Model 3 streamlines	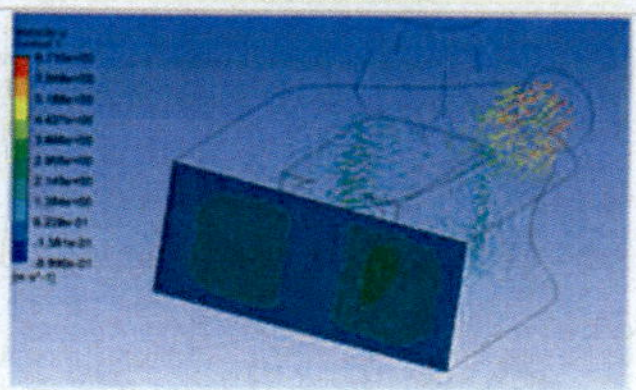 **Fig. 4 f** Model 3 velocity contour (exit) with vectors from inlet
4.	 **Model 4**	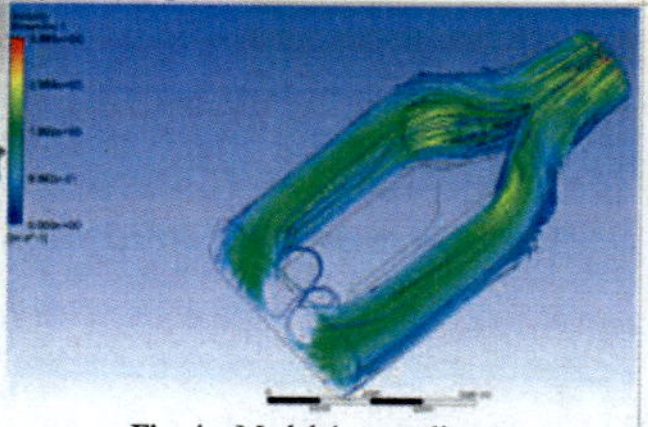 **Fig. 4 g** Model 4 streamlines	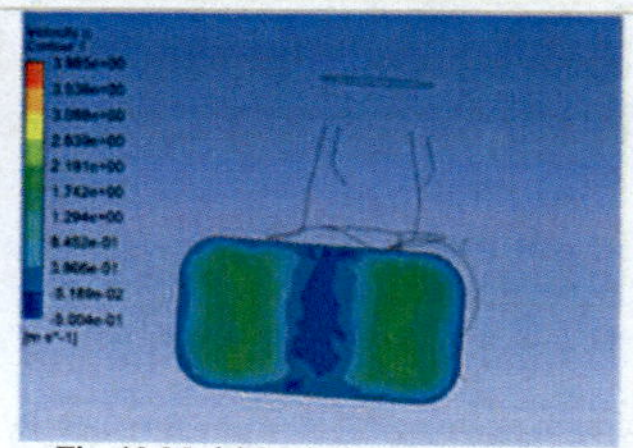 **Fig. 4 h** Model 4 velocity contour (exit)

we have also discovered the dependence of various parameters on output. Finally, we have successfully achieved a 3D printable streamline designs for Venturi valve.

The Charlotte valve needs to be streamlined which is successfully achieved at the end of the iterations and losses have been reduced by 3–4%.

3D printing technology has already been integrated in the medical domain but low-cost desktop 3D printers have fewer quality approvals. So, this research is preeminent for hospitals that are yet to implement 3D printing facilities.

References

1. Paladino L, Silverberg M, Charchaflieh JG, Eason JK, Wright BJ, Palamidessi N, Arquilla B, Sinert R, Manoach S (2008) Increasing ventilator surge capacity in disasters: ventilation of four adult-human-sized sheep on a single ventilator with a modified circuit. Resuscitation
2. Chen GH, Hellman S, Irie T, Downey RJ, Fischer GW (2020) Regulating inspiratory pressure to individualise tidal volumes in a simulated two-patient, one-ventilator system. Br J Anaesthesia
3. Lewith H, Pandit JJ (2020) Lung ventilation and the physiology of breathing. Surgery
4. Zuckerberg J, Shaik M, Nelin TD, Widmeier K, Kilbaugh T (2020) A lung for all: novel mechanical ventilator for emergency and low-resource settings. Life Sci 257:118113. ISSN 0024-3205
5. Titheradge PJ, Robergs R (2018) Venturi method concept for human pulmonary ventilation. Crimson Publishers. ISSN 2576-8816

Analysis of Wave Breaking in Pipe Flow Using Image Processing Technique

Digpriya Chaudhary, Sunny Saini, and Jyotirmay Banerjee

Nomenclature

$A_{\text{Pipe}}, A_{\text{f}}$	Cross-sectional area of the pipe, cross-sectional area of fluid flow
ρ_{f}	Dynamic viscosity of the fluid
$\dot{m}_{\text{f}}$	Mass flow rate of fluid
Re_{SL}	Superficial Reynolds number of liquid
Re_{SG}	Superficial Reynolds number of gas
$U_{\text{SF}}, U_{\text{bf}}, U_{\text{AVG}}, U_{\text{Wave}}$	Superficial velocity of the fluid, bulk velocity of the fluid, average velocity of fluids, wave speed
Crest $\|_x$	Location of wave crest at point x
t_n	Time at point n ($n = 1, 2, 3, \ldots$)

1 Introduction

Gas–liquid two-phase flow has attention in petrochemical and nuclear industries where simultaneously both the fluids are transported several kilometers through horizontal or near horizontal pipelines. Since the two fluids flow in a closed conduit, gas is compressible, while the liquid is incompressible and influenced due to pipe orientation, and alteration in inlet flow rate leads to complex flow hydrodynamics. These complex hydrodynamics are two-phase flow regimes, demarcated as stratified flow,

D. Chaudhary (✉) · S. Saini · J. Banerjee
Department of Mechanical Engineering, SVNIT, Surat, India
e-mail: p19tm018@med.svnit.ac.in

J. Banerjee
e-mail: jbaner@med.svnit.ac.in

J. Banerjee et al. (eds.), *Recent Advances in Fluid Dynamics*, Lecture Notes in Mechanical Engineering, https://doi.org/10.1007/978-981-19-3379-0_23

plug flow, slug flow, and dispersed bubbly flow [6]. Several researchers [1, 2, 10, 14–17] suggested that intermittent flow associated with vigorous aeration called slug which can cause pressure fluctuation that further leads to pipe failure due to erosion, and shear stress in the pipe flow leads to pipe wall thinning. Moreover, intermittent flow is a discontinuous periodic flow that increases the pipeline's vibration and cyclic loading (fatigue strength) and results in pipe failure. Hence, to avoid such catastrophic failure due to slug flow, it is required to analyze the evolution of slug flow. Slug flow is generally evolved from either plug flow or wavy flow. Associated to slug development, a study is [11] reported the mechanism of slug development and strong aeration inside the slug body due to plunging wave breaking.

Wave breaking is a phenomenon observed in oceans or seashores and pipe flows. In nature, the wave break happens either at the seashore due to the slope of the beach or at the deep seawater, shown in Fig. 1a [13]. Seawater wave breaking has different types, i.e., spilling, plunging, collapsing, and surging [4], which many researchers have discussed. In two-phase pipe flows, the wave-breaking phenomenon is observed, which is different from seawater wave breaking due to the influence of pipe curvature [11]. In the case of two-phase pipe flow, the wave-breaking phenomenon causes slug formation [8, 11, 12]. The wave energy transforms into turbulent kinetic energy during wave breaking and induces multiple waves to break further. Sometimes, it leads to a cascade of wave breaks which finally culminate into the formation of slug associated with the strong aeration [11] in the pipe flow. The study of wave breaking can also throw light on the type of slug formed in a pipeline. For the prediction of slug, it is essential to study the forces involved in the wave-breaking phenomenon.

(a)

(b)

Fig. 1 **a** Wave breaking in deep seawater waves [breaking wave] and **b** wave breaking in pipe of 50 mm diameter

Most of the wave breaking are observed during experiments, in large amplitude roll waves at Re_{SG} of air greater than 2000. Wave breaking refers to the tipping over (overturning of wave crest) of wave peaks. Wave breaks are widespread, and it is an outcome of the K-H instability mechanism [7] and the curvature effect of pipe; a captured image of a wave break observed in our experiment is shown in Fig. 1b. In literature, the microscale wave breaking in pipe flow has been discussed by [13, 18], which is a weak form of wave breaking in pipe flow without air entrainment. On the other hand, [11] reported the strong form of wave breaking (plunging), which causes an augmentation of aeration in the slug.

In this work, interfacial analysis of wave breaking is done using the image processing technique. Wave breaking is a nonlinear and turbulent flow phenomenon, so the image processing technique (a non-intrusive technique) gives an edge without disturbing the flow. Using an image processing algorithm, the signal of the wave breaking is obtained to understand the mechanism and prediction of wave breaking.

2 Experimental Setup

The experiment for two-phase flow in a horizontal pipe is performed in Advanced fluid dynamics Laboratory SVNIT Surat. The two-phase experimental flow facility used for the present work has two significant parts, a two-phase flow test rig (TPFTR) and a high-speed photography system (HSPS).

The two-phase flow test facility is shown in Fig. 2 [15]. The test rig (TPFTR) contains an airflow loop, a water flow loop, a flow visualization section, supervisory control and data acquisition) (SCADA), etc. The analysis presented in this work is based on experiments carried out on 50 ± 0.2 mm diameter and 14-m-long Perspex transparent glass pipe. The mass flow rate of air and water is controlled by Coriolis mass flow meter. The working fluids air and water flow inside the horizontal pipe with a range of mass flow rate combinations is controlled by the SCADA system during the experiment.

The flow visualization is done using the HSPS, which consists of two main parts Photron-made FASTCAM camera with a tripod and Photron FASTCAM Viewer (PFV) software shown in Fig. 3. Other equipment like a portable whiteboard, light, and scale is also used. The Photron camera is mounted on the tripod stand, facilitated with a spirit-level arrangement for its orientation and height adjustment from ground level in the desired direction. The camera is controlled by PFV software, which contains the frame per second and resolution. The camera is set in front of the test section to capture the different flow regimes of the two-phase flow. The scale is placed behind the test section for the geometrical estimation of the two-phase flow characteristics of captured images.

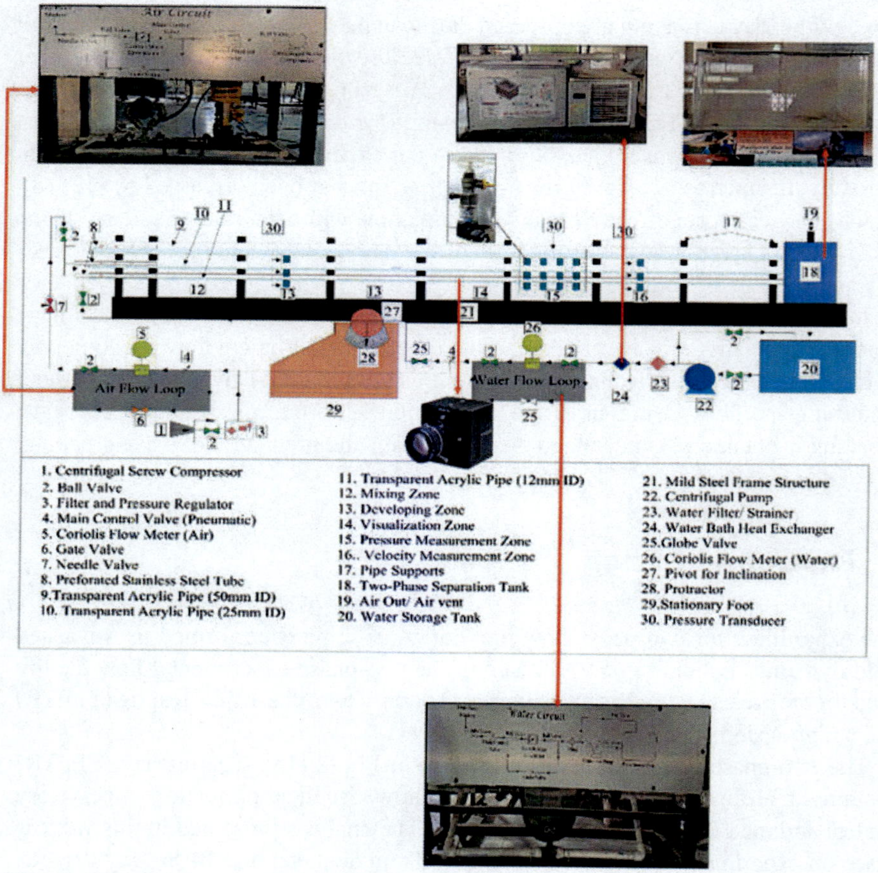

Fig. 2 Schematic view of the experimental test facility

3 Image Processing Technique

Image processing is a technique in which algorithms are performed on an image to extract information [5]. In the first step, the captured image (shown in Fig. 4) is converted into a binary (shown in Fig. 5) or grayscale image to find the edge of the air–water interface of the two-phase flow. Then, the next step is to find out the actual coordinates of the detected edge and get wave properties, such as wave frequency, amplitude, wavelength, and wave speed [9]. Using the coordinate of the interface, the signal of a different wave can also be obtained. The present work only employs image processing techniques to conduct a two-phase flow interfacial analysis.

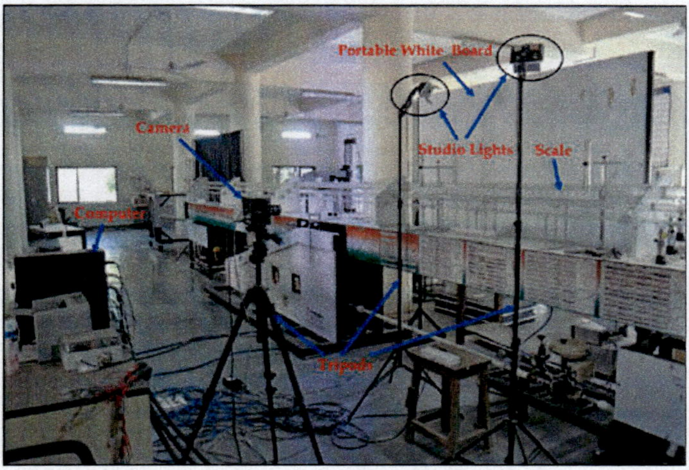

Fig. 3 Experimental setup for high-speed photography system

Fig. 4 Raw image of wave break (at $Re_{SG} = 7000$ and $Re_{SL} = 8000$)

Fig. 5 Processed binary image of the captured wave break

4 Results and Discussion

4.1 Mechanism of Wave Breaking

To explain the wave-breaking phenomenon, we take the help of Fig. 6, in which wave breaking of roll waves has been shown. It is observed that due to the large amplitude of the roll waves, the area inside the pipe behaves like a venturi tube. This means that when the air flows through this area, it will accelerate as it moves toward the wave's peak. Such an air acceleration further reduces the static pressure and sucks

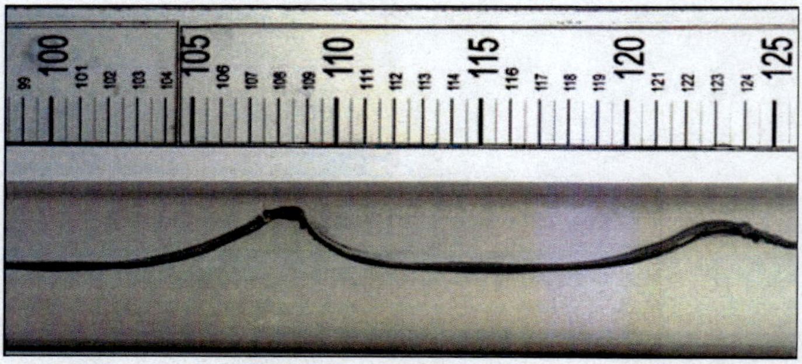

Fig. 6 Roll wave before wave breaking, here the cross-sectional area near the crest and trough is marked with arrows ($\mathrm{Re_{SG}} = 4000$ and $\mathrm{Re_{SL}} = 9000$)

the interface even further. An increase in air velocity also increases shear stresses at the wave's peak. This increased shear stress and increase in the dynamic pressure on the back of the wave create a net moment in the clockwise direction. When this clockwise moment becomes stronger than the capillary forces, the wave tips over, and a wave break is witnessed.

Further, the formation of venturi acts as an obstruction to the flow of air, and consequently, waves ahead of this wavelet experience less airflow. However, the flow is restored after wave breakage with a momentarily high-pressure fluctuation. These pressure and flow velocity fluctuations initiate wave breakage in other wavelets and cause a cascade.

To verify our hypothesis, we calculate different velocities before the phenomenon of wave breaking in Fig. 6 (the captured image just before wave break), which helps in quantitative analysis of the contribution of the Beroulli effect and interfacial shear stress in wave break. Our results (velocity calculation using image processing algorithm) indicate. Superficial velocity of air (U_{SG}) is 1.27 m/s [3], and superficial velocity of water (U_{SL}) is 0.14 m/s. These values are obtained from the inlet mass flow rate values, fixed for an experiment. The mean velocity of the two-phase flow was found to be 0.71 m/s. Interestingly, the wave speed was also 0.71 m/s. The formula used in the calculation of the above velocities is given below.

$$U_{\mathrm{SF}} = \frac{\dot{m}_{\mathrm{f}}}{\rho_{\mathrm{f}} \cdot A_{\mathrm{Pipe}}}$$

$$U_{\mathrm{Avg}} = \frac{U_{\mathrm{SL}} + U_{\mathrm{SG}}}{2}$$

$$U_{\mathrm{Wave}} = \frac{\mathrm{Crest}|_{x2} - \mathrm{Crest}|_{x1}}{t_2 - t_1},$$

In the above expressions, liquid height can be obtained at any two points of the pipe with an image processing algorithm, which further helps in calculating A_{f} (cross-sectional area of fluid flow). If we calculate the bulk velocity of air at two stations in the venturi, indicated by arrows in Fig. 6, we observe that near the trough (at point

1 in Fig. 6), the velocity is around 2 m/s, and near the crest (throat), air velocity is around 6.2 m/s. The bulk velocity of air has been calculated using the formula below,

$$U_{\mathrm{bf}} = \frac{\dot{m}_{\mathrm{f}}}{\rho_{\mathrm{f}} \cdot A_{\mathrm{f}}}$$

The calculated velocity of air near the peak is approximately three times greater than the velocity at the trough. It implies that the shear stresses are almost three times stronger at the height and lead to wave breakage. Here, the increased velocity near the wave crest increases the dynamic pressure of air due to the Bernoulli effect, which pushes the wave crest in the flow direction. This phenomenon also causes the wave to overturn and break.

4.2 *Classification of Wave Breaking*

Analysis of image processing results and flow visualization of wavy flow regime shows that the wave breaking occurs at the different liquid levels in the influence of several mass flow rates of air and water, which causes various kinds of slug formation. Based on the above argument, the wave breaking in the horizontal pipe can be divided into two types discussed below.

4.2.1 Strong Wave Breaking

At a high air superficial Reynolds number ($\mathrm{Re}_{\mathrm{SG}}$) ranges from 5000 to 10,000, the wave breaks at a low liquid level, which means that the wave crest does not touch the pipe's upper wall when it breaks. This kind of wave break causes highly aerated slug and pseudo-slugs. Due to the high flow rate of air, the wave breaks at a low liquid level, and the highly aerated slug forms. In Fig. 7, the captured image is shown, where wave breaking occurs without being in contact with the upper wall of the pipe, and it further forms a strong aerated slug.

In Fig. 8, the obtained signal of liquid height before and after wave breaking (Fig. 7) is shown in which the crest height at the time of the wave break is about

Fig. 7 Wave break captured in $\mathrm{Re}_{\mathrm{SG}} = 7000$ and $\mathrm{Re}_{\mathrm{SL}} = 8000$

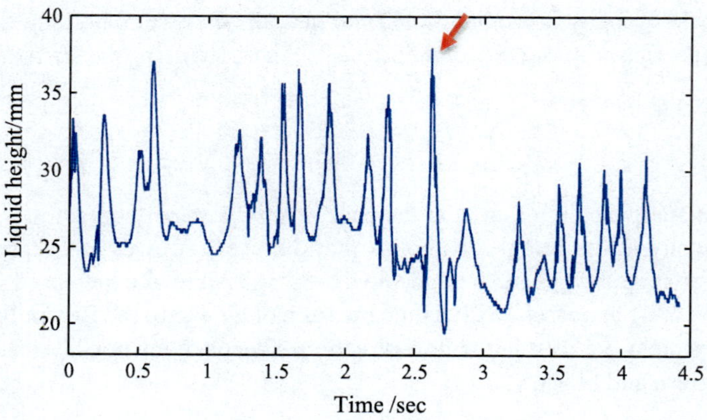

Fig. 8 Variation of interfacial height with time at fixed a point ($Re_{SG} = 7000$ and $Re_{SL} = 8000$)

38 mm, causing slug formation with strong air entrainment. The wave crest during wave break does not appear smooth in the plot as it is a turbulent and nonlinear process with air entrainment [13]. The red arrow shows the wave crest, which breaks, and after the wave break, the height of the wave drops suddenly due to slug formation (Fig. 8).

The captured image of another wave-breaking phenomenon shown in Fig. 9, which falls under similar category of wave-breaking (strong wave breaking), where the wave breaks at a lower liquid level (32 mm) than in Fig. 7. The red arrows in Fig. 10 show the wave peaks, which break simultaneously due to the high flow rate of air. In Fig. 10, the plot of liquid height with time is calculated by the developed image processing algorithm. The wave crest is near 32 mm of pipe diameter when the wave breaks, and it causes a pseudo-slug, not a slug, due to low liquid level and high gas flow rate. And that is the reason why the liquid height of the next wave crest after these wave

Fig. 9 Wave break captured in $Re_{SG} = 9000$ and $Re_{SL} = 7000$

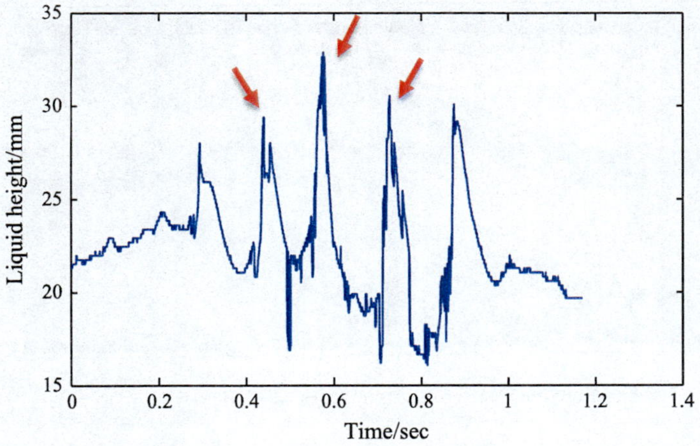

Fig. 10 Variation of interfacial height with time at fixed a point ($Re_{SG} = 9000$ and $Re_{SL} = 7000$)

breaks (shown in Fig. 10) is not dropped significantly. The plot of the wave crest is not smooth as it breaks at high Re_{SG}, making the flow turbulent.

4.2.2 Weak Wave Breaking

At a low air superficial Reynolds number (Re_{SG} from 1000 to 4000), the wave breaks at a high liquid level, and in this case, the wave breaks when wave comes in contact with the pipe's upper wall. Thus, more stable-slug (less aerated and long slug) formation due to surface tension, which prevents the air entrainment in the liquid phase, is obtained, which signifies the weak form of the wave breaking. The slug formation due to weak wave breaking has lesser velocity than that of strong wave breaking.

The weak wave-breaking phenomenon discussed in this article is not microscale wave breaking [18], as the wave crest at the time of weak wave breaking also overturns and forms a slug.

In Fig. 11, the captured image of weak wave breaking is shown; in this, during the wave breaking phenomenon, the wave crest overturns and then breaks while touching the upper wall of the pipe. In the developed plot (shown in Fig. 12), we can see that the wave peaks are about 44 mm of pipe diameter, and the wave breaks near 42 mm of pipe diameter are shown by the last peak in the plot and indicated by the red arrow in Fig. 12. The wave break plot is relatively smooth here as there is less air entrainment due to low air Re_{SG}.

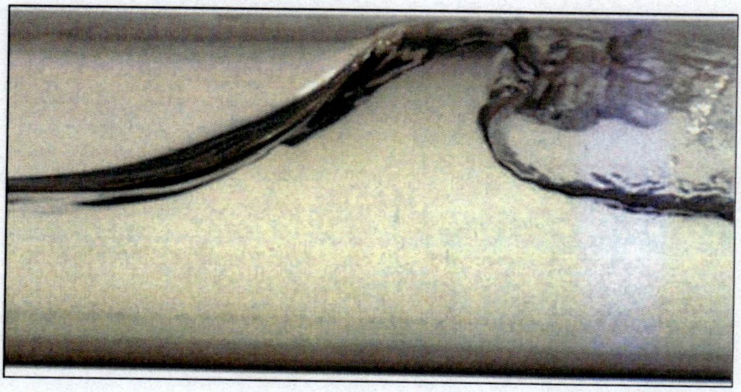

Fig. 11 Wave break captured in $Re_{SG} = 4000$ and $Re_{SL} = 9000$

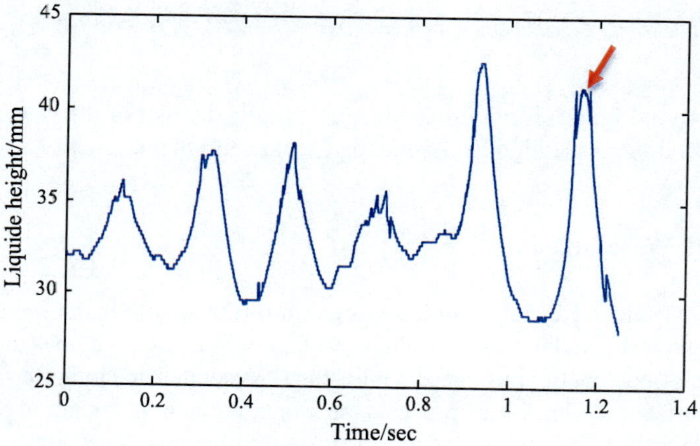

Fig. 12 Variation of interfacial height with time at fixed a point ($Re_{SG} = 4000$ and $Re_{SL} = 9000$)

4.3 Prediction of Wave Breaking

Large amplitude roll waves have a high tendency of wave breaking. Thus, we can infer the large amplitude roll wave signal as an indication of wave breaking. The large amplitude roll wave signal is shown in Figs. 13 and 14. The roll waves have a steeper front and gradually sloping back, and they are found at air superficial Reynolds numbers greater than 2000 (Re_{SG}). The signal of large amplitude roll waves has a sharp crest and broad trough, and they are very regular.

The signal in Figs. 13 and 14 is shown for large amplitude roll waves at different superficial Reynolds numbers, and these kinds of signals can be used to predict the wave-breaking phenomenon in two-phase pipe flow.

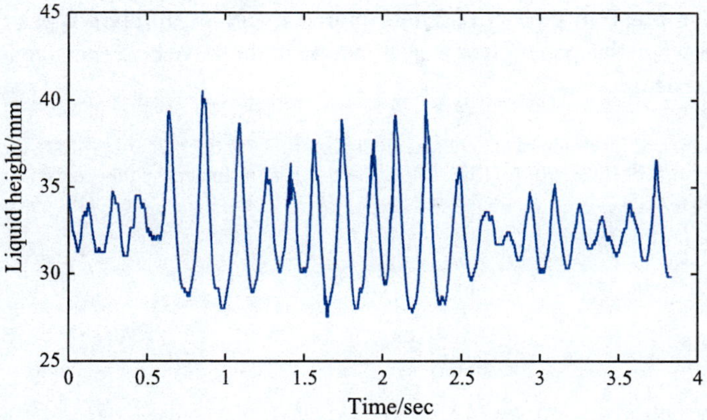

Fig. 13 Variation of interfacial height with time at fixed a point (at $Re_{SG} = 4000$ and $Re_{SL} = 9000$)

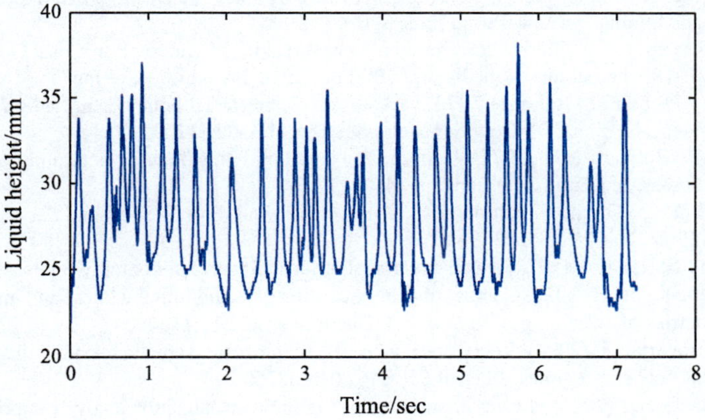

Fig. 14 Variation of interfacial height with time at fixed a point (at $Re_{SG} = 6000$ and $Re_{SL} = 8000$)

5 Conclusion

In this study, the wave-breaking phenomenon has been analyzed using the image processing technique. Two different types of wave breaking in pipe flow have been discussed here: strong wave breaking and weak wave breaking. The influence of the Bernoulli effect and increase in interfacial shear stress causes the waves to break (which is an integral part of K-H instability), shown by bulk velocity calculation before wave breaking. During experiments, the wave-breaking phenomenon occurs when the superficial Reynolds number of water reaches 7000 (Re_{SL}) for 9000 (Re_{SG}) and 10,000 (Re_{SG}) superficial Reynolds number of air in 50-mm-diameter pipe. Thus, it could be summarized that specific liquid film thicknesses are needed for a wave to break. Here, it is also described that the signal of large amplitude roll waves can indicate wave break in pipe flow. The captured image of different types of wave breaking shown in this paper gives a clear picture of the wave breaking phenomenon for the visual aspect.

Acknowledgements The support of Science and Engineering Research Board (SERB), India (sanction letter no SB/S3/MIMER/0111/2013 dated 23-05-2014) in financing this study is gratefully acknowledged.

References

1. Ahmed WH, Bello MM, Al-Sarkhi A, El Nakla M (2012) Flow and mass transfer downstream of an orifice under flow accelerated corrosion conditions. Nucl Eng Des 252:52–67
2. Arabi A, Salhi Y, Zenati Y, Si-Ahmed EK, Legrand J (2020) On gas–liquid intermittent flow in a horizontal pipe: influence of sub-regime on slug frequency. Chem Eng Sci 211:115251
3. Ayati AA (2018) Experimental characterization of non-linear interfacial wave interaction in stratified gas-liquid pipe flow. Phys Fluids 30(6):063305
4. Bosco DR (1985) A qualitative description of wave breaking. J Waterw Port Coast Ocean Eng 111(2):171–188. https://en.wikipedia.org/wiki/File:Large_breaking_wave.jpg
5. Chaudhary D, Saini S, Banerjee J (2021) Analysis of interfacial behavior in two-phase flow using image processing. IOP Conf Ser Mater Sci Eng 1146(1):012002
6. Dukler AE, Hubbard MG (1975) A model for gas-liquid slug flow in horizontal and near horizontal tubes. Ind Eng Chem Fundam 14(4):337–347
7. Jeffreys H (1925) On the formation of water waves by wind. Proc R Soc Lond Ser A Contain Pap Math Phys Char 107(742):189–206
8. Kordyban ES, Ranov T (1970) Mechanism of slug formation in horizontal two-phase flow
9. Liu L, Wang K, Bai B (2020) Comparative investigation of liquid film thickness and interfacial wave properties of swirling gas-liquid flows. Chem Eng Sci 213:115407
10. Saini S, Banerjee J (2021) Recurrence analysis of pressure signals for identification of intermittent flow sub-regimes. J Petrol Sci Eng 204:108758
11. Saini S, Banerjee J (2021) Physics of aeration in slug: flow visualization analysis in horizontal pipes. J Visualization 24:917–930
12. Saini S, Banerjee J (2021) Recognition of onset of slug using recurrence analysis of pressure signal. Nucl Eng Des 381:111325
13. Smith L (2018) Experiments of breaking waves in pipes and flumes
14. Sun JY, Jepson W (1992) Slug flow characteristics and their effect on corrosion rates in horizontal oil and gas pipelines. In: SPE annual technical conference and exhibition
15. Thaker J, Banerjee J (2016) Influence of intermittent flow sub-patterns on erosion-corrosion in horizontal pipe. J Petrol Sci Eng 145:298–320
16. Thaker J, Saini S, Banerjee J (2021) On instantaneous pressure surges and time averaged pressure drop in intermittent regime of two-phase flow. J Petrol Sci Eng 205:108971
17. Vaze M, Banerjee J (2011) Experimental visualization of two-phase flow patterns and transition from stratified to slug flow. Proc IME C J Mech Eng Sci 225(2):382–389
18. Vollestad P, Ayati AA, Jensen A (2019) Experimental investigation of intermittent airflow separation and microscale wave breaking in wavy two-phase pipe flow. J Fluid Mech 878:796–819

Performance Evaluation of Porous Layer in Jet Impingement Heat Transfer

Abdul Rahman Ansari and Vipul M. Patel

Nomenclature

B	Nozzle width (m)
C_f	Forchheimer coefficient (m^2)
h	Porous layer height (m)
H	Channel width (m)
K	Thermal conductivity of fluid (W/mK)
L	Channel length (m)
Nu	Nusselt number
Re	Reynolds number
T	Temperature (K)

Greek Symbols

ρ	Density (kg/m^3)
Φ	Porosity
μ	Viscosity (Ns/m^2)

A. R. Ansari (✉) · V. M. Patel
Department of Mechanical Engineering, Sardar Vallabhbhai National Institute of Technology, Surat, Gujarat 395 007, India
e-mail: p19tm004@med.svnit.ac.in

J. Banerjee et al. (eds.), *Recent Advances in Fluid Dynamics*, Lecture Notes in Mechanical Engineering, https://doi.org/10.1007/978-981-19-3379-0_24

1 Introduction

Heat transfer enhancement techniques have an essential role in a variety of applications including refrigeration, aerospace automotive, process industries and solar energy heaters. Using appropriate procedures for heat transfer, technical advantages and cost reductions might be met. A cold fluid jet impingement is an effective method of cooling explained by Zackerman and Loir [1]. A sudden release of liquid or gaseous flow against the surface can transfer a large amount of thermal energy between liquid and surface. The application of jet impingement includes cooling of electronic components, heating of optical surfaces for defogging and cooling of turbine components such as combustor case, turbine case and turbine blades. Porous medium, as their name suggests, they have a number of interconnected pores. Porous materials are all around us and play a role in a variety of sectors, including energy management, vibration suppression, heat insulation, sound absorption and fluid filtering. Metallic foams are one example of the porous medium. They are very light; approximately, 75–95% volume fraction is void, good energy absorption with high compression strength, low thermal conductivity and high strength. Dutta et al. [2] performed numerical computation of turbulent jet impingement for two nozzles to flat plate spacing (H/B) 4 and 9.2 with Reynolds number at 20,000. Various RANS-based models like k-ε and k-ω were tested. With little modification for transitional flow, the standard SST k-ω model shows best agreement with the experimental results in terms of fluid flow and heat transfer and secondary peak. Ashforth and Jambunathan [3] performed experimental analysis to investigate the effect of nozzle geometry and confinement on the potential core. Here, four cases such as flat and fully developed flow at nozzle exit with and without semi-confinement were investigated. It was found that the length of the potential cores for flat unconfined jet, fully developed unconfined jet, flat semi-confined jet and fully developed semi-confined jet was 4.5d, 4.8d, 5.3 and 5.8, respectively. Numerical modelling of heat transfers of the circular jet for confined and unconfined cases was carried out by Behnia et al. [4]. It was found that the confinement has little effect on heat transfer. It tends to decrease the average heat transfer rate, but no change was observed in the local stagnation heat transfer coefficient. Katti and Prabhu [5] experimentally studied the jet impingement heat transfer to understand the effect of jet to plate spacing and Reynolds number on the local distribution of the heat transfer coefficient and wall static pressure. It was found that with the increase in the nozzle to plate spacing up to 6d, the stagnation point Nusselt number increases. Graminho et al. [6] carried out the numerical investigation of jet impingement on the flat plate covered with a porous material. The results obtained explain the flow behaviour in terms of the velocity profiles, contour of pressure, streamlines and coefficient of friction along the target wall. Fu et al. [7] numerically studied the three types of porous blocks, i.e. rectangle, convex and concave for laminar-convective heat transfer. Their results showed that the thermal performance is mainly enhanced by the more fluid flowing through the porous medium near the heated region. Fischer et al. [8] investigated the heat transfer results, obtained by the turbulent jet impingement on the heated plate covered with the porous material. They considered various parameters of porous material such as porosity, thickness, permeability and thermal conductivity to see their effect on the local Nusselt number distribution on the heated wall. Buonomo et al. [9] carried out an experimental investigation for the jet impingement on the wall heated at constant heat flux and covered with aluminium foam having the pores per inch (PPI) 40. It was found that the use of metal foam improves the heat transfer on the wall. The increase in Re shows the increase in the Nu and for high nozzle diameter to slot height (D/H) 1.2 and low heat flux 1400 w/m^2. To understand the effect of the local non-thermal equilibrium (LNTE) model for energy transport, numerical analysis of the impinging jet against the heated target plate covered with the porous medium was performed by Dorea and de Lemos [10]. Their results indicate that the porous material eliminates the secondary peak and allow controlled heat transfer from a wall. de Lemos and Fischer [11] numerical findings for a jet impinging on a flat plate coated with a porous material that was kept at a higher temperature than the incoming jet fluid temperature were obtained. It was found that the porosity only affects the stagnation Nu, whereas porous layer thickness against the flat plate shows more variation along the wall jet direction.

The present study is focussed on the jet impingement cooling with and without the porous medium to validate the effect of the porous medium on the heat transfer characteristics. Important parameters of the porous medium such as porosity and porous layer thickness are varied to understand their effect on the transfer of heat for the flat plate. The main motive of the study is to enhance the heat transfer characteristics of jet impingement using porous layer on flat plate.

2 Problem Description

The problem consists of a laminar jet with uniform velocity v_0 enters through a slotted nozzle of width B into the channel of height H as shown in Fig. 1a. The temperature of the incoming jet is constant and maintained at T_0 The length of the plate on which the jet is impinging is $2L$, and it is maintained at constant temperature T_1 which is 10 °K above the incoming jet temperature T_0. As shown in Fig. 1b for the porous channel, the bottom plate is coated with a porous medium of height h.

The flow considered here is two-dimensional, laminar, steady and incompressible and porous medium homogeneous in nature. Properties of the fluid are assumed to be constant, and effect of gravity is neglected.

3 Methodology

At first, the Nusselt number is obtained on the bottom wall for clear channel, i.e. without porous media for four different Reynolds number (Re = 250, 500, 750 and 1000). The Nusselt numbers are then compared with the work of de Lemos and

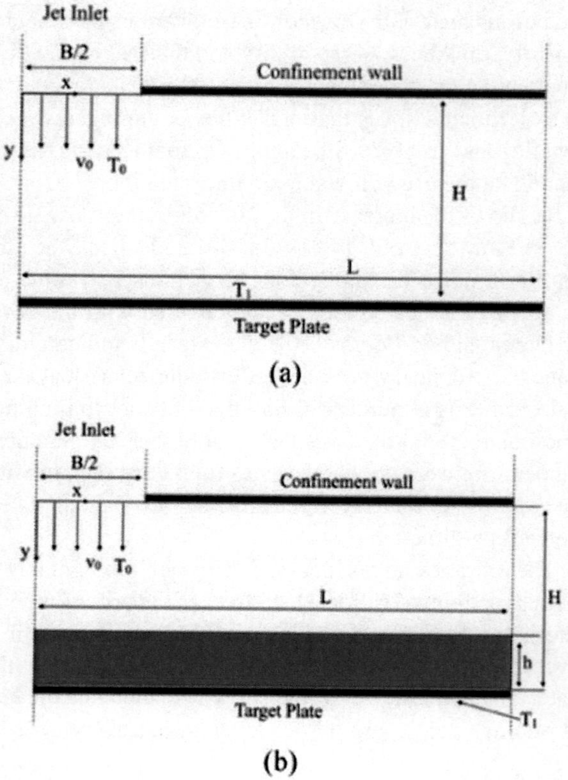

Fig. 1 **a** Confined jet impingement channel-clear channel case, **b** confined jet impingement channel (channel filled with porous medium of height *h*)-porous case

Fischer [11]. For the porous medium jet impingement case, the porous layer of height h is applied on the bottom wall of the channel. The Nusselt number is obtained on the bottom wall by varying the parameters of the porous medium such as porosity (ϕ) and porous layer height (h).

3.1 Computational Domain

Numerical simulation is performed using commercial CFD software Ansys workbench 16.2. 2D geometry is designed in the design modular, and Ansys Fluent is used to set up the problem. The connection between pressure and velocity is handled using SIMPLE algorithm, and the convective terms are handled using a second-order upwind technique (Fig. 2).

Fig. 2 **a** Geometry in design modular, **b** mesh grid

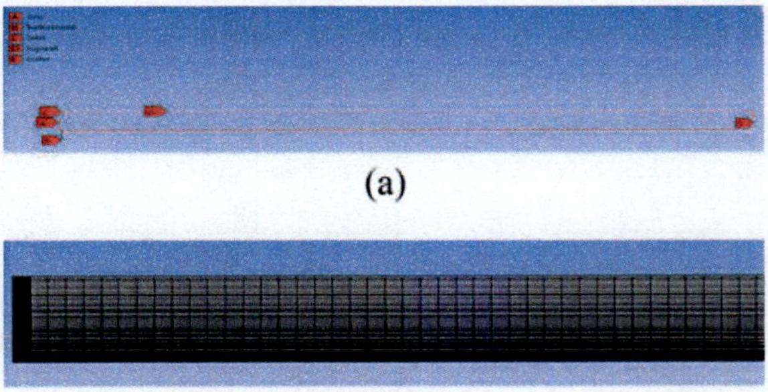

(a)

(b)

3.2 Governing Equations

In the presence of the porous media, the governing equations for the jet impingement are modified as follows.

Macroscopic continuity equation:

$$\nabla \cdot u_D = 0$$

u_D is average surface velocity. For a continuity equation, Dupuit–Forchheimer relationship is used. $u_D = \phi \langle u \rangle^i$ where ϕ, $\langle u \rangle^i$ is porosity and intrinsic velocity average, respectively.

Macroscopic momentum Equation:

$$\rho \nabla \cdot \frac{u_D u_D}{\phi} = -\nabla \phi \langle P \rangle^i + \mu \nabla^2 u_D - \left(\frac{\mu \phi u_D}{K} + \frac{C_F \phi \rho}{\sqrt{K}} |u_D| u_D \right)$$

The last two terms are the Darcy and Forcheimer contributions, respectively. K represents permeability of the porous material; $C_F\phi = 0.55$ is the coefficient of form drag (Forchheimer coefficient); $\langle P \rangle^i$ is the intrinsic fluid pressure; ρ and μ are density and the fluid viscosity, respectively. At the interface velocity and pressure, condition is as follows

$$u_{D|0<\phi<1} = u_{D\phi=1}$$
$$\langle P \rangle^i_{0<\phi<1} = \langle P \rangle^i_{\phi=1}.$$

Macroscopic Energy Equations

For the local thermal equilibrium hypothesis (LTE), the macroscopic energy equation is as follows:

$$(\rho C_p)_f \nabla \cdot \left(u_D \langle T \rangle^i \right) = \nabla \cdot \left\{ K_{eff} \cdot \nabla \langle T \rangle^i \right\}.$$

where $\langle T\rangle^i$ and K_{eff} are the average temperature (fluid and solid) and the conductivity tensors (effective), respectively.

K_{eff} can be written as

$$K_{\text{eff}} = \lfloor \phi k_{\text{eff}} + (1-\phi)k_{\text{s}} \rfloor I + K_{\text{tor}} + K_{\text{disp}}$$

where K_{tor} and K_{disp} are defined as tortuosity and thermal, dispersion.

Tortuosity: $\left[\frac{1}{\Delta V} \int\limits_{A_i} n(k_{\text{f}}\overline{T_{\text{f}}} - k_{\text{s}}\overline{T_{\text{s}}})\text{d}S \right] = K_{\text{tor}} \cdot \nabla \langle \overline{T} \rangle^i.$

Thermal dispersion

$$-(\rho C_{\text{p}})_{\text{f}}(\phi \langle u^i \overline{T_{\text{f}}} \rangle^i) = K_{\text{disp}} \cdot \nabla \langle \overline{T} \rangle^i$$

De-composing the convection term in space creates dispersion, and then, volume averaging will give

$$(\rho C_{\text{p}})_{\text{f}} \nabla \cdot (\phi \langle uT \rangle^i) = \left\{ \phi \left(\langle u \rangle^i \langle T_{\text{f}} \rangle^i + \left\langle \overline{u}^{i} \, \overline{T}^{-i}_{\text{f}} \right\rangle^i \right) \right\}$$

Convection and dispersion terms are the last two terms in the parenthesis of the above equation. The physical significance of those terms is (1) Convection heat flux due to macroscopic time mean temperature and velocity that after using $u_D = \phi \langle u \rangle^i$ and (2) thermal dispersion related to the variation in mean temperature and velocity.

3.3 *Geometric Configuration and Boundary Conditions*

The results are obtained by simulating the problem with geometric configurations described in Fig. 1. The considered boundary conditions are (i) constant velocity and temperature profile at jet inlet, (ii) no slip conditions for the walls, (iii) symmetry condition for $x = 0$, (iv) at outlet fully developed flow $x = L$. (v) Bottom wall is maintained at constant temperature, and top confinement wall is maintained at null heat flux. The jet inlet temperature T_0 is maintained at 300 K; the bottom wall temperature T_1 is maintained at constant temperature of 310 K. The nozzle width is 1×10^{-2} m; the non-dimensional distance between the nozzle to plate is $\frac{H}{B} = 2$, and length of the plate is $L = 0.5$ m. The convergence criterion for mass momentum and energy is set as 10^{-6}.

4 Results and Discussions

The distribution of the Nusselt number is obtained on the bottom plate for the different Reynolds number without porous medium. After understanding the effect of Reynold number on the heat transfer characteristics of jet impingement, porous layer of height *h* is introduced in the channel. By varying the parameters of the porous medium, the Nu numbers are obtained on the bottom wall.

4.1 Jet Impingement Without Porous Media

For Reynolds number 250, 500, 750 and 1000, on the bottom wall, the Nusselt number is calculated and compared to the numerical results of de Lemos and Fischer [11], as shown in Fig. 3a, b. The present results show good agreement with the published results.

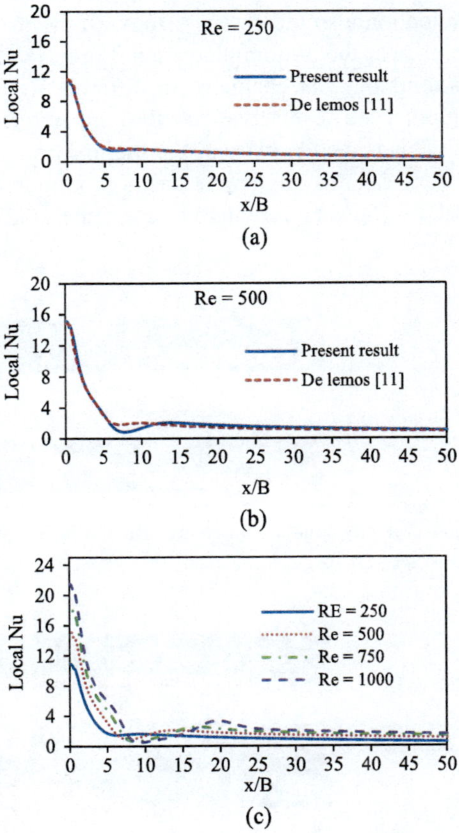

Fig. 3 Nusselt number distribution on the flat plate for clear channel **a** Re = 250, **b** Re = 500, **c** effect of Reynolds number on Nusselt number

In Fig. 3, local Nusselt number is obtained against the non-dimensional position (x/B) on the flat plate for increasing Reynolds number. It has been observed that the local Nusselt number on the target plate also increases with an increase in the Reynolds number. The observed value stagnation Nu (i.e. $X/B = 0$) at the Re = 250 is 11. As Re increases to 500, 750 and 1000, the local Nu also increases to the values of 15, 18 and 21, respectively (see Fig. 3c). Also, the secondary maxima in the Nu can be observed for the flow having a Re greater than 250. The secondary maxima in Nu occurs due to recirculation of the flow. When flow impinges on the flat plate at the stagnation point, the flow velocity comes to rest, but as jet proceeds along with the target plate, its velocity increases, and recirculation occurs which causes more convective heat transfer between the heated plate and fluid. Hence, the secondary maxima were observed. For the application of homogeneous cooling, these re-circulations are not desirable; hence, porous materials are being used to bring the homogeneity in the heat transfer.

Figure 4 shows the velocity contours for various Re for the clear channel. When the jet enters into the channel, it interacts with the stagnant fluid present in the channel because of which primary circulation is observed. With the increase in the Re, the primary circulation increases, and hence, the size of the bubble relevant to primary circulation also increases. Also, with the increase in Re, the secondary circulation arises near the bottom plate, and it grows with the increase in Re. The temperature contours of clear channels for different Re are shown in Fig. 5. At the stagnation region, it can be observed that the temperature gradient increases with the increase in Re. This is mainly because the thermal boundary layer is reduced with the increase in the Re. Since higher Re has high kinetic energy which causes more convective heat transfer therefore high temperature gradient is achieved.

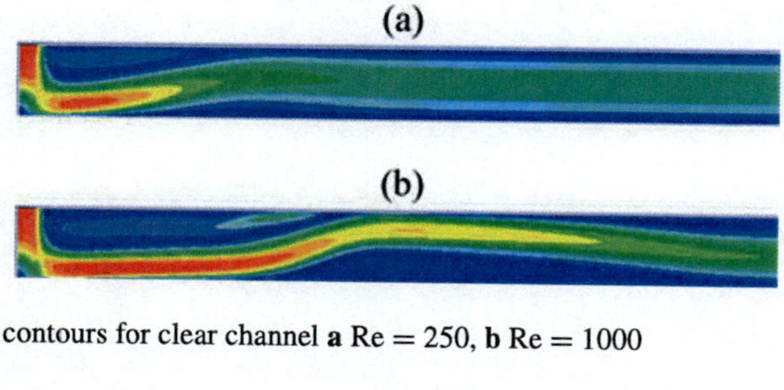

Fig. 4 Velocity contours for clear channel **a** Re = 250, **b** Re = 1000

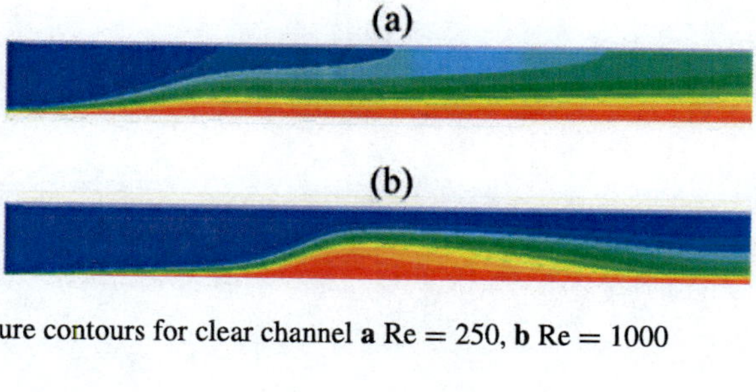

Fig. 5 Temperature contours for clear channel **a** Re = 250, **b** Re = 1000

4.2 *Jet Impingement with Porous Medium*

In this case, the channel is partially filled with the porous material, and numerical analysis is performed to investigate the characteristics of heat transfer on the flat plate. The obtained results are compared with the work of de Lemos and Fischer [11]. Parameters like porosity and porous layer height are adjusted in a channel filled with porous material to see how they affect heat transmission properties.

4.2.1 Effect of Porosity

In Fig. 6, the distribution of Nusselt number is plotted along the plate length. For this case, the Re = 750, Da = 8.28×10^{-3} and porous layer height $h/H = 0.5$ are considered. The obtained results are compared with the work of de Lemos and Fischer [11].

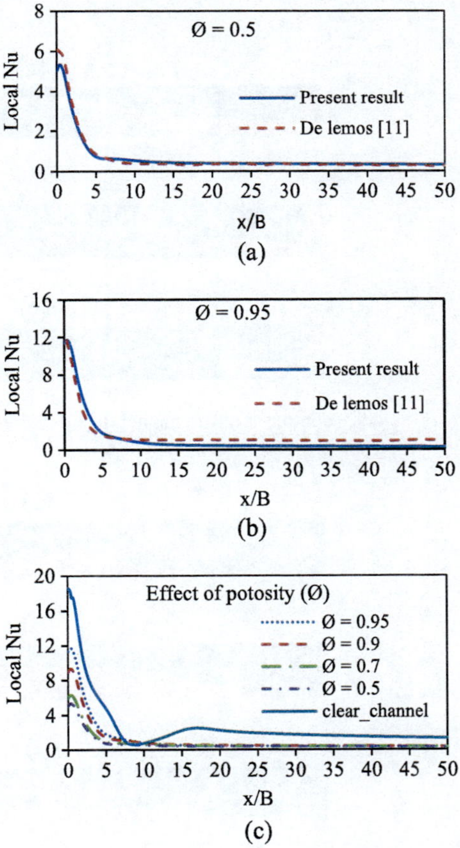

Fig. 6 Nusselt number distribution on the flat plate for the porous channel **a** $\phi = 0.5$, **b** $\phi = 0.95$, **c** effect of porosity on Nusselt number

From the graph of porosity variation (see Fig. 6), it is observed that with insertion of porous layer the secondary peak in Nusselt number distribution is eliminated. Further, for $x/B \geq 10.$, the Nu decreases to a minimum point, and from there, it remains almost constant throughout the length of the bottom wall. With increase in porosity, the voids in the porous material also increase. This makes it easy for the incoming jet to pass through the porous material which results in the increase in the stagnation Nu. By incorporating porous material, the distribution of the Nu becomes homogeneous. This can be observed in the comparison plot Fig. 6c. This indicates the enhancement in heat transfer distribution along the target plate length.

From the velocity contour Fig. 7, it is observed that the variation in porosity does not influences the flow behaviour. The flow behaviour is almost identical for the range of porosity considered in this study. The increase in porosity from 0.5 to 0.95 has no significant effect on the primary and secondary circulation. However, from temperature contours (Fig. 8), it can be seen that as the porosity increased, the temperature field soon homogenised. In addition, when the porosity of the porous medium rises, the jet penetrates more easily and reaches to the bottom wall. As the porosity increases, the thermal boundary layer compresses to the wall at the

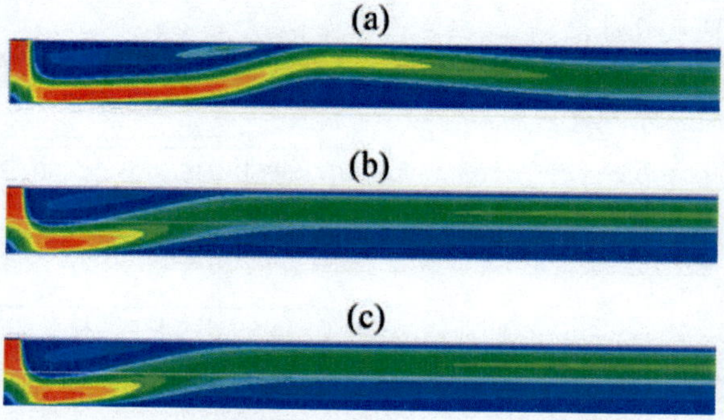

Fig. 7 Velocity contour **a** clear channel, **b** $\phi = 0.5$, **c** $\phi = 0.95$

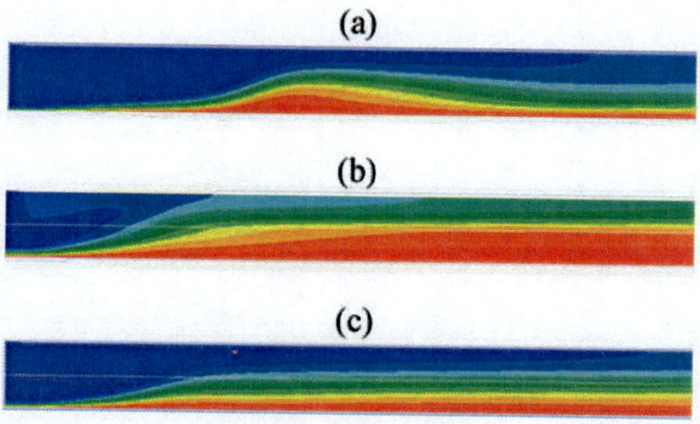

Fig. 8 Temperature contour **a** clear channel, **b** $\phi = 0.5$, **c** $\phi = 0.95$

stagnation point, resulting in a high temperature difference at the wall, and heat transfer is enhanced.

4.2.2 Channel Blockage Effect (h/H)

In order to understand the effect of porous layer height on the jet impingement heat transfer characteristics, the height of the porous layer is changed in the range of $h/H = 0.25$–0.75. The other parameters, such as Re, Da and porosity, are kept constant. For the present case, Re = 750, Da = 8.28×10^{-3} and porosity ($\varnothing$) = 0.9 are considered.

Figure 9 shows the effect of variation in porous layer height (h) on local Nusselt number. The addition of the porous layer appears to suppress the Nusselt number's secondary maxima. For porous layer thickness less than $h = 0.4H$, the secondary Nusselt number peak is observed. For the case of porous layer thickness $h = 0.25H$ and Reynolds number of 750, the porous layer height is not sufficient to diminish the secondary maxima. One can observe that the variation in the stagnation Nu is not as

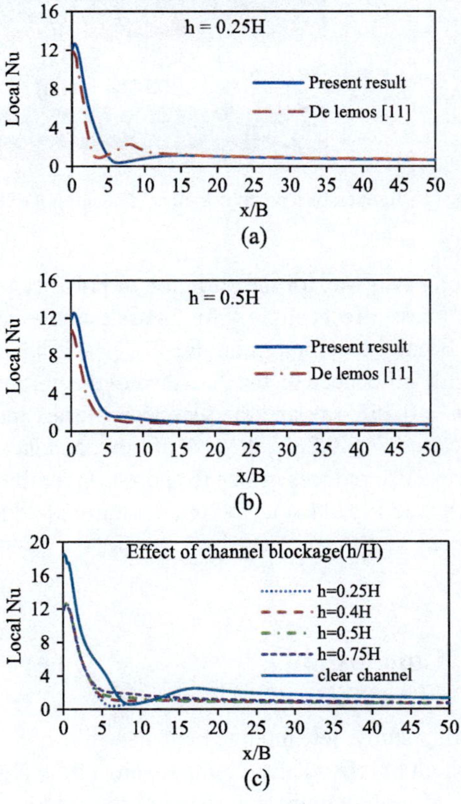

Fig. 9 Nusselt number distribution on the flat plate for the porous channel **a** $h = 0.25H$, **b** $h = 0.5H$, **c** effect of porous layer height on Nusselt number

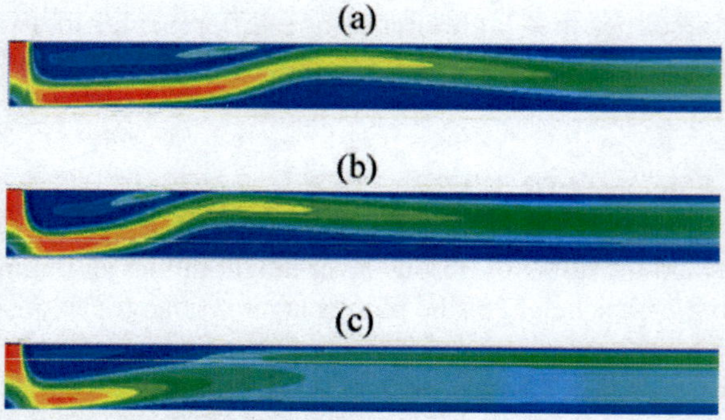

Fig. 10 Velocity contour **a** clear channel, **b** $h = 0.25H$, **c** $h = 0.75H$

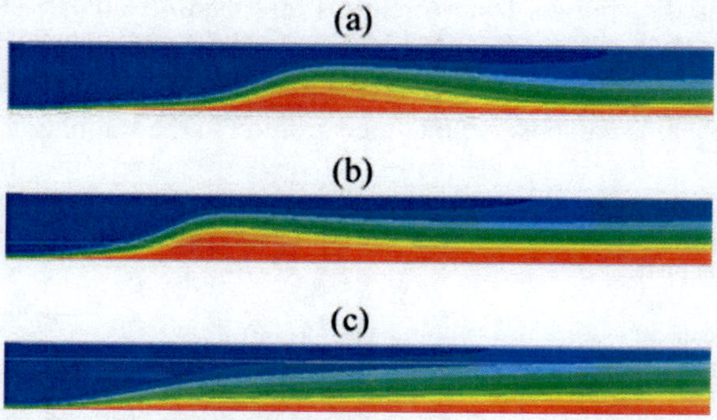

Fig. 11 Temperature contour **a** clear channel, **b** $h = 0.25H$, **c** $h = 0.75H$

strong as it was for the variation of porosity (see Fig. 9c). Variation in porous layer thickness affects slightly for Nusselt number distribution in the stagnation region.

From the velocity contour Fig. 10, it has been observed that flow behaviour is highly influenced by the thickness of the porous layer. The primary vortex diminishes at $h = 0.25H$ (a), and the secondary vortex increases near the wall. As porous layer thickness $h \geq 0.25H$, the size of the secondary vortex gets disappeared, and primary vortex also reduces in size (b) porous layer thickness also influences the temperature field (see Fig. 11a) as the temperature contour varies more along the length of the plate.

5 Conclusion

In this study, jet impingement heat transfer is studied for the flat plate with and without porous media. Comparisons of the Nusselt number are made to understand the different parameters of the porous medium on the heat transfer characteristics of the jet impingement. (i) Nusselt number on the flat plate increases with increase in the Reynolds number since kinetic energy of fluid increases, and it enhances the convective heat transfer. (ii) Porous medium decreases the stagnation Nusselt number when compared with the clear channel jet impingement but homogenise the heat transfer of the jet impingement. (iii) Insertion of porous layer on the target plate allows the controlled heat transfer by eliminating the secondary peak in the Nusselt number. (iv) Porosity strongly influences the stagnation Nusselt number values, whereas porous layer height mostly affects its distribution on the flat plate.

References

1. Zuckerman N, Lior N (2006) Jet impingement heat transfer: physics, correlations, and numerical modeling. Adv Heat Transf 39:565–631
2. Dutta R, Dewan A, Srinivasan B (2013) Comparison of various integration to wall (ITW) RANS models for predicting turbulent slot jet impingement heat transfer. Int J Heat Mass Transf 65:750–764
3. Ashforth-Frost S, Jambunathan K (1996) Effect of nozzle geometry and semi-confinement on the potential core of a turbulent axisymmetric free jet. Int Commun Heat Mass Transfer 23(2):155–162
4. Behnia M, Parneix S, Shabany Y, Durbin PA (1999) Numerical study of turbulent heat transfer in confined and unconfined impinging jet. Int J Heat Fluid Flow 20(1)
5. Katti V, Prabhu SV (2008) Experimental study and theoretical analysis of local heat transfer distribution between smooth flat surface and impinging air jet from a circular straight pipe nozzle. Int J Heat Mass Transf 51(17–18):4480–4495
6. Graminho DR, de Lemos MJS (2008) Laminar confined impinging jet into a porous layer. Numerical Heat Transf Part A Appl 54(2):151–177
7. Fu W-S, Huang H-C (1997) Thermal performances of different shape porous blocks under an impinging jet. Int J Heat Mass Transf 40(10):2261–2272
8. Fischer C, de Lemos MJS (2010) A turbulent impinging jet on a plate covered with a porous layer. Numerical Heat Transf Part A Appl 58(6):429–456
9. Buonomo B (2006) Confined impinging jets in porous media. J Phys Conf Ser 745(3)
10. Dórea FT, le Lemos MJS (2010) Simulation of laminar impinging jet on a porous medium with a thermal non-equilibrium model. Int J Heat Mass Transf 53(23–24):5089–5101
11. de Lemos MJS, Fischer C (2008) Thermal analysis of an impinging jet on a plate with and without a porous layer. Numerical Heat Transf Part A Appl 54(11):1022–1041

Simple Analytical Model for Mass Transport Resistance for Passive DMFC

Seema S. Munjewar, Rohan Pande, and Arunendra K. Tiwari

Nomenclature

T_1	Initial thickness of GDL
T_C	Operating thickness of GDL
A_C	Area of GDL under current collector
A_O	Open area or active area of GDL
A	Total area of GDL ($A_C + A_O$)
OR	Open ratio

Greek Symbols

Φ	Compression ratio of GDL
Ω_1	Resistance to fuel flow through opened GDL
Ω_2	Resistance to fuel flow through compressed GDL
Ω_{eq}	EMTR

S. S. Munjewar (✉)
VNIT, Nagpur, India
e-mail: seemamunjewar@gmail.com

R. Pande
SVNIT, Surat, India

A. K. Tiwari
SPRERI, Anand, India

J. Banerjee et al. (eds.), *Recent Advances in Fluid Dynamics*, Lecture Notes in Mechanical Engineering, https://doi.org/10.1007/978-981-19-3379-0_25

1 Introduction

A fuel cell covert chemical energy of the reactants continuously into the electrical energy as long as reactant is supplied [1–3]. Thus, fuel cells are the most promising energy device for portable applications, stationary applications as well as for transport applications [2–6]. Nowadays, direct methanol fuel cells (DMFCs) have been attracted much with several advantages. The DMFC is also subcategorised into active and passive DMFCs. The main difference between active and passive DMFCs is the nature of supply of reactants (oxygen and methanol). Reactants are supplied by external means in active DMFCs and by natural means in passive DMFC.

Design of anode current collector (ACC) and cathode current collector (CCC) such as open ratio, shape structure and their material are important factor which contributes in performance of passive DMFC [7]. Both current collectors have different function [7, 8]. The major literatures are found on material of current collector, open ratio of current collector and also on manufacturing process of current collectors [9–12]. The most preferable channel for current collector is perforated type [13–19]. The tremendous study has also been done on passive fuel cell with different current collector structure, open ratio and material but no one studied surface roughness of current collector. Therefore, an attempt has been made to study the effects of CC roughness on the passive DMFC performance using polarization and EIS test in our previous study [20]. GDL of membrane electrode assembly (MEA) also contributes more in DMFC performance. The performance of passive DMFC also depends on compression of GDL. Contact between the current collector and GDL decided the contact resistance at the interfaces. The fuel cell performance affected with ohmic résistance and mass transfer resistance [21, 22]. Electrochemical impedance spectroscopy is a non-destructive technique and contributes information for passive DMFC without disturbing system. The aim of this paper is to study the effect of CCR, open ratio and compression ratio of diffusion layer on the equivalent mass transport resistance mathematically.

2 Construction and Working of Passive DMFC

The sequential component of passive DMFC is as shown in Fig. 1.

Anode end plate with methanol tank, ACC, MEA, CCC and cathode end plate and nut and bolt are required to tighten cell [13, 14, 15, 28–32]. MEA is the important components in passive DMFC placed in between current collectors. The MEA is composed of anode diffusion layer (ADL), anode catalyst layer (ACL), polymer electrolyte membrane (PEM), cathode catalyst layer (CCL) and cathode diffusion layer (CDL) as shown in Fig. 2. Both the electrodes are supported on GDL backing. GDL gives channels to flow reactants, oxidants and products.

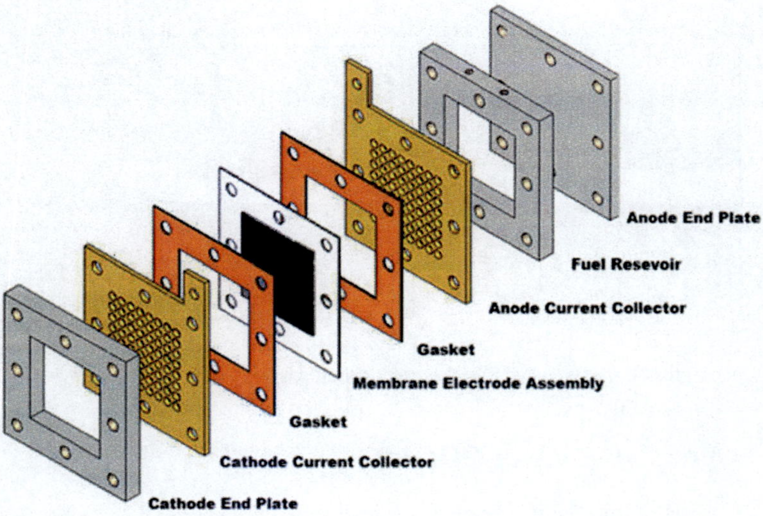

Fig. 1 Passive DMFC

Fig. 2 Electrode membrane assembly

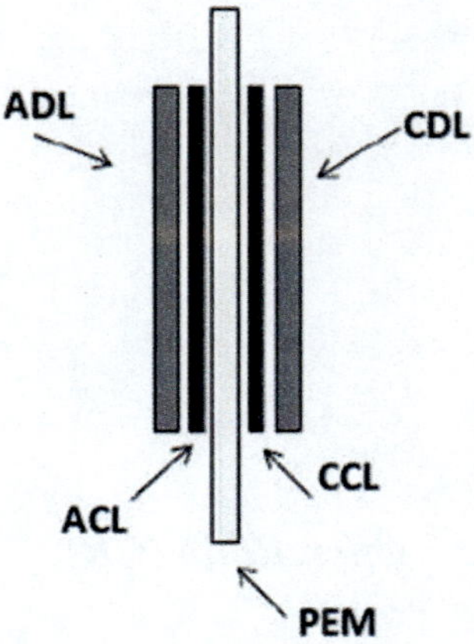

3 Mathematical Prediction

Figure shows 1D steady state mathematical model of passive DMFC to develop an equivalent mass transfer resistance. Figure 3a shows uncompressed passive DMFC, and Fig. (b) shows compressed passive DMFC. This 1D model helps to identify the effect of open ratio of CC, compression due tightening of fuel cell and CCR on transport resistance.

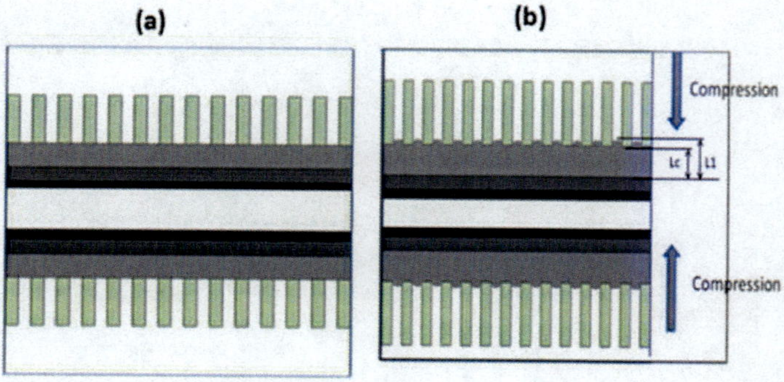

Fig. 3 **a** Uncompressed passive DMFC and **b** compressed passive DMFC

3.1 *Compression Ratio of GDL (Φ)*

$$\emptyset = \frac{\text{initial thickess of DL} - \text{operating thickess of DL}}{\text{initial thickess of DL}}$$

Thus mathematically

$$\emptyset = \frac{T_1 - T_c}{T_1}$$

Thus

$$T_c = (1 - \emptyset)T_1 \quad (1)$$

3.2 *Open Ratio (OR)*

$$\text{OR} = \frac{\text{Active area or open area of GDL}}{\text{total area of GDL}}$$

Let

A_c Area of GDL under current collector
A_0 open area of GDL and
$A_c + A_o$ total active area of GDL (A)

The open ratio is given by

$$\text{Open ratio (OR)} = \frac{A_o}{A_o + A_c} \quad (2)$$

$$\text{Open ratio (OR)} = \frac{A_o}{A} \tag{3}$$

The open ratio, compression ratio and CCR affect the EMTR thus affects the reactant flow to the electrodes. Thus, fuel cell performance also due to restriction of reactant supply. Thus, this is essential to predict mass transport resistance for passive DMFC.

Figure 3a shows uncompressed passive DMFC, and Fig. (b) shows compressed passive DMFC, i.e. compressed diffusion layer. GDL under current collector gets compressed to T_C due to the tightening effect. Some portion of GDL is uncompressed and opened for reactant supply as shown in Fig. 3

Thus, it quite obvious that $T_C < T_1$, increase in compressive force decreases the diffusion layer thickness.

Some important assumptions to develop model are

1. $D_C < D_O$ Here, D_C is the diffusivity of compressed GDL and D_O is the diffusivity of opened GDL.
2. It is assumed that diffusivity is a function of compression ratio (Ø), and the compression ratio depends on the operating tightening pressure.
3. It is also assumed that diffusivity of compressed GDL is a function of the current collector roughness.
4. The relation is obtained between D_C and D_O

$$D_C = (1 - Ø)^n D_O \tag{4}$$

Here, n is any index which is depending on current collector roughness and which is always greater than zero. Increased in the value of CCR increases the value of index n.

The reservoir supplies fuel to the catalyst electrode through diffusion layer. So it is clearly understood that fuel experience more resistance when passed through compressed path than the uncompressed path.

Figure 4 shows the resistance circuit.

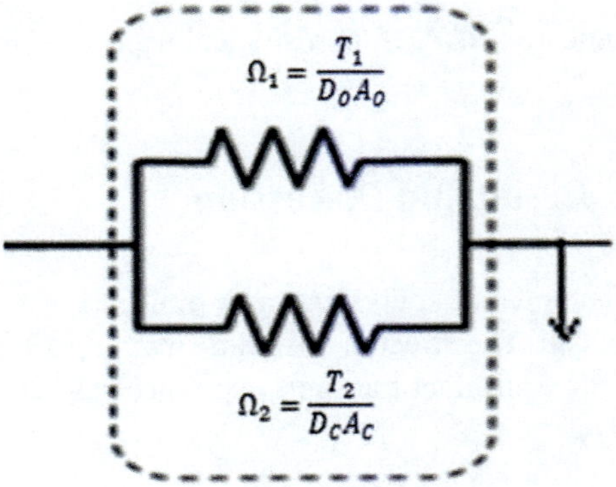

Fig. 4 Resistance circuit for EMTR

Here, Ω_1 is the resistance to fuel flow through opened diffusion layer and Ω_2 is the resistance to fuel flow through compressed diffusion layer.

It is assumed that the methanol concentration gradient is equal for the compressed GDL and opened GDL. We can write

$$\frac{1}{\Omega_{eq}} = \frac{1}{\Omega_1} + \frac{1}{\Omega_2} \tag{5}$$

where Ω_{eq} is the EMTR. As per fick's law, we can write

$$\Omega_1 = \frac{T_1}{D_O A_O} \quad \text{and} \quad \Omega_2 = \frac{T_2}{D_C A_C}$$

Thus, we get

$$\Omega_{eq} = \frac{\Omega_1 \Omega_2}{\Omega_1 + \Omega_2} \tag{6}$$

$$\Omega_{eq} = \frac{\frac{T_1}{D_o A_O} \times \frac{T_C}{D_C A_C}}{\frac{T_1}{D_o A_O} + \frac{T_C}{D_C A_C}} \tag{7}$$

Put all the values of T_C, A_C, A_O and D_C, we get

$$\Omega_{eq} = \frac{T_1}{AD_O\left[(1 - \Phi)^{n-1}(1 - OR) + OR\right]} \tag{8}$$

Here, $n = 1, 2, 3 \ldots$ and $D_O = D_{MeOH}^{ADL}$ diffusion coefficient methanol through the anode GDL is given by

$$\mathbf{D_{MeOH}^{ADL}} = \epsilon_{\mathbf{ADL}}^{2.5}\left[\left(\frac{7.608 \times 10^{-7} \times \mathbf{T}}{\mu_{H_2O} \times 9.485}\right)\right] \text{cm}^2\text{s} \tag{8}$$

where $\epsilon_{ADL}^{2.5}$ is ADL porosity and μ_{H_2O} is water viscosity (at $T = 25$ °C).

4 Result and Discussion

This model of equivalent mass transport resistance can be analysed passive DMFC fuel cell. The effect of open ratio, compression ratio and current collector roughness on the equivalent transport resistance has been discussed below.

Table 1 Percentage increment in EMTR

S. No.	n	% Ω_{eq}
1	1	–
2	2	20.55
3	3	39.15
4	4	55.33
5	5	68.12

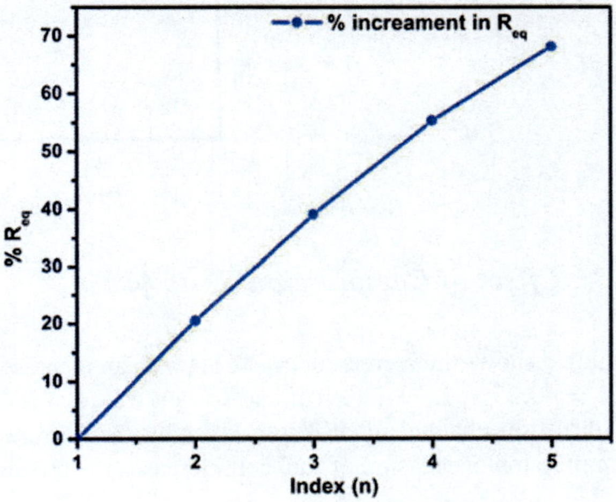

Fig. 5 Index n versus equivalent mass transport resistance

4.1 *Effect of Index* (N)

Table 1 and Fig. 5 show the percentage increment in EMTR for different values of index (n) when $\phi = 34\%$, $L_1 = 340$ μm and OR = 49.25% if the index n increases (CCR increases), the percentage EMTR (% Ω_{eq}) increases. Thus, diffusion of methanol decreases and thus decreases the power density. The qualitative nature between CCR and n is obtained. The actual relation between CCR and n is desirable for proper evaluation. This is left as scope for future work. It is noticed that EMTR of passive DMFC is depend on CCR.

4.2 *Effect of Open Ratio (OR)*

Figure 6 shows the decrement in EMTR with an increase in open ratio of the current collector, when $\phi = 34\%$, $L_1 = 340$ μm and for any value n. This is the obvious and expected trend for an open ratio of the current collector. Increase in the open ration increases the diffusion rate of fuel thus decreases the transport resistance.

Fig. 6 Open ratio versus equivalent mass transport resistance

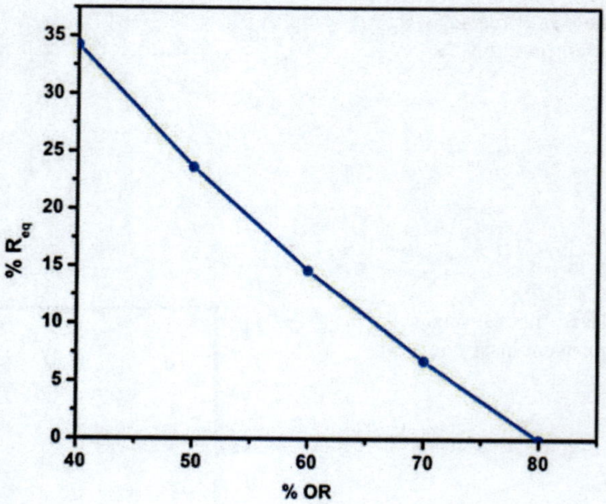

4.3 Effect of Compression Ratio (Φ)

Figure 7 shows the increment in EMTR with an increase in compression ratio, when OR = 49.25%, L_1 = 340 μm and for any value of *n*. This is due to the blockage of diffusion channel of diffusion layer by the current collector. This increase in compression or pressure of fuel cell increases the transport resistance of fuel cell.

Fig. 7 Compression ratio versus equivalent mass transport resistance

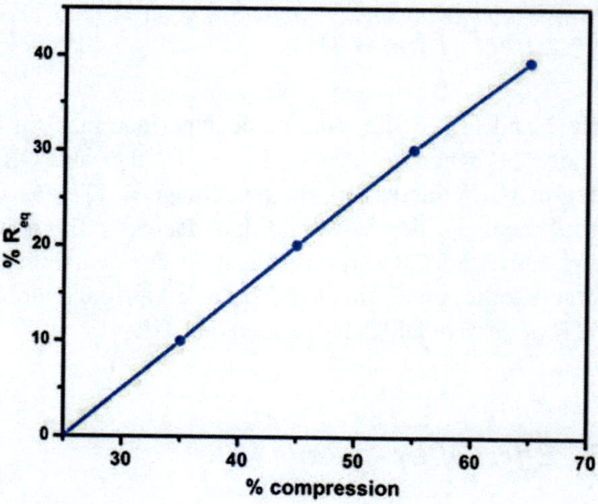

5 Conclusion

In this study, mathematical model with calculation EMTR is developed which is otherwise obtained from EIS test. Thus, it is noticed that Eq. 8 analytically apprehended all the effect of compression ratio of GDL, open ratio of current collector and CCR.

References

1. Hart D (2000) Sustainable energy conversion: fuel cells—the competitive option? J Power Sour 86(1):23–27
2. Stambouli AB (2011) Fuel cells: the expectations for an environmental-friendly and sustainable source of energy. Renew Sustain Energy Rev 15(9):4507–4520
3. Mench MM (2008) Fuel cell engines, 1st ed. Wiley, New Jersey
4. Pollet BG, Staffell I, Shang JL (2012) Current status of hybrid, battery and fuel cell electric vehicles: from electrochemistry to market prospects. Electrochim Acta 84:235–249
5. Mekhilef S, Saidur R, Safari A (2012) Comparative study of different fuel cell technologies. Renew Sustain Energy Rev 16(1):981–989
6. Andújar JM, Segura F (2009) Fuel cells: history and updating. A walk along two centuries. Renew Sustain Energy Rev 13(9):2309–2322
7. Hsieh S-S, Huang C-F, Feng C-L (2008) A novel design and micro-fabrication for copper (Cu) electroforming bipolar plates. Micron 39(3):263–268
8. Lin Y et al (2008) Surface-modified Nafion membranes with mesoporous SiO_2 layers via a facile dip-coating approach for direct methanol fuel cells. J Power Sour 185(2):904–908
9. Gholami O, Imen SJ, Shakeri M (2013) Effect of non-uniform parallel channel on performance of passive direct methanol fuel cell. Int J Hydrog Energy 38(8):3395–3400
10. Wang Y, Northwood DO (2007) An investigation into TiN-coated 316L stainless steel as a bipolar plate material for PEM fuel cells. J Power Sour 165(1):293–298
11. Chan YH et al (2008) A small mono-polar direct methanol fuel cell stack with passive operation. J Power Sour 178(1):118–124
12. Yang WM, Chou SK, Shu C (2007) Effect of current-collector structure on performance of passive micro direct methanol fuel cell. J Power Sour 164(2):549–554
13. Mallick RK, Thombre SB, Shrivastava NK (2015) A critical review of the current collector for passive direct methanol fuel cells. J Power Sour 285(Supplement C):510–529
14. Tsujiguchi T et al (2010) Development of a passive direct methanol fuel cell stack for high methanol concentration. J Power Sour 195(18):5975–5979
15. Zhang Y et al (2010) Development of an air-breathing direct methanol fuel cell with the cathode shutter current collectors. Int J Hydrog Energy 35(11):5638–5646
16. Shrivastava NK, Thombre SB, Motghare RV (2014) Wire mesh current collectors for passive direct methanol fuel cells. J Power Sour 272:629–638
17. Chan YH et al (2008) A self-regulated passive fuel-feed system for passive direct methanol fuel cells. J Power Sour 176(1):183–190
18. Zhang Y et al (2011) Design and fabrication of a silicon-based direct methanol fuel cell with a new cathode spoke structure. J Power Sour 196(6):3015–3025
19. Munjewar SS, Thombre SB, Mallick RK (2017) A comprehensive review on recent material development of passive direct methanol fuel cell. Ionics 23(1):1–18
20. Munjewar SS, Thombre SB (2019) Effect of current collector roughness on performance of passive direct methanol fuel cell. Renew Energy 138:272–283

21. Asghari S, Mokmeli A, Samavati M (2010) Study of PEM fuel cell performance by electrochemical impedance spectroscopy. Int J Hydrog Energy 35(17):9283–9290
22. Wu J et al (2008) Diagnostic tools in PEM fuel cell research: part I electrochemical techniques. Int J Hydrog Energy 33(6):1735–1746

CFD Analysis and Validation of Airflow Pattern in Typical Indian Rooms

Shivam Prajapati, Anuj Sahu, Ankit Kumar, Mitesh Gandhi, and Ashis Acharjee

1 Introduction

The science that deals with creating and retaining definite temperature, relative humidity, and air purity environment in indoor spaces, typically used to maintain a level of personal comfort in indoor space, is referred to as air conditioning. One of the major changes in the twentieth century is the usage of environmental calmness and dynamic air conditioning systems. At present, air conditioners and fans have become a necessity. In India, people utilize both ceiling fans and air conditioners together, which helps them to reduce electricity consumption while maintaining the same thermal comfort. While it looks unnecessary to be using electricity to power two ways to remain yourself cool, the reality is that fans can keep your air conditioner more efficient with something called the "wind chill" effect. Just as the wind can lead cold weather seem colder, a fan blowing in a house can make air-conditioned air seem chillier. This incident happens because your fans take heat away from your body, making you feel cooler and ideally more comfortable. In the context of room air conditioners, many published pieces of literature were reviewed. There has been a lot of research for the flow pattern of air and heat ventilation and air conditioning. Each previous research paper focused only on airflow patterns from air conditioners to different locations such as bedrooms, lecture halls, and schools.

This paper has a slight difference from the previous articles because we have considered a typical Indian study room, bedroom, and saree shop for simulation with both the air conditioner and the ceiling fan. Dipendrasingh et al. [1] presented a study on "Quantification of Air Flow Pattern in Air-Conditioned Room—A Review." In this paper, he studied room airflow patterns regarding temperature and velocity

S. Prajapati (✉) · A. Sahu · A. Kumar · A. Acharjee
Department of Mechanical, NIT Agartala, Agartala, India
e-mail: prajapatishivam64@gmail.com

M. Gandhi
Department of Mechanical, SVNIT, Surat, India

J. Banerjee et al. (eds.), *Recent Advances in Fluid Dynamics*, Lecture Notes in Mechanical Engineering, https://doi.org/10.1007/978-981-19-3379-0_26

along with quantifying the comfort from thermal and environment and by varying for clothing indices in the CFD study. He concluded that the placement of units for cooling should be decided according to the particular environment situation. Patel et al. [2] presented a study on the CFD analysis of air conditioning in room using ansys fluent. The main target of this study was to examine the cooling by air and the distribution of temperature in the room. The air conditioner's performance is analyzed for various locations of duct positions. Yongsan et al. [3] presented a study on analysis of flow of air in an air-conditioned room. In this paper, he analyzed for more comfort of the occupants in the room. He got that the occupants would feel the most comfortable when the placement of the blower of air conditioner blower is on location. Hurak et al. [4] presented a study to examine distinct ventilation configurations in a room. He concluded that the overall system's effectiveness could be improved with no need of more efficient/expensive components. Detaranto et al. [5] presented a CFD analysis of airflow patterns and heat transfer in small, medium, and large structure. He concluded that designing buildings in such a way that they are more efficiently ventilated can help take down the overall cost of energy.

With the increase of developed populations, energy usage in buildings is also going to rise. Economically strong regions use 25 times more energy than other regions. Rahman et al. [6] presented a study on "Analysis of Lecturing Room using Computational Fluid Dynamics Natural and Forced Ventilation." He carried out a CFD simulation of natural and forced ventilation in a lecturing room and the thermal comfort of the human being. Kumar1 et al. [7] presented "a study on CFD Analysis of room with Air Conditioner by using Ansys Workbench." He experimentally analyzed the performance within two equal varieties of room for air conditioning.

2 Methodology

2.1 Geometry

We have created the framework to optimize the air conditioning and ceiling fan analysis using ANSYS Fluent. The engineer starts with the symbolization of the layout done with the aid of geometry, which helps in the engineering simulations. Every aspect including that of structural analysis, air amount of the fluid, or electromagnetic area is taken into consideration. The geometry is either provided with computer aided design CAD or the geometry is made from scratch.

The solidworks software is used to model the room's furniture, and ANSYS Design Modeler is used to model the geometry of the room domain. Since we are only interested in the analysis of the fluid domain, only the aero solid model is modeled, leaving the rest of the domain like wall and furniture as vacant space inside the room domain. First, all the three models, bedroom, study room, and narrow long shop, have been created in solidworks, and then the aero solid model is created in ANSYS Design Modeler.

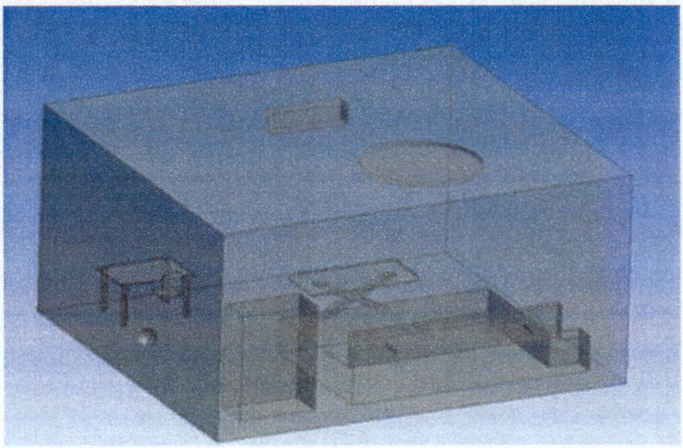

Fig. 1 Diagram of bedroom domain (isometric view)

The dimensions of all rooms are given below (Figs. 1, 2, and 3; Tables 1, 2, and 3).

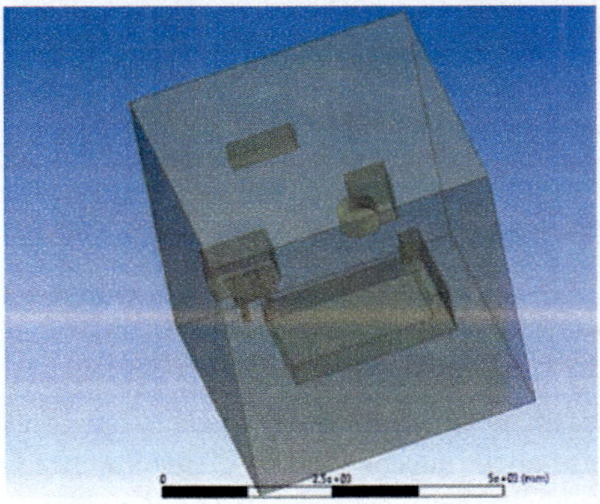

Fig. 2 Diagram of study room domain (isometric view)

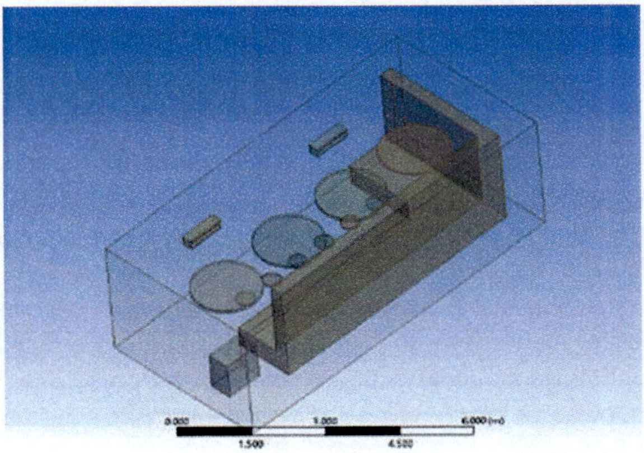

Fig. 3 Diagram of narrow long shop domain (isometric view)

Table 1 Details of bedroom

Bedroom parameters	Size (m)
Room domain	5 × 5.5 × 3
Fan (diameter)	1.5
A.C. (inlet)	1 × 0.3
A.C. (outlet)	1 × 0.2

Table 2 Details of study room

Study room parameters	Size (m)
Room domain	5 × 5.5 × 3
Fan(diameter)	1.5
A.C. (inlet)	1 × 0.3
A.C. (outlet)	1 × 0.2

Table 3 Details of narrow long shop

Narrow long shop parameters	Size (m)
Room domain	8.52 × 3.9 × 3.74
Fan (diameter)	1.5
A.C. (inlet)	0.82 × 0.07
A.C. (outlet)	0.82 × 0.14

2.2 *Mesh Adoption*

Mesh generation is one of the tedious and time-consuming simulation events in CFD. It discretizes the computational domain. Analytical equations of Navier–Stokes exist for only the simplest of flows under ideal conditions. A numerical approach must be adopted to obtain solutions for real flows whereby the equations are changed with the algebraic approximations, which can be solved using numerical methods. These numerical methods at first discretize the governing equations and then fragment the volume domain into small finite control volumes using a mesh. The discretized governing equations are integrated over each control volume, in order to conserve energy, mass, momentum, etc., and are stored appropriately for each control volume. The quality of the computational fluid dynamics (CFD) results is highly dependent on the quality and size of mesh. Many cells in the domain might lead to long solver runs, and very few might lead to an appropriate set of results. Meshing is done by default ANSYS Mesher. After generation of the mesh, it can be transferred to solution mode using the Mode toolbar or the command switch-to-solution-mode. Defining the boundary conditions, the fluid properties, getting the solution, along with post-processing of results are performed in solution mode.

Fig. 4 Mesh view of enclosure

2.3 *Inflation*

In the ANSYS Fluent, cell stacking normal to the boundary is achieved by inflation. Here, the mesh with multiple layers from the boundary surface has inflated until boundary layer thickness is fully covered. The inflation feature is utilized to set up the growth of all five layers of inflation from the boundary surface. If the inflations are used, it becomes very easy to capture the effect of the boundary layer more precisely. It is observed that inflation layers contribute to less computational time because of lesser element and node count (Fig. 4).

2.4 *Grid Independence Study*

Three different grid sizings are analyzed. First one has 437,832 elements in it, second one has 488,962 elements contained, while the third model has 537,996. We have selected the geometric model which has least number of elements, i.e. first model with 437,832 elements as it would be least computationally expensive.

2.5 *Models Used*

The k-ϵ model is used for the analysis. This model is a turbulence model and is used since the flow can be turbulent. Two-equation models of turbulence help calculate both a time scale and turbulent length by solving two different transport equations. The standard k-ϵ model is derived from model transport equations which include the parameters such as the rate of dissipation (ϵ) and the turbulent kinetic energy (k). While the model transport equation for k is obtained from the equation, physical reasoning is used to acquire the model transport equation for ϵ and hence, has little resemblance to its mathematical counterpart. There are two assumptions in the k-ϵ

model, the flow is fully turbulent and molecular viscosity repercussions are insignificant. For mitigation and improving its efficiency, there have been amendments in the standard k-ϵ model. ANSYS Fluent contains two such alterations of it. They are renormalization group (RNG) k-ϵ model and the realizable k-ϵ model where k and ϵ are obtained from the transport equations. The turbulent (or eddy) viscosity, μ_t, is computed by combining k and ε. To solve conservation equations for chemical species.

2.6 Simulation Procedures

The assumptions while simulating in ANSYS Fluent are as follows: (a) The air conditioning and ceiling fan configuration in selected bedrooms are operational and working correctly. (b) Excluding doorways and exhaust duct, the rooms are properly confined. (c) External room surface temperature is not faltering (d) There is trivial effect on temperature from internal digital devices and light sources (Tables 4 and 5).

Table 4 Outline of analysis variables

Items	Bedroom
Energy conserve	ON
Viscous model	*k*-epsilon (2 eq)
Variant of *k*-epsilon	Standard
Mesh statistics	
Nodes	115,316
Elements	437,832
Turbulent kinetic energy	1st order upwind
Rate of dissipation	1st order upwind

Table 5 Boundary conditions of bedroom

Boundary conditions of bedroom			
	Mass flow specification method	Mass flow rate (kg/s)	Temperature (K)
A.C._inlet	Mass flow rate	0.569625	288.15
A.C._outlet	Mass flow rate	0.569625	–
FAN_inlet	Mass flow rate	7.155	303.15

2.7 Boundary Conditions

The whole room domain is initially patched to be containing air at a pressure of 101,325 Pa and a temperature of 303.15 K. Initial boundary conditions are given to be the mass flow inlet and convection with 5 w/m^{2}k heat coefficient at fluid domain walls.

2.8 Solver Used

The pressure-based solver uses an algorithm that belongs to a general class of methods known as the projection method. In this projection method, the constraint of conservation of mass (continuity) of the velocity field leads to some achievement by solving a pressure (or pressure correction) equation. The equation of pressure achieved from the momentum and the continuity equations is so that the velocity field, obtained by the pressure, satisfies the continuity. Since the governing equations used in this method are nonlinear and are coupled to one another, the solution process has the involvement of iterations wherein the complete set of governing equations is solved in *k*-omega repeatedly until and unless the solution converges (Tables 6 and 7).

Table 6 Boundary conditions of study room

Boundary conditions of study room			
	Mass flow specification method	Mass flow rate (kg/s)	Temperature (K)
A.C._inlet	Mass flow rate	0.569625	288.15
A.C._outlet	Mass flow rate	0.569625	–
FAN_inlet	Mass flow rate	7.155	303.15

Table 7 Boundary conditions of narrow long shop

Boundary conditions of narrow long shop			
	Mass flow specification method	Mass flow rate (kg/s)	Temperature (K)
A.C._inlet	Mass flow rate	0.569625	288.15
A.C._outlet	Mass flow rate	0.569625	–
FAN_inlet	Mass flow rate	7.155	303.15

2.9 Flowchart for Modeling and Analysis

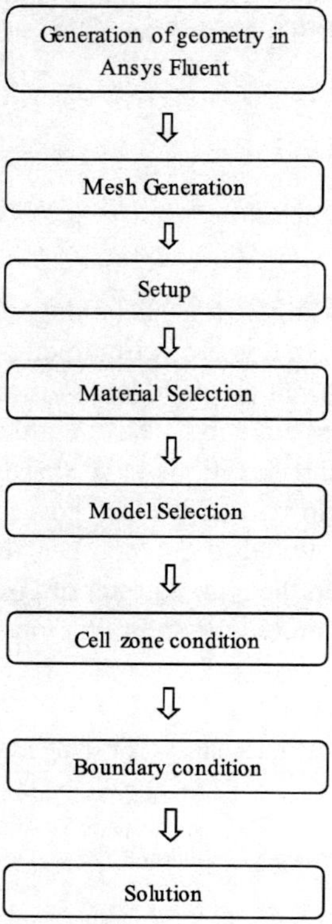

2.10 Governing Equations

The k-ϵ model was used for analysis. The kinetic energy k and dissipation rate ϵ can be extracted from:

$$\rho \frac{\mathrm{D}k}{\mathrm{D}t} = \frac{\partial}{\partial x_i}\left[\left(\mu + \frac{\mu_{\mathrm{t}}}{\sigma_{\mathrm{k}}}\right)\frac{\partial k}{\partial x_i}\right] + G_{\mathrm{k}} + G_{\mathrm{b}} - \rho\varepsilon - Y_{\mathrm{M}} \tag{1}$$

$$\rho \frac{\mathrm{D}\varepsilon}{\mathrm{D}t} = \frac{\partial}{\partial x_i}\left[\left(\mu + \frac{\mu_{\mathrm{t}}}{\sigma_{\varepsilon}}\right)\frac{\partial \varepsilon}{\partial x_i}\right] + C_{1\varepsilon}\frac{\varepsilon}{k}(G_{\mathrm{k}} + C_{3\varepsilon}G_{\mathrm{b}}) - C_{2\varepsilon}\rho\frac{\varepsilon^2}{k} \tag{2}$$

In the k-ϵ model, parameters of convective heat and mass are provided by the following equation:

$$\frac{\partial}{\partial t}(\rho E) + \frac{\partial}{\partial x_i}[u_i(\rho E + p)] = \frac{\partial}{\partial x_i}\left[\left(k + \frac{c_p \mu_t}{Pr_t}\right)\frac{\partial T}{\partial x_i} + u_j(\tau_{ij})_{\text{eff}}\right] + S_h \tag{3}$$

Transport equations for the transport of the Reynolds stresses, $\overline{u_i u_j}$, are

$$\frac{\partial}{\partial t}(\rho\overline{u_i u_j}) + C_{ij} = D_{ij}^{T} + D_{ij}^{L} + P_{ij} + G_{ij} + \phi_{ij} + \varepsilon_{ij} + F_{ij} \tag{4}$$

The turbulent kinetic energy, k, and its rate of dissipation, ϵ, are obtained from the following transport equations.

$$\rho\frac{Dk}{Dt} = \frac{\partial}{\partial x_j}\left[\left(\mu + \frac{\mu_t}{\sigma_k}\right)\frac{\partial k}{\partial x_i}\right] + \frac{1}{2}(P_{ii} + G_{ii}) - \rho\varepsilon(1 + 2M_t^2) \tag{5}$$

$$\rho\frac{D\varepsilon}{Dt} = \frac{\partial}{\partial x_j}\left[\left(\mu + \frac{\mu_t}{\sigma_\varepsilon}\right)\frac{\partial \varepsilon}{\partial x_j}\right] + C_{\varepsilon 1}\frac{1}{2}[P_{ii} + C_{\varepsilon 3}G_{ii}]\frac{\varepsilon}{k} - C_{\varepsilon 2}\rho\frac{\varepsilon^2}{k} \tag{6}$$

3 Results and Discussion

3.1 Results of Bedroom

The case of bedroom contains various furniture like in a typical Indian bedroom. Figure 5 shows the velocity contours of the fan for different diagonal views. They represent the region where the velocity is high. Figure 6 shows the temperature contour for isometric view having maximum temperature in the upper part of the

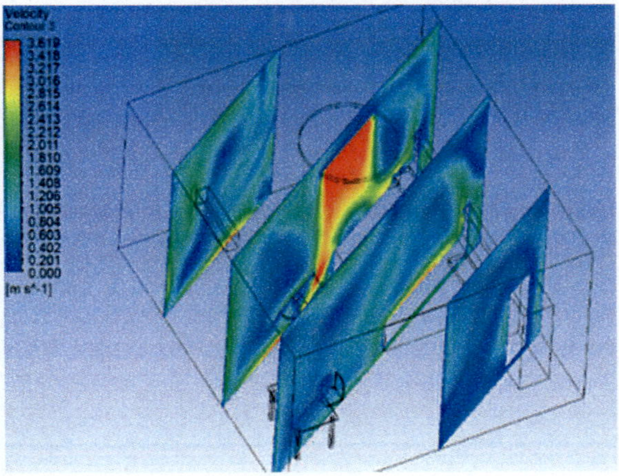

Fig. 5 Velocity contour *in different layers of bedroom*

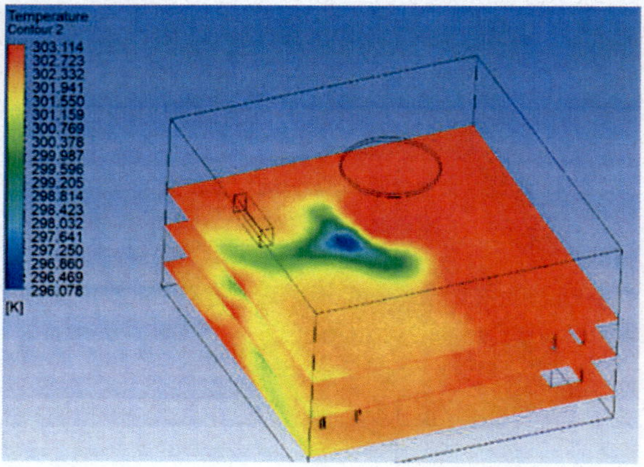

Fig. 6 Temperature contour in horizontal layers of bedroom

room as 303.114 K. It is seen that there are different velocities at different zones of the room. Velocities vary from 0 to 3.62 m/s.

3.2 Results of Study Room

Case of study room contains various furniture like in a typical study room in India. Figure 7 shows the velocity contour of the fan, and Fig. 8 shows temperature contours. Velocities vary from 0 to 4.06 m/s at different zones of room. The combined effect of the air conditioner and the fan could be seen in Fig. 7.

Figure 8 shows the temperature variation on the study room walls, which concludes that the temperature range is from 27 to 29 °C after A.C. and fans both work combined in a closed space.

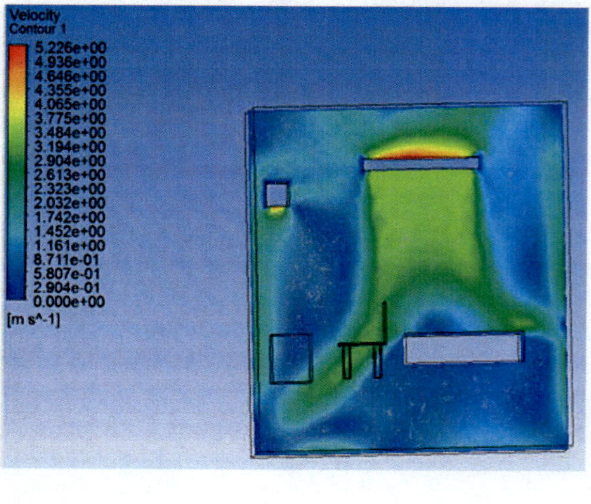

Fig. 7 Velocity contour showing the combined effect of A.C. and fan

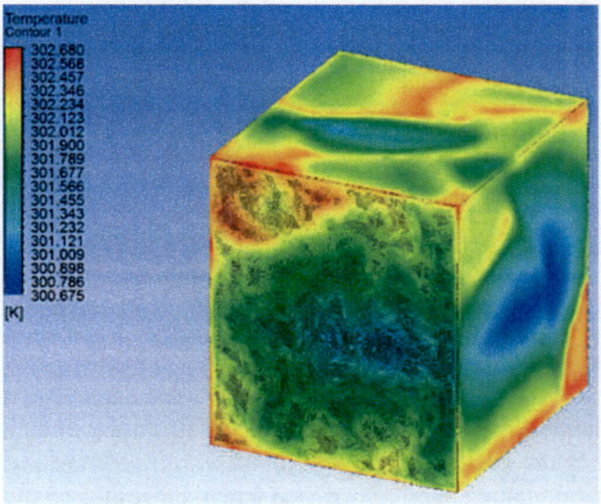

Fig. 8 Temperature contour showing the combined effect of A.C. and fan

Fig. 9 Velocity contour of the shop having the effect of fan

3.3 *Results of Long Narrow Shop*

The long narrow shop contains various pieces of furniture like a typical physical shop in India. It is seen that there are different velocities at different zones of the room. Velocities vary from 0 to 6.338 m/s. The combined effect of A.C. and fan could be seen in Fig. 9.

4 Validation

The simulations are done for different fan speeds and the graphs were plotted for PMV index. It was observed that the PMV values rise when the A.C. set temperature elevates. Growth in PMV index with growth A.C. set temperature of lecture hall signifies that dwellers are perceiving warmer conditions. In contrast to the lecture hall, where the airflow conditions were good, conditions with no exterior airflow

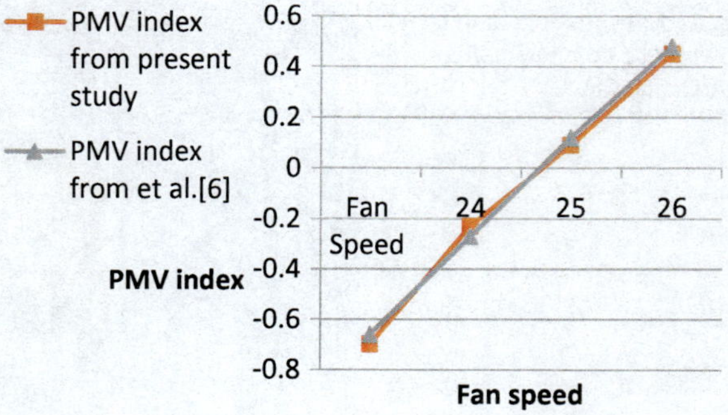

Fig. 10 Fan speed and PMV index results

exhibited greater PMV index. According to the findings of Muhieldeen et al. [8], occupants feel colder in conditions with higher airflow. The results obtained from the simulations closely matched with their results as shown in Fig. 10.

Later efficiency of cooling when both A.C. and fan are ON are also calculated. It was slightly higher than the efficiency obtained by Muhieldeen et al. [8] and Stefano et al. [9]. The reason for this can be due to different room configuration.

5 Conclusion

An important goal for HVAC design engineers is to maintain the standard of thermal comfort for residents of buildings. Fulfillment with the thermal environment is critical since some thermal conditions can be fatal to people. If the body heat is above 37.5–38.3 °C, it is said to be in a condition of hyperthermia or hypothermia below 35.0 °C. The airflow was maximum below the fan, so maximum comfort would be provided in that region. The rate of cooling increases with respect to the speed and during the start of the devices, the level of comfort of the occupant starts to increase too. The airflow of air leads to lesser cooling rate at locations away from device and where the air experiences drag due to objects ultimately leading to a temperature gradient across the room. Airflow can provide direct cooling to occupants, especially if they are not wearing much clothing. Speed of air up to 0.8 m/s is permitted, and 1.2 m/s is permissible with any local control. This elevated airflow increases the maximum temperature for a room in the summer to 30.0 °C from 27.50 °C. It was observed that the most uniform or constant pattern of temperature and velocity was observed when both fan and A.C. were kept ON condition. It was recommended that lower fan speed provides a better airflow pattern which depends on the location of the AC.

Acknowledgements We, the authors of this paper, would like to thank the Director of NIT Agartala and the HOD of the mechanical department for extending all possible help and support for carrying out this research work as a minor project under the supervision of Dr. Ashis Acharjee, Assistant professor of the mechanical engineering department. We also acknowledge all of the other pupils who have, directly and indirectly, help us to complete our minor project and this research paper accordingly as an outcome of our minor project. We also acknowledge our parents and family member who has help and support for making this research paper in reality. Last but not least authors also thank the editor of the journal for accepting us to publish our paper in the reputed journal.

References

1. Dipendrasingh T, Patel P, Shah PB. Quantification of air flow pattern in air conditioned room—a review. ISSN(O): 2348-4470, p-ISSN(P): 2348-6406
2. Patel A, Dhakar PS. CFD analysis of air conditioning in room using ansys fluent. Rajiv Gandhi Proudiyogika Vishwavidyalaya, Bhopal (M.P.)
3. Yongsona O, Badruddina IA, Zainala ZA, Aswatha Narayanab PA (2007) Airflow analysis in an air conditioning room. Build Environ 42:1531–1537
4. Hurak BS, Mazumder S, Guezennec Y. Computational fluid dynamics analysis of air flow and temperature distribution in buildings
5. Detaranto MF. CFD analysis of airflow patterns and heat transfer in small, medium, and large. Virginia Polytechnic Institute and State University
6. Rahman MA, Narahari GA (2014) Analysis of lecturing room using computational fluid dynamics natural and forced ventilation, vol 3, issue 3. ISSN: 2278-0181
7. Kumar A, Bartaria VN, Mechanical Engineering Department, LNCT Bhopal, Mechanical Engineering Department, LNCT Bhopal. CFD analysis of room with air conditioner by using ansys workbench
8 Muhieldeen MW, Kuang YC (2019) Saving energy costs by combining air-conditioning and air-circulation using CFD to achieve thermal comfort in the building. J Adv Res Fluid Mechan Thermal Sci 58(1):84–99
9. Schiavon S, Melikov AK (2008) Energy saving and improved comfort by increased air movement. Energy Build 40(10):1954–1960

Shape and Size Effects of Glass Mini-Channels on Infrared Sensors in Air–Water Two-Phase Flow

N. Mithran, K. Sowndarya, and M. Venkatesan

Nomenclature

r	Reflection coefficient
t	Transmittance coefficient
n_1, n_2	Refractive index of first medium and second medium
I	Intensity of rays coming out of the medium
I_0	Intensity of incident ray
x	Photon path
q	Ray position
k	Wave vector

Greek Symbols

α	Absorption coefficient of the medium
θ_t, θ_1	Angle of transmittance, incidence
ω	Angular frequency

N. Mithran
Department of Mechatronics Engineering, Sri Krishna College of Engineering and Technology, Coimbatore, India

K. Sowndarya
Department of Mechanical Engineering, SVNIT, Surat 395007, India

M. Venkatesan (✉)
School of Mechanical Engineering, SASTRA Deemed University, Thanjavur 613401, India
e-mail: mvenkat@mech.sastra.edu

J. Banerjee et al. (eds.), *Recent Advances in Fluid Dynamics*, Lecture Notes in Mechanical Engineering, https://doi.org/10.1007/978-981-19-3379-0_27

1 Introduction

Simultaneous flow of two different phases of same fluid or different fluids is called two-phase flow. Gas–liquid two-phase flow occurs in bubble column reactors, microfluidic systems, electronic chip cooling, atomization in combustion processes, droplet formation, impaction in material processing, coating, mini heat exchanger tubes, and in power plants.

Void fraction plays a significant role in determining flow boiling heat transfer characteristics in mini/micro-channels. The effectiveness of compact heat exchangers with mini-channels depends upon the two-phase flow regimes, pressure drop, and void fraction as detailed by Ide et al. [5]. These parameters also play a vital role in designing of evaporators and condensers. Most importantly the void fraction which depends upon the flow regimes is a significant parameter which often varies along the length of the test section. Development of sensors for measuring void fraction is still a challenging task. One of the reasons is that the interface between the two phases in a gas–liquid two-phase regime is composed of a compressible and incompressible phase and exhibits different flow behavior.

The effect of channel size on two-phase flow patterns and void fraction in microchannels was studied by Sur and Liu [16]. A wire mesh sensor to determine void fraction in the vertical pipeline was developed by Prasser [12]. The results were compared with gamma-ray densitometry measurements. The bubble size distributions were measured. Rocha et al. [14] developed a conduction probe sensor to measure the instantaneous signal generated by the capacitance probe. The signal could be used for measuring two-phase flow void fraction. Lawal [9] used electrical impedance method for measuring void fraction in a multiphase flow. The capacitance value was reduced, while solid phase was added to a two-phase gas/liquid flow. Rocha and Moreira [15] experimentally measured Taylor bubble velocity and length. A multielectrode impedance sensor was used in which a rotating electric field sweeping across the test section was used to measure the necessary parameters involved.

Barreto et al. [3] measured void fractions using resistive impedance sensor. The influence of liquid flow regime on flow pattern and pressure drop was detailed. Cartellier [4] used optical probes to measure void fractions employing piercing methods and found that the results were in good agreement with direct visualization. Juliá et al. [7] measured void fractions in a bubbly flow using an optical probe. It was found that for non-perpendicular piercing, probe inclination leads to the additional source of underestimation. Jagannathan et al. [8] utilized laser pointer as a monochromatic coherent source of light and employed refraction phenomenon. Using Fresnel reflectance and transmittance coefficients [13], two-phase flow regime identification and slug velocity measurement were done.

All materials emit radiation if their temperature is above absolute zero in the form of electromagnetic waves or photons. If the emitted wavelength ranges between 700 nm and 1 mm, then it is termed as infrared radiation. When the temperature of a solid body increases, radiation of shorter wavelength is emitted. When ferrous materials are heated to a very high temperature, incandescence is in the visible region and starts to glow in red color [6]. Identification of two-phase flow regime and void fraction can be examined by a source which emits thermal radiation. Infrared transmitter and reflector is such a source which employs refraction phenomenon. The photons from IR LED (transmitter) when excited get scattered. But most of them get collimated as the construction of LED is such that it has a lens at its front end. The receiving end also has the same construction. The amount of scattered photon rays may be neglected in the absence of any medium. Adhavan et al. [1] determined the size and velocity of a slug using IR sensors without any complex filtering or processing circuit and validated with high-speed photography. Arunkumar et al. [2] identified the two-phase flow regime using infrared sensor and compared with CFD results employing a volume of fluids method. The results were in good agreement with high-speed photography. Recently, Mithran and Venkatesan [10] analyzed the effect of IR transceiver orientation on gas/liquid two-phase flow. It was noticed that the IR sensors were sensitive to angles at which the transceiver is placed. The thickness and shape of tubes varied the signal output which was not reported in the literature for multiphase flows.

Above literature shows that limited works are reported on using IR sensors for two-phase flow. When compared to other sensor arrangements, which require a separate signal processing circuit, IR sensors are simple in construction. The transceiver does not need any complicated signal processing techniques. The objective of the present work is to analyze the effect of wall thickness and shape of the channel on irradiation behavior of the infrared sensor during two-phase flow. The effect of various dimensional ratios and shapes (circular, square, and triangular) of test sections on the signal output characteristics of IR rays is analyzed for bubble and slug flow regimes. A numerical model is developed using COMSOL package to predict the irradiation behavior during single-phase flow. An IR transceiver circuit with the aid of NI-DAQ is used for measuring the intensity of scattered infrared rays in a gas/liquid two-phase flow. The results obtained are related with image processing techniques with good agreement. The developed numerical model will be useful to predict irradiation behavior of IR sensors for various shapes and sizes. The developed sensor may be further used along with arrays to determine volumetric void fraction.

2 Experimental Setup

The experimental setup for the present work is shown in Fig. 1. The setup is arranged in such a way that both IR sensing and imaging using high-speed camera is done simultaneously. Experiments are done in borosilicate glass test sections of internal diameters 2.50, 3.50, 5.20, and a 2 mm side square channel. The wall thickness of the test sections mentioned above is 0.3, 0.6, 0.6, and 0.5 mm, respectively.

Two separate E-spin SPK101 single channel syringe pumps with flow rate capacity of 0.4 nl/min–13.6 ml/min are used for air and water supply. Flow rates are controlled using a digital controller which is an integral part of the syringe pump. Water with

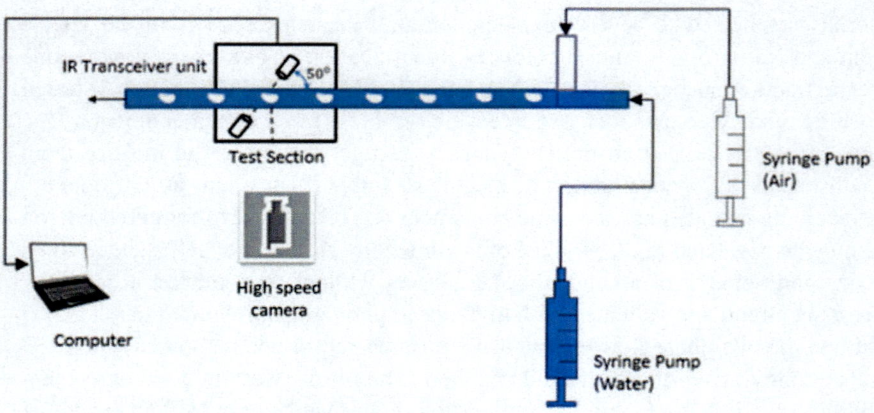

Fig. 1 Schematic diagram of the experimental setup

TDS concentration of 250 ppm is used for the experiments. The temperature of water used is 29 °C. Air and water are mixed in a T-section which is an integral part of the test section and has a vertical length of 5 cm. The IR transceiver circuit is positioned at a distance of 5 cm away from the mixing T-section so that the regime is fully developed. The IR transmitter and receiver face each other close to the tube in a straight line horizontally at an angle of 50° with the vertical. The angle chosen is based on the experiments initially conducted for single phase fluid so that maximum rays from the transmitter fall on the receiver unit. Two-phase flow is from right to left concerning the incident rays of the sensor. The length of the test section is 150 mm (uncertainty for length of the test section is in the order of $\pm$ 0.6 mm for all the tubes). BASLER CMOS acA2000 monochrome high-speed camera which can record at 340 fps at 2 MP resolution is positioned next to the sensor. Navitar zoom 7000 lens is used with the camera. The high-speed video and the sensor data are simultaneously recorded. IR transceiver signal is recorded using an NI-DAQ 9174 chassis with NI 9203 module with a maximum sampling rate of 200 kS/s using LabVIEW software. The image obtained is processed using MATLAB image processing technique which detects the edges of the flow regime and measures the dimensions of the flow regime.

3 Design and Simulation

Infrared rays are invisible to the naked eye at ambient temperature. It is electromagnetic radiation with a wavelength of the order of more than 700 nm. IR rays with longer wavelength are less scattered than visible light and are not obscured by gas or dust particles. All objects emit infrared radiation at all temperatures except at absolute temperature. The amount of radiation emitted is inversely proportional to the temperature of the object. The IR sensor used is a commercially available IR333A model. The IR transmitter and receiver are 5 mm in diameter and 7.6 mm length. IR

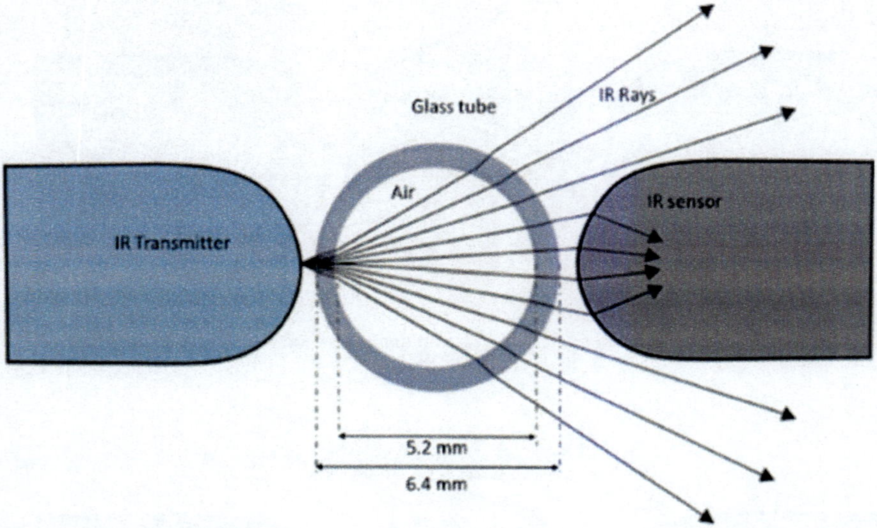

Fig. 2 IR rays for 5.2 mm diameter tube with 0.6 mm thickness (only air)

rays emitted from IR diode form a 30° conical shape at the exit of the transmitter, and after making contact with the glass tube, the rays diverge as shown in Fig. 2.

Focused conical rays emitted from IR diode are passed through the test section made of borosilicate glass. Initially, the intensity of IR rays is measured when only air is present inside the tube. The refractive indices of air, glass, and water are 1, 1.517, and 1.33, respectively. As the refractive index ratio for glass-air is 1.517 and for glass-water is 1.14, based on Snell's law, IR rays get refracted more while passing from glass to air than glass to water. The intensity of rays after divergence from the glass tube is relatively less, and only a small part of the incident IR rays reaches the receiver.

When the experiment is done with the tube filled with water, the diverging rays start to converge. A large part of incident IR ray reaches the receiver as shown in Fig. 3. As shown in Fig. 4, when a gas slug enters the test section, due to changes in the refractive index, the converged rays toward the sensor gets deflected. This will result in a reduction of measured current in IR receiver. Snell's law of refraction is used for depicting the above phenomena which are as follows

$$n_1 \sin \theta_1 = n_2 \sin \theta_2 \tag{1}$$

As the rays from IR transmitter are projected at 30° angles, the ray will meet at an incidence angle of θ_1 when compared with a normal tangent drawn on the circular boundary of the test section. Snell's law can be used to determine the angle of refraction θ_2. Similarly, all other angles of refraction can be calculated by considering the angle of refraction as the angle of incidence for other medium change.

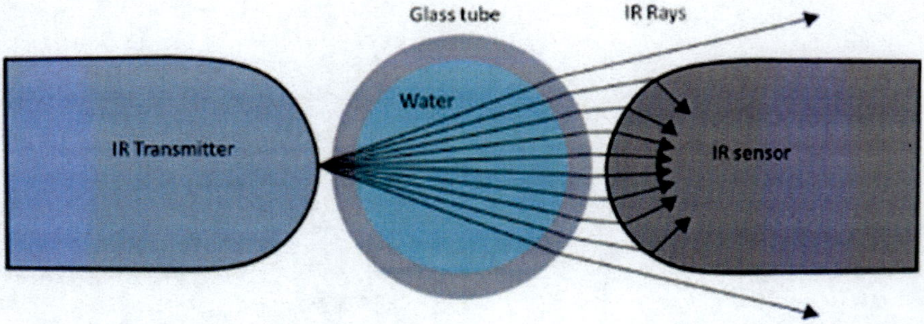

Fig. 3 IR rays for 5.2 mm diameter tube with 0.6 mm thickness—only water in the test section

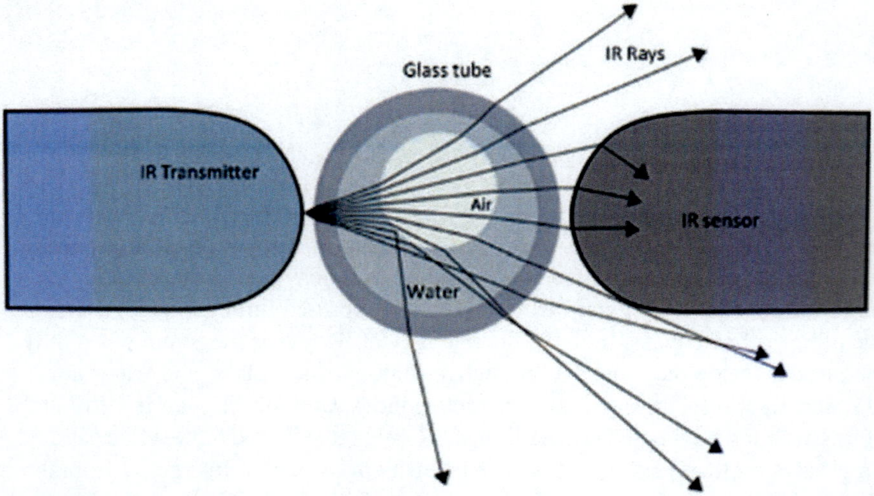

Fig. 4 IR rays for 5.2 mm diameter tube with 0.6 mm thickness—water and air in the test section

3.1 Method of Measurement Using IR Sensor

The limited region of the infrared wavelength spectrum with specific bandwidth is sensed by commercially available sensors. In the present work, the intensity of IR rays from IR transmitter is adjusted using a 10 k variable resistor from 0 to 3 V in IR transmitter circuit. A constant DC source of 5 V is used for power supply. The IR receiver is directly connected to an NI-cDAQ 9174 chassis with NI 9203 module (mA measurement) with its negative terminal as input with a 5 k variable resistor. The current variation in the sensor is measured using NI-cDAQ with 50 samples at 100 Hz for varying two-phase flow regimes of interest. The module is capable of measuring—20 to 20 mA. High-speed video image is instantaneously captured for the same flow regime. The images are captured using a high-speed camera at 300 frames per second. The captured image is processed by image processing techniques described in [11]. After the image is processed, the Euclidian pixel length is obtained. The size of the test section is already known, and hence, width and length of the bubbles are calculated.

3.2 Modeling in COMSOL Multiphysics

COMSOL Multiphysics software with ray optics module is used to solve the first order ordinary differential equations of Snells's law. The current generated from IR photodiode is related to a total number of photons entering the IR receiver Mithran and Venkatesan [10]. The governing equations for predicting ray direction in COMSOL ray optics are

$$\frac{\mathrm{d}q}{\mathrm{d}t} = \frac{\partial \omega}{\partial k} \tag{2}$$

The numerical model is developed with the test section, IR LED, and photodiode of dimensions as same as that of the experiment. The IR transceiver setup is positioned at an angle of 50° with respect to the test section for three different cross-sectional channels as shown in Fig. 5.

Horizontal circular tubes of internal diameter 2.50, 3.50, and 5.20 mm and thickness 0.3, 0.6, 0.6, and 0.5 mm, square channel of cross section—2 × 2 mm with 0.5 mm thickness, and an equilateral triangular test section of side 3.04 mm with 0.50 mm thickness is modeled for single phase water and air flow separately. The rays transmitted from LED incident on IR photodiode through a borosilicate glass test section and water with refractive index 1.517 and 1.33, respectively, are modeled using COMSOL package. The scale of IR rays is adjustable in terms of a number of rays. In the present case, it is chosen as 50 to have better visibility of the ray. Figure 5 shows the irradiation behavior of IR rays for various shapes of the test section. It can be observed from Fig. 5 that the divergence of IR rays in a triangular channel is more and fewer photons reaches the IR photodiode when compared to circular or square test sections. Due to the shape of the triangular test section, three different orientations with the IR transceiver are numerically studied as shown in Fig. 6. It is observed from the simulations that the percentage of rays reaching the IR photodiode is more in case 1 compared to other cases. In case 2 and case 3, the percentage of rays received for single phase air in test section is relatively higher, and percentage of rays received for a single phase of water in test section is less. While, in case 1, square channel and circular tubes the percentage of rays received for single phase water in test section is higher. The variation is observed as the divergence of IR rays has increased in case 2 and case 3 as transmitted photon rays are directed toward the corner of the triangular channel.

Hence, the orientation of triangular test section in case 1 is preferred as the divergence patterns of IR rays in this orientation is similar to the test sections of other

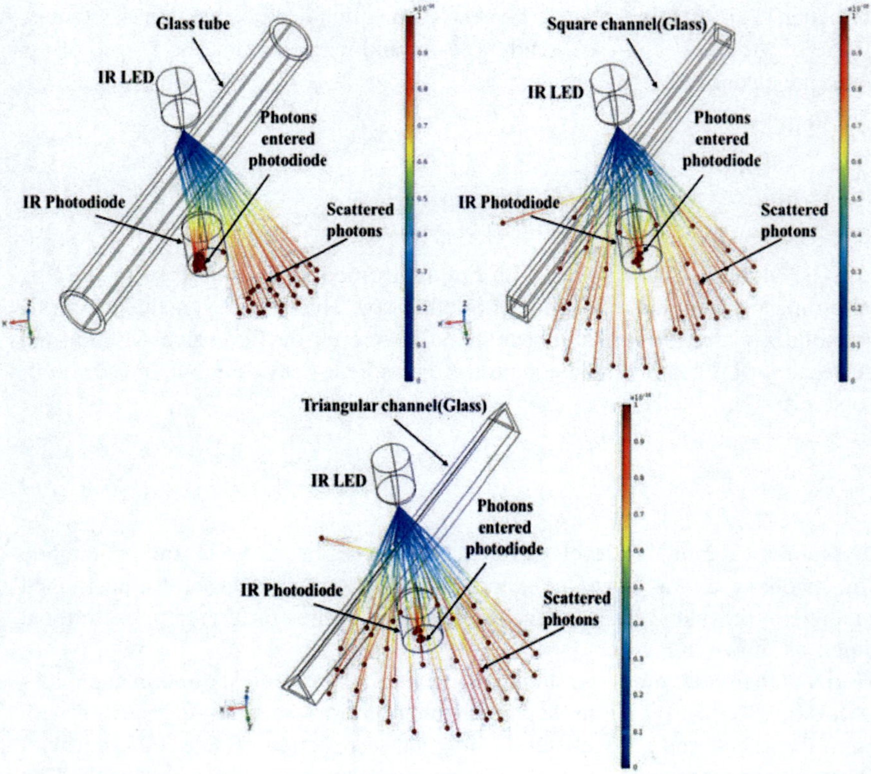

Fig. 5 Simulated COMSOL model for test sections of different shapes with IR transceiver setup

cross-sectional shapes. The other orientations of the square channel are not considered as the same phenomena would be observed, and there will not be any difference in signal output because of orientation. Two-phase flow experiments are planned on triangular channels separately for varying thickness and are scope for future work.

4 Results and Discussions

The various flow regimes visualized using high-speed photography and the corresponding current–time output obtained using IR sensor is compared for the square and circular test section. The experiments show that the cross-sectional void fraction (ratio of air to water at that cross-sectional area) is directly proportional to the amplitude difference of the current measured by the sensor. Initially, the experimental variation of current values is measured for single-phase flow of air and water on the same test section separately. The numerical results obtained from the model

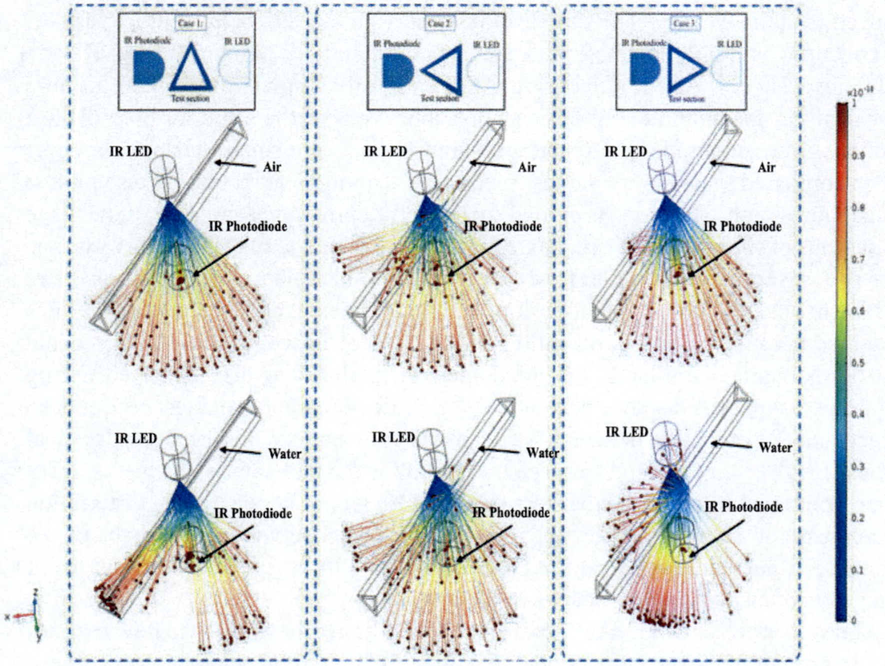

Fig. 6 Simulated COMSOL model for different orientation of triangular test sections with IR transceiver setup

developed using COMSOL package as in previous section is compared against the experiments and is shown in Fig. 7.

The current measured experimentally and the predicted percentage of photons numerically using COMSOL Multiphysics package is compared. From Fig. 7, it can be inferred that the measured mA value is lower for air and higher for water. The value of current measured with water for 2.5 mm tube is higher as the wall thickness of the tube is less than other test sections leading to increasing in the

Fig. 7 Current measured for air and water in different circular test sections

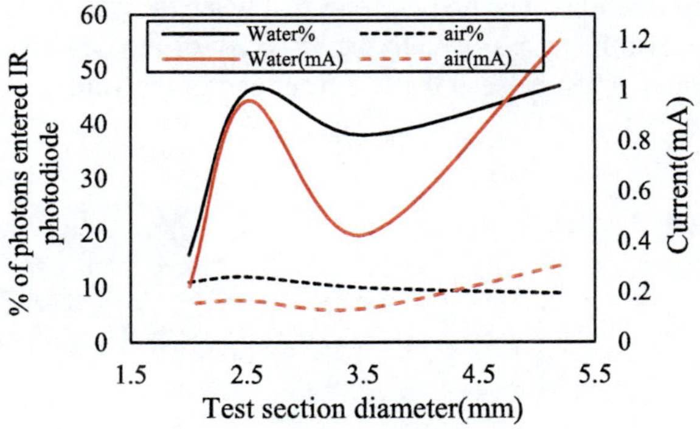

convergence of IR rays. The dimensional ratio (internal diameter to thickness) of test sections with inlet diameter 2.5 mm, 3.5 mm, and 5.2 mm are 0.12, 0.18, and 0.11, respectively. As the dimensional ratio value is similar for 2.5 and 5.2 mm test sections, the amplitude difference also appears to be the same for both the test sections. The amplitude difference measured for a 2 mm square channel is lower when compared to other circular test sections. The numerical results are also plotted in the same graph. The graph is plotted such that the current measured and percentage of photons of similar scale lie in y-axis for the same diameter of test section in x-axis. The patterns formed are similar for experiments as well as numerical data's, but there is a slight deviation of numerical results from experimental results. The difference is observed because the exact positioning of the transceiver with the test section could not be maintained similar to that of numerical model. It is also observed during experiments that the distance between IR LED, IR photodiode and test section also affects the IR irradiation behavior. The same phenomena are observed in numerical simulation. The amplitude difference measured for a 2 mm square channel is lower when compared to other circular test sections. Hence, only one square test section is compared to show the difference of signal pattern observed for the shape of the regime. Further, studies are required for bubble and slug regimes in various sizes of square channels and various shapes of the test section.

The values are shown in Fig. 7 and are taken as limits for two-phase flow regimes for the present work since the two-phase flow mixture will generate a signal whose current value will be between these limits. Bubbly and slug flow regimes are the two-phase flow regimes considered in the present work since accurate measurement of these regimes can be done with image processing. The bubble and slug regimes occurring in the four test sections are compared. High-speed photographs of bubble regimes for a 5.2 mm round tube and its processed image using the proposed algorithm are measured for superficial velocity of $U_g = 0.0116$ m/s and $U_l = 0.0081$ m/s. The superficial velocities are measured based on the data from syringe pump. The width and length of the bubbles in Fig. 8 are calculated as 4.54 and 10.92 mm using image processing technique. The signal output of the same bubbles measured using an IR transceiver is shown in Fig. 9. The variation in mA as a function of time is plotted to show the refracted IR rays. The direction of two-phase flow is from right to left. The measured mA as a function of time plotted represents the cross-sectional void fractions. The graph shows that when the size of the air void increases, drop in mA also increases. Due to the viscosity of the denser medium and velocity of the bubble, it can be seen that the front end of the void is sharper compared to its back end.

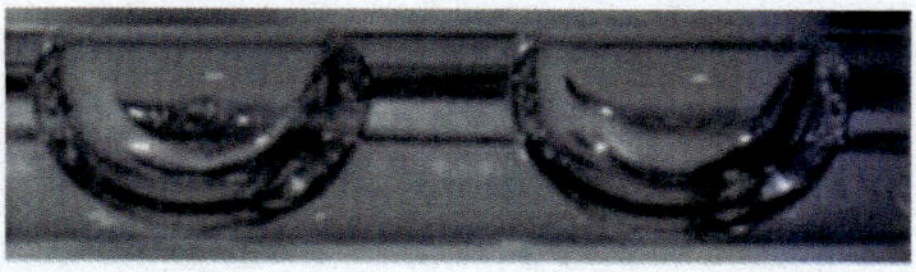

Fig. 8 Bubble in 5.2 mm tube

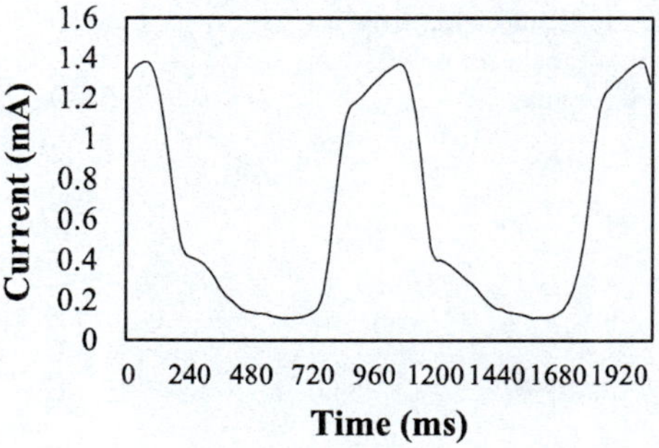

Fig. 9 IR measurements in 5.2 mm tube

The phenomena can also be observed from the plot where the drop of mA is stable and then drops further. The amplitude of current mA measured at the mid-section of the bubbles of size 4.59 mm and 4.54 mm are 0.112 mA and 0.114 mA, respectively. It may be recalled that the amplitude value for only water in the 5.2 mm inner diameter tube is 1.4 mA as detailed in Fig. 7. The difference in current value between only water and at mid-section of the bubble will give the current amplitude difference for cross-sectional void at the mid-section. The values recorded are 1.288 mA and 1.286 mA, respectively, which is proportional to the cross-sectional void fraction at the mid-section of the mentioned bubbles.

The observation is similar for regimes in a 3.5 mm tube and 2.50 mm tube with reduced amplitude of mA as shown in Figs. 10 and 11. The bubbles formed in this flow for given superficial velocities are almost identical in front and rear ends. Hence, the plot formed is mostly curved.

However, for 2 mm square section, the refraction of IR rays is different. It can be visualized from the plot that from the reference line (water only value of 1.70 mA), the value of current initially increases and then reduces which appears to be different from that of circular tubes. Further, the amplitude of mA is also reduced. The increase in mA is because of the flat section of the channel. In round tubes, the IR rays from the

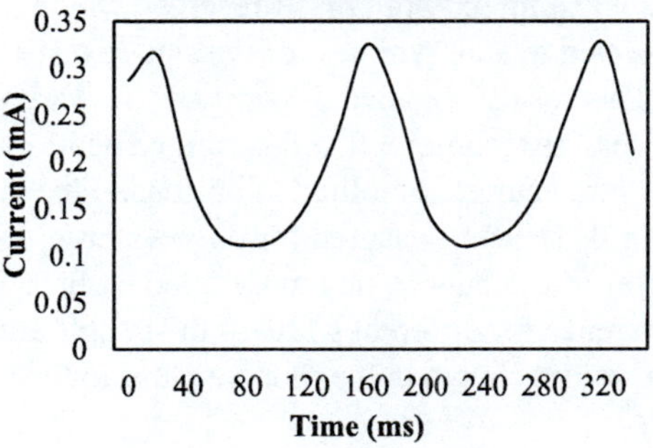

Fig. 10 IR measurements in 3.5 mm tube

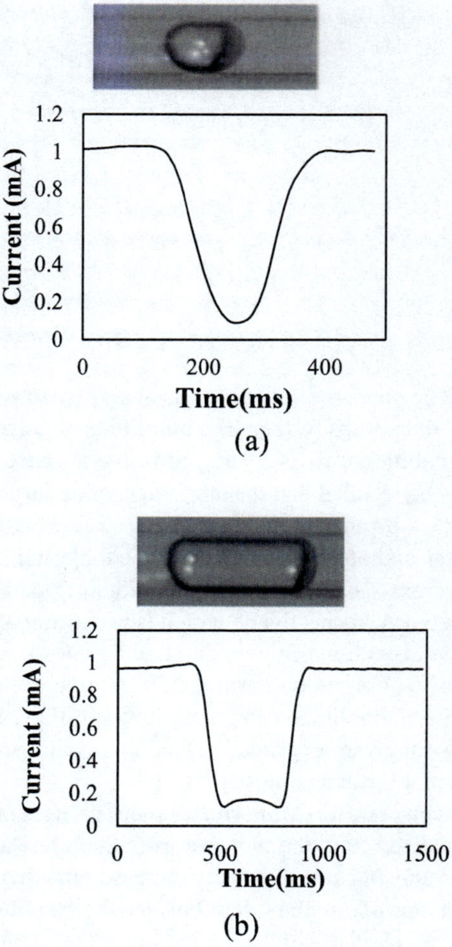

Fig. 11 IR measurements in 2.5 mm tube **a** for bubble and **b** for slug

transmitter are refracted more because of curvature shapes. In a square test section, since the section is flat, more photons reach the receiver without getting scattered. The surface of the glass is flat, and the refracted IR rays do not converge or diverge much and directly fall on IR receiver, which is shown in Fig. 12. A similar pattern is observed in slug regimes for the square test section.

The result obtained is verified for various sizes of bubble/slug using image processing technique. The determined width of the regime is compared with change in measured current amplitude difference. The difference is between the current output from IR sensor measured for only water in test section and current measured in the center of the bubble/slug as detailed earlier. Figure 13 shows the current amplitude difference for different width of the bubble and slug regimes in various test sections. The values shown in the plot are close to the current measured for only water in test

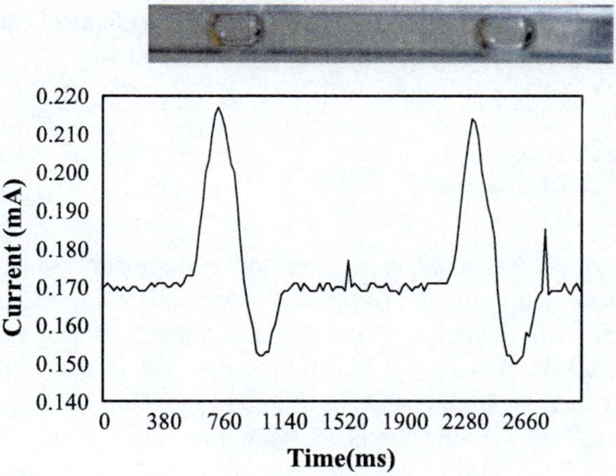

Fig. 12 IR measurements in 2 mm square tube

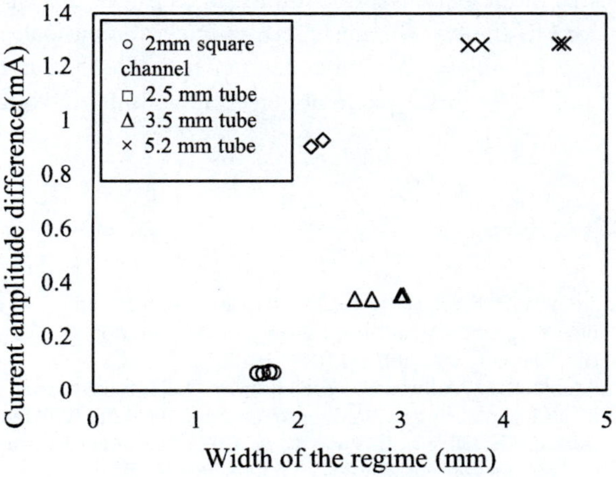

Fig. 13 Current amplitude difference for the width of the regimes in different test sections

section as shown in Fig. 6. Hence, the diametrical ratio of test section influences the irradiation behavior of IR rays.

The increase in current amplitude difference is observed with increase in width of the regime. For test sections, 2.5, 3.5, and 5.2 mm circular tubes, the amplitude difference values obtained are within the current range measured for only water and only air in the test section. However, the amplitude difference for 2 mm square test section is lower than the current measured for only air in the test section. The phenomena observed in the square channel is different from round tubes due to the curved shape of nose and tail of the regime and orientation of the IR transceiver setup, and hence, further analysis is required for varying thickness ratios of square channels and is scope of future work. The above results show that the thickness and size of the test sections play a vital role in the output of IR sensor. Comparison of the square, round and triangular (numerical) channels indicates that the output of the sensor depends on the shape of the channel also.

5 Conclusions

In the present study, two-phase flow experiments are conducted in the small round, square and triangular (numerical) channels of varying diametrical ratios which are kept in horizontal position. IR irradiation behavior is studied numerically using COMSOL package for the different dimensional ratio (thickness/cross section) of the test section. High-speed camera and image processing algorithm are used to determine the size of the bubble and slug flow regimes. The IR transceiver circuit with the help of NI-cDAQ is used for measuring the amount of scattering of IR rays for the same flow regimes. The intensity of scattering of IR rays is directly proportional to the size of the bubble/slug but influenced by the dimensional ratio (thickness/cross section) of the test section. No filtering or amplification circuit is required in the present technique. The numerical model developed can be further used to design an array of IR sensors capable of detecting volumetric void fraction.

References

1. Adhavan J, Balachandar C, Arunkumar S, Venkatesan M (2016). Determination of two phase flow slug velocity and length using infrared sensor. In: Fluid mechanics and fluid power—contemporary research, pp 1011–1017
2. Arunkumar S, Adhavan J, Venkatesan M, Das S, Balakrishnan A (2016) Two phase flow regime identification using infrared sensor and volume of fluids method. Flow Meas Instrum 51:49–54
3. Barreto EX, Oliveira JLG, Passos JC (2015) Frictional pressure drop and void fraction analysis in air-water two-phase flow in a microchannel. Int J Multiph Flow 72:1–10
4. Cartellier A (1992) Simultaneous void fraction measurement, bubble velocity, and size estimate using a single optical probe in gas–liquid two-phase flows. Rev Sci Instrum 63(11):5442–5453
5. Ide H, Kariyasaki A, Fukano T (2007) Fundamental data on the gas–liquid two-phase flow in minichannels. Int J Therm Sci 46(6):519–530
6. IR Sensing Catalog (2015) 132.55 revision 3.2, 15-04
7. Juliá E, Harteveld W, Mudde R, Van den Akker H (2005) On the accuracy of the void fraction measurements using optical probes in bubbly flows. Rev Sci Instrum 76(3):035103
8. Jagannathan N, Chidambaram B, Seshadri A, Muniyandi V (2015) Characterization of gas–liquid two-phase flows using laser patterns. Can J Chem Eng 93(9):1678–1685
9. Lawal DU (2014) Void fraction measurement using electrical impedance techniques. Int J Adv Eng Technol 7(5):1539–1548
10. Mithran N, Venkatesan M (2017) Effect of IR transceiver orientation on gas/liquid two-phase flow regimes. Flow Meas Instrum 58:12–20
11. Mithran N, Venkatesan M (2020) Volumetric reconstruction of taylor slug gas flow using IR transceiver in minichannels. IEEE Trans Instrum Meas 69(6):3818–3825
12. Prasser HM (1999) Wire-mesh sensors for two-phase flow investigations. Inst Safety Res 31(18):23–28
13. Peatross J, Ware M (2013) Physics of light and optics, 7th edn. Brigham Young University, Provo, UT, USA

14. Rocha MS, Cabral ELL, Simões-Moreira JR (2009) Capacitance sensor for void fraction measurement in a natural circulation refrigeration circuit. In: International nuclear Atlantic conference. 978-85-99141-03-8
15. Rocha MS, Moreira JRS (2008) Void fraction measurement and signal analysis from multiple-electrode impedance sensors. Heat Transf Eng 29(11):924–935
16. Sur A, Liu D (2012) Adiabatic air–water two-phase flow in circular microchannels. Int J Therm Sci 53:18–34

Comparative Analysis of Drag Force in a Deep-Water Wading Simulation of Ahmed Body with Different Commercial Automobile Models

Shivam Prajapati, Shivam Gupta, and Nishi Mehta

Nomenclature

k	Turbulence kinetic energy
t	Span
U_j	Internal energy
x_j	Initial position
P_k	Production Limiter
β^o	Coefficient of thermal expansion
ω	Specific turbulent dissipation rate (Omega)
υ	Viscosity
υ_T	Kinematic eddy viscosity
α	Constant
S	Vorticity
σ_ω	Turbulent Prandtl number
F_1	Blending function
y	Distance to the next surface
$\mathrm{CD}_{k\omega}$	Positive portion of cross—diffusion term of the Eq. (2)
ρ	Density
a_1	Constant
F_2	Second blending function
F_d	Downforce
C_d	Drag coefficient

S. Prajapati (✉)
NIT Agartala, Agartala, India
e-mail: prajapatishivam64@gmail.com

S. Gupta
ISM (IIT) Dhanbad, Dhanbad, India

N. Mehta
SVNIT, Surat, India

J. Banerjee et al. (eds.), *Recent Advances in Fluid Dynamics*, Lecture Notes in Mechanical Engineering, https://doi.org/10.1007/978-981-19-3379-0_28

v Velocity
A^w Area of reference

1 Introduction

General: Considering the advancements in computational power, CFD simulations have become a reliable solution to predict the flow in the fluidic system. Computational fluid dynamics can be broadly used for fluid–structure interaction and can provide valuable insight into the fluid flow around an object. CFD is used by automotive industries for the vehicle fuel efficiency and drivability affecting forces like drag and turbulence forces on external flow exposed components like car body. CFD, which stands for computational fluid dynamics, is the method that uses numerical analysis to calculate and prophesy the fluid behavior with the power of computation. CFD methods have been extensively applied in aerodynamic design and optimization of both aircraft and automobiles. Thanks to CFD, various engineering problems can be considerably solved, such as aerodynamics, gas turbines, turbomachinery, multiphase modeling, and ship hydrodynamics. Many of the engineering problems presented above demand experimental results that are often unfeasible to realize or economically non-viable. Owing to these reasons, exponential growth of computational power as well as low cost and time consumption in CFD, it has become an integral part of solving fluid flow problems.

Recently, CFD is being used in both industry as well as academia to deal with fluid flow problems, owing to its ability to solve them quicker and obtain more precise results than traditional wind tunnel methods. To ensure the reliability of CFD results, their validation and verification are required. The codes used in CFD are based on numerical algorithms. They basically consist of three main steps: Pre-processing, processing, and post-processing.

In the first step, geometry and mesh are generated corresponding to the problem domain. For the second step, initial and boundary conditions are provided to the system, and pertinent physical models, numerical schemes, and solvers are selected for the calculation. In the last step, using post-processing software, the required results are visualized that are obtained from the calculation executed in the previous step. Anbarsooz et al. [1] presented "A numerical study on drag reduction of underwater vehicles using hydrophobic surfaces". In this study, he used a hydrophobic surface for an underwater vehicle with a body profile having unseparated flow. In the results, it was witnessed that the drag force and drag coefficient can be decreased using hydrophobic surfaces. He also found that drag force can be lower than supercavitating hull.

Problem Description: CFD being a research rather than design tool, vehicle aerodynamics are still being understood by coupling CFD with wind tunnel results. As a road vehicle is bluff body having complex geometry, it has a fully three-dimensional flow, turbulent boundary layers, and common flow separation. Moreover, with the current development of CFD's application in increasing vehicle performance, scenarios like water wadding are also trending with traditional aerodynamics study. There has been a little bit exciting work in this field which has led to arise more new questions than ever. Also, there has been no single study analyzing and comparing geometrically different vehicle's performance in deep wading scenarios using CFD. The main aim of this work is to address the aforementioned research gap.

Literature Review: On the motion of any vehicle in a fluid, exemplifying the motion of car in water in the monsoon, the action of drag force on it prevails. The vehicle's performance is affected by the drag force caused by fluids such as air and water. A numerical study on the reduction of drag on underwater vehicles consisting of hydrophobic surfaces was presented by Anbarsooz et al. [1]. The coefficients of total friction and pressure volumetric were the basis of this idea. Slip velocity is changed by changing sliding coefficient. The slippage of fluid on the hydrophobic surface reduced the skin friction drag up to a large extent. They numerically investigated the reduction of drag force in an underwater hull having an unseparated flow profile. When the sliding coefficients were smaller than 10, the level of drag was quite noticeable. It was noticeable to such an extent that an underwater vehicle consisting of an unseparated flow profile was fairly comparable to a supercavitating hull. A drag coefficient estimation model was proposed by Tan et al. [2] for the simulation of dynamic control of autonomous underwater vehicle (AUV). It was based mainly on the trial runs' frequency and the onboard inertial sensors of the vehicle. The results obtained in this paper through the proposed approach on ANSYS CFX were quite close to the real drag coefficient. The robustness as well as the accuracy of this model were investigated by exemplifying the basic models of cylinders. More research was suggested in the direction of other fitting methods like locally weighted smoothing regression for the accurate prediction of drag coefficient. The CFD simulation of flow around the body of Ahmed, an external vehicle, was presented by Banga et al. [3]. The paper basically focused on the inefficiencies' and losses' reduction which are caused due to the forces considered in the road vehicles' aerodynamics like drag and lift forces. The variation of the rear slant angle of the body of the Ahmed vehicle and its impact on the coefficient of drag and lift were investigated through numerical methods in order to accomplish the purpose of this paper. Also, they simplified the geometry of the vehicle. By plotting the coefficients of drag and lift on a graph, some observations were made related to their trend with respect to the rear slant angles. They were:

- An increasing linear trend is followed by lift coefficient for the rear slant angle between the range of 0–200.
- The minimum positive value of the lift was 0.0292, and it was obtained for the value of 7.50° of the rear slant angle.
- The drag coefficient showed decrement with the rise in rear slant angle in the range of 0°–7.50°, the minimum value of drag coefficient being 0.2346.
- After the rear slant angle reached to 7.50°, the drag coefficient gradually increased up to the maximum value of 300.

- After the drag coefficient reached to 300, the flow separation occurred along with a random dispersal of the coefficients of drag and lift.

The shallow water effect was numerically investigated by Nakisa et al. [4] on the resistance of multipurpose amphibious vehicle. Water pumps were employed by them for the control of the water ingress during the mission of river crossing. Reynolds-average Navier–Stokes (RANS) equation-based finite volume Method (FVM) solver in ANSYS CFX was used for modeling and simulation. The ship resistance was significantly impacted by the hydrodynamic pressure distribution on the hull of the ship. The ascent of the flow under the keel may lead to pressure reduction that can cause buoyancy to decrease. As a result, the sinkage of the ship can occur due to which wetted surface area will increase, and thus, viscous drag gets increased. Khapane et al. [5] presented "Deep-water wading simulation of Automotive Vehicles". In other similar studies, there were limitations. One of them was the computation of the vehicle's inertial field during wading. In this study, the wading at different depths with varying velocities was calculated. Buscariolo et al. [6] presented "Multiphase water flow simulation of vehicle's roof". In this research work, a method for simulation for verification of the behavior of the water flow from the vehicle's roof to its side ditch was developed. Further, they manufactured a device for the simulation of rain, which was serving a dual purpose of simulating as well as testing. They inserted virtual parameters in it for reproducing the same effect.

In all the researches, there were many studies related to drive through water using computational fluid dynamics, but it was noticed that there was no proper comparison of commercial cars and Ahmed body. This simulation was done to fill that gap. In the paper, a comparison of two commercial vehicles (shown in Figs. 3 and 4) was made with the Ahmed body. We have compared the drag, lift coefficients, as well as downforce produced by the walls of the car.

Vehicle Wading: It refers to a situation where a vehicle travels through a road where water is flooded at a certain height and thus suffers drag as a combined effect of air and water. Vehicle wading at various speeds and different depths of water is a principal test to check the efficiency of automobiles in the flood-prone region. After massive rain events, sometimes water level exceeds the critical height, and the wading drag becomes sufficient enough to halt the car. When the water level surpasses a required height, the underwater components of the automobile like bumper, engine undertray, transmission scoop, radiator, plastic sills, and electronic circuits become sensitive to damage. Thus, to observe the effect on internal components of the body requires more vigorous CFD analysis, including dynamic mesh though such situations are not considered in the current study. During the high-speed wading, water splashing is observed. It is quite complex for modeling. Because of the wheels' rotation as well as vehicle floor's impact on the surface of water, a hydrodynamic inertial force field is generated. Such effects are not considered in the present study of standalone CFD simulation.

Ahmed Body: The generic Ahmed body as shown in Fig. 1 is taken as a reference model in the current study to compare the drag force experienced by various

Fig. 1 Ahmed body

commercial automobile's models. The body was originally developed by Ahmed [7] in 1984 in his research "Some Salient Features of the Time-Averaged Ground Vehicle Wake" to mimic the conventional flow field around a bluff car body. Aerodynamics study is used in reducing drag coefficient while designing the car, and Ahmed body has become a benchmark for aerodynamic simulation analysis. Ahmed body is a simple bluff body that allows for accurate flow simulation through its simple enough shape but retains essential practical phenomenon significant to automobile bodies. The Ahmed body consists of a rectangular box, which connects the round front part and the movable rear slant plane placed at the back part of the body. The Ahmed body can be used as a standardization model in aerodynamic designing of automobiles, and thus, in the present study, author has used this model to produce a constructive comparison report of drag force with other two commercial models. More information on the geometry will be imparted in the latter section.

Scope of the paper: In the above literature review section, one can find many research works accomplished in aerodynamics designing of a model and modeling of automobiles in water wading scenarios. However, these researches lack a proper comparison of the various automobiles and the effects of drag suffered by the vehicles. In this paper, CFD analysis has been performed on two unique commercial car models and standard Ahmed body under similar deep-water wading conditions and physical setup. The main aim of this analysis is to provide a comparison report of total drag force suffered by each individual model. Furthermore, the current study also investigates the fluid behavior around the car model to provide better insight into the external fluid flow. The total drag is the combined drag force offered to the model due to change in pressure and viscosity of both fluid media (air and water).

2 Methodology

The approach followed in this paper is based on the comparative assessment of drag effects on selected car models which are similar to the typical commercial vehicles. Computational fluid dynamics can be broadly used for the analysis of airflow around a body. Each of the two chosen models and the Ahmed body is examined in detail using CFD analysis. This analysis is wholly based on multiphase flow, using volume of fluid method (VOF) and having two phases (air as primary phase and water as secondary phase). K-omega SST model consisting of open channel flow is used for the simulation. The boundary conditions of the simulation were set to a constant velocity of '8 m/s' at the inlet and a zero-gauge pressure at outlets. Steady-state simulation was carried out. Moreover, before running the calculations, the hybrid initialization was carried out.

The methodology opted in this paper is similar to the typical approach used in most CFD analysis. First, the physical problem is transformed into a mathematical model by converting the continuous domain into a discrete domain, as shown in Fig. 2 (this process is known as meshing). Then by employing boundary conditions and various models, the mathematical model is then solved using a relevant solver to obtain a numerical solution (required variables at selected points). Using post-processing software, these numerical solutions can be visualized, and thus, can be utilized to study the physical flow as needed.

The criteria for assessing the performance, respectively, the 'effectiveness' of a commercial car model in a deep-water wading situation are vague. Many objective factors, as well as manufacturing cost, play a role. Therefore, the main criteria of assessment of the vehicle's performance focused here are the amount of drag force experienced by it during wading. In order to keep the results to be presented later on

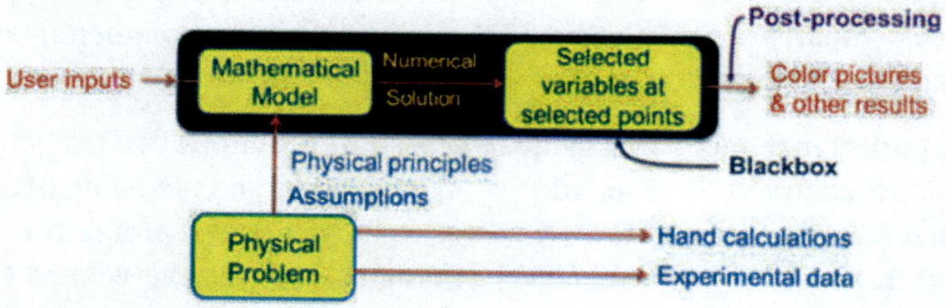

Fig. 2 Methodology

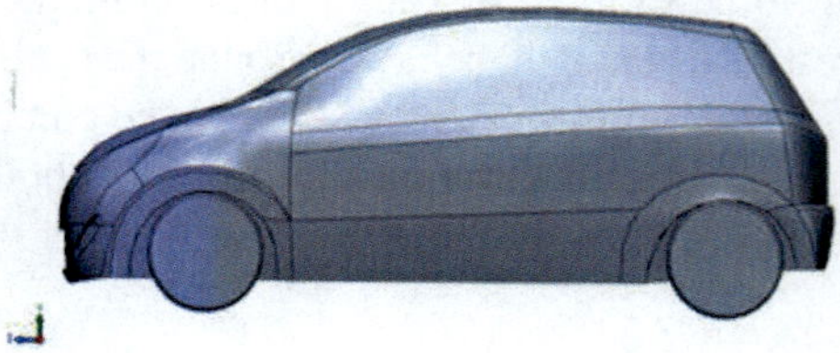

Fig. 3 First vehicle geometry

Fig. 4 Second vehicle geometry

as comparable as possible, all the simulations are performed in an environment as identical as possible. This means especially that the same water wading height was initialized in all simulations.

CAD Geometry: We used two simplified models of commercial cars and an Ahmed body which were designed as per the standard dimensions. The three-dimensional CAD models of the commercial cars were developed using solidworks 2016. The first model shown in Fig. 4 was intended to have a streamlined depicting figure, while the second model shown in Fig. 3 was developed in a way to have a certain extruding figure in the rear. Furthermore, windshield and front wing were designed in order to facilitate the airflow pattern around these sections. In the whole simulation, functioning and effects on various internal parts of the body are not taken into the interpretation in the results. Hence, we have assumed the complete model of this vehicle to be a single individual solid rigid body. The geometry of the body of Ahmed vehicle is considerably different from regular vehicles. The fundamental aerodynamic properties of the vehicle are represented by the vehicle's body, especially its rear part. The rear slant angle ϕ has a significant impact on the aerodynamic drag and lift coefficients, as they change quite abruptly when its value becomes 30°, also known as the value of critical slant angle. It is important to mention that dimensions of all three models, including Ahmed body, are scaled likewise in order to get reliable comparative results.

Meshing: Firstly, the geometry designing of the car was carried out. Then, an enclosure in which the vehicle model was contained was constructed in order to define the domain of finite volume, as shown in Figs. 5 and 6. The simulation was done as shown in Fig. 6. The meshing tool of ANSYS Workbench was used to discretize the domain. Although a flow domain larger in size is generally preferred for the CFD studies, it is limited owing to the computational load and the solved problem's flow

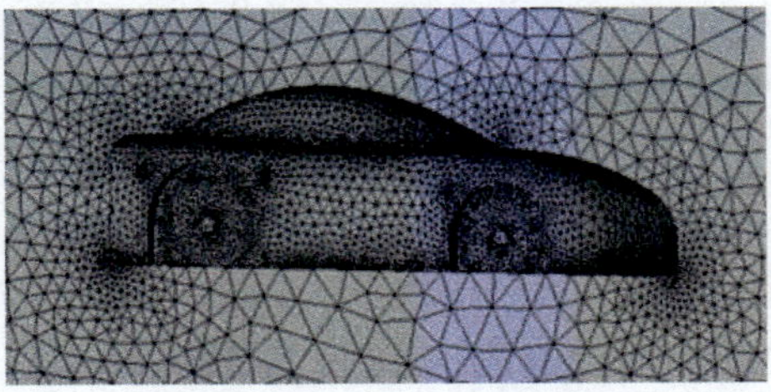

Fig. 5 Meshing for car body

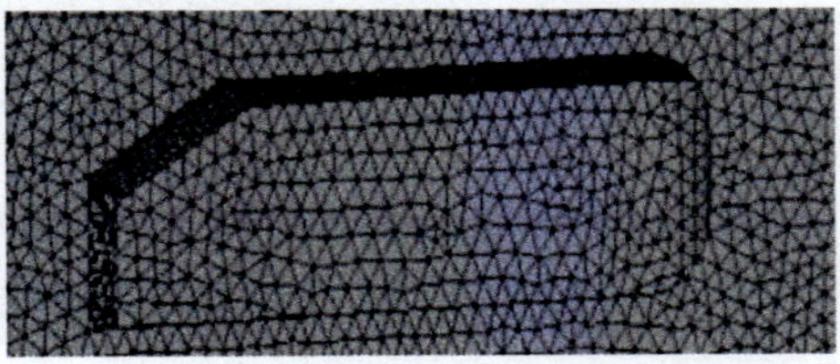

Fig. 6 Meshing of Ahmed body

nature. The flow domain generated around all the vehicles considered in this study has the following dimensions:

- In the vehicle's front, there is 2*L*
- Behind the vehicle, there is 5*L*
- Above the vehicle, there is 1*L*
- There is 1.5*L* towards the vehicle's side.

Here, *L* is considered to be the car's length parallel to the direction of flow. Half symmetric model was used for the analysis for reduction of time of computation with least impact on accuracy. A symmetry plane which was longitudinally passing through the middle of the vehicle body bifurcated the entire domain comprising of the vehicle's body.

Meshing software uses various algorithms for different types of mesh generation. In this study, patch conforming meshing method is used to discretize the computational domain into tetrahedral elements. A local sphere of influence method for body sizing is applied around the vehicle for concentrating the cell count close to the vehicle where the checking of the results is required. The element size in the refinement sphere was kept as 60 mm, while for the remaining domain it was approximately 120 mm. During the meshing of the models of commercial cars and Ahmed body, it was ensured that their orthogonal quality is above 0.1 while skewness was kept below 0.95. The number of elements in first, second, and Ahmed body models was 1,738,046, 1,636,950, 1,541,436, and respectively.

Modeling: The physical modeling described in this section is similar for all the three-simulations carried out in this study for each respective geometry. Each simulation consists of two phases: air and water. Multiphase modeling is accomplished by the open channel flow volume of fluid (VOF) model formulated using implicit parameters. This is an Eulerian type model in which governing equations are solved for each separated phase, and it does not deliberate the heat and mass transfer phenomena. For each model, pressure inlet boundary condition is applied for the mixture at the inlet, which requires the velocity magnitude to calculate the dynamic pressure for the total pressure calculation. The flow inlet comprises of two main parts, which were in shallow water area zone's water inlet and the air inlet which was above the free surface level. The difference between the free surface and bottom road level can be considered as the water wading height. The velocity, $v = 8$ m/s, free surface level, and $y_{\text{local}} \sim 0.67$ m (for deep-water wading) were imparted in the inlet boundary conditions of each model. Pressure outlet by prescribing the free surface level was provided at the outlet boundary, and the bottom road was defined as a stagnant wall with the no-slip condition. Symmetry condition was assigned to the symmetry wall to replicate the calculated results to the other side of the bounding box as well.

The model used for analysis was shear stress transport (SST) k-ω turbulence. The reason being, its ability of prediction of flows having adverse pressure gradient as well as flows having separation and reattachment. The SST k-ω model was developed by Menter et al. [8] in 1994 as well as it was modified by Menter et al. [9] in 2003. Basically, it is a two-equation, eddy-viscosity as well as fully turbulent model.

Moreover, it is the combination of the two famous turbulence models, which are the k-ω model in the sublayer of the viscous boundary and the k-ϵ model which is in the field which is quite distant from the wall.

Governing Equations: Owing to the accuracy of SST k-ω turbulence model, it is used for the CFD analysis in this paper.

A blending function is used between the equations of this model to gradually transition between the different k-ω models near the wall and the outer portion of the boundary layer. A standard model is used near the wall, whereas a model with high Reynolds number is used toward the outer portion. A modified turbulent viscosity formulation is also used in them to consider the transport effects of the principle turbulent shear stress.

Turbulence Kinetic Energy

$$\frac{\partial k}{\partial t} + U_j \frac{\partial k}{\partial x_j} = P_k - \beta^o \kappa\omega + \frac{\partial}{\partial x_j}\left[(\upsilon + \sigma_k \upsilon_T)\frac{\partial k}{\partial x_j}\right] \quad (1)$$

Specific Dissipation Rate

$$\frac{\partial \omega}{\partial t} + U_j \frac{\partial k}{\partial x_j} = \alpha S^2 - \beta\omega^2 + \frac{\partial}{\partial x_j}\left[(\upsilon + \sigma_\omega \upsilon_T)\frac{\partial \omega}{\partial x_j}\right] + 2(1 - F_1)\sigma_{\omega 2}\frac{1}{\omega}\frac{\partial k}{\partial x_j}\frac{\partial \omega}{\partial x_j} \quad (2)$$

F_1 (Blending Function)

$$F_1 = \tanh\left\{\left\{\min\left[\max\left(\frac{\surd k}{\underset{*}{\beta}\,\omega y}, \frac{500\upsilon}{y^2\omega}\right), \frac{4\sigma_{\omega 2} k}{\mathrm{CD}_{k\omega} y^2}\right]\right\}^4\right\} \quad (3)$$

Note: $F_1 = 1$ inside boundary layer and 0 in the free stream

$$\mathrm{CD}_{k\omega} = \max\left(2\rho\sigma_{\omega 2}\frac{1}{\omega}\frac{\partial k}{\partial x_j}\frac{\partial \omega}{\partial x_j}, 10^{-10}\right) \quad (4)$$

Kinematic eddy viscosity

$$\upsilon_T = \frac{a_1 k}{\max(a^1\omega, \mathrm{SF}_2)} \quad (5)$$

F_2 (Second Binding Function)

$$F_2 = \tanh\left[\left[\max\left(\frac{2\lfloor k}{\underset{*}{\beta}\,\omega y}, \frac{500\upsilon}{y^2\omega}\right)\right]^2\right] \tag{6}$$

P_k (Production Limiter)

$$P_k = \min\left[\tau_{ij}\frac{\partial \mathrm{U}_i}{\partial \mathrm{x}_j}, 10\,\underset{*}{\beta}\,k\omega\right] \tag{7}$$

where the velocity in x_j direction is represented by x_j, fluid pressure is represented by p, shear stress tensor is represented by τ_{ij}, and the body force term is represented by f_{bi}. Equations (3) and (4) are collectively known as Navier–Stokes equations.

Coupled pressure–velocity coupling scheme with second-order spatial discretization for momentum and pressure were used, while for turbulence kinetic energy, specific dissipation rate first-order upwind schemes were utilized. Least squares cell-based method was adopted for the calculation of gradients. Lastly, hybrid initialization was performed before starting the calculations.

3 Results and Discussion

The results after simulating all three geometries vehicle body 1, vehicle body 2, and Ahmed were analyzed after simulations. It was found that the overall drag of the vehicle body 1 was far less than that of the vehicle body 2. Since the former is more streamlined than the latter. Figure 10 presents the velocity streamlines of the geometry of the first vehicle. Figure 11 presents the velocity streamlines in the case of Ahmed body. All the results are found to be nicely correlating with the literature. The streamline velocity contours of all the considered bodies are displayed in Figs. 7 and 8. The contour of volume fraction of water as shown in Fig. 9 has helped us to understand that how the structure of a vehicle is affecting the water wadding flow. This will help us to analyze the contribution of each external parts of the body to the total drag suffered by the vehicle. Thus, we can consider vehicle body 1 to be the most optimum design for deep-water wading condition.

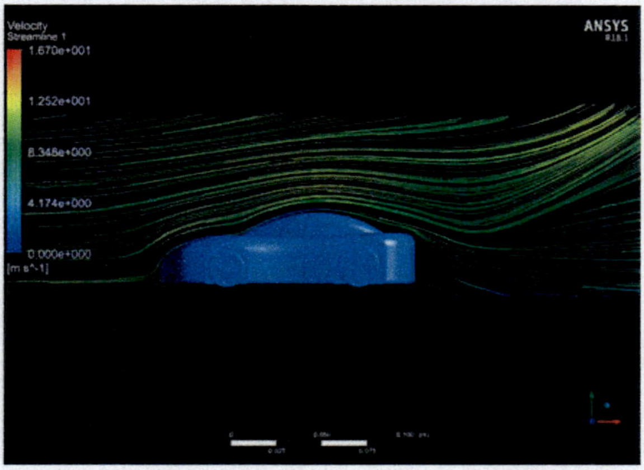

Fig. 7 Velocity streamline

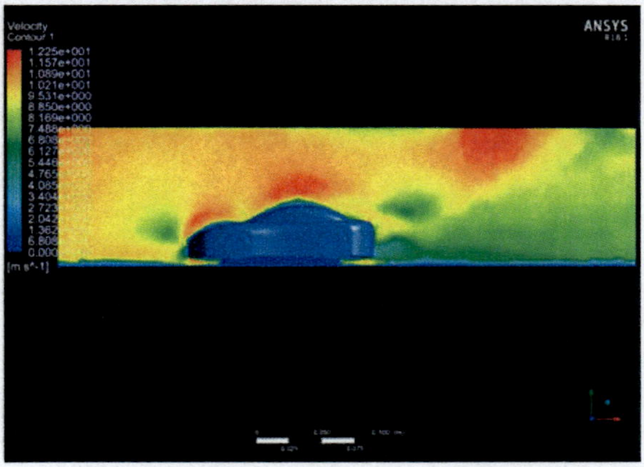

Fig. 8 Velocity contour of second vehicle geometry

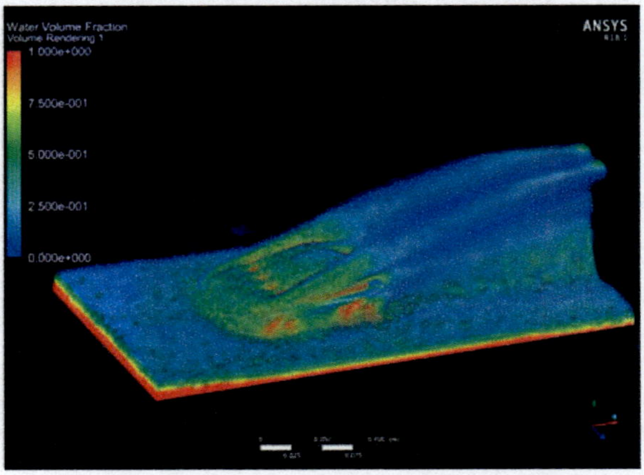

Fig. 9 Water volume fraction in the second vehicle geometry

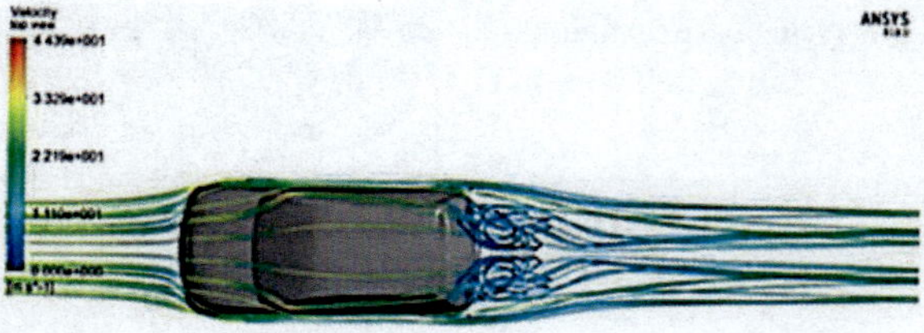

Fig. 10 Streamlines in the first vehicle geometry

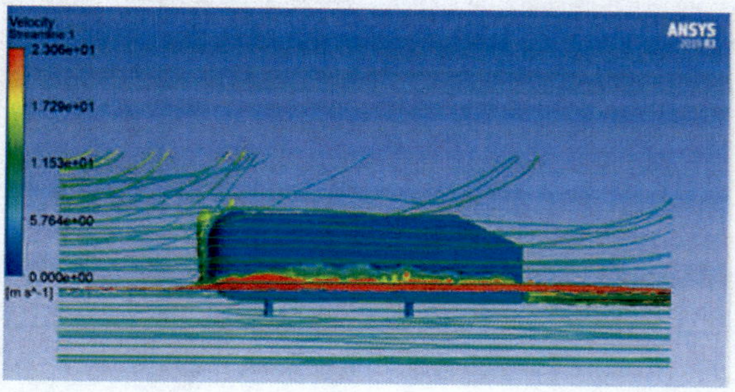

Fig. 11 Velocity streamline in the Ahmed body

4 Conclusion

Table 1 concludes that the second vehicle is more vulnerable to the wading conditions than the first vehicle. The results can also be seemed to be validated by the results from Ahmed body. Furthermore, the variations of drag on both the vehicle with the increasing velocity are also plotted in Fig. 12. The graph as shown in Fig. 12 is nicely correlating with the hand calculations and literature. Thus, from our study, we conclude that the vehicle body 1 is more appropriate for the deep-water wading scenarios because the drag force experienced by this body is the least.

Table 1 Drag values for different velocity values

Velocity (KMPH)	Drag in first vehicle geometry (hatchback)	Drag in second vehicle geometry (Streamlined)	Drag in Ahmed body
20	279.42	219.1	217.1
40	302.5	248.6	244.7
60	337.8	297.8	288.1
80	401.6	306.6	300.6
100	497.8	356.1	346.9

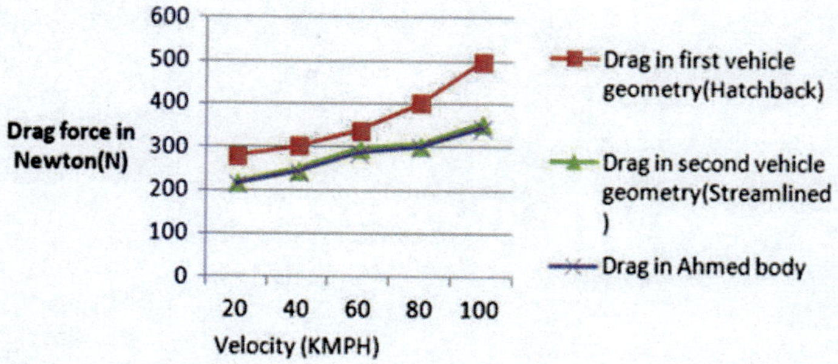

Fig. 12 Graph showing the comparison of drag forces between all three geometries

References

1. Anbarsooz M (2019) A numerical study on drag reduction of underwater vehicles using hydrophobic surfaces. Proc Inst Mechan Eng Part M J Eng Maritime Environ 233(1):301–309
2. Tan KM, Lu TF, Anvar A (2013) Drag coefficient estimation model to simulate dynamic control of autonomous underwater vehicle (AUV) motion. In: 20th international congress on modelling and simulation, pp 1–6
3. Banga S, Zunaid M, Ansari NA, Sharma S, Dungriyal RS (2015) CFD simulation of flow around external vehicle: Ahmed body. IOSR J Mechan Civ Eng 12(4):87–94
4. Nakisa M, Maimun A, Ahmed YM, Behrouzi F, Tarmizi A (2017) Numerical estimation of shallow water effect on multipurpose amphibious vehicle resistance. J Nav Archit Mar Eng 14(1):1–8
5. Khapane P, Ganeshwade U, Senapathy J, Kalmykov DII, Bayrasy IP, Wolf K (2015) Deep water wading simulation of automotive vehicles. In: Proceedings of NAFEMS world congress
6. Buscariolo FF, Budavari A, Mantovani D, Almeida E, Rossi G, Bigarella R, Ramos RP (2014) Multiphase water flow simulation of a vehicle's roof (no. 2014-36-0256). SAE technical paper
7. Ahmed SR, Ramm G, Faltin G (1984) Some salient features of the time-averaged ground vehicle wake. SAE Trans:473–503
8. Menter FR (1994) Two-equation eddy-viscosity turbulence models for engineering applications. AIAA J 32(8):1598–1605
9. Menter FR, Kuntz M, Langtry R (2003) Ten years of industrial experience with the SST turbulence model. Turbul Heat Mass Transf 4(1):625–632

Improving Efficiency of Diesel Engine by Oxygen Enrichment Using Pressure Swing Adsorption Technique

Mahaveer Vindhyachal Jaiswal and P. R. Dhamangaonkar

Nomenclature

C	Concentration of nitrogen in gas phase (mol/m^3)
D_x	Axial dispersion coefficient for nitrogen (m^2/s)
U	Velocity of air flow (m/s)
ρ_p	Density of zeolite particle used (zeolite 13X)
$\in_{bed}$	Porosity of adsorbent bed
q	Average amount of nitrogen in outlet (mol/g)
q^*	Equilibrium amount of nitrogen adsorbed (mol/g)
k	Mass transfer coefficient for LDF
k_1, k_2, k_3 and k_4	Are parameters of Runge–Kutta 4th order
d_p	Adsorbent (zeolite) particle diameter
μg	Gas viscosity
ρg	Gas density

1 Introduction

Today, diminishing fuel utilizations and exhaust outflows become an extensive matter since, supposing that these things are done then cost of energy or fuel will be diminished and an ecological issue will be settled part of the way as well. Taking care of unadulterated oxygen amounts to the blend prior to entering the ignition chamber

M. V. Jaiswal (✉)
College of Engineering Pune, Pune, India
e-mail: mahaveerjaiswal2616@gmail.com

P. R. Dhamangaonkar
Department of Mechanical Engineering, College of Engineering Pune, Pune, India

J. Banerjee et al. (eds.), *Recent Advances in Fluid Dynamics*, Lecture Notes in Mechanical Engineering, https://doi.org/10.1007/978-981-19-3379-0_29

can be a successful method to do as such. Diesel motor producers face significant difficulties to meet the force necessity with high ignition productivity. In addition, the abatement of fuel utilization has constrained the vehicle business to deliver better motors with new innovation, prompting advancement of new ignitions frameworks. Heaps of exploration were done and various test contemplates exhibited the benefits of applying oxygen improved ignition innovation (OECT) in diesel motors. In the current work, separate oxygen detachment arrangement is utilized to improve the oxygen level in the admission air. A little blending chamber can be given before intake manifold.

Effect of utilization of oxygen enhanced air can be compared at various loads and diverse degree of oxygen advancement. Different parts of oxygen improvement like decrease in particulates, smoke and unburned hydrocarbon (HC) were excluded from this work. Under the consistent gulf absolute, oxygen enhanced ignition can further develop the burning effectiveness, and the discharges of residue can diminish enormously. At the point when the oxygen fixation is more than 24%, the impact of oxygen focuses on the burning of the chamber is bit by bit debilitated. The chamber divider heat misfortune increments incredibly, which upsets the improvement of the viable warm productivity of the diesel motor [1].

Fuel transformation proficiency improved marginally with the 25% oxygen content air contrasted and ordinary air, however, that improvement vanished when oxygen content was expanded to 28%. At and under an oxygen content of 28%, the vehicle gives brilliant reaction, however, after 28% oxygen improvement, the motor started to thump [2].

There is a large mass transfer zone (MTZ) for zeolite 13X. Thus, the adsorption pace of nitrogen on zeolite 13X is higher than that on zeolite 5A. The main drop of nitrogen concentration in the outlet of zeolite 13X occurs at the time of about 125 s [3, 4].

The mass stream pace of fuel with oxygen taking care of is not exactly that of with no oxygen taking care of at some specific values of motor rates and exactly the same thing was found for air mass stream rate. The air–fuel proportion additionally is less with impressive qualities for the situation with oxygen taking care of than that with no oxygen samples [5].

1.1 Methodology

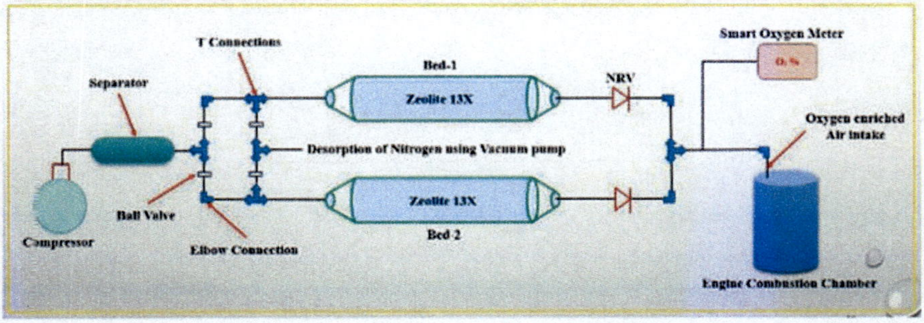

Pressure swing adsorption is preferred over temperature swing adsorption because pressure can be changed substantially more quickly than the temperature, consequently making it conceivable to work a PSA interaction on a lot quicker cycle, subsequently expanding the throughput per unit of adsorbent bed volume.

Two types of zeolites which are available as commercial molecular sieves are zeolite 5A and zeolite 13X. But, the most well-known sort of commercial zeolite for oxygen concentration measure is zeolite 13X because of its extraordinary nitrogen to oxygen adsorption selectivity. There is a huge mass transfer zone (MTZ) for zeolite 13X and subsequently, the adsorption pace of zeolite 13X is high [4].

Compressed air at pressure of 4 bar is passed through separator containing silica gel, where moisture adsorption takes place. Demoisturized air is then passed through bed 1 containing zeolite 13X, where some amount of nitrogen is adsorbed. Before this, processed air enters in engine cylinder smart oxygen meter measures oxygen in processed air to understand level of enrichment.

Factors affecting adsorption process are;

a. Design parameters—Bed size, physical properties of adsorbent and number of adsorption bed.
b. Operational variables—Temperature, pressure and velocity of air intake and retention time.
c. BSF—Decreasing process duration for a PSA interaction normally brings about a gain in adsorbent usage, frequently addressed in industry by the bed size factor (BSF) [6]. Increase in adsorbent usage is addressed by a decline in BSF [7] (Graph 1).

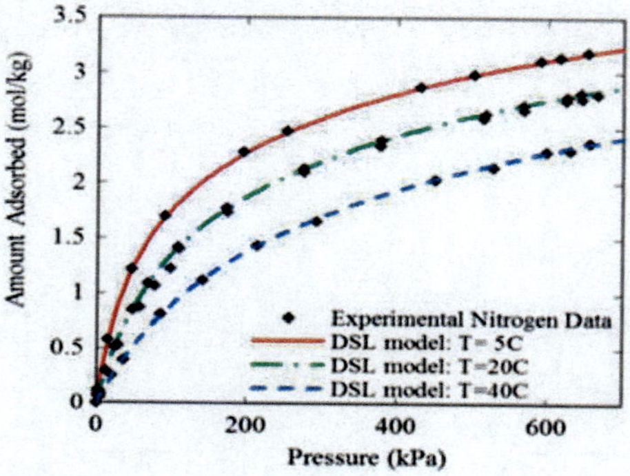

Graph 1 Nitrogen adsorption in mol/g on zeolite 13X versus pressure at different temperature [7]

2 Mathematical Modeling

$$\frac{\partial C}{\partial t} = Dx\frac{\partial^2 C}{\partial x_2} - \frac{\partial (C * U)}{\partial x} - \frac{\rho_{\text{par}}(1 - \epsilon_{\text{bed}})}{\epsilon_{\text{bed}}} * \frac{\partial q}{\partial t}$$

where $C = \frac{q * P}{\text{RT}}$ [3, 8].

$$D_x = 10^{-9}\,\text{m}^2/\text{s}$$
$$\rho_\text{p} = 1.170\,\text{g/m}^3$$
$$\epsilon_{\text{bed}} = 0.37$$

B. Linear Driving Force (LDF) [3]

$$\frac{\partial q}{\partial t} = k(q^* - q)$$
$$k = 0.1971/\text{s}$$

C. Dual-Site Langmuir Isotherm [3]

$$q^* = \frac{qmB * P}{1 + B * P} \quad \left(q_m = k_1 + k_2 * T \& B = k_3 * e^{\frac{k4}{T}}\right)$$
$$k_1 = 12.52\,\text{mol/g},\ k_2 = -1.785 * 10^{-2}\,\text{mol/g K},$$
$$k_3 = 2.154 * 10^{-4}\,\text{atm}^{-1}\ \text{and}$$
$$k_4 = 2333\,\text{K}$$

D. Moment Balance Equation [9]

$$-\frac{\partial P}{\partial x} = \frac{150\mu g(1 - \epsilon_{\text{bed}})U}{d_\text{p}2 * \epsilon_{\text{bed}}\ 2} + \frac{1.75(1 - \epsilon_{\text{bed}})}{d_\text{p}\ \epsilon_{\text{bed}}} * U2\rho g$$

$$d_p = 0.00149\,\text{m}$$
$$\mu g = 1.813 * 10^{-5}\,\text{Ns/m}^2$$
$$\rho g = 1.2\,\text{kg/m}^3$$

E. Simplified Equations

$$\frac{\partial C}{\partial t} = Dx\frac{\partial 2C}{\partial x2} - \frac{\partial (C * U)}{\partial x} - \frac{\rho_p(1- \in b)}{\in b} * \frac{\partial q}{\partial t}$$

$$\frac{\partial (q * P)}{\text{RT}\partial t} + \frac{\partial (q * P * U)}{\text{RT}\partial x} + \frac{\rho_p(1- \in b)}{\in b} * \frac{\partial q}{\partial t} = 0$$

$$P * \frac{\partial q}{\partial t} + q * \frac{\partial (P)}{\partial t} + UP * \frac{\partial (q)}{\partial x} + qU * \frac{\partial (P)}{\partial x}$$
$$+ \text{RT} * \frac{\rho_p(1- \in b)}{\in b} * \frac{\partial q}{\partial t} = 0$$

$$\left[P + \text{RT} * \frac{\rho_p(1- \in b)}{\in b}\right] * k(q^* - q) + q\left[\frac{\partial (P)}{\partial t} + U * \frac{\partial (P)}{\partial x}\right] = 0$$

$$q = \frac{k(q*) * \left[P + \text{RT} * \frac{\rho_p(1-\in b)}{\in b}\right]}{k * \left[P + \text{RT} * \frac{\rho_p(1-\in b)}{\in b}\right] - \frac{\partial (P)}{\partial t} - U * \frac{\partial (P)}{\partial x}}$$

Equation of q is to be simulated in the time domain.

2.1 MATLAB Simulation

MATLAB Simulink model built in MATLAB which;

1. Validates reduction in nitrogen adsorption with respect to time.
2. Identifies adsorption bed saturation time.
3. Validates increase in nitrogen adsorption with increase in air pressure.
4. Validates decrease in nitrogen adsorption with increase in air temperature.

Simulation is done at $P = 4$ bar, $T = 298$ K and $U = 0.1$ m/s.
From Graph 2 it is clear that

1. Nitrogen adsorption in zeolite bed decreases with increase in time and reaches to zero in 60 s. Hence, saturation time for 1 bed can be concluded as 60 s.
2. Maximum adsorption during cycle time of 60 s is found to be in first 20 s.
3. Total amount of nitrogen adsorbed in cycle time of 60 s is 0.8025 mols which is equivalent to 0.8025 * 24, i.e., 19.26 g.

From Graph 3 it is clear that, nitrogen adsorption continuously increases with increase in pressure but, at the same decreases with increase in time.

From Graph 4 it is clear that, nitrogen adsorption increases till 308 K and then

Graph 2 Nitrogen adsorption versus time

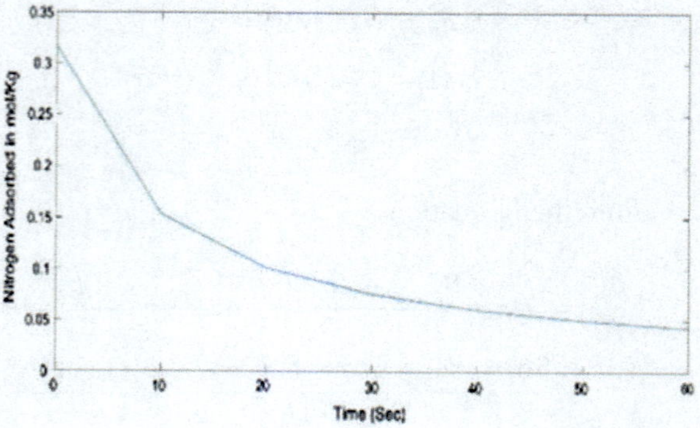

Graph 3 Nitrogen adsorption versus pressure

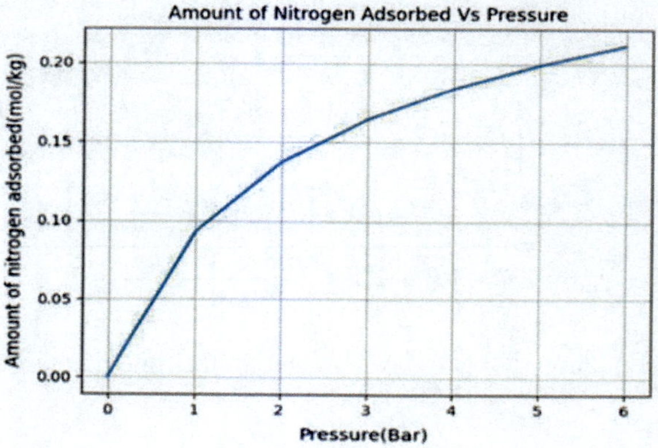

Graph 4 Nitrogen adsorption versus temperature

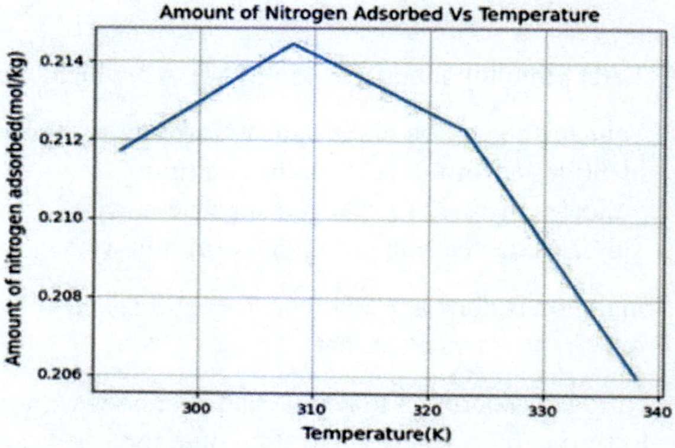

starts decreasing after 308 K. Maximum nitrogen adsorption is found between 295 and 323 K.

Numerical calculations and GT suit simulation are done for 1-cylinder, 4-stroke diesel engine with,

Bore size 87.5 mm.

Stroke 110 mm.
Rated power, $P = 3.5$ kW.
Speed, $N = 1500$ RPM.
Air–fuel ratio = 16.
Orifice diameter = 20 mm.

3 Results

A. Oxygen enrichment calculation

1. When 1 kg air passed through bed, 19.26 g nitrogen get adsorbed resulting in 980.74 g processed air available at bed outlet.
2. To complete 1 kg of air, additional 19.26 g processed air is added in already available 980.74 g processed air.
3. From above to steps amount of nitrogen in 1 kg processed is,
   = 770 (Initial amount of nitrogen in fresh air)—19.26 − 19.26 * 19.26/770 = 750.26 g.
4. Therefore, **percentage oxygen enrichment achieved** can be calculated as,
   (1000 − 750.26) * 21/230 = **22.8%** (Initially it was 21% in fresh air).

B. GT suit simulation results

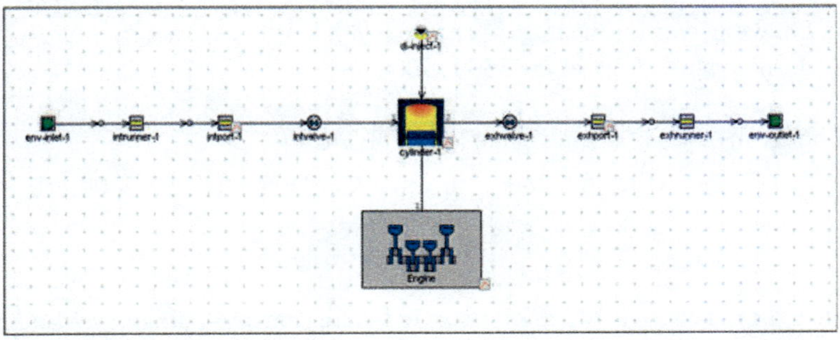

GT Suit Model for 1-cylinder 4-stroke Diesel Engine

C. Calculation for change in mass of fuel required to maintain A/F ratio for 22.8% enriched oxygen (Table 1)

Combustion reaction for diesel fuel can be written as;

$$C_{12}H_{22} + 17.5 * ((79/21) * N_2 + O_2) = 12CO_2 + 11H_2O)$$

Here, mass of fuel is 12 * 12 + 22 * 1 = 166 g…1.
According to above reaction, air–fuel ratio can be calculated as,

$$\text{A/F} = [17.5 * (100/21) * 32]/[12 * 12 + 22 * 1] = 16.07$$

Now, to maintain A/F ratio of 14.48 with 22.8% oxygen in air,

$$\text{Mass of fuel} = [17.5 * (100/22.8) * 32]/16.07 = 153\,\text{g}\ldots 2$$

From 1 and 2 its clear that, there is 13 g/min, i.e., 7.83% of reduction in mass of fuel required for 22.8% oxygen.

D. Calculation for change in brake-specific fuel consumption for 22.8% enriched oxygen and 3.5 kW rated power

Rated brake power for mentioned engine is 3.5 kW,

$$\begin{aligned} \text{Change in BSFC} &= 2 * \left(m_{f(22.8\%\text{O}_2)} - m_{f(21\%\text{O}_2)}\right)/(\text{B.P.} * 60) \\ &= 2 * (153 - 166)/(3.5 * 60) \\ &= -0.123\ \text{g/kWh} \end{aligned}$$

Hence, with 22.8% O_2 in air, there occurs 0.123 g/kWh, i.e., 7.83% of reduction in BSFC.

E. Calculation for change in brake thermal efficiency ($\eta_{B\text{th}}$) for 22.8% enriched oxygen and 3.5 kW rated power

Rated brake power for mentioned engine is 3.5 kW,
Calorific value of diesel is 45500 kJ/kg.

$$\begin{aligned} \%\ \text{Change in}\ \eta_{B\text{th}} &= (\text{B.P.} * 3600)/\left(m_f * 60 * \text{CV}\right) \\ &= [(3.5 * 3600)/(0.153 * 45{,}500)] - [(3.5 * 3600)/(0.166 * 45{,}500)] \\ &= 0.142\% \end{aligned}$$

Hence, there occurs 0.142% increase in brake thermal efficiency with 22.8% O_2 in air.

4 Conclusions

1. Oxygen content in air has been increased to 22.8% from 21%.
2. Theoretically, 7.83% decrease in brake-specific fuel consumption has been calculated, while by simulation, 12.6% decrease in brake-specific fuel consumption is recorded.

Table 1 Variation in performance parameters with variation of O_2 content in air

Parameter	For 21% O_2	For 22.8% O_2	Change
Brake torque (Nm)	98.3	98.2	− 0.1 Nm
IMEP (bar)	20.11	20.09	− 0.02 bar
BSFC (g/kWh)	233.3	233.1	− 0.2 g/kWh
Volumetric efficiency (%)	81.5	81.7	+ 0.2%
A/F ratio	16.07	16.1	+ 0.03
Brake efficiency (%)	35.8	35.9	+ 0.1%

3. Theoretically, 0.142% increase in brake thermal efficiency has been calculated, while by simulation, 0.1% increase in brake thermal efficiency is recorded.
4. Theoretically, decrease of 7.83% in fuel requirement is observed with 22.8% O_2 content in air.
5. According to simulation, 0.2% increase in volumetric efficiency is recorded for 22.8% O_2 content in air.

Acknowledgements I would like to express my profound gratitude and deep regards to my guide Dr. P. R. Dhamangaonkar, Dean of Student Affairs, COEP, for his valuable guidance, monitoring and constant encouragement throughout this project work. I am thankful to Dr. M. R. Nandgaonkar, Head of Department, Department of Mechanical Engineering, COEP, and Dr. K. C. Vora, Head of Department, ARAI academy, for their cooperation and motivation.

References

1. Wang H, Liu W (2015) Simulation studies of diesel engine combustion characteristics with oxygen enriched air. J Power Energy Eng 3:15–23
2. Kraft SW, Ng HK, Sekar RR. The potential benefits of intake air oxygen enrichment in spark ignition engine powered vehicle. In: SAE technical paper series_932803
3. Kakavandi IA, Shokroo EJ (2017) Dynamic modeling of nitrogen adsorption on zeolite 13X bed
4. Kakavandi IA, Shakroo EJ (2016) Dynamic survey of nitrogen adsorption on zeolite 13X bed. ResearchGate
5. Momani W (2009) Effects of oxygenated gasoline on fuel and air mass flow rates and air-fuel ratio. Am J Appl Sci 6(5):974–977
6. Lin L. Book on 'numerical simulation of pressure swing adsorption process'
7. Sinha P, Padhiyar N (2019) Optimal startup operation of a pressure swing adsorption. IFAC PapersOnLine 52(1):130–135
8. Moran AA. Book on 'limits of small-scale pressure swing adsorption'
9. Mofarahi M (2013) Comparison of two pressure swing adsorption processes for air separation using zeolite 5A and zeolite 13X. ResearchGate
10. Hamed HH (2015) Oxygen separation from air using zeolite type 5A. IJSER 6

11. Cho S, Nettem VC (2002) Adsorption of nitrogen, oxygen and argon on Na-CaX zeolites. ResearchGate
12. Pan M, Omar HM (2017) Application of nanosize zeolite molecular sieves for medical oxygen concentration. Nanomaterials
13. Mathias PM, Kumar R (1996) Correlation of multicomponent gas adsorption by the dual-site Langmuir model. Application to nitrogen/oxygen adsorption on 5A-zeolite. Ind Eng Chem Res 35:2477–2483
14. Liu Z, Grande CA. Adsorption and desorption of carbon dioxide and nitrogen on zeolite 5A. Journal ISSN: 0149-6395, 1520-5754 (online)
15. Rao VR, Kothare MV, Sircar S (2013) Numerical simulation of rapid pressurization and depressurization of a zeolite column using nitrogen.https://doi.org/10.1007/s10450-013-9548-x,9@SpringerScience
16. Melissa Magee H. Nitrogen gas adsorption in zeolites 13X and 5A. Walla Walla University, 204 S. College Ave., College Place, WA 99324
17. Razmus DM, Hall CK (1991) Prediction of gas adsorption in 5A zeolites using monte Carlo simulation. AICHE 37(5)
18. Tan W, Wang Q (2011) Adsorption of nitrogen and phosphorus on natural zeolite and its influencing factors. In: 2011 third international conference on measuring technology and mechatronics automation
19. Mofarahi M, Shakroo EJ (2011) Numerical simulation of a pressure swing adsorption for air separation. In: 7th international chemical engineering congress and exhibition Kish, Iran
20. Mehta B, Patel H (2015) Opportunity to improve the engine performance and emission characteristics by using oxygen enriched combustion. IJIRAE 6(2). ISSN: 2349-2163
21. Rajkumar K, Govindarajan P (2011) Impact of oxygen enriched combustion on the performance of a single cylinder diesel engine. Front Energy 5(4):398–403
22. Maxwell TT, Setty V, Jones JC, Narayan R. The effect of oxygen enriched air on the performance and emissions of an internal combustion engines. In: SAE technical paper series_932804
23. McCroskey T, Rhee KT. A spark ignition engine operated by oxygen enriched air. In: SAE technical paper series_922174
24. Wu Y-Y, David Huang K. Improving the performance of a small spark-ignition engine by using oxygen-enriched intake air. 20076504 (JSAE), 2007-32-0004 (SAE)
25. 'Zeolite a versatile air pollutant adsorber' by U.S. Environmental Protection Agency Research Triangle Park, North Carolina 27711, EPA-456/F-98-004 July 1998
26. Pierotti RA, Rouquerol J, Siemieniewska T (1985) Reporting physisorption data for gas/solid systems with special reference to the determination of surface area and porosity. Pure Appl Chem 57(4):603—619
27. Yates DJC (1968) Studies on the surface area of zeolites, as determined by physical adsorption and X-ray crystallography. Can J Chem 46:1695–1701
28. Chao CC, Milwood (1989) Process for separating nitrogen from mixture with less polar substances. U.S. Patent-4859217
29. McKee DW (1964) Separation of an oxygen-nitrogen mixture. U.S Patent-3140933
30. Abdul Mujeebu M, Zulkifly Abdullah M, Mohamad AA (2010)Trends in modeling of porous media combustion. Progr Energy Combustion Sci 36:627–650
31. Siriwardane R, Biegler LT (2003) Optimization of a pressure-swing adsorption process using zeolite 13X for CO_2 sequestration. Ind Eng Chem Res 42:339–348
32. Shi Y, Liu Y (2015) Simulation study of oxygen-enriched cylinder combustion in the zero emissions internal combustion engine. In: 5th international conference on education, management, information and medicine

Qualitative Study on Parameters Affecting the Structure of Sprays and Its Atomization

Abhishek Bhupendra Gade and Nikhil A. Baraiya

Nomenclature

Oh	Ohnesorge number
We_c	Critical Weber number
t_{hyb}	Hybrid time scale
B_1	Primary breakup parameter
C_{RT}	Secondary breakup diameter
PDA	Phase Doppler Anemometry
LPT	Lagrangian Particle Tracking
VOF	Volume of fluid
SMD	Sauter mean diameter

1 Introduction

A lot of research has been done in the field of primary and secondary atomization to understand the breakup of sprays although the fact that atomization process is more complex to understand. There are two important dimensionless number [8] in the study of spray formation, namely Weber number and Ohnesorge number. The Weber number is the ratio of aerodynamic force (disruptive type) to surface tension force (restorative type). The velocity of droplet is higher for high Weber number which fastens the process of droplet breakup. The Ohnesorge number is the ratio of viscous force of drop to the surface tension forces acting on it. The breakup is dependent on Ohnesorge number only if its value is greater than 0.1 [19]. The

A. B. Gade · N. A. Baraiya (✉)
Sardar Vallabhbhai National Institute of Technology, Surat, India
e-mail: nikhil@med.svnit.ac.in

J. Banerjee et al. (eds.), *Recent Advances in Fluid Dynamics*, Lecture Notes in Mechanical Engineering, https://doi.org/10.1007/978-981-19-3379-0_30

manner in which the fragmentation of drops takes place is decided by the Weber number [12]. Several transitional modes are classified on the basis of Weber, namely vibrational, bag, multimode, sheet thinning and catastrophic mode. A lot of research is done for Newtonian fluids, although the complexity of non-Newtonian fluids [23] has limited the research, where the rheological characteristic is considered. Having understood these basic breakup modes, many research studies have been carried out in the last decade to understand about the changes happening to Sauter mean diameter, penetration length and the way a droplet undergoes atomization by means of simulations and experiments. A variety of surrounding parameters affecting the spray formation and its initiation have been taken into account and results have been established for different cases.

1.1 *Qualitative Literature of the Advancement of Studies of Spray and Its Properties*

The bag breakup regime in a continuous cross-flow air jet was examined by Zhao et al. [29] using a high-speed camera. The critical Weber number was found to be strongly influenced by the viscosity of the liquid. This also affected the Oh which increases quickly with the increment in We_c (Fig. 1). The test liquids used in the experiments are glycerol solutions, water and ethanol.

The paper by Omidvar and Khaleghi [22] suggested a new mathematical model based on t_{hyb} which depends on both effective turbulence time scale and breakup time scale and takes into account the ambient turbulence effects. The model is based on Taylor analogy breakup model which uses a second-order differential equation of forced damped spring-mass system and relates it with the oscillating droplet. If the turbulence intensity is increased, the critical Weber number is reduced (Fig. 2). At

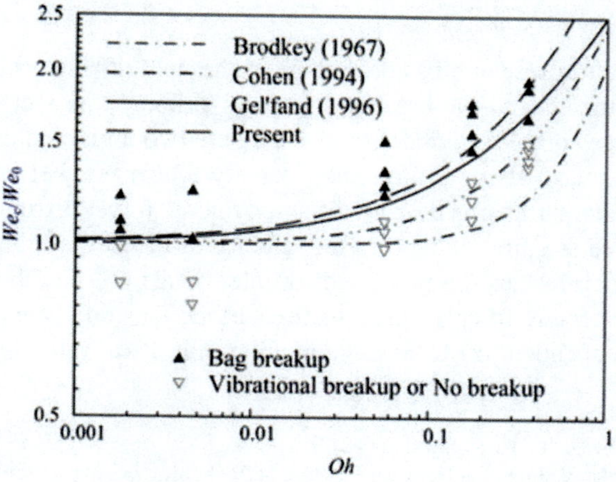

Fig. 1 Critical we at different oh [29]

Fig. 2 Variation of critical Weber number with relative turbulent intensity [22]

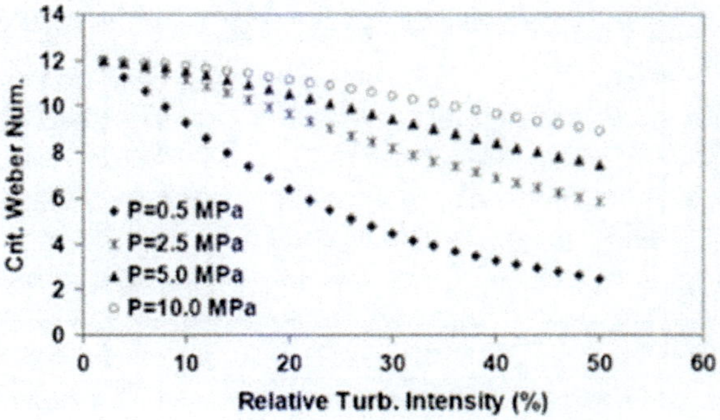

high pressure, the free-stream turbulence effect on droplet breakup is also observed, though it is less.

To investigate the effects of ambient pressure and injection pressure on penetration length, [15] performed a numerical simulation of spray in OPENFOAM. It is difficult for the droplet to diffuse in the surrounding gaseous medium if the ambient pressure is high. As the injection pressure is increased, the spray penetration length increases due to the increased momentum of the spray droplet (Fig. 3).

The difference in atomization properties between diesel and biofuel was studied by Boggavarapu and Ravikrishna [4]. The liquid length of biodiesel is higher and hence ensures a good amount of penetration. The diameter of droplets is higher for biodiesel as compared to diesel spray. The atomization process is greatly affected by the turbulence, effect of cavitation and geometry inside the nozzle.

Naz et al. [20] observed that the quantity of thermal energy linked with the spray had a significant effect on its width, length and tip penetration. The value of Sauter mean diameter reduces as the heating temperature is increased (Fig. 4) and is not much affected by the orifice diameter at higher temperatures.

Fig. 3 Spray penetration length v/s time for different injection pressures [15]

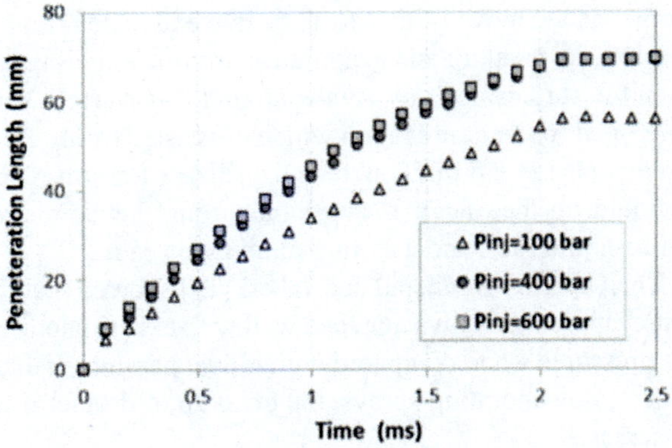

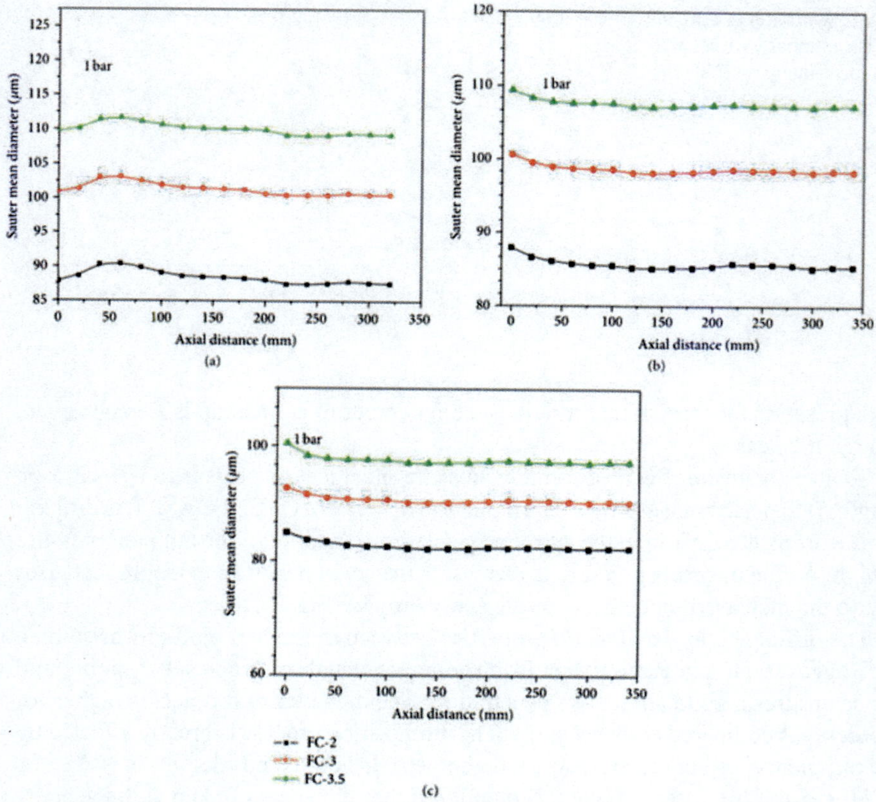

Fig. 4 Variation of SMD with axial distance at different fixed temperatures **a** 20 °C, **b** 60 °C, **c** 90 °C [20]

In the studies done using PDA, it was observed that there was reduction in size droplet as we move farther from nozzle exit. The reason for this is the spreading of droplets, its breakup and evaporation into surrounding atmosphere.

In the studies done by Avulapati and Rayavarapu Venkata [2], a new strategy is developed for jet atomization which is assisted by air. It is observed that the effect of viscosity is more prominent than the surface tension. There is no effect of properties of liquid on the Sauter mean diameter and hence the atomization beyond critical gas-to-liquid ratio and a certain momentum ratio.

The paper by Irannejad and Jaberi [13] showed that the large eddy simulation of non-evaporating sprays matched well with experimental results for different ambient gas pressures when compared with global parameters like liquid penetration length. In case of evaporating sprays, the breakup of droplet increased if the density of gas is increased.

Kourmatzis and Masri [17] studied about how primary atomization is affected by gas phase velocity fluctuations. The increment in turbulent intensity resulted in smaller size droplets with more deformation and an effective primary atomization.

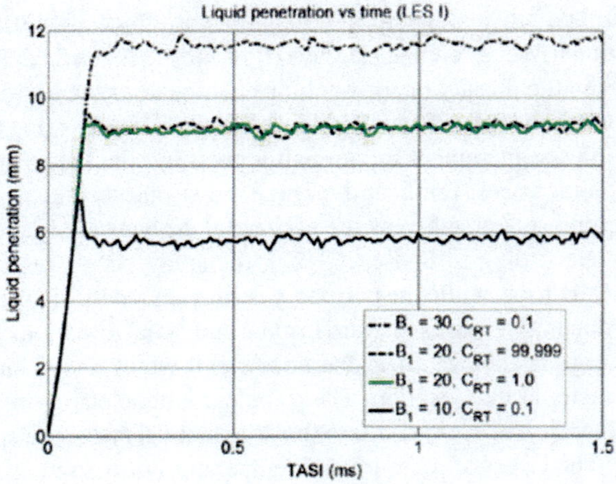

Fig. 5 Variation of liquid penetration with different breakup parameters [7]

The turbulent Weber number should also be used along with mean droplet Weber number to correctly identify the level of droplet deformation.

The studies by Devassy et al. [6] suggested a new model based on two-surface density and Eulerian-Eulerian two-fluid approach. This model takes into account the formation of ligament structures in the beginning and further breakup of them to form droplets. The disintegration is observed to occur at jet tip, where a spray cap is observed. This disintegration is a result of both gas and liquid turbulences.

The paper by Elmtoft et al. [7] provided results for diesel spray using Reynolds-averaged Navier–Stokes (RANS) and large eddy simulation (LES) models for turbulence. The size of the droplet injected has a significant impact on liquid and vapor penetration. Also, the liquid penetration was found to be dependent on the B_1 and C_{RT} (Fig. 5).

The impact of high ambient pressure on the features of swirl injector spray qualities was investigated by Fu and Yang [9]. When the pressure drop across the injector was increased, the liquid sheet breakup length was reduced and the spray cone angle was increased in the scenario when ambient pressure was applied. The length of the liquid sheet breakup and the spray cone angle both decreases as ambient pressure rises. When the value of nozzle diameter is raised, the angle of the spray cone rises.

The paper by Grosshans et al. [11] used the VOF model to show that there is very less effect on aerodynamic breakup by varying the liquid–gas density ratio from 10 to 30. On other hand, if the liquid–gas viscosity is reduced from 7 to 1, the resulting droplets were smaller in size and hence effective dispersion.

The primary atomization of water jet at high Reynold number was studied by Saeedipour et al. [24] using Eulerian–Lagrangian model. The combination of LPT with VOF was used to track the location of droplets moving in small scale. The Sauter mean diameters and mean drop velocity measured through experiments matched well with the simulation results.

The research of injection settings on the structure of spray of current GDI injectors was the focus of [18]. The spray atomization was effective for fuel having lower surface tension. There is no significant effect of fuel and injection pressure once critical Weber number at the exit of nozzle is reached. The size of droplet is bigger when the density of surroundings is high after the termination of atomization.

Urbán et al. [28] studied the dynamics of spray using a plain-jet air blast atomizer. With the increment in atomizing pressure, the velocity of droplets in radial direction increases significantly. There is a strong secondary atomization observed in the central region, whereas the peripheral droplets are found to be highly stable.

According to Sun et al. [27], the spray cone's middle is primarily made up of small drops, while the periphery is surrounded by big drops. The velocity of droplets reduces as it moves forward in axial and radial direction. The highest turbulent kinetic energy is achieved near the nozzle exit which shows that there is high momentum transfer at that location. The turbulent kinetic energy reduces at other places due to expansion of sprays, where the drag force reduces the speed of the droplets.

The effect of pressure and temperature of incoming air was studied by Bazdidi-Tehrani and Abedinejad [3] using numerical simulation. The incoming air pressure had no relation with droplet size distribution. When the temperature of the incoming air is raised from 373 to 573 K, the depth of penetration decreases (Fig. 6).

The effect of ambient density on the parameters of hollow-cone sprays created with a GDI injector was studied by Jeon and Moon [14]. Higher needle lifts result in shorter breakup lengths than lower needle lifts. In the case of low needle lift, the breakup length is reduced by increasing the density of surrounding air. In the event of high needle lift, ambient air density has no effect on breakup length. The spray atomization stops near the nozzle if the ambient density is high.

The paper by Omidvar [21] proposed new model that considers effective surface tension. This new idea was useful in capturing the impact of carrier phase turbulence on droplet breakup. It was checked with the experimental results by implementing the idea into a two-phase flow CFD code. The original Pilch and Erdman model is modified to include the effective surface tension. This brings a change in the value

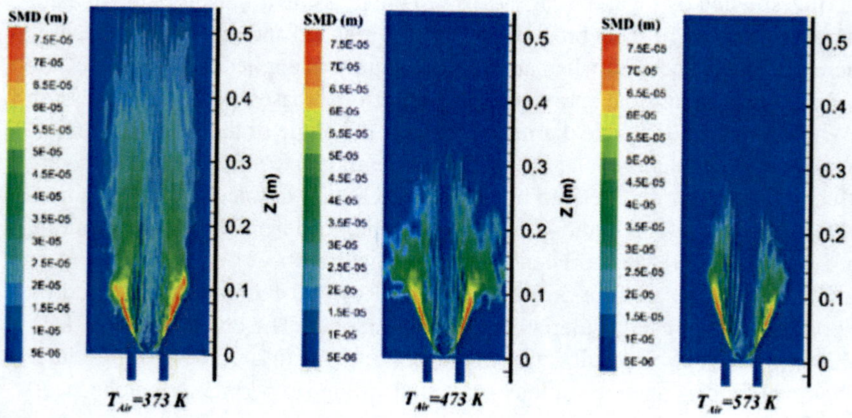

Fig. 6 Penetration depth and SMD distribution for different ambient temperature [3]

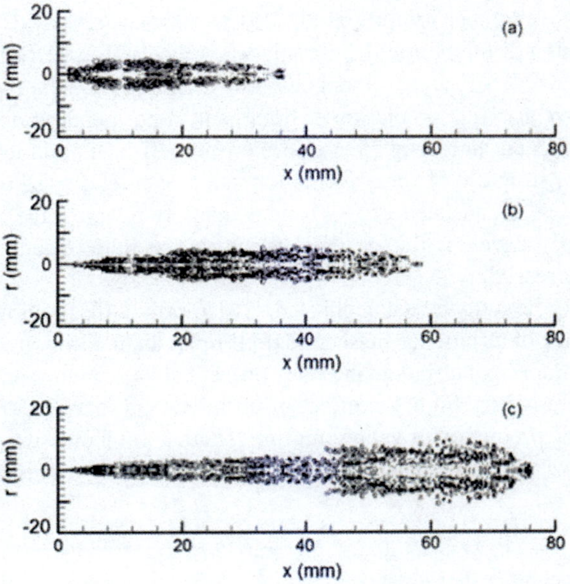

Fig. 7 Liquid spray structure for **a** 0.4 ms, **b** 1.12 ms, **c** 1.75 ms [21]

of Weber number and Ohnesorge number, and now, a new corrected value of stable droplet diameter is obtained.

The liquid penetration length obtained by observation of the formation of spray (Fig. 7) at 0.4, 1.12 and 1.75 ms is found to be in good agreement with the available data from experiments. The same positive results are observed while comparing the SMD at 15 and 30 mm far distance from nozzle exit.

The paper by Zhao et al. [30] showed how the breakup of droplets was affected by turbulence in surroundings with counter air flow using high-speed camera. Particle image velocimetry was utilized to investigate nine different turbulent counterflow setups. The turbulent intensity is highest for plate having high solidity keeping the mean velocity as constant. The bag structure as well as the stamen structure increases with the increment in turbulent intensity.

Sharma et al. [25] showed that there is considerable effect on turbulence intensity and swirl velocity if the swirl angle is increased in a simplex atomizer. The multimodal droplet distribution is observed near the axis and at long distance from the atomizer. This is a direct consequence of secondary atomization process and the recirculatory flow owing to the interplay between incoming spray and flow of swirling air. When a curved swirler is utilized instead of a flat swirler, the droplet sizes are smaller as the flow will be smoothly guided.

The study on variation of Sauter mean diameter of droplets injected using swirl atomizer at different locations was done by Amedorme [1]. The Sauter mean diameter of droplets increased moving along the center line in axial direction. But if the injection pressure is high, then this value of Sauter mean diameter decreases. The nozzle with smaller exit orifice diameter out of the two compared has smaller Sauter mean diameter.

The paper by Sula et al. [26] performed large eddy simulations of liquid spray using different atomization models, namely Reitz-Diwakar, Taylor analogy breakup and Pilch-Erdman models. Out of the three models, the TAB model is modified to give accurate prediction of liquid and vapor penetration length (Fig. 8).

Chen and Tang [5] performed numerical simulation and experiments to study the structure of spray formation using open-end swirl injector. The spray formation process consisted of four stages, namely pencil stage, onion stage, tulip stage and fully developed stage (Fig. 9) in an ascending order as the injection pressure is increased.

The study by Khaleghi et al. [16] improves the previous model to include the influence of turbulence on secondary droplet disintegration. This is because of the higher influence of turbulent intensity on critical Weber number. When the surrounding pressure is very high, the effect of turbulence is very less (Fig. 10) and can be ignored. The prediction of vapor pressure length at high pressure using the new model almost gives the same results and does not indicate any significant difference.

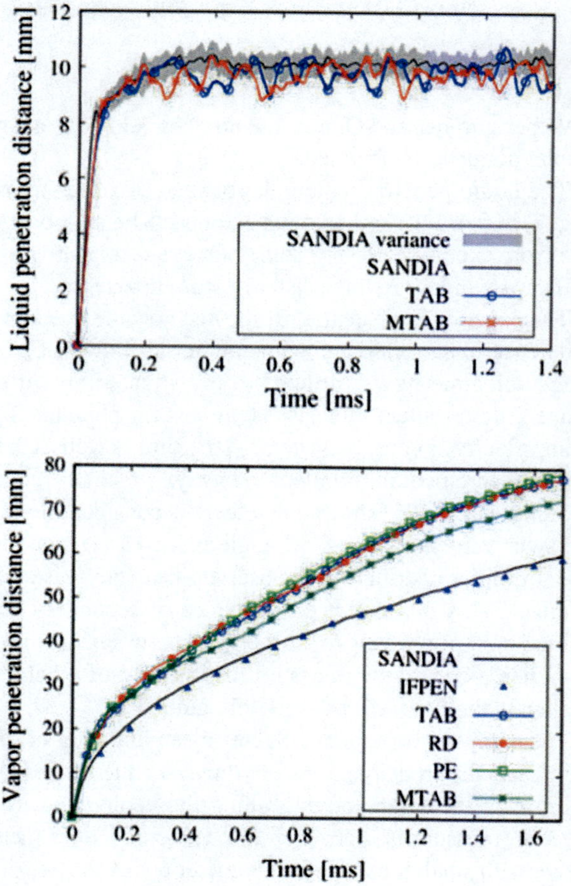

Fig. 8 Plot of liquid and vapor penetration distance for different breakup models [26]

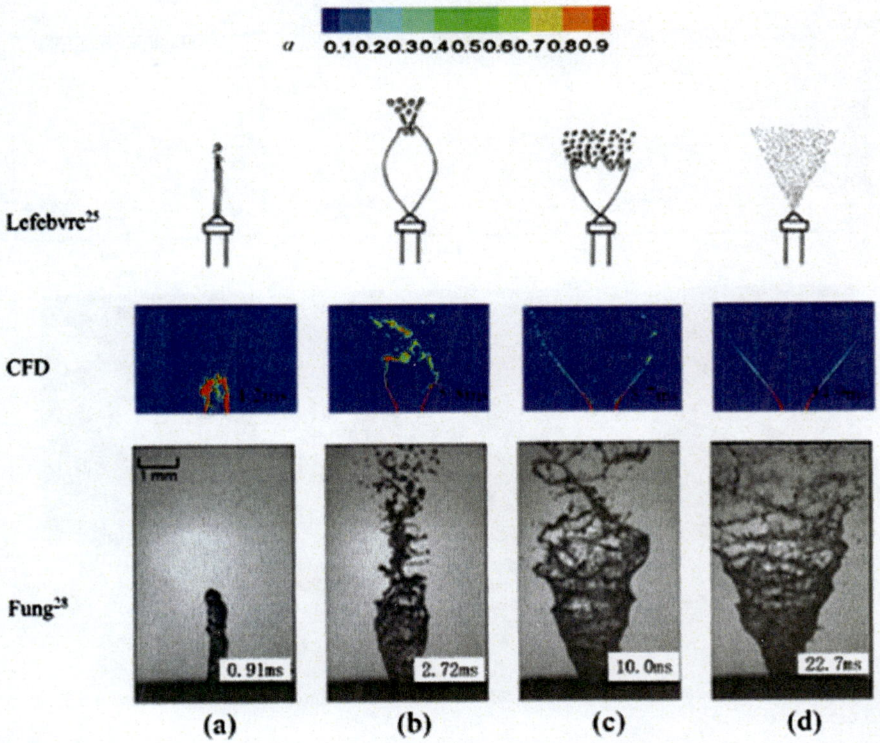

Fig. 9 Formation of stages **a** pencil stage, **b** onion stage, **c** tulip stage, **d** fully developed stage [5]

Gai et al. [10] studied about the turbulence properties for mitigation of fire using industrial nozzles. The strength of the turbulent flow is highest at the nozzle and lowers as we move away from it. The diameter of droplets has strong effect on the turbulent intensity which is maximum inside the spray cone and decreases along the vertical length moving away from spray nozzle.

2 Conclusions

The breakup of spray depends on different parameters and representative properties out of which the following are listed based on the studies in this review paper:

(1) Density of gas: The breakup frequency increases, and droplet size is big after atomization completes at higher value of gas density. If the ambient density is high, then it hampers the reach of atomization process.
(2) Viscosity of the fluid: The property of viscosity has higher effect on spray parameters than the surface tension. Higher the viscosity, higher is the critical Weber number.

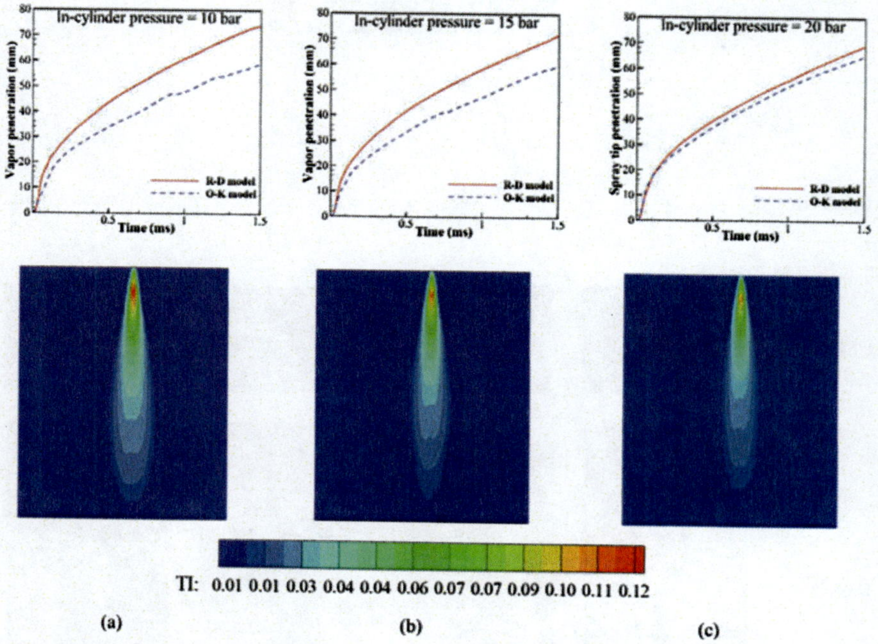

Fig. 10 Spray tip penetration and turbulent intensities for different cylinder pressure **a** 10 bar, **b** 15 bar, **c** 20 bar [16]

(3) Ambient pressure: As ambient pressure rises, the distance traveled during liquid sheet breakup and the spray cone angle both decreases.

(4) Injection pressure: If the injection pressure is increased, the penetration length increases due to higher kinetic energy of droplets. Another effect is the decrement in Sauter mean diameter.

(5) Temperature of fluid: The penetration depth is higher for lower value of incoming fluid temperature.

(6) Type of swirler: The size of droplet is smaller and is guided smoothly when the swirler is curved instead of flat.

(7) Solidity of plate: The turbulent intensity can be increased by increasing the solidity of plates and the size of the droplets reduces.

(8) Type of fuel: Biodiesel has an advantage on conventional fuel in terms of its environmental impacts and longer penetration depth is a plus advantage.

(9) Turbulence: It plays a significant role in breakup of droplets due to its solid effect on critical Weber number and still a number of experiments need to be performed for different conditions to capture the physics of transition during breakup and global spray parameters. The size of the droplet determines turbulent intensity at a given location.

The combination of all these and many such parameters decides the droplet diameter at different locations, the liquid and vapor penetration lengths and whether the atomization will be effective or not for certain application. In the future, more studies can be done with simulation and advanced experimental tools to understand the turbulence nature of flow and understand the physics of complex flow of fluids yet to be discovered.

Acknowledgements The author wishes to thank his colleagues Vijender Singh and Nishantt N. who constantly helped him while making this review paper.

References

1. Amedorme S (2020) Experimental study of mean droplet size from pressure swirl atomizer. Int J Eng Sci Technol 4(6):49–59. https://doi.org/10.29121/ijoest.v4.i6.2020.124
2. Avulapati MM, Rayavarapu Venkata R (2013) Experimental studies on air-assisted impinging jet atomization. Int J Multiph Flow 57:88–101. https://doi.org/10.1016/j.ijmultiphaseflow.2013.07.007
3. Bazdidi-Tehrani F, Abedinejad MS (2018) Influence of incoming air conditions on fuel spray evaporation in an evaporating chamber. Chem Eng Sci 189:233–244. https://doi.org/10.1016/j.ces.2018.05.046
4. Boggavarapu P, Ravikrishna RV (2013) A review on atomization and sprays of biofuels for IC engine applications. Int J Spray Combustion Dyn 5(2):85–121. https://doi.org/10.1260/1756-8277.5.2.85
5. Chen C, Tang Z (2020) Investigation of the spray formation and breakup process in an open-end swirl injector. Sci Prog 103(3):1–19. https://doi.org/10.1177/0036850420946168
6. Devassy BM, Habchi C, Daniel E (2015) Atomization modelling of liquid jets using a two-surface-density approach. Atom Sprays 25(1):47–80. https://doi.org/10.1615/AtomizSpr.2014011350
7. Elmtoft E et al (2015) Injected droplet size effects on diesel spray results with RANS and les turbulence models. In: SAE technical papers. https://doi.org/10.4271/2015-01-0925
8. Faeth GM, Hsiang L-P, Wu P-K (1995) Structure and breakup properties of sprays. Int J Multiphase Flow
9. Fu QF, Yang LJ (2015) Visualization studies of the spray from swirl injectors under elevated ambient pressure. Aerosp Sci Technol 47:154–163. https://doi.org/10.1016/j.ast.2015.09.027
10. Gai G et al (2021) Numerical study of spray-induced turbulence using industrial fire-mitigation nozzles. Energies 14(4):1–20. https://doi.org/10.3390/en14041135
11. Grosshans H et al (2016) Sensitivity of VOF simulations of the liquid jet breakup to physical and numerical parameters. Comput Fluids 136:312–323. https://doi.org/10.1016/j.compfluid.2016.06.018
12. Hsiang LP, Faeth GM (1992) Near-limit drop deformation and secondary breakup. Int J Multiph Flow 18(5):635–652. https://doi.org/10.1016/0301-9322(92)90036-G
13. Irannejad A, Jaberi F (2014) Large eddy simulation of turbulent spray breakup and evaporation. Int J Multiph Flow 61:108–128. https://doi.org/10.1016/j.ijmultiphaseflow.2014.01.004
14. Jeon J, Moon S (2018) Ambient density effects on initial flow breakup and droplet size distribution of hollow-cone sprays from outwardly-opening GDI injector. Fuel 211:572–581. https://doi.org/10.1016/j.fuel.2017.09.016
15. Kayhani M, Aghaie A, Razavi M (2012) Investigation of different numerical models in spray behavior simulation in order to predict the spray tip penetration. Psrcentre.Org. Available at http://psrcentre.org/images/extraimages/25.212171.pdf
16. Khaleghi H et al (2021) Effects of turbulence on the secondary breakup of droplets in diesel fuel sprays. Proc Inst Mechan Eng Part D J Automob Eng 235(2–3):387–399. https://doi.org/10.1177/0954407020958581

17. Kourmatzis A, Masri AR (2014) The influence of gas phase velocity fluctuations on primary atomization and droplet deformation. Exper Fluids 55(2). https://doi.org/10.1007/s00348-013-1659-3
18. Moon S et al (2017) Governing parameters and dynamics of turbulent spray atomization from modern GDI injectors. Energy 127:89–100. https://doi.org/10.1016/j.energy.2017.03.099
19. Muston P et al (2015) Droplet breakup in turbulent counterflow. In: 7th Australian conference on laser diagnostics in fluid mechanics and combustion Melbourne, December
20. Naz MY et al (2013) Investigation of vortex clouds and droplet sizes in heated water spray patterns generated by axisymmetric full cone nozzles. Sci World J 2013.https://doi.org/10.1155/2013/796081
21. Omidvar A (2019) Development and assessment of an improved droplet breakup model for numerical simulation of spray in a turbulent flow field. Appl Therm Eng 156:432–443. https://doi.org/10.1016/j.applthermaleng.2019.04.090
22. Omidvar A, Khaleghi H (2012) An analytical approach for calculation of critical weber number of droplet breakup in turbulent gaseous flows. Arab J Sci Eng 37(8):2311–2321. https://doi.org/10.1007/s13369-012-0319-x
23. Ortiz C, Joseph DD, Beavers GS (2004) Acceleration of a liquid drop suddenly exposed to a high-speed airstream. Int J Multiph Flow 30(2):217–224. https://doi.org/10.1016/j.ijmultiphaseflow.2003.11.004
24. Saeedipour M et al (2017) Multiscale simulations and experiments on water jet atomization. Int J Multiph Flow 95:71–83. https://doi.org/10.1016/j.ijmultiphaseflow.2017.05.006
25. Sharma S et al (2019) Effects of air swirler geometry on air and spray droplet interactions in a spray chamber. Adv Mech Eng 11(5):1–15. https://doi.org/10.1177/1687814019850978
26. Sula C, Grosshans H, Papalexandris MV (2020) Assessment of droplet breakup models for spray flow simulations. Flow Turbul Combust 105(3):889–914. https://doi.org/10.1007/s10494-020-00139-9
27. Sun Y et al (2018) Numerical and experimental study on the spray characteristics of full-cone pressure swirl atomizers. Energy 160:678–692. https://doi.org/10.1016/j.energy.2018.07.060
28. Urbán A et al (2017) Droplet dynamics and size characterization of high-velocity airblast atomization. Int J Multiph Flow 95:1–11. https://doi.org/10.1016/j.ijmultiphaseflow.2017.02.001
29. Zhao H et al (2011) Breakup characteristics of liquid drops in bag regime by a continuous and uniform air jet flow. Int J Multiph Flow 37(5):530–534. https://doi.org/10.1016/j.ijmultiphaseflow.2010.12.006
30. Zhao H et al (2019) Effect of turbulence on drop breakup in counter air flow. Int J Multiph Flow 120.https://doi.org/10.1016/j.ijmultiphaseflow.2019.103108

Comparative Assessment of Various Turbulent Models for 2D Flow Over an Airfoil

Aakash Desai, Abhishek Desai, Vikas Bhenjaliya, and Hemantkumar B. Mehta

Nomenclature

CFD	Computational Fluid Dynamics
AoA	Angle of Attack
EWT	Enhanced Wall Treatment
C_L	Coefficient of Lift
C_D	Coefficient of Drag
c	Chord Length

Greek Symbols

ε	Rate of dissipation of Turbulent Energy
ω	Specific rate of dissipation of Turbulent Energy

1 Introduction

Airfoils are basically the structures that consist of curved surfaces which are designed for providing the most favourable lift-to-drag force ratio. In order to predict the aircraft performance, the simulation of flow over an airfoil plays an important role.

A. Desai (✉) · A. Desai · V. Bhenjaliya · H. B. Mehta
Department of Mechanical Engineering, SVNIT, Surat, India
e-mail: desaiaakash08@gmail.com

H. B. Mehta
e-mail: hbm@med.svnit.ac.in

J. Banerjee et al. (eds.), *Recent Advances in Fluid Dynamics*, Lecture Notes in Mechanical Engineering, https://doi.org/10.1007/978-981-19-3379-0_31

The methodology which is developed to analyse the flow over an airfoil which reduces cost of experiments is computational fluid dynamics (CFD). In the present study, flow is under turbulence region. Turbulence is a three-dimensional property occurring due to the random and unsteady motion of particles in the fluids at moderate to high Reynolds number. The fluctuations which occur in turbulent flow occur over a wide range of length scale. Also, they possess high frequency, and hence, they are computationally very expensive to simulate directly in physical calculations. To overcome this problem, the governing equations are modified and due to this some additional variables come into existence. For this, various turbulence models were developed to determine these unknown variables in terms of the known quantities.

Eleni et al. [1] presented the computational fluid dynamics (CFD) calculations for the subsonic flow over a NACA 0012 airfoil. The results were compared with the experimental data, and it was found that predicted drag coefficient was higher than that of the experimental value. Later on, they calculated the transition point and then carried out the simulations, and the results were in good agreement with the experimental results.

Sarsaf et al. [2] developed a procedure to numerically model airflow over NACA 4412 airfoil using Gambit and Fluent. They solved the governing equations of conservation of continuity and conservation of momentum in steady state along with the additional equation of turbulence models and validate the numerical model by comparison of predicted results and experimental results of the airfoil. Calculations were carried out for constant air velocity altering only AoA for every turbulence model tested. In this work, calculations showed that at high AoA, accurate results were not obtained using the turbulence models used in commercial CFD codes.

Rao and Sahitya [3] had presented both numerical and experimental models for flow over an airfoil NACA 0012. Experimental data was obtained by performing experiments in wind tunnel whereas numerical model was simulated in Ansys. Experimental data was obtained by establishment of flow over airfoil of 0.1 m chord length with 12 pressure gauge setup along the profile, and velocity was kept around 45 m/s. Numerical solution was obtained by performing various simulations on various turbulent models. Through numerical solutions, it was found out that only at low AoA the dimensionless quantity coefficient of lift force was linearly proportional to AoA and the flow was attached with the airfoil throughout its regime. But when the AoA was near 15°–16°, the flow separated from the surface and condition of stall began to develop.

From the literature review, it was observed that the main focus was on comparing turbulent models for the flow over an airfoil. The analysis was carried on various operational parameters like low Reynolds number, low velocity, etc. The comparison was also made between the turbulent model, but the comparison was limited to some extend only that is all the turbulence models are not compared together along with the comparison with experimental data [4]. Main aim of the present paper is to carry out CFD simulations using various turbulent models at different conditions keeping inlet velocity constant (0.2 Mach number) and verifying it to find which is the best suitable turbulence model.

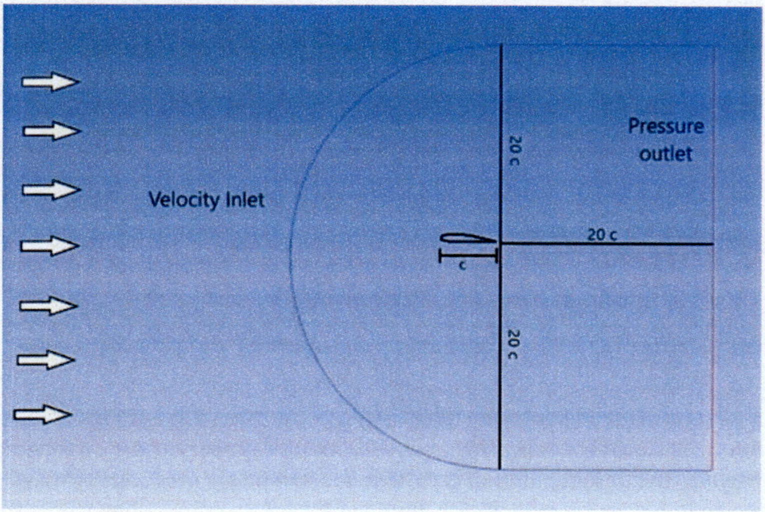

Fig. 1 Schematic of physical domain and its boundary conditions

2 Mathematical Formulation

2.1 Physical Domain

The schematic of the physical domain and its associated boundary conditions is shown in Fig. 1. Lift and drag forces get influenced by the size of the computational domain. The C-type domain is selected as computational domain. The domain is extended up to 20c in upstream and 20c in downstream of the airfoil trailing edge so that complete vortex effects of the flow can be captured in the downstream of the airfoil. In the transverse direction at the downstream, the domain size is 20c to eliminate the effect of the boundaries on the solution. The airfoil wall is prescribed as no-slip boundary condition. Velocity inlet and pressure outlet are specified at inlet and outlet of the computational domain.

3 Computational Method

For the computational work, NACA 0012 airfoil is used. The first digit of airfoil determines the camber in percentage of chord length, while the second digit represents the location of camber in tenths of the chord length. The last two digits represent the maximum thickness of the airfoil in percentage of the chord length. NACA 0012 is symmetrical airfoil as first two digits '00' represent that there is no camber present in the airfoil, whereas the last two digits '12' represent that the maximum thickness of the airfoil is 12% of the chord. The calculations were carried out at high Reynolds

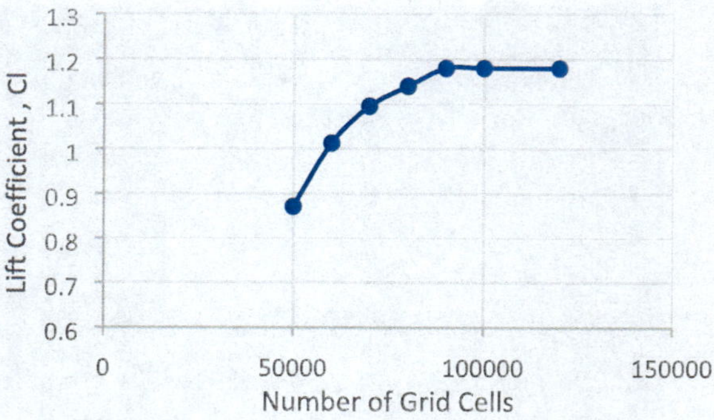

Fig. 2 Grid validation at 16° AoA

number (4.7 × 10^6), and the free stream temperature was kept similar to that of the environment, i.e. 300 k. The flow at this Reynolds number can be considered as incompressible, and this assumption is close to reality and is not necessary to resolve the energy equation. The calculations were done for AoA ranges from -10° to 16° at steady-state conditions.

The most important and foremost step in any CFD calculations is to determine the effect of size of mesh on the solution results. Normally, it is said that the results are more accurate if the mesh contains a greater number of nodes, but more the number of nodes more is the computational time. Therefore, appropriate number of nodes should be determined by increasing the number of nodes till the mesh shows a constant result and further refining would not generate any effective change in the results. Figure 2 shows the variation of coefficient of lift (C_L) with respect to number of grid cells, and it was found that after 90,000 cells there is almost negligible deviation, and hence, this grid size was considered to be appropriate for the calculation process.

From the study, it was observed that a C-type grid topology with 90,000 quadrilateral cells would be adequate enough to generate a grid independent solution. The height of first layer adjacent to the airfoil was set to nearly 10^{-5} which corresponds to average y^+ of approximately 1. The average skewness of the mess is in the order of 10^{-2}. The skewness and value of y+ of this size would be sufficient enough to properly resolve the inner components of the boundary layers. There is more refinement of the mesh in a region where more computational accuracy is required, such as nearer to the airfoil surface (Fig. 3).

For the simulation process, the pressure-based solver is used in most of the simulations, but in some cases, density-based solver was used. In some of the models at high AoA, the computational results were not accurate using pressure-based solver due to flow separation and wake generations. Thus, for some of the models at high AoA the density-based solver was used. In pressure-based solver, the semi implicit method for pressure linked equations (SIMPLE) scheme was used as pressure velocity coupling scheme and momentum equations were solved as per second order upwind solutions. In density-based solver, the implicit scheme was used as pressure–velocity coupling scheme and the Courant number was set as 5. In both the solvers, the boundary

Fig. 3 Mesh refinement near airfoil (NACA 0012)

Table 1 Operating parameters

Operating parameter	Value
Velocity of flow	0.2 Mach Number
Operating pressure	101,325 Pa
Operating temperature	300 k
Density of fluid	1.225 kg/m^3
Viscosity of fluid	1.7894 × 10^{-5} kg /ms
Angle of attack	−10° to 16°
Fluid	Air

conditions were given same and operational parameters were given similar to that as shown in Table 1.

4 Validation

The calculations using the above numerical model (except inlet velocity) were carried out and the results of the simulations were compared with the results of Eleni et al. [1]. The comparison was made in order to validate the numerical model. The results of the numerical model developed were in good agreement with the results of the paper, and hence, model was validated. The simulations were carried on using the k-ω SST Model, and comparison of C_L was made with the results of Eleni et al. [1]. The average MARD for the C_L was nearly about 2.4%. The comparison of C_L is shown below and it was cleared that the results were correct, and hence, the numerical model developed for simulation of flow on NACA 0012 airfoil was validated (Fig. 4).

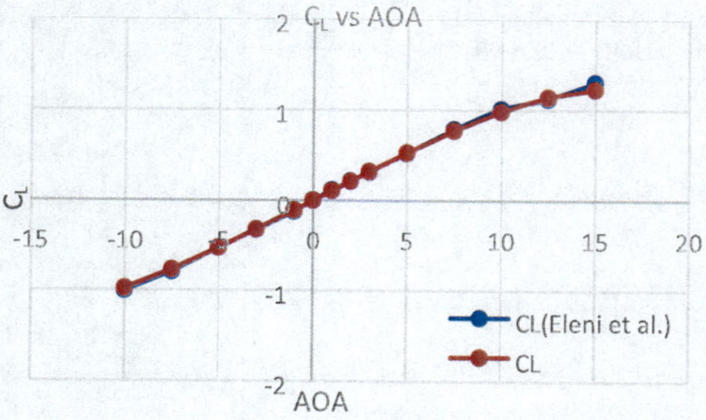

Fig. 4 Validation curve of C_L against AoA

5 Results and Discussions

Spallart–Allmaras is based on single transport equation [5] whereas k-ω and k-ε are two equation models. k-ε standard model is used to simulate far away condition, whereas k-ω gives better estimate for both near wall and far away conditions. k-ε realizable model with enhanced wall treatment is modified from k-ε to effectively capture near wall condition (y + ~1).

Coefficient of Drag (C_D) The resultant force on the airfoil is basically divided into two components, i.e. lift force and drag force. The component of the net force acting normal to the incoming flow stream is known as the lift force, and the component of the net force acting parallel to the incoming flow stream is known as the drag force. Following curves show the relation between coefficient of drag (C_D) with various AoA for various turbulence models. It is observed that the drag coefficient varies at slow rate for small AoA, but when the AoA reaches towards stall angle, the drag coefficient increases at a very high rate (Figs. 5, 6, 7, and 8).

Coefficient of Lift (C_L) Following Figs. 9, 10, 11, and 12 show the variation of lift coefficient (C_L) with various AoA for different turbulence models at 0.2 Mach number. It is clearly observed that at low AoA the lift coefficient varies linearly with AoA, whereas at high AoA, the linearity is not seen.

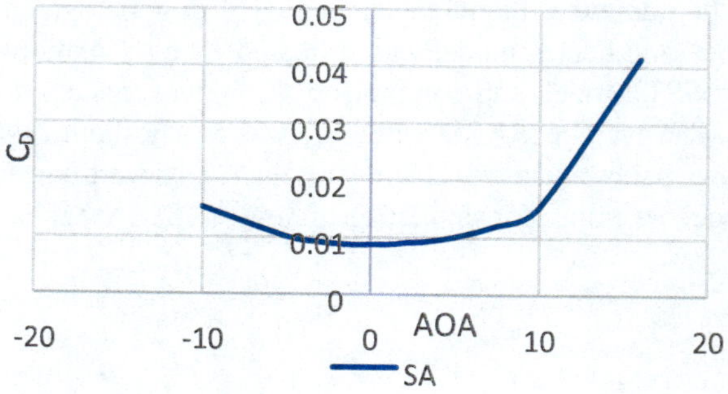

Fig. 5 Curve of C_D against AoA (SA Model)

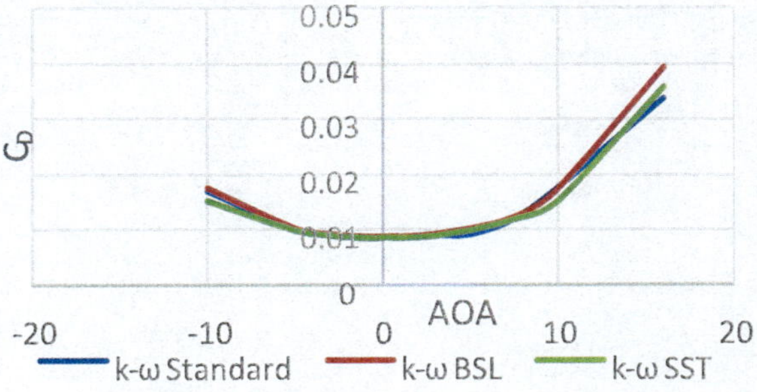

Fig. 6 Curve of C_D against AoA (k-ω Model)

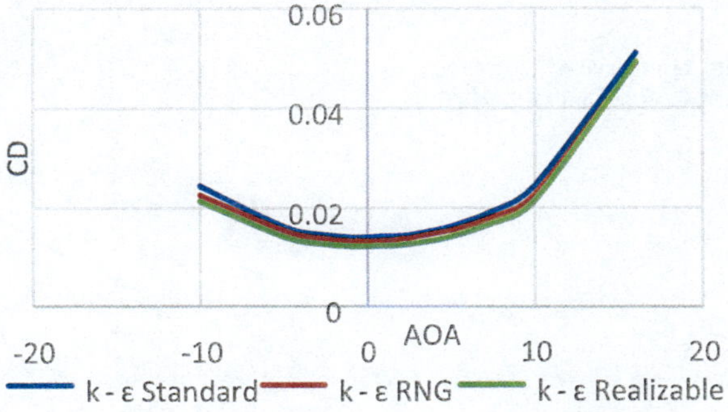

Fig. 7 Curve of C_D against AoA (k-ε Model with SWF)

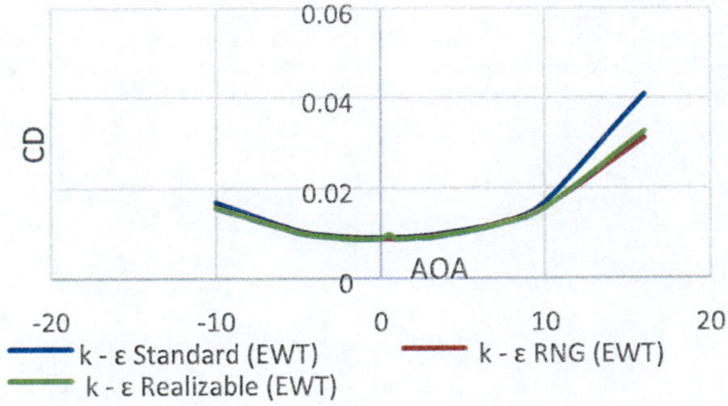

Fig. 8 Curve of C_D against AoA (k-ε Model with EWT)

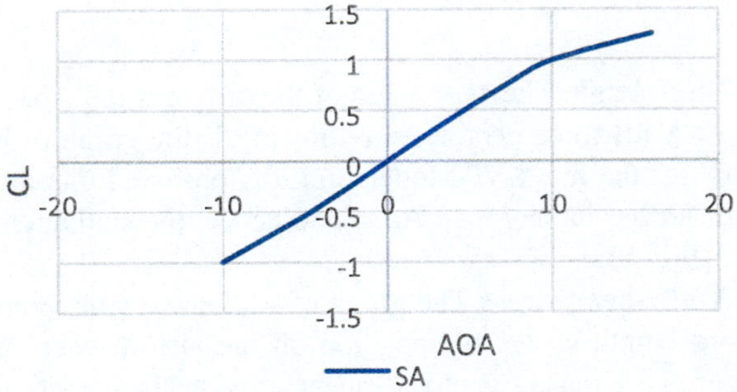

Fig. 9 Curve of C_L against AoA (SA Model)

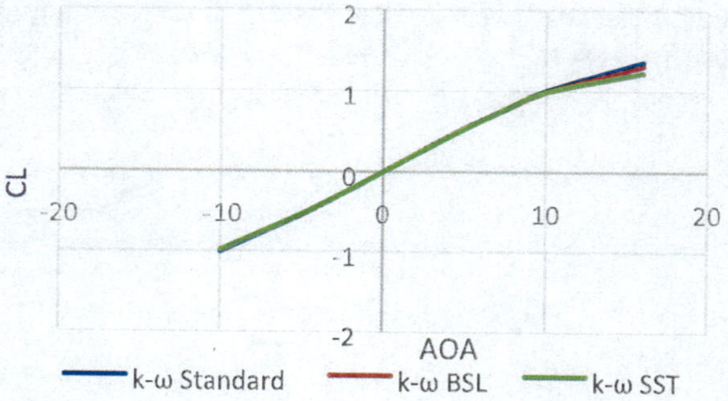

Fig. 10 Curve of C_L against AoA (k-ω Model)

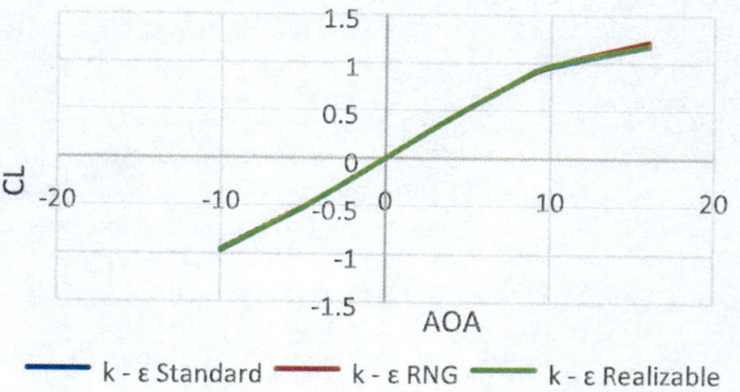

Fig. 11 Curve of C_L against AoA (k-ε Model with SWF)

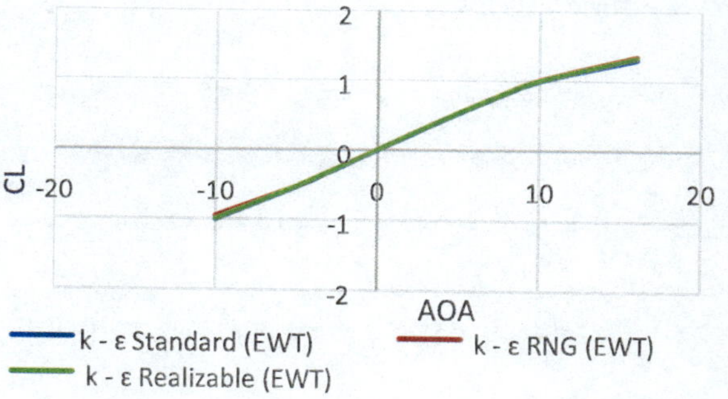

Fig. 12 Curve of C_L against AoA (k-ε Model with EWT)

Stall Angle The stall angle of the airfoil is the AoA after which with increases in AoA lift force decreases. Following is the graph of lift coefficient (C_L) against AoA for the k-ω SST Model, and it is observed that after 16° AoA, C_L decreases with further increases in AoA, and hence, the stall angle for NACA 0012 airfoil is 16° (Fig. 13).

Wall shear stress The graph of wall shear stress against the position along the chord length shows the position on the airfoil where the flow starts to separate. Figure 14 is the graph of wall shear stress against position along chord length for the k-ω SST model at stall angle (16°), and it is clearly observed that the flow starts to

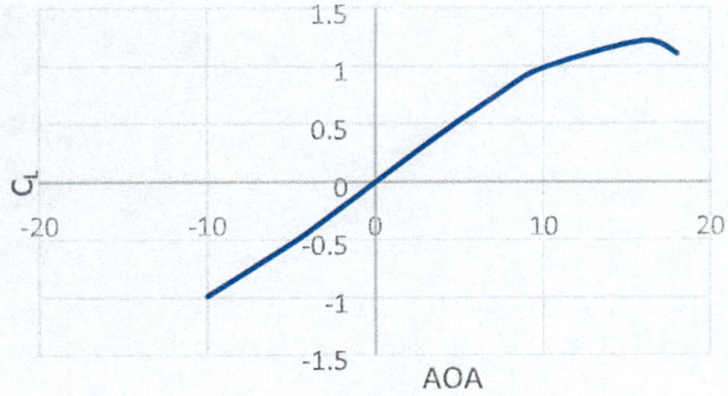

Fig. 13 Curve of C_L against AoA

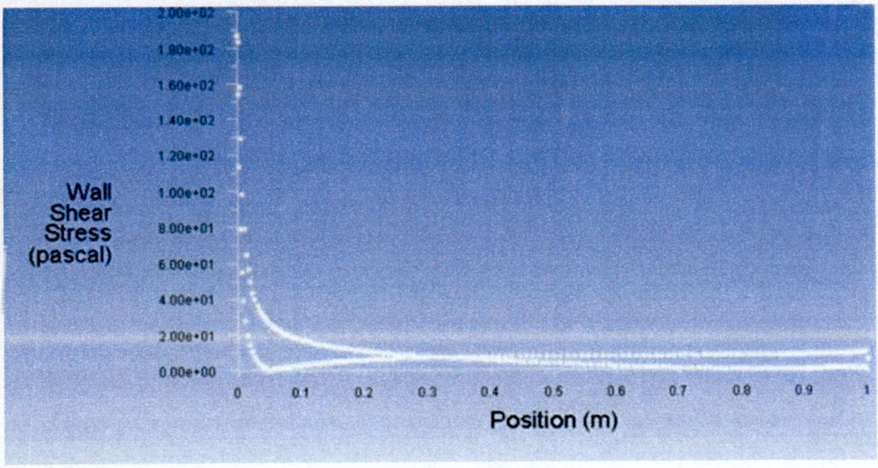

Fig. 14 Graph of Wall Shear Stress against position along chord length (k-ω SST Model)

separate at nearly 0.65c on the airfoil. Figure 15 shows the graph corresponding for k-ε realizable model with EWT at stall angle (16°) and for realizable model, a delay is observed in the flow separation, i.e. the flow starts to separate at nearly 0.8c on the airfoil.

Curve of C_L/ C_D against AoA Following curve shows the variation of C_L/C_D vs AoA for the k-ε realizable model with enhanced wall treatment. It shows that at low AoA the graph is dominated with lift force, whereas at high AoA the graph is dominated with drag force. It is observed that the ratio increases with AoA from 0° to 7.5°. After this, with further increase in AoA, the ratio decreases. The maximum value of ratio is obtained at 7.5° AoA and its value is 64.758 which means that at 7.5° AoA, NACA 0012 airfoil will work more efficiently than other AoAs at 0.2 Mach number. (Fig. 16).

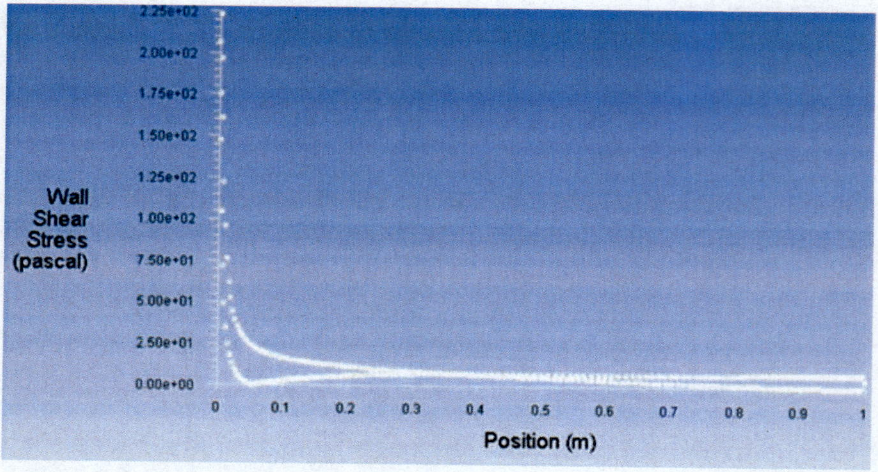

Fig. 15 Graph of Wall Shear Stress against position along chord length (k-ε Realizable Model with EWT)

Fig. 16 Curve of C_L/C_D against AoA

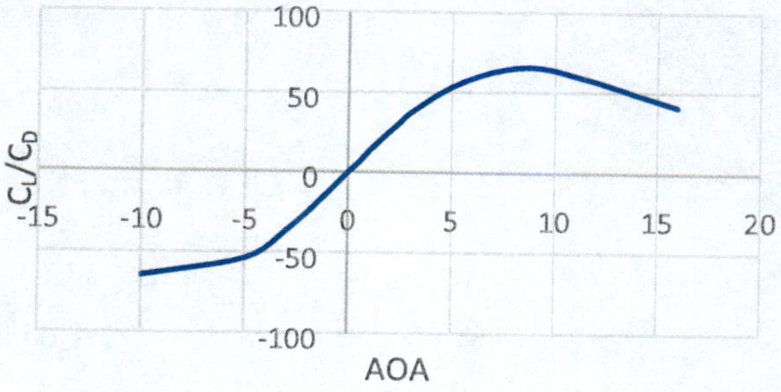

6 Conclusion

The calculations were carried out at 0.2 Mach number which is nearer to the compressible zone and hence at high AoA, accurate results were not obtained from the pressure-based solver and hence density-based solver was used in some models at high AoA. It was also concluded that the stall angle is 16° as after this angle the lift coefficient decreases and hence lift decreases. It is clear from the wall shear stress versus. x/c graph of k-ω SST and k-ε realizable model with enhanced wall treatment that the latter model effectively captures the delay in the flow separation. As compared to other models, delay in flow separation observed by k-ε realizable model with enhanced wall treatment accurately calculated the high lift coefficient (C_L) and low drag coefficient (C_D). These were the primary reasons for concluding that the k-ε realizable model with enhanced wall treatment is best suited for assessment of flow over a 2D airfoil at 0.2 Mach number.

References

1. Eleni DC, Athanasios TI, Dionissios MP (2012) Evaluation of the turbulence models for the simulation of the flow over a National Advisory Committee for Aeronautics (NACA) 0012 airfoil. Journal of Mechanical Engineering Research 4(3):100–111
2. Saraf AK, Singh M, Kumar A (2012) Analysis of the Spalart-Allmaras and k-ω standard models for the simulation of the flow over a National Advisory Committee for Aeronautics (NACA) 4412 airfoil. Int J Scientific & Engineering Res 3(8)
3. Rao BR, Sahitya R (2015) Numerical and experimental investigation of the flow field around NACA 0012 Aerofoil. 4(2). ISSN (Print): 2319-3182
4. Sadikin A, Yunus NAKM, Abd Hamid SA, Ismail AE, Salleh S, Ahmad S, Rahman MNA, Mahzan S, Ayop SS (2018) A comparative study of turbulence models on aerodynamics characteristics of a NACA0012 Airfoil. Int J Integrated Engineering 10(1):134–137
5. Spalart PR, Allmaras SR (1992) A one-equation turbulence model for aerodynamic flows. AIAA Paper, pp 92–439

Numerical Validation of Power Take-Off Damping in An Owc Chamber and Modifications for Increased Efficiency

Dasadia Kush, Mitesh Gandhi, and Jyotirmay Banerjee

Nomenclature

NWT	Numerical Wave Tank
OWC	Oscillatory Water Column
PTO	Power Take-Off
T	Time period
h	Water depth
H	Wave height
λ	Wavelength
k	Wavenumber
BC	Boundary Conditions
C	Permeability coefficient

1 Introduction

The concept of the OWC device is utilized to collect energy from waves of ocean. Oscillating movement of free surface in a subsection in OWC transfers the wave energy to the air inside the section. A power take-off arrangement converts pneumatic energy into electrical energy or another form that can be used. Wave parameters

D. Kush (✉) · M. Gandhi · J. Banerjee
Mechanical Engineering, SVNIT, Surat, India
e-mail: kushdasadia51@gmail.com

M. Gandhi
e-mail: miteshagandhi1@gmail.com

J. Banerjee
e-mail: jbaner@med.svnit.ac.in

J. Banerjee et al. (eds.), *Recent Advances in Fluid Dynamics*, Lecture Notes in Mechanical Engineering, https://doi.org/10.1007/978-981-19-3379-0_32

such as wave height, wavelength, time period, depth of water, depth of immersion, a bottom profile of OWC chamber, dimensions of chamber, distance from wave generation zone, number of OWC chambers in rows, the distance between them, and phase control. affect the efficiency of OWC device. Considering all these parameters for the analysis of OWC is difficult. This analysis reported in the literature mainly focuses on a few of these parameters [7] performed tests to evaluate the impact of thickness of walls, front wall shape, and front lip submergence on the hydrodynamic efficiency of the OWC chamber for various wave parameters. They also discovered front wall that has smoother geometry and minimizes entry losses, enhancing OWC device efficiency [5] and used porous-medium flow theorem to induce the PTO damping. We have generated a numerical model consisting of a PTO damping in ANSYS Fluent and validate the results with the numerical model of Morris-Thomas et al. [5] and experiments of Morris-Thomas et al. [5]. Modifications in the OWC chamber are done in two parts. Firstly, the vent location is varied at three positions in left, mid, and right locations in the OWC chamber. The side supporting walls to the vent, which were earlier horizontal, are now given specific slopes. The reason to do this was to increase efficiency to absorb the incident wave energy without significant seaward leakage of said energy so that generation of negative power is minimized. The pressure to velocity conversion efficiency is four vertical plates at equal distances and constant thickness are placed inside the former modified shapes. According to Masoomi et al. [6], introducing four vertical plates led to much more uniform oscillatory motion inside the chamber and greatly increased the efficiency.

2 Problem Description

This research aims to compare the experimental results of a small-scale prototype of OWC with numerical results. In a 2D NWT, ANSYS Fluent solver is used for simulation of an OWC. Firstly, we compare the variations of interior chamber pressure, velocity of free surface, and free surface elevation with experimental outcome of Morris-Thomas et al. [7] and Kamath et al. [4] numerical model due to which our numerical model results are confirmed. Later, the efficiency and power output were calculated for both the cases. Increasing the efficiency of the OWC is the last part of the problem.

3 Methodology

The continuity equation is the governing equation for two-phase fluid flow. During the simulations, we used the homogeneous volume of fluid multiphase model. An implicit scheme was applied while keeping the global courant number as 0.1 for better accuracy—numerical modal consists of solving NavierStokes set of equation

Table 1 Solution methods used

Gradient	Green Gauss-cell based
Pressure	Body force weighted
Momentum	2nd order upwind
Fraction of volume	Compressive
Kinetic energy of Turbulence	2nd order upwind
Specific rate of dissipation	2nd order upwind

Table 2 Wave parameters

λ (m)	*T* (s)	*h* (m)	*H* (m)
4.07	1.7019	0.92	0.12

with the equation of transport. The turbulent model of SST k-ω was adopted in simulations.

Methods & Discretization schemes For solution methods, the PISO method is applied for pressure and velocity coupling. For space–time discretization, the following strategies are applied (Table 1):

Wave Generation For generating waves in NWT, the Fluent solver's integrated open channel flow and open channel wave boundary conditions were applied as a part of VoF's submodels. Similarly, the open channel wave B.C was applied to inlet velocity through which shallow/intermediate waves were generated following Stokes 5th order wave theory [3]. All the wave parameters are constant and listed below (Table 2).

4 Hydrodynamic Efficiency

The OWC section's hydrodynamic efficiency determines how much power is available on the vent at any given timestep, which can be used to generate energy using any PTO mechanism. Efficiency is the parameter utilized to find and compare the effects on the output power of numerical studies one and two with the results of Kamath et al. [4] and Morris-Thomas et al. [7]. Inside the OWC chamber, the free surface is raised as a result of the incident wave energy. Pressure is created, and the energy is transported to the top of the chamber via the vent. This output power is estimated via the free surface oscillating motion. The corresponding incident wave power is

$$P_i = \left(\frac{1}{8}\rho g H_{\text{inc}}^2\right)\frac{\omega}{k}\left[\frac{1}{2}\left(1 + \frac{2kh}{\sinh(2kh)}\right)\right] \tag{1}$$

The hydrodynamic power from the oscillating movement of the free surface is:

$$P_{\text{hyd}} = pAu \tag{2}$$

where p denotes the pressure in the OWC section, along with A denoting the chamber's area, and u denoting the free surface average velocity. The ratio of output power to incident power is the efficiency which is calculated at each time step [1, 2].

$$\varepsilon = \frac{\mathrm{P_{hyd}} * 100}{\mathrm{Pi}} \tag{3}$$

5 Model of the Numerical Tank

For Validation: The model of the numerical tank has the identical dimension as in the experimental setup of Morris-Thomas, et al. [7]. The OWC chamber is located 37.5 m from the zone of wave generation. Water depth is taken as $h = 0.92$ m, the height of the NWT is 2.2 m, the height of the chamber is 1.275 m, front wall draught is $a = 0.15$ m, chamber length is $b = 0.64$ m, and the wall thickness is 0.04 m. The vent size is taken as 0.05 m, and a porous media of certain permeability constant is applied in that vent area for PTO damping (Fig. 1).

For Modifications: The dimensions and the parameters of the model were kept constant, and the modifications include the change in the vent location and the upper wall shape for modification 1. Vent size for modified models is kept 5 mm (Fig. 2).

Modification 2 includes adding four vertical plates inside the chambers of the former modified models. These plates were of thickness of 0.02 m and had a height of 0.22 m with the bottom collinear with the front wall bottom face. The plates had a spacing of 0.1122 m.

Meshing: For meshing, GAMBIT software is used. Along the free surface of water, around chamber's front wall, and in the air section within the chamber, a small mesh of 0.01 m was constructed. Since the precision of the results in these crucial regions is critical. The remaining domain had a 0.025 m mesh size. The grid size for modification 1 and modification 2 models is 0.025 m. A bit of non-uniformity in the mesh is present for the modified models due to the sloping surface of the walls (Figs. 3, 4, 5, 6, 7, 8, and 9).

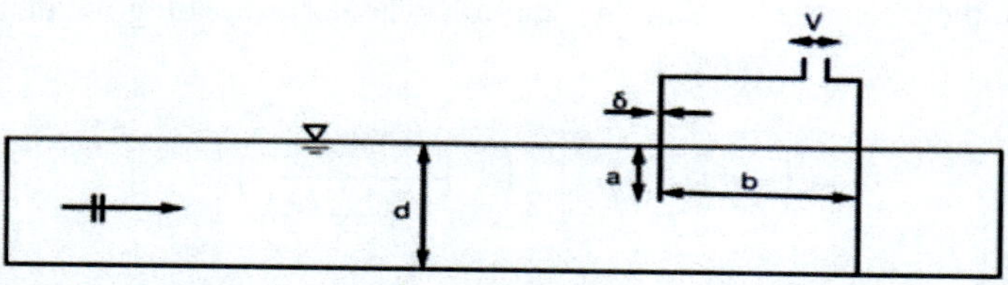

Fig. 1 Dimensions of the NWT

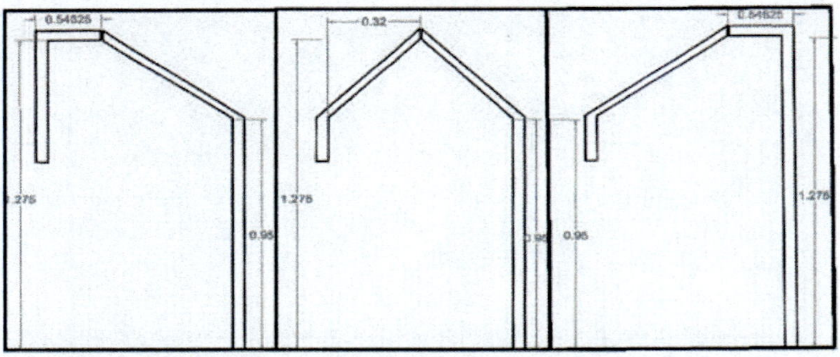

Fig. 2 OWC models with modified wall shape and vent location (all units in m)

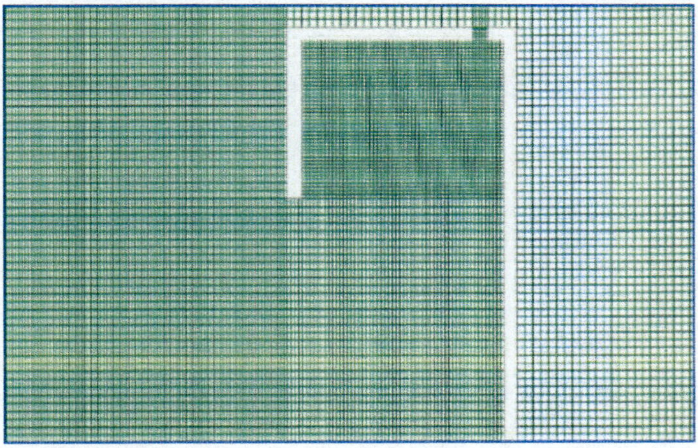

Fig. 3 Mesh generated for validation model

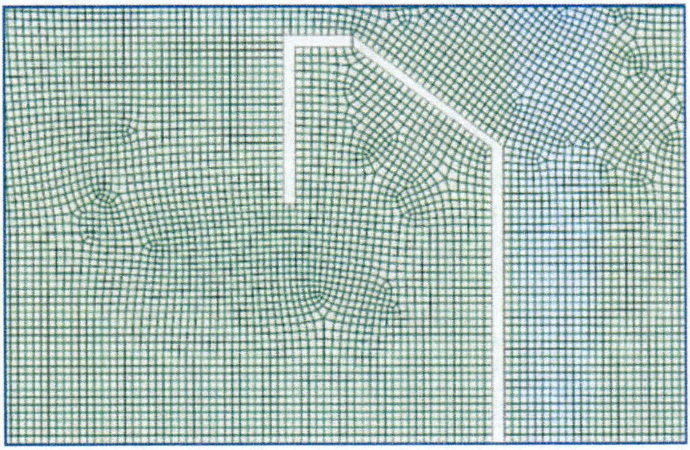

Fig. 4 Mesh generated for modification 1 model (left vent)

Fig. 5 Mesh generated for modification 1 model (mid-vent)

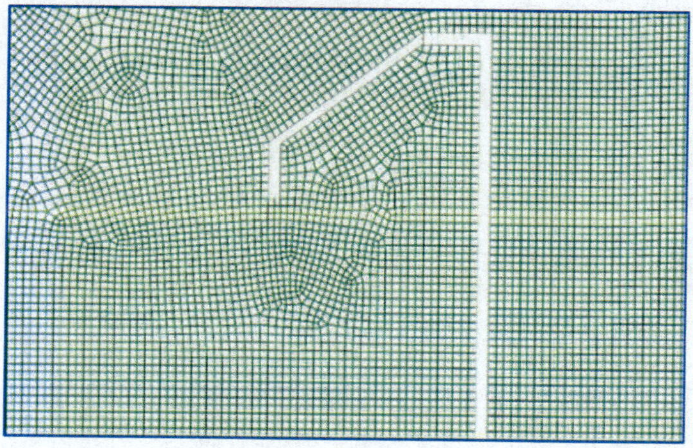

Fig. 6 Mesh generated for modification 1 model (right vent)

Porous Media Modelling The vent size is taken as 0.05 m which is comparatively large, so it will not create sufficient pressure to transform the energy incidented on the vent. So, a porous media is applied at the vent area, which generates a PTO damping and can give sufficient power output for the same. Darcy's law of flow through porous medium is used to estimate the permeability coefficient $C = 1/k_p$:

$$q = \frac{-k_p A}{\mu} \frac{\Delta P}{L} \tag{4}$$

Fig. 7 Mesh generated for modification 2 model (left vent)

Fig. 8 Mesh generated for modification 2 model (mid-vent)

Here, q represents flow rate, k_p denotes intrinsic permeability, and A represents area of cross section. The cross-section area, the flow-through vent, and pressure are known from experiments. So, the permeability coefficient is calculated as 3.6×10^8 for the given length of porous media (Fig. 10).

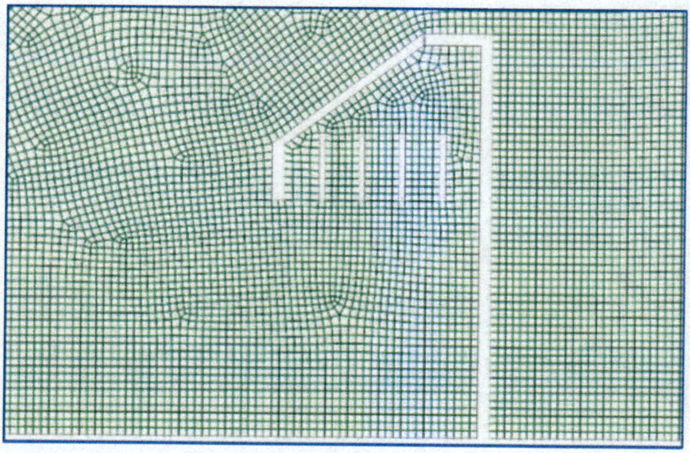

Fig. 9 Mesh generated for modification 2 model (right vent)

Fig. 10 Mesh in the vent region (validation model)

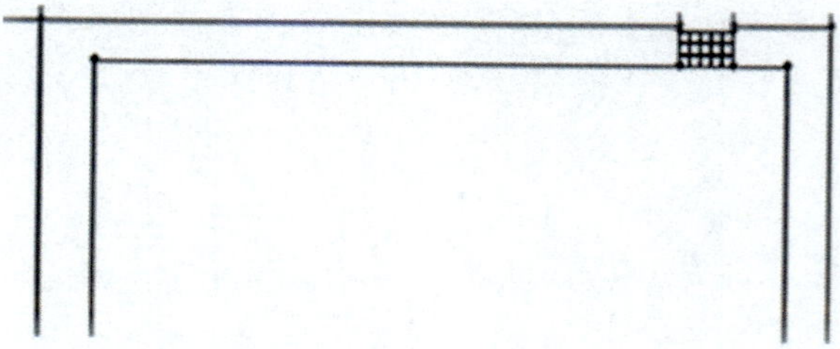

6 Results and Discussion

The experimental setup of Thomas is simulated with the PTO damping in the vent. Calculations are carried out in this study utilizing the incident wave whose physical parameters fulfil Stokes fifth-order wave theory. As previously stated, the OWC chamber is located 37.5 m away from the wave generating zone. The similar configuration is recreated in simulations along with a small alteration in PTO unit, at which vent size is altered as 0.05 m. Using the experimental data of Morris-Thomas et al. [7] $C_{exp} = 3.6 \times 10^8$ m^2 is calculated in the numerical model to give similar drop of pressure and flow rate across a vent with a thickness of 0.05 m for the corresponding length of the porous layer. The Fluent solver assumes the unit width in the case of 2D simulations. In experiments, the width of the chamber is 1.37 m.

The pressure fluctuation in OWC section, elevation of free surface, and velocity of free surface are calculated via custom field functions implemented in Fluent solver. Here are the custom field functions:

For pressure it is,

$$(1 - \text{VOF}_{\text{water}}) \times \text{ total pressure} \tag{5}$$

For free surface velocity, it is,

$$\text{VOF}_{\text{water}} \times V_y \tag{6}$$

For free surface elevation, it is,

$$\frac{(Y - \text{ coordinate } \times \text{VOF}_{\text{water}}) - 0.92}{A_0} \tag{7}$$

Then, these solutions are differentiated with the experimental outcome of Morris-Thomas et al. [7] and Kamath et al. [4]). Here are the comparison data (Figs. 11, 12, 13, 14, 15, 16, 17, 18, and 19).

In these figures, the numerical outputs and the experimental findings are in very much good agreement. Similarly, the power output and efficiency were calculated using custom field functions. Incident power was calculated as 38.34 W (Figs. 20, 21, 22, 23, 24, 25, 26, 27, Table 3).

The peak efficiency and average efficiency observed are 66.5% and 28.8%, respectively. In the modified models, we applied the same conditions as validation models, and the efficiency of each modified model is observed.

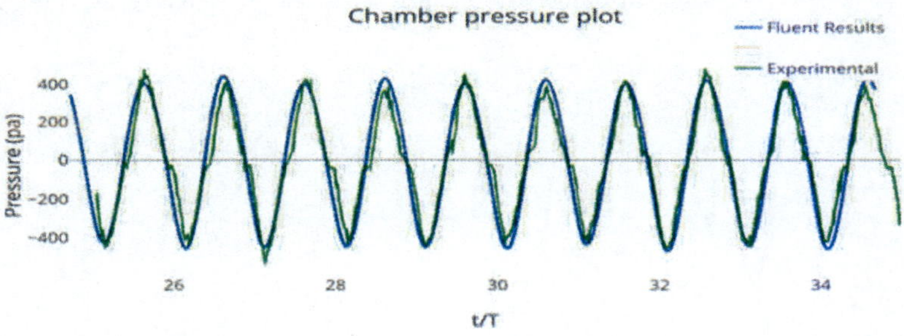

Fig. 11 Pressure results with reference to the experimental outcome of Thomas et al.

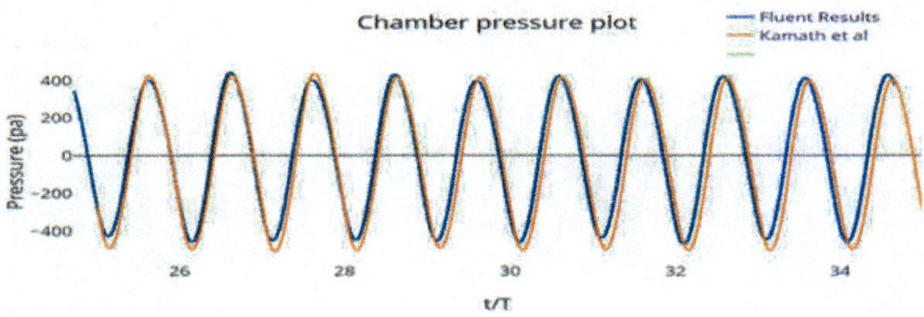

Fig. 12 Pressure comparison with solution of Kamath et al.

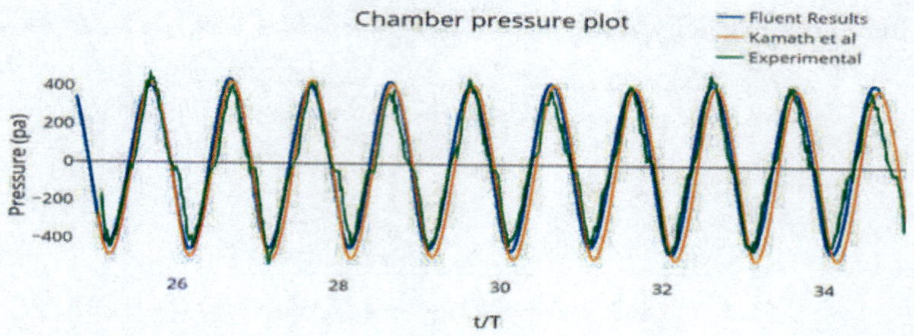

Fig. 13 Combined pressure solution with reference to the experimental outcome of Thomas et al. and results of Kamath et al.

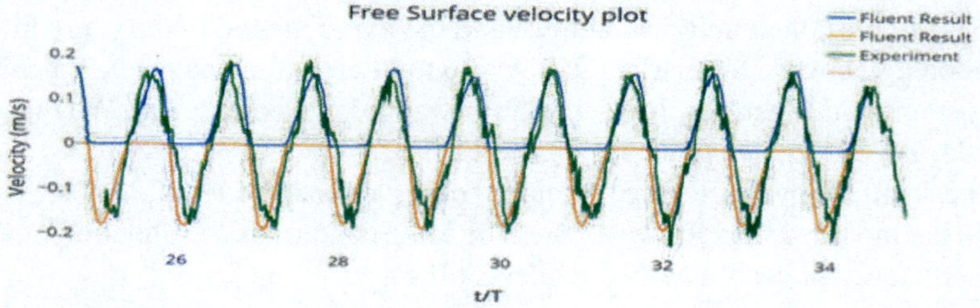

Fig. 14 Velocity results with reference to the experimental outcome of Thomas et al.

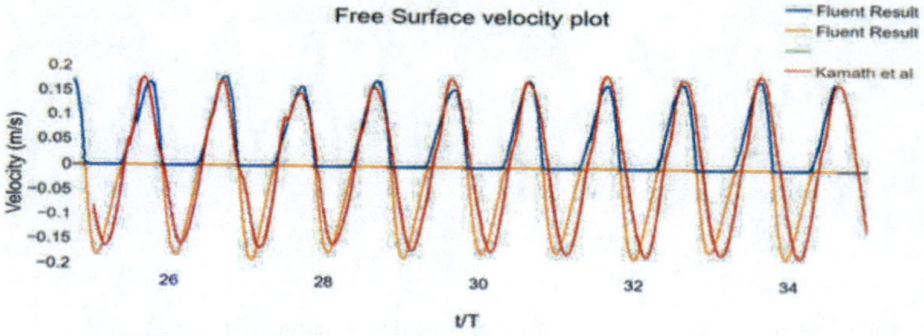

Fig. 15 Velocity comparison with solution of Kamath et al.

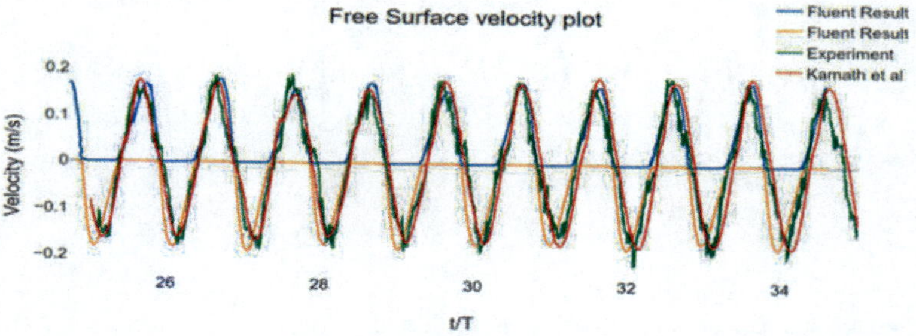

Fig. 16 Combined velocity comparison with solution of Kamath et al. and experimental outcome of Thomas et al.

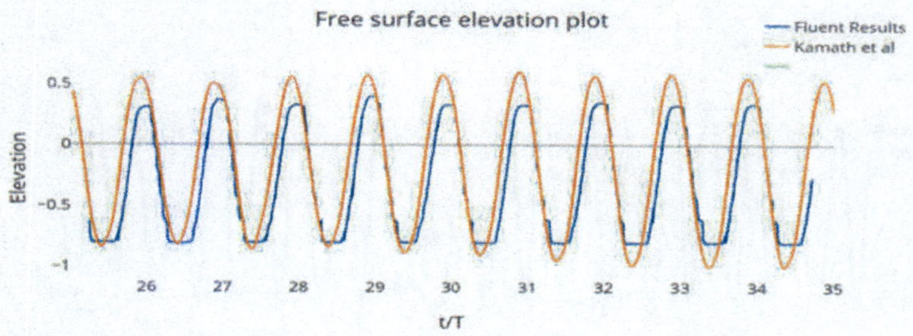

Fig. 17 Elevation of free surface compared with the results of Kamath et al.

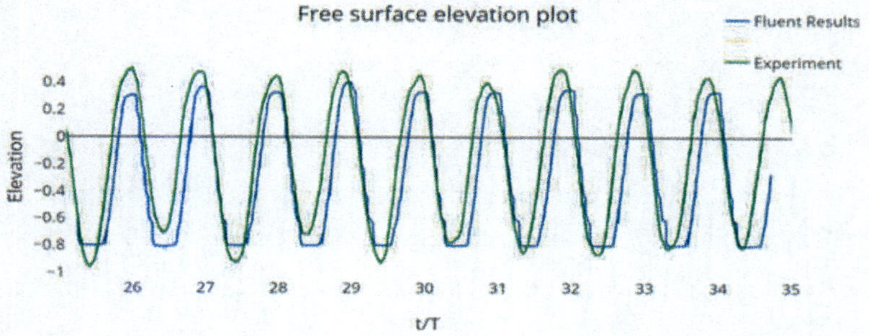

Fig. 18 Elevation of free surface with reference to experimental outcome of Thomas et al.

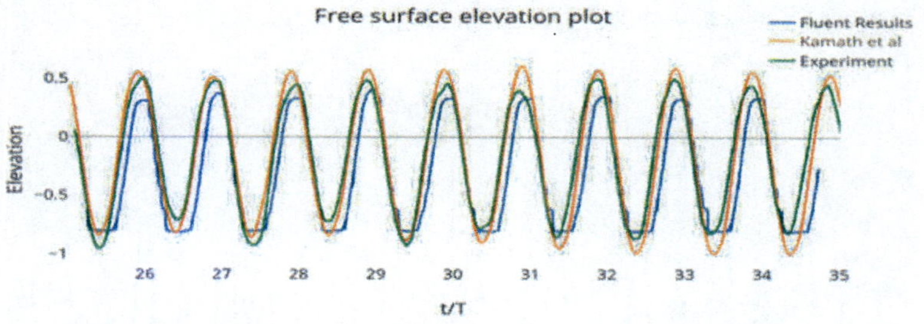

Fig. 19 Combined free surface elevation comparison with experimental outcome of Thomas et al. and Results of Kamath et al.

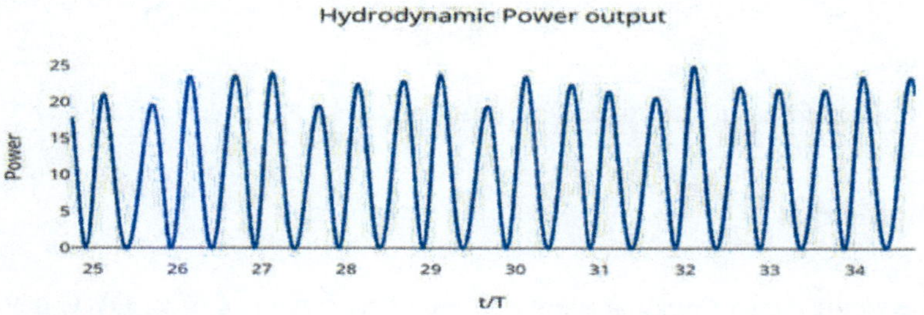

Fig. 20 Efficiency calculated

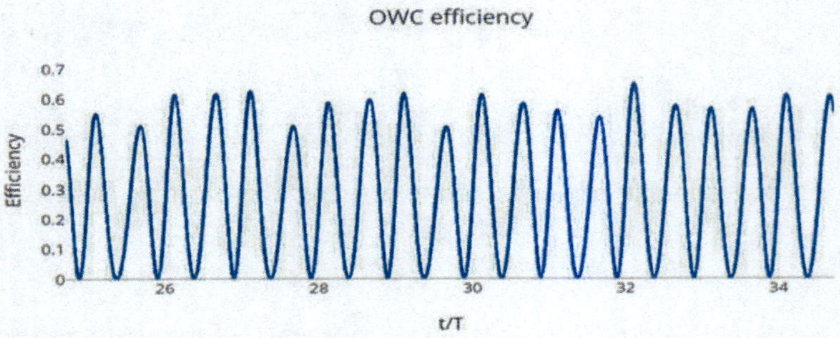

Fig. 21 Efficiency ratio of obtained and maximum power output

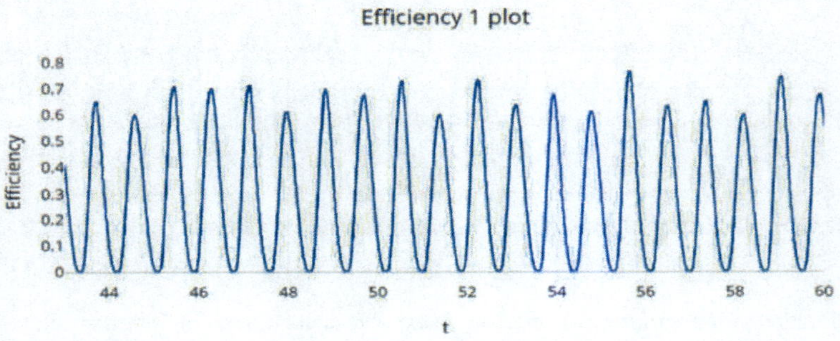

Fig. 22 Efficiency for modification 1 (left vent)

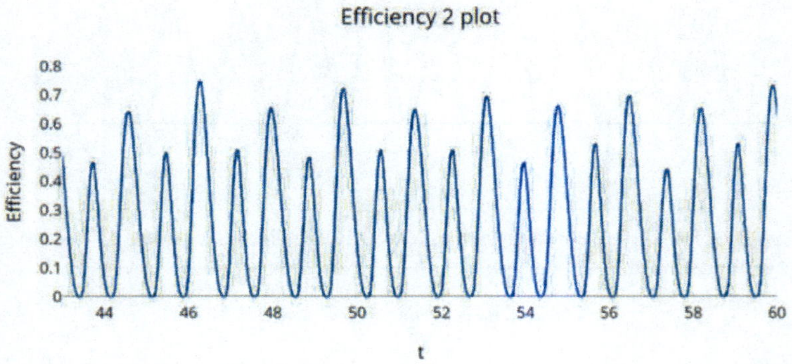

Fig. 23 Efficiency for modification 1 (mid vent)

7 Conclusions

The influence of incident regular waves on the hydrodynamics of an OWC device was investigated using a 2-D NWT. It has been determined that PTO damping may be used to mimic the OWC chamber as viewed in experiments. The water velocity at

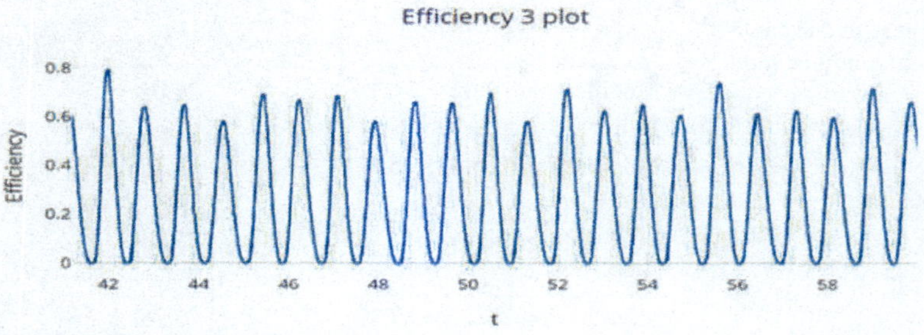

Fig. 24 Efficiency for modification 1 (right vent)

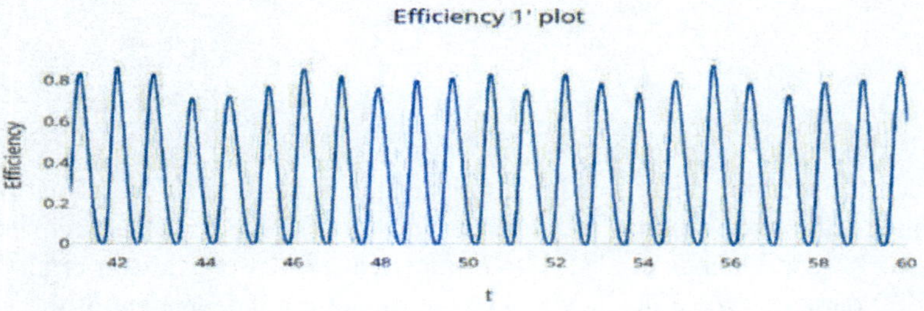

Fig. 25 Efficiency for modification 2 (left vent)

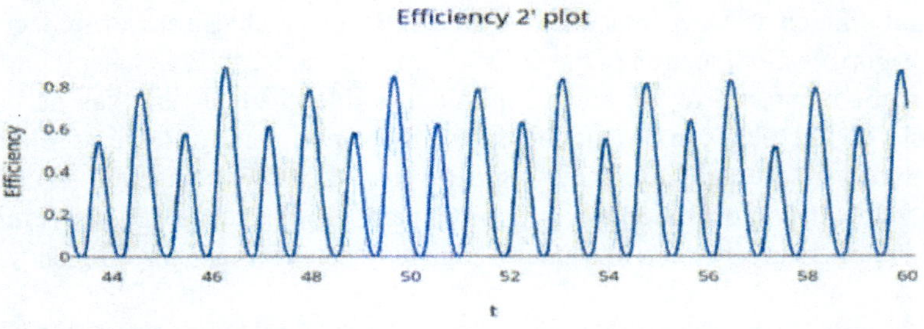

Fig. 26 Efficiency for modification 2 (mid vent)

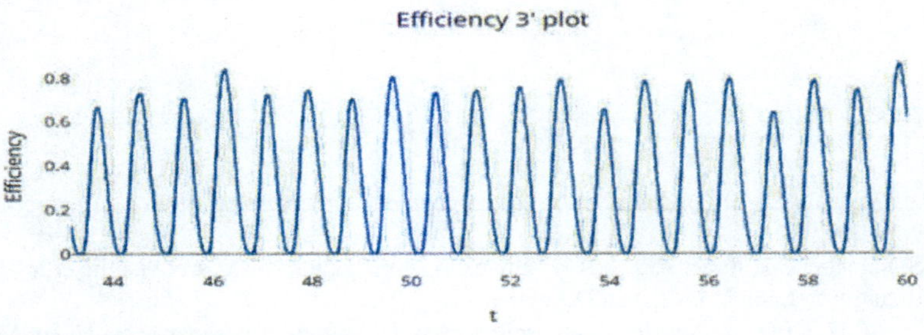

Fig. 27 Efficiency for modification 2 (right vent)

Table 3 Peak and average efficiency of modified models

	%Peak efficiency	%Average efficiency
Modification 1 left vent	76.76	34.04
Modification 1 mid vent	75.02	27.35
Modification 1 right vent	74.64	34.44
Modification 2 left vent	87.82	40.75
Modification 2 mid vent	90.27	34.03
Modification 2 right vent	86.85	37.87

free surface, the pressure created within the air section owing to pressurization of air, and the fluctuation of elevation of free surface were all compared to the experimental outcome of Morris-Thomas et al. [7] and numerical results of Kamath et al. [4]. Due to PTO damping, the efficiency and power output are independent of the vent dimensions with corresponding porous layer areas.

It is preferable that the free surface motion within the chamber must be piston-type for optimal efficiency. Variations in water free surface can have a major impact on efficiency. Further modification of the OWC chamber is done, and average efficiency values can be compared to get an optimum and stable chamber that can be used commercially. The plates in the modified designs help in getting stable oscillatory motion, and their average efficiency is greater than those without plates. Further study on the design optimization can be done on various other vent locations. A more complex upper chamber shape like a bell shape may help increase the efficiency.

Acknowledgements The authors are grateful to Suraj Garad for the aid and insights of their NWT modelling work.

References

1. Boualia B, Larblb S (2013) Contribution to the geometry optimization of an oscillating. s.l., s.n., p 9
2. Garad S, Saincher SA, Banerjee J (2017) Numerical investigations on hydrodynamic efficiency of oscillating water column. s.l., s.n, December
3. Horko M (2007) CFD optimisation of an oscillating water column wave energy converter. M.S. Thesis, The University of Western Australia. s.l.:s.n
4. Kamath A, Arntsen ØA, Bihs H (2015) Numerical investigations of the hydrodynamics of an oscillating water column device. Ocean Eng, 40–50
5. Kamath A, Bihs H, Arntsen ØA (2015) Numerical modeling of power take-off damping in an oscillating water column device. Int J Marine Energy, 1–16

6. Masoomi M, Yousefifard M, Mosavi A (2021) Efficiency assessment of an amended oscillating water column using OpenFOAM. Sustainability, 13
7. Morris-Thomas M, Irvin R, Thiagarajan K (2007) An investigation into the hydrodynamic efficiency of an oscillating water column. J Offshore Mechanics Arctic Eng, pp 273–278

Parametric Investigation to Improve Blending Performance

Deepak Kumar and Mahesh Vaze

Nomenclature

A	Nozzle/jet cross-sectional area,m^2
c^*	Degree of mixing
c_p	Correlation constant, (Fosset, 1951)
C	Tracer concentration, $kmol/m^3$
C_d	Coefficient of drag
C_μ, $C_{1\varepsilon}$, $C_{2\varepsilon}$, C_1, C_2	'K-ε' model constant
d_j	Diameter of nozzle/jet, m
d_z	Diameter of jet at free jet path end, m
u_z	Jet velocity at free jet path end, ms^{-1}
L_j	Jet path length, m
n_j	Number of jets
u_j	Jet velocity, m/s
Re_j	Jet Reynolds number $(\rho d_j u_j/\mu)$
T_{mix}	Time of mixing, s
T_{inj}	Injection timing, s
ε	Turbulent energy dissipation rate
μ_t	Eddy viscosity
D_o	Diameter of outlet, m
D_t	Diameter of the tank, m
H_l	Height of liquid, m
k	Turbulent kinetic energy

D. Kumar (✉)
University of Petroleum and Energy Studies, Dehradun, India
e-mail: dc98083@gmail.com

M. Vaze
ABB Global Industries and Services Private Ltd, Bangalore, India
e-mail: mahesh.vaze@in.abb.com

J. Banerjee et al. (eds.), *Recent Advances in Fluid Dynamics*, Lecture Notes in Mechanical Engineering, https://doi.org/10.1007/978-981-19-3379-0_33

K_c	Correlation constant
P_j	Jet power input, W
q_j	Jet flow rate, m^3s^{-1}

Greek Symbols

ρ	Density of the liquid, kg/m^3
θ_m	Time of mixing, s
μ	Molecular viscosity, kg/m.s

1 Introduction

Jet mixers are the simplest and important mixing devices to achieve mixing in many chemical processes. In many applications, side jet mixers are the most commonly used mixers to achieve mixing. Jet mixers are cheaper than agitator mixers and easy to install. Jet mixers have become a very important mixing device in many chemical processes because of their many advantages, including low-operating cost, easy installation and maintenance, no. of vibration because of moving parts, and simple design as compared to agitator mixers. In jet mixers, a high-velocity fluid is drawn with the help of a pump and recirculates into the tank with the help of a jet. The recirculation of fluids using a jet mixer was first introduced by [2].

Over the years, many experimental investigations were done by the researchers on the jet mixers by using different geometries of the tanks. Most of the experimental studies end up with empirical correlations to predict mixing and mixing time. Table 1 gives the previous experimental work given by many researchers on the jet mixers with the correlations given for the estimation of mixing time. These correlations are case-specific and do not helpful in understanding the complex three-dimensional phenomena of mixing.

Hence, our objective is to develop a CFD model to understand the mixing phenomena. There are certain parameters in jet mixing like velocity and jet angle which have a significant effect on mixing and concentration profile. This would be the further objective of this paper.

2 Problem Description

The schematic diagram of the jet mixing tank [9] is shown in Fig. 1. The flat bottom cylindrical tank with the diameter (D) of 0.5 m in which the water is filled with the height (H) 0.5 m ($H/D = 1$). The outlet with a diameter of 0.0381 m was located at

Table 1 Proposed correlations from previous experimental work on jet-mixed tanks

Authors	Geometry and dimensions	Correlations
Fossett and Prosser [2]	Inclines side entry jet, $D_t = 1.524$ m, $Re_j = 4500–80{,}000$, $\theta = 40^0$, $n_j = 1$ and 2, $D_o = 2.54$ cm, $H_l = 0.9144$ m, $d_j = 1.9$ mm	$\theta_{mix} = c_p \frac{D_t^2}{u_j d_j}$ $c_p = 4.5$, and $c_p = 9$ when $T_{inj} > \theta_{\text{mix}}/2$
Fox and Gex [3]	$H_l/D_t = 0.25–2$, $D_t = H_l = 0.15–4.27$ m $D_t/d_j = 45–180$, $u_j = 0.6–11$ m/s $d_j = 0.159–3.81$ cm	$T_{\text{mix}} = C \frac{D_t H_l^{0.5}}{(u_j d_j)^{4/6} g^{1/6} \text{Re}_j^{1/6}}$
Vergleichende Ruhrversuche zum Mischen loslicher Flussigkeiten in einem 12 000-m3-Behalter [12]	Fox and Gex [3] data re-analysis	$T_{mix} = 3.68 \frac{D_t^2}{(u_j d_j)}$
Okita and Oyama [7]	Side entry jets with the inclination	$T_{mix} = 5.5 \frac{D_t^{1.5} H_l^{0.5}}{(u_j d_j)}$
Lehrer [6]	Compared model predictions show good agreement with experimental data	$T_{\text{mix}} = \frac{0.658}{u_j} \left(\frac{\rho_c}{\rho_d}\right)^{5/8} d_j^{0.25} \times \left(\frac{u_j}{n_j A}\right)^{3/4}$ (-log(1-c^*))
Yianneskis [15]	$D_t = H_l = 294$ mm, $d_j = 4, 8$ mm, $\theta = 45°$ side entry jet with 10 mm clearance (from bottom and sidewall)	$T_{\text{mix}} = (C\varepsilon)^{-0.33}$ $\text{Re}_j > 30{,}000$
Orfaniotis et al. [8]	D_t, $H_l = 500$ mm, $d_j = 9, 15$ mm Horizontally located two jets at $\frac{H_l}{3}$ and $\frac{H_l}{2}$	$T_{\text{mix}} = C \frac{D_t^{1.82}}{u_j^{0.82} d_j^{0.82} H_l^{0.09}}$
Grenville and Tilton [5]	$D_t/H_l = 0.2–1.0$, $D_t = 0.61–36$ m $d_j = 5.8–50$ mm	$T_{mix} = K_c \frac{D_t^2 H_l}{u_j d_j L_j}$

(continued)

Table 1 (continued)

Authors	Geometry and dimensions	Correlations
Patwardhan and Gaikwad [10]	Inclined side entry jet $\theta = 45\,°$, $D_t = 0.5$ m, $H_l = 0.5$ m,	$P_j = \left(\frac{1}{2}\rho {u_j}^2\right) q_j = \frac{\pi}{8}\rho d^2 {u_j}^3$
Wasewar [13]	Fossett and Prosser [2] geometry	$T_{\text{mix}} = c_p \frac{D_t}{d_j u_j}$ $c_p = 9$, when $T_{inj} > \theta_{\text{mix}}/2$, $c_p = 4.5$, when $T_{inj} < \theta_{\text{mix}}/2$,
Wasewar and Sarathi [14]	$D_t = H_l = 0.4$ m, $d_j =$ 0.01 m	$\frac{[T - T_{mean}]}{T_{mean}} = m$ $m = 1$, when mixing starts, $m = 0$, when complete mixing is achieved
Grenville and Tilton [4]	Experiment was done for various liquid depth to vessel diameter ratios, jet velocity, and jet nizzle diameters	$\theta_{\text{mix}} = 32.37 \left[\left[\frac{d_z Z^2}{{u_z}^3}\right]\right]^{1/3}$

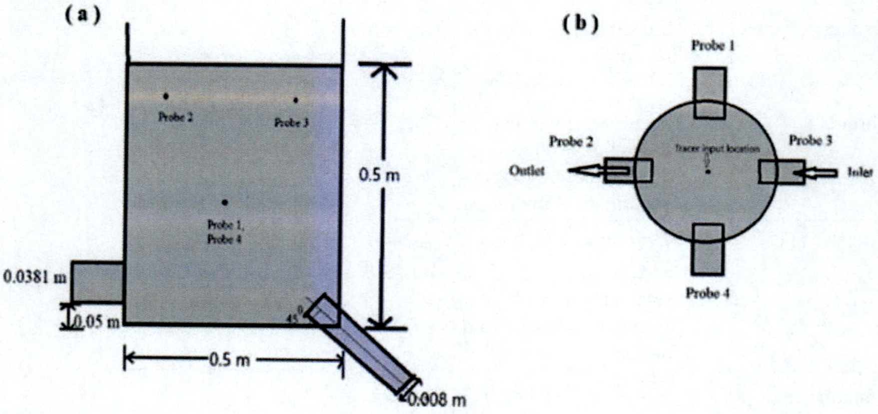

Fig. 1 Schematic diagram of **a** Jet mixing tank side view and **b** Jet mixing tank top view

0.05 m from the base of the tank. To understand the effect of the inclination of the nozzle jet, it was oriented at 30°, 45°, and 60° with reference to the base of the tank. Figure 2 shows the three-dimensional solid jet mixing tank model.

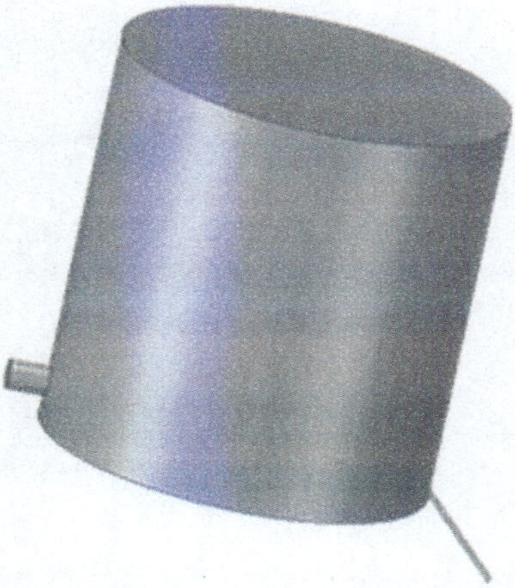

Fig. 2 Jet mixing tank three-dimensional model

3 Grid Generation

Grid generation was done using ANSYS Fluent mesher to establish the polyhedral mesh. Grid was further refined in the viscous-dominated region by creating the inflation layers. The poly-hexcore grid was carefully generated with the domain decomposition technique. The grid of the jet mixing tank is shown in Fig. 3.

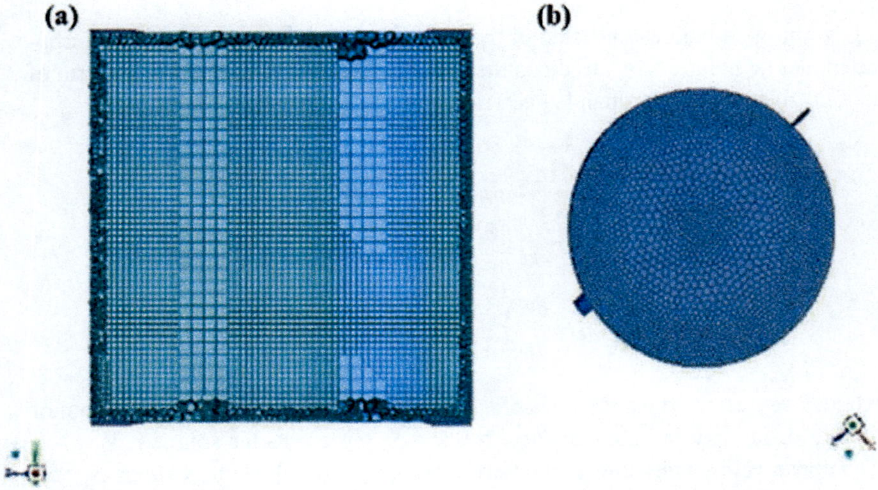

Fig. 3 Jet mixing tank grid generation **a** inside view and **b** top view

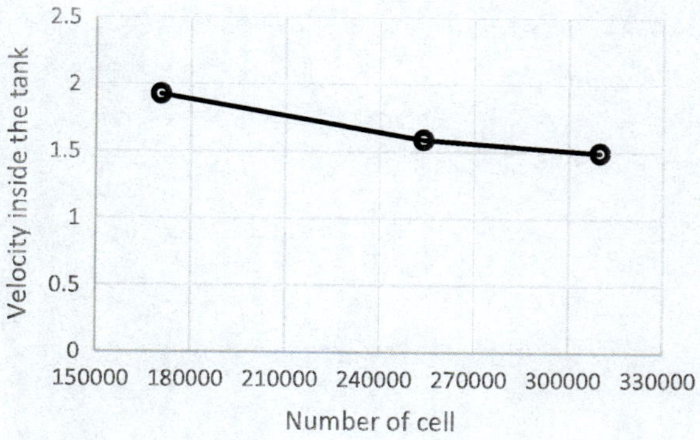

Fig. 4 Grid independence results concerning the velocity change inside the tank

4 Grid Independence Study

Three grid sizes with 1,70,000 cells, 2,53,000 cells, and 3,10,000 cells were created to investigate the effect of grid size. In Fig. 4, it can be seen that the velocity variation inside the tank is more from 1,70,000 cells to 2,53,000 cells but there is a very slight variation in the velocity inside the tank from 2,53,000 cells to 3,10,000 cells. So, for further study and validation, purposes grid 2 were used.

5 Governing Equations

The governing equations consisted of the Reynolds transport equation and a turbulence model (standard 'k-ε') to close the set of equations. The generalized form of equations given by Patwardhan [9] is written as,

$$\begin{aligned} &\frac{1}{r}\frac{\partial}{\partial r}(ru_r\emptyset) + \frac{1}{r}\frac{\partial}{\partial \theta}(u_\theta\emptyset) + \frac{\partial}{\partial z}(u_z\emptyset) \\ &= \frac{1}{r}\frac{\partial}{\partial r}\left(r\Gamma_{eff}\frac{\partial\emptyset}{\partial r}\right) + \frac{1}{r}\frac{\partial}{\partial \theta}\left(\frac{\Gamma_{eff}}{r}\frac{\partial\emptyset}{\partial \theta}\right) \\ &+ \frac{\partial}{\partial \theta}\left(\Gamma_{eff}\frac{\partial\emptyset}{\partial z}\right) + S_\emptyset \end{aligned} \tag{1}$$

where Ø represents transport variable, Γ_{eff} stands for effective diffusivity (sum of molecular diffusivity and eddy diffusivity), and $S_\emptyset$ represents the source [11].

The conservation equation for an inert tracer given by [9] can be written as,

$$\frac{\partial C}{\partial t} + \frac{1}{r}\frac{\partial}{\partial r}(u_r r C) + \frac{1}{r}\frac{\partial}{\partial \theta}(u_\theta C) + \frac{\partial}{\partial z}(u_z C) = \frac{1}{r}$$
$$\frac{\partial}{\partial r}\left(r\Gamma_{\text{eff}}\frac{\partial C}{\partial r}\right) + \frac{1}{r}\frac{\partial}{\partial \theta}\left(\Gamma_{\text{eff}}\frac{\partial C}{r\partial \theta}\right) + \frac{\partial}{\partial z} \tag{2}$$
$$\left(\Gamma_{\text{eff}}\frac{\partial C}{\partial z}\right) + S_c$$

In the above equation of inert tracer, the eddy diffusivity determines the dispersive transport of mass, and predicted mean velocity fields determine the convective transport of mass. For the mixing time estimation, the concentration profiles can be used.

6 Boundary Conditions

For steady-state and transient calculation, velocity inlet and pressure outlet boundary conditions were imposed at inlet and outlet, respectively. Standard wall function and no-slip condition were adopted at tank base and boundary. The top surface of the flow domain is simulated by assuming a flat surface for case T1 (Table 2). This implies normal gradients and normal velocity of all variables are zero.

This is assumed for computational convenience and to reduce the computational cost.

Turbulence is very important parameter, and hence, the simulation was initialized by imposing k and ε values at the jet inlet [1].

Table 2 gives the inlet boundary conditions for different cases which were adopted for the parametric investigation.

Table 2 Inlet boundary conditions for parametric investigation

Case	Nozzle angle	Velocity (m/s)	Turbulent intensity	TKE $k = \frac{3}{2}(u_j I)^2$ (m^2s^{-2})	TDR $\epsilon = C_D \frac{k^{\frac{3}{2}}}{l}$ (m^2s^{-3})
T1	45°	4.4	10%	0.2904	22.35614
T2		4.8		0.3456	29.02436
T3		5.4		0.4374	41.32570
T4	30°	4.4		0.2904	22.35614
T5	60°	4.4		0.2904	22.35614

7 Numerical Methods

In the present work, double precision pressure-based solver was used to omit the round-off error. SIMPLE algorithm was used during the simulation as a pressure–velocity coupling scheme. Second-order upwind spatial discretization scheme with first-order implicit temporal discretization scheme is used for the prediction of tracer mixing time.

8 Results and Discussion

The initial investigation was done by setting the concentration zero throughout the flow domain. NaCl solution is used as a tracer. For investigating, the mixing time water was used as a primary fluid and the properties of tracer and primary fluids were assumed identical because tracer NaCl solution was completely dilute. The concentration of tracer was monitored and recorded at four locations during this calculation. Overall mixing time was determined by the time required for the conductivity to reach 95% of the fully mixed concentration value. Convergence criteria 0.005 with the time scale of 0.0025 s was chosen for this simulation. The simulation was performed for case T1(Table 2) in which a 45-degree nozzle angle with 4.4 m/s velocity was used for 8 mm jet diameter. The results of which were compared with the experimental data from literature [9] in Fig. 5. It can be observed that the start of the concentration profile is earlier than the concentration profile observed by the experiment. Moreover, it was observed that the peak value is also well predicted in the concentration profile. It was also observed that the predicted concentration profile exactly follows the experimental profile after 15 s. In the predicted concentration profile, the peak is higher than that observed experimentally. This peak might be formed because the initial turbulent conditions and inaccurate measurement locations which were not mentioned in the literature. Because of which the values of turbulent kinetic energy 'k' and turbulent dissipation rate 'ε' were calculated and directly imposed at the jet inlet.

Fig. 5 Validation of CFD results with experimental results [9]

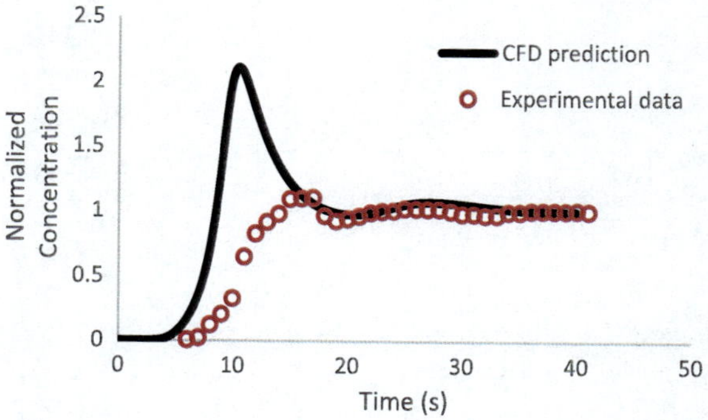

After the validation of the CFD code, the parametric investigation was carried out as mentioned in Table 2 to improve the performance of jet mixing tanks. This parametric investigation was based on the change of various aspects like velocity, inclination angle, and their effects on mixing time. The graphs are also plotted and explained in a detailed manner in the following paragraphs.

We considered three velocities as 4.4 m/s, 4.8 m/s, and 5.4 m/s for 45° of nozzle inclination as mentioned in Table 2. It can be observed from Fig. 6 that for the case T1 (velocity 4.4 m/s) predicted concentration peak is higher compared to all other cases. The concentration profile follows the exact trend for all cases after 15 s.

To understand the variation of mixing time according to the change in the velocity, the graph is shown in Fig. 7. From which, we can easily predict that as we are increasing the velocity from 4.4 m/s to 4.8 m/s, and 5.4 m/s, the mixing time decreases.

So, from here, we can conclude that there is an effect of change in velocity on mixing time of jet mixing tank, and the increase in velocity reduces the mixing time.

The other parameter that we have investigated was the jet angle. The result of this parametric investigation was plotted in Fig. 8 to understand its effect on concentration profile. The highest peak was observed for case T4 (nozzle angle 30°). The

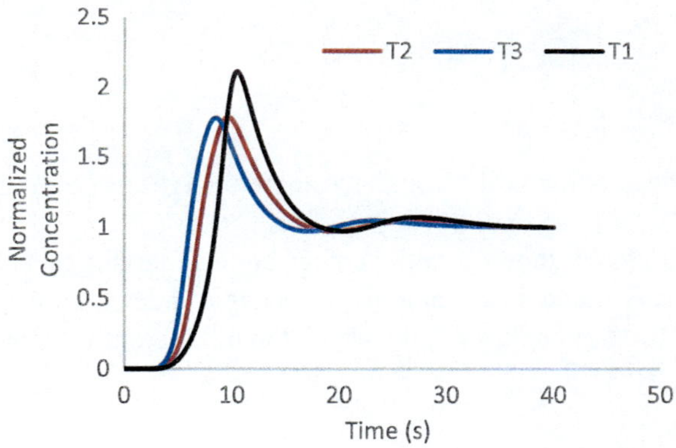

Fig. 6 Concentration profile for case T1, T2, and T3

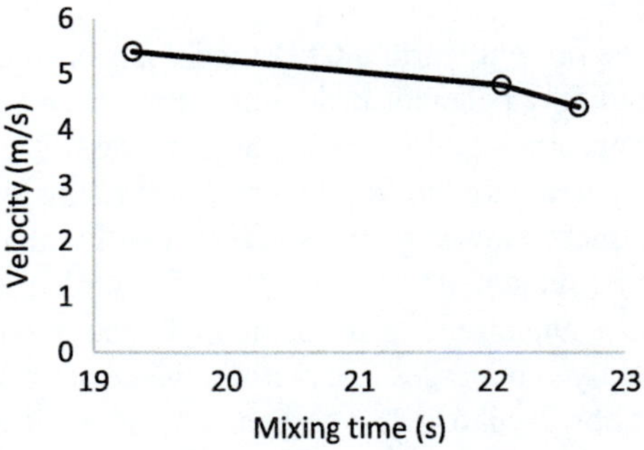

Fig. 7 Effect of increase in velocity at mixing time

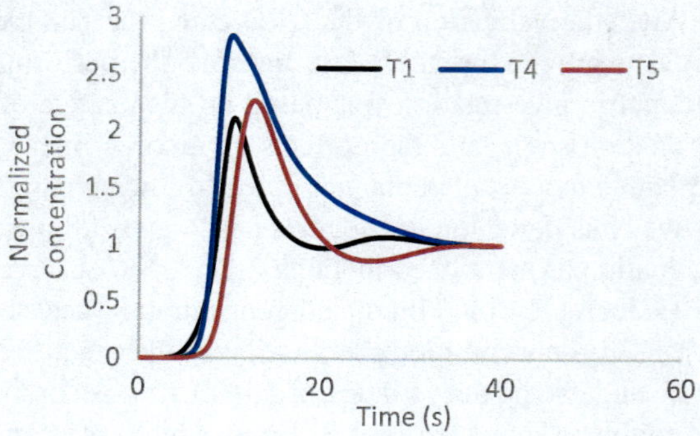

Fig. 8 Comparison of CFD concentration profiles for different jet angles

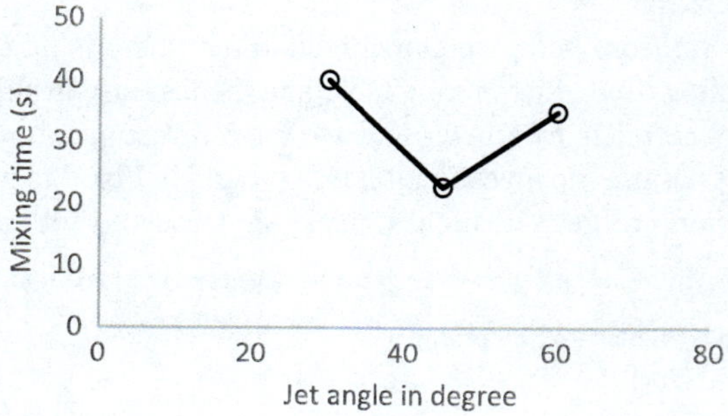

Fig. 9 Effect of change in jet angle at mixing time

concentration profile for all the angles merges to a common concentration value after 30 s.

Figure 9 shows the effect of jet angle on mixing time. It was observed that mixing time is affected by changing the jet angle and optimum angle observed to be the case T2 (nozzle angle 45°); for which the mixing time obtained was 22.

9 Conclusions

In the present work, a CFD model has been developed for the jet mixers to predict the mixing behavior inside the jet mixer. A parametric investigation has been done to improve the performance of jet mixers. The predicted concentration profiles were compared with the experimental data of the literature which shows good agreement. Parametric investigation suggests that by increasing the velocity, the concentration profile remained the same after 15 s while the mixing time reduces. The change in nozzle angle reflects the same fact about the concentration profile which fluctuates initially and merges to a constant value after 30 s. The mixing time for 45° jet angle was observed to be less compared to other jet angles. Further work would be directed toward predicting the concentration profile peak value with appropriate turbulence conditions.

References

1. Bumrungthaichaichan E, Namkanisorn A, Wattananusorn S (2018) CFD modelling of pump-around jet mixing tanks: a discrepancy in concentration profiles. J Chinese Institute of Engineers, Trans Chinese Institute of Engineers, Series A 41(7):612–621. https://doi.org/10.1080/02533839.2018.1530956
2. Fossett H, Prosser LE (1949) The application of free jets to the mixing of fluids in bulk. Proceedings of the Institution of Mechanical Engineers 160(1):224–232. https://doi.org/10.1243/PIME_PROC_1949_160_024_02
3. Fox EA, Gex VE (1956) Single-phase blending of liquids. AIChE J 2(4):539–544. https://doi.org/10.1002/aic.690020422
4. Grenville RK, Tilton JN (2011) Jet mixing in tall tanks: Comparison of methods for predicting blend times. Chem Eng Res Des 89(12):2501–2506. https://doi.org/10.1016/j.cherd.2011.05.014
5. Grenville R, Tilton JN (1997) Turbulence for flow as a predictor of blend time in turbulent jet mixed vessels, Recent Progress en Genie des Procedes. Proceedings of the Ninth European Conference On Mixing, France 11(51):67–74
6. Lehrer IH (1981) A new model for free turbulent jets of miscible fluids of different density and a jet mixing time criterion. Transactions of the Institution of Chemical Engineers 59:247–252
7. Okita N, Oyama Y (1963) Mixing characteristics in jet mixing. Chem Eng 27(4):252–260. https://doi.org/10.1252/kakoronbunshu1953.27.252
8. Orfaniotis A et al (1996) Experimental study of the fluidic mixing in a cylindrical reactor. The Canadian J Chemical Eng 74:203–212. https://doi.org/10.1002/cjce.5450740205
9. Patwardhan AW (2002) CFD modeling of jet mixed tanks. Chem Eng Sci 57(8):1307–1318. https://doi.org/10.1016/S0009-2509(02)00049-0
10. Patwardhan AW, Gaikwad SG (2003) Mixing in tanks agitated by jets. Chem Eng Res Des 81(2):211–220. https://doi.org/10.1205/026387603762878674
11. Sahu AK, Joshi JB (1995) Simulation of flow in stirred vessels with axial flow impellers: effects of various numerical schemes and turbulenee model parameters. Ind Eng Chem Res 34(2):626–639. https://doi.org/10.1021/ie00041a025
12. Vergleichende Ruhrversuche zum Mischen loslicher Flussigkeiten in einem 12 000-m3-Behalter* (1959),pp 583–587
13. Wasewar KL (2006) A design of jet mixed tank. Chem Biochem Eng Q 20(1):31–46
14. Wasewar KL, Sarathi JV (2008) CFD modelling and simulation of jet mixed tanks. Engineering Appl Comp Fluid Mechanics 2(2):155–171. https://doi.org/10.1080/19942060.2008.11015218
15. Yianneskis M (1991) The effect of flow rates and tracer insertion time on mixing times in jet-agitated vessels. In: 7th European Conference on Mixing Brugges Belgium, 1, pp 121–128

Review on Micro Hydro Power Plant

Anupkumar Chaudhari and Gaurang C. Chaudhari

1 Introduction

Renewable energy resources can be the alternative source to decrease the dependency on fossil fuel. Among all renewable energies, hydroelectricity is pollution free and economical. Rise of demand for energy and diminishing sources of fossil fuel, energy from water in mini/micro hydro power plant are considered an attractive cause of renewable energy. Micro hydro plant (10 kW and 200 kW) can best way to extracting energy from streams and small rivers to electrify rural region in developing countries [1]. It refers to the conversion of energy from flowing water into electrical energy. Hydro energy is a cheap and environmental friendly energy source has a great sustainable future. India is endowed with economically exploitable and viable hydro power potential, which can be utilized to fulfill energy requirement to remote areas away from the grid [2]. According to the report by Ministry of New and Renewable Energy, small hydro power has contribution of 13% from the total amount of grid-connected power generation [3].

2 Energy Generation in India

India is still facing problem to meet its energy demands even if it has an immense source of renewable energy. As of 2020, India's electricity generation is about 76% (55% coal, 14% fuel, and 7%gas) from fossil fuels and still renewable energy sources

A. Chaudhari (✉)
Gujarat Technological University, Ahmedabad, India
e-mail: anup6607@gmail.com

G. C. Chaudhari
C.K. Pithawalla College of Engineering and Technology, Surat, India
e-mail: gaurang.chaudhari@ckpcet.ac.in

J. Banerjee et al. (eds.), *Recent Advances in Fluid Dynamics*, Lecture Notes in Mechanical Engineering, https://doi.org/10.1007/978-981-19-3379-0_34

are still less utilized. It is expected that power generation from renewable energy resources contribute around 44.6% of the total electricity generation by the year 2029–2030.Rise of electric power plant from 2010 to 2020 succeeding 12.2% per year in average [4]. The energy consumption in India grew by 2.4% in 2019. The total primary energy consumption from coal is 55.88%, from fuel is 30.64%, from gas is 6.17%, and from renewable power 7.31% in the 2018.India's energy requirements are undergoing major alterations that are forming challenges but also some opportunities due to rise in population and requirements from numerous energy-intensive equipment. The difficult task is to meet growing demand and require the use of renewable resources to generate electricity and increased energy efficiency [5].

India is located in the middle of latitudes 8°4'N and 37°6'N and longitudes 68°7'E and 97°25'E, lying entirely in northern side of hemisphere. In term of geographic position, it is located in Southeast Asia. India shares its territorial boundaries with several countries; it shares land borders with China, Afghanistan, Bhutan, Pakistan, Nepal, Myanmar and Bangladesh. It is bounded by the Bay of Bengal, the Arabian Sea, and the Himalayan mountain ranges. India's total area is 3.287 million km^2 and consists of 572 islands [6].

This tropical country hosts two climatic types: tropical monsoon climate, tropical wet, and dry climate. India has winter season, summer season, monsoon season, and a post-monsoon period. India has advantages geographical positions, which motivate to develop renewable energy source such as MHP. Below table represents hydro power plant installed capacity, potential, and under implementation in India as per Annual Report 2019–20, Ministry of New and Renewable Energy, Govt. of India [6]. The Ministry has to still take series of steps to promote growth of MHP in a well-planned way and improve quality and reliability of the projects. By providing various types incentives and benefits have been attracted in commercial MHP projects apart from financially assisting State Governments to set up micro hydro projects. These projects may help to taken up with the involvement of local organizations such as the village energy cooperatives, and State Nodal Agencies, Water Mills Associations, cooperative societies, and registered NGOs.

No of states	Installed capacity		Potential capacity		Under capacity implementation	
	Number	(MW)	Number	(MW)	Number	(MW)
30	1127	4671.557	7133	21,133.62	109	529.240

3 Hydro Power Plant

Hydropower plays a major role in electricity generation, it generate up to 1000 s of MW. To meet energy demand, mega projects of hydropower plants need to be developed which flooded big area of land, due to which, it becomes difficult to build

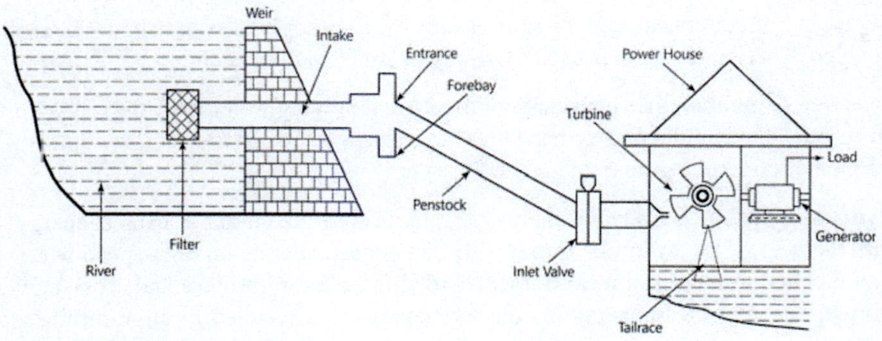

Fig. 1 Schematic diagram of MHP

new dams for power generation [7]. Therefore, smaller scale hydroelectric power plants need to be developed. Hydroelectric power plants categorized as small and large hydro projects based on power generation. In India, plants of equal to or less than 25 MW are small hydro, which can be more categorized into 00 kW or below as micro, 101 kW–2 MW as mini and 2–25 MW as small hydropower plants [8]. Typical MHP is shown in Fig. 1.

4 Components of MHP

The micro hydro power plant comprises weir, intake, penstock, forebay, turbine, and generator. The river water distracted at the weir through an intake; filter is attached at entry level to remove any impurities present in the water. Then the water is stored in forebay tank, which is located in the middle of penstock and the intake. Water is supplied through penstock to the turbine. The turbine which is installed in the power house generates mechanical energy from the potential energy of the water. Turbine will help to generate mechanical energy which is further converted to electric energy through generator.

Water Intake: In a hydropower system, water intake placed at the highest point from which water feeds to the turbine. Filter is provided at the intake pipeline to remove dirt and debris. Water is diverted from the source and transport to the turbine. Diversion systems can be grouped into open systems and closed system. In an open diversion system, large volume of water is directly imparted above the turbine inlet. In a closed diversion system, water is transported using sealed pipes helps to develop high-pressure head.

Forebay: For continues generation of electricity, the supply of water should be not troubled. The forebay tank is used to supply of water to the system.

Penstock (Pipeline): The penstock not used for supplying water from the intake to the turbine, but parallel creates head pressure. They can be installed at low points in the pipe or under the ground to get the flow going.

Hydraulic Turbines: Potential energy of water is converted into mechanical energy with the help of the hydraulic turbine. Turbines those operate on high-speed water jet are called impulse turbines, further divided into Pelton turbines and cross-flow turbine. Turbines those operate on the water pressure are called reaction turbines, further divided into Francis turbines (radial flow) and Kaplan turbine (axial-flow). The power generated from stream flow in a turbine is given as,

$$P = \rho \eta_t g Q_n$$

where,

P = power generated (Watt)
ρ = density of Water (Kg/m^3)
H_n = net head (m)
Q = flow rate of water (m^3/s)
g = gravity (m/s^2)
η_t = turbine efficiency (%)

The turbine efficiency (η_t) is the ratio of mechanical power generated in the turbine to the hydraulic power.

Drive System: One end of the drive system is the turbine and the other end is the generator, which produces voltage and frequency at the velocity that delivers the best efficiency. Direct drive system (one to one coupling between the turbine and generator) is most favorable in micro hydro power plant to avoid losses caused by gears, chains, or belt.

Generator: The generator converts velocity of turbine into constant or variable frequency. The outcome of generator depends on operations mode of turbine and the specific speed. The energy generates from the generator can be store in isolated or battery systems directly transmitting various types of loads. Generators used in power station are working at low speed and of salient-pole type, having a numerous of poles, a large diametric short rotor.

5 Review on Small Hydropower Plant

Sr No	Title (Year)	Author	Summary
1	Study of silt erosion mechanism in Pelton turbine buckets (2012) [9]	M.K. Padhy	– Partials which traveling at higher velocity created pits on the bucket at larger impact angle – Abrasive type of erosion caused by smaller particles at the outlet of the bucket [9]
2	Analysis of stress on Pelton turbine blade due to jet impingement (2014) [10]	Varun Sharma	– To reduce the stress in turbine blade water should move out in the flow direction – When water jet strikes the turbine blade at 0°, ma maximum stress will be 113 MPa and the minimum value of stress will be 0.027 MPa [10]
3	Energy generation from gray water in high-raised buildings: The case of India (2014) [11]	Prabir Sarkar	– New possibility of energy harnessing drifts down over high-raised buildings was discussed in this paper from water. To fit the necessary requirements usage of micro hydro turbine was proposed to utilize energy of gray water tumbling from floors above – The proposed design shows commercial viability for major cities of India [11]
4	Effects of impeller diameter and rotational speed on performance of pump running in turbine mode (2014) [12]	Sanjay V. Jain	– The experiments were performed on impellers after blade rounding with speed range of 900 to 1500 rpm – The maximum working efficiency of 76.93% with 10% of trimmed impeller at speed of 1100 rpm [12]

(continued)

(continued)

Sr No	Title (Year)	Author	Summary
5	Developing a test rig to measure hydro abrasive erosion in Pelton turbine (2015) [13]	Anant Kr. Rai	– To predict erosion occurs in the prototype plants was addressed and proposed the erosion relations for 3D digitization of Pelton buckets [13]
6	Effect of temperature, suction head, and flow velocity on cavitation in a Francis turbine of small hydro power plant (2015) [14]	Pankaj P Gohil	– It was found that cavitation rate and efficiency loss increase with temperature and suction head. With flow rate, cavitation rate increases – Initially, efficiency loss decreases then reaches minima ends as increasing with flow velocity [14]
7	Design and implementation of micro hydro turbine for power generation and its application (2016) [15]	N. J. Kumbhar	– The mathematical design calculation for hydro-electric power was presented in this paper [15]
8	Design of a Pelton wheel turbine for a micro hydro power plant (2016) [16]	N. Manjunatha	– The Pelton wheeled turbine is essential component for micro hydro type of power plants which can run on the maximum efficiency of 96% for different of head [16]
9	Small hydro development in the Indian Himalaya: implications for environmental assessment Reform (2016) [17]	Alan Paul Diduck	– For successful implementation of small hydro power plant locals should have to participate for improving the project-level assessment [17]
10	A review on turbines for micro hydro power plant (2017) [18]	C.P. Jawahar	– Turbine is very essential component of this system, in this paper, author had tried to cover all the aspect of selection of turbine with different specification [18]

(continued)

(continued)

Sr No	Title (Year)	Author	Summary
11	Design of 15 kW micro hydro power plant for rural electrification at Valara (2017) [19]	Prawin Angle Michael	– The installation and designing of micro hydro power plant is proposed to meets the energy requirement of 120 tribal families of Valara [19]
12	Energy situation, current status, and resource potential of run of the river (RoR) large hydro power projects in Jammu and Kashmir: India (2017) [20]	Ameesh Kumar Sharma	– Private investors should invite for maximum harnessing of hydro power in specified time frame [20]
13	Selection of best location for small hydro power project using AHP, WPM, and TOPSIS methods (2018) [21]	Rana, Shilpesh	– For appropriate site selection of micro hydroelectric power plant, TOPSIS method and weighted product method should be used [21]
14	Renewable energy potential in India and future agenda of research (2019) [22]	Santosh Singh Raghuwanshi	– Private company shows the interest in the development of micro hydro-electric power plant projects [22]
15	Risk management in small hydro power projects of Uttarakhand: An innovative approach (2019) [23]	Neha Chhabra Roy Sr	– Before and after installation of power plant always the risk management policy should be effectively

6 Potential Research Area in Small Hydropower Plant

Turbine Design: Selection of turbine depends on water flow rate and available pressure head. Turbines of regulated type can make movement of their inlet guide vanes to vary the quantity of flow they will draw [25].

Turbine Materials: Altering materials to form lightweight of turbines could help us to increase the efficiency of production. Aluminum can be more efficient because of its being lightweight. This helps to conclude that material can affect efficiency of turbine [26].

Variable parameters: Parameters likes shape of spear, mass flow rates, nozzle opening, no. of bucket, bucket radial position, angular position, size of bucket, and angle attack affect the performance of turbine [27].

Bucket design: The density of bucket materials also an important parameter because that effect on mass moment of inertia of system so material for bucket should be selected carefully. Number of buckets is one of the important factor while enhancing a Pelton turbine runner [28].

7 Conclusion

Renewable energy is the only option to avoid the catastrophic situation of supply and demand of electricity. One of the renewable energy sources is hydro emery which gives promising results in different aspect. The hydro power plant should be used to generate the electricity from the potential energy of water. Site selection and huge investment is major drawbacks for huge hydro power plant. Instead of creating centralize source of energy, we can discretize the source by using small or micro hydropower plant on the local river. India has rich source of river water and location for installation of small hydro power plant. We can conclude that for the design and components of small hydro power plant, there is a wide range of possibility to improve the performance efficiency of small hydro power plant.

References

1. Erinofiardi et al (2016) A review on micro hydropower in Indonesia. Energy Procedia
2. Nasir, BA (2014) Suitable selection of components for the micro-hydro-electric power plant. Adv Energy Power
3. Barnett SK, Andrew (2000) Best Practices for sustainable development of micro hydro power in developing countries
4. Chan Z (2019) Design calculation of penstock and nozzle for 5kW Pelton turbine micro hydropower plant. Int J Trend in Scientific Res Develop 3
5. Elbatran AH, et al (2015) Operation, performance and economic analysis of low head micro-hydropower turbines for rural and remote areas: A review. Renewable Sust Energy Rev 43
6. Sarkar P, et al (2014) Energy generation from grey water in high raised buildings: The case of India. Renewable Energy 69
7. Ratnata IW, Saputra WS, Somantri M, Mulyana E, Ardhika A (2017) Preliminary study of micro-hydro power plant (MHPP) in the rural area. In: International Symposium on Materials and Electrical Engineering
8. Gao OD, Sarsing (2020) An overview of small hydro power development in India. AIMS Energy
9. Padhy MK, Saini RP (2012) Study of silt erosion mechanism in Pelton turbine buckets. Energy, 286–293
10. Dhama VS, Sanjeevkumar (2014) Analysis of stress on Pelton turbine blade due to jet impingement. Int J Current Engineering Tech 4
11. Sarkar P, Sharma B, Malik U (2014) Energy generation from grey water in high raised buildings: The case. Renewable Energy
12. Jain SV, et al (2015) Effects of impeller diameter and rotational speed on performance of pump running in turbine mode. Energy Conv Manag 89
13. Rai AK (2015) Developing a test rig to measure hydro-abrasive erosion in Pelton turbine. In: International Conference on Hydropower for Sustainable Development

14. Gohil PP (2015) Effect of temperature, suction head and flow velocity on cavitation in a Francis turbine of small hydro power plant. Energy
15. Kumbhar NJ, et al (2016) Design and implementation of micro hydro turbine for power generation and its application. Int Res J Engineering Technology 3
16. Manjunatha N (2016) Design of a Pelton wheel turbine for a micro hydro power plant. In: International Conference on Science, Technology and Management, New Delhi
17. Diduck AP, et al (2016) Small hydro development in the Indian Himalaya: implications for environmental assessment reform. J Environmental Assessment Policy Management 18
18. Jawahar CP, Micheal PA (2017). A review on turbines for micro hydro power plant. Renewable Sust Energy Rev 72
19. Michael PA, et al (2017) Design of 15 kW micro hydro power plant for rural electrification at Valara. Energy Procedia 117
20. Sharma AK, Thakur NS (2017) Energy situation, current status and resource potential of run of the river (RoR) large hydro power projects in Jammu and Kashmir: India. Renewable Sust Energy Rev 78
21. Patel SCR, Jayantilal N (2018) Selection of best location for small hydro power project using AHP, WPM and TOPSIS methods. ISH J Hydraulic Eng
22. Arya SSR, Rajesh (2019) Renewable energy potential in India and future agenda of research. Int J Sustain Eng
23. Roy NC et al (2019) Risk management in small hydro power projects of Uttarakhand: an innovative approach. IIMB Management Review
24. Nouni MR, Mullick SC, Kandpal TC (2005) Techno-economics of micro-hydro power plants for remote villages in Uttaranchal in India. Int J Global Energy 24
25. Nasir BA (2014) Design considerations of micro-hydro-electric power plants. Energy Procedia
26. Sritram P, Treedet W, Suntivarakorn R (2015) Effect of turbine materials on power generation efficiency from free water vortex hydro power plant. In: 4th Global Conference on Materials Science and Engineering
27. Gudukeya LK, Mbohwa C (2017) Improving the efficiencies of Pelton wheel in micro-hydro power plants. In: Colombia: Proceedings of the International Conference on Industrial Engineering and Operations Management
28. Audrius Židnois GAA (2016) Pelton turbine: Identifying the optimum number of buckets using CFD. J Journal Hydrodynamics 28

A CFD Analysis of Closed Loop Pulsating Heat Pipe Using Fourth-Generation Refrigerant

Sagar M. Asodiya, Kalpak R. Sagar, and Hemantkumar B. Mehta

Nomenclature

CLPHP	Closed-loop pulsating heat pipe
VOF	Volume of fluid
HFCs	Hydro fluorocarbons
HFOs	Hydro fluoro-olefins
GWP	Global warming potential
ODP	Ozone depletion potential
FRs	Filling ratios
T_C	Critical temperature
P_C	Critical pressure
M	Molecular mass
k	Thermal conductivity (W/m K)
C	Specific heat (J/kg K)
t	Time (sec)

Greek Symbols

ρ	Density (kg/m^3)
μ	Dynamic viscosity (Pas)
α	Volume fraction
σ	Surface tension coefficient (Nm)

S. M. Asodiya (✉) · K. R. Sagar · H. B. Mehta
Sardar Vallabhbhai National Institute of Technology, Surat, India
e-mail: sagarasodiya2210@gmail.com

H. B. Mehta
e-mail: hbm@med.svnit.ac.in

J. Banerjee et al. (eds.), *Recent Advances in Fluid Dynamics*, Lecture Notes in Mechanical Engineering, https://doi.org/10.1007/978-981-19-3379-0_35

1 Introduction

Pulsating heat pipe (PHP) was introduced by Akachi in 1990. PHP is a passive heat transfer device that transfers heat from the heating section to the cooling section by continuously changing the phase of working fluid from liquid to vapour and vice versa. PHP is used at a location where high heat flux is removed from smaller space, so it applies cooling microelectronic devices, Solar applications, Cryogenics, Space, and Fuel cell [1]. It comprises continuous long capillary tubes bent in the number of turns and divided into three parts viz. evaporator, adiabatic, and condenser section.

The greenhouse gases (HFC refrigerants) operating in current refrigeration and air conditioning systems are under the time-barred permission period under the Kyoto Protocol (1997). So, to diminish ozone depletion and reverse climate change effects, the phasing out of these harmful refrigerants is a must. Suitable refrigerants required to eliminate these harmful refrigerants, which can deliver similar performance with negligible environmental impact. The GWP value of HFO-1234yf is 4, and it is significantly less compared to the GWP value of HFC-134a, which is 1430 [2]. In the present numerical study, HFO-1234yf is considered to be a possible alternative to the refrigerant HFC-134a. A 2-D CLPHP model is developed and investigated with ANSYS Fluent.

2 Mathematical Modeling

Governing Equations In computational fluid dynamics (CFD), the VOF model used to model the interface between the multiple phases. In the VOF method, a single set of momentum equation shared by fluids/phases and volume fraction of each phase in computation cell tracked throughout the domain [3]. The volume fraction of liquid and vapour represented by α_l and α_v. Now, if the value of $\alpha_l = 1$ then it means that computational domain filled with liquid and value of $\alpha_l = 0$, then it means that the computational area filled with vapour. The volume fraction of the computational domain shows by Eq. (1).

$$\sum_{q=0}^{n} \alpha_q = 1 \text{ Where } \alpha_q \text{ is the volume fraction} \tag{1}$$

The Energy equation shared between the two phases is shown in Eq. (2).

$$\frac{\partial(\rho E)}{\partial t} + \nabla * \left(\vec{v}(\rho E + p)\right) = \nabla * (k_{\text{eff}} \nabla T) + S_h \tag{2}$$

The VOF model treats energy E and temperature T as mass-averaged variables and can calculate by Eq. (3):

$$E = \frac{\sum_{q=1}^{n} \alpha_q \rho_q E_q}{\sum_{q=1}^{n} \alpha_q \rho_q} \tag{3}$$

where E_q for each phase is based on the specific heat of that phase and the shared temperature. The properties ρ and k_{eff} (effective thermal conductivity) are shared by the phases. The source term S_h contains contributions from radiation, as well as any other volumetric heat sources.

Mass continuity equation satisfied by liquid and vapour phase shown in Eqs. (4) and (5)

$$\frac{\partial \alpha_l}{\partial t} + \nabla(\vec{v}\alpha_l) = \frac{S_{m,1}}{\rho_1} \tag{4}$$

$$\frac{\partial \alpha_v}{\partial t} + \nabla(\vec{v}\,\alpha_v) = \frac{S_{m,v}}{\rho_v} \tag{5}$$

where $\vec{v}$ represents the velocity (m/s), t represents time, $S_{m,1}$ is the source term of mass transfer between the liquid and vapour phase, which can find out by Eqs. (4) and (5), ρ_v and ρ_l represent the density (kg/m^3) of the vapour phase and liquid phase, respectively.

$S_{m,1}$ is the source term of mass transfer between liquid and vapour phase, which can be loaded into Boundary Condition by UDF (user-defined function), and it can be finding out by Eqs. (6) and (7) [4].

$$S_{m,1} = \begin{cases} -0.1\alpha_l \rho_l \left| \frac{T_{mix} - T_{sat}}{T_{sat}} \right| & \text{when } T \geq T_{sat} \\ 0.1\alpha_v \rho_v \left| \frac{T_{mix} - T_{sat}}{T_{sat}} \right| & \text{when } T \leq T_{sat} \end{cases} \tag{6}$$

$$S_{m,v} = \begin{cases} 0.1\alpha_l \rho_l \left| \frac{T_{mix} - T_{sat}}{T_{sat}} \right| & \text{when } T \geq T_{sat} \\ -0.1\alpha_v \rho_v \left| \frac{T_{mix} - T_{sat}}{T_{sat}} \right| & \text{when } T \leq T_{sat} \end{cases} \tag{7}$$

where T_{mix} represent the temperature of the mixture of two phases,T_{sat} represent the saturation temperature.

The Density, Dynamic viscosity and Thermal conductivity of a mixture of two phases can be calculated by the following Eqs. (8), (9) and (10)

$$\rho = \alpha_l \rho_l + \alpha_v \rho_v \tag{8}$$

$$\mu = \alpha_l \mu_l + \alpha_v \mu_v \tag{9}$$

$$k = \alpha_l k_l + \alpha_v k_v \tag{10}$$

The momentum equation can be defined as (11) [4, 5],

$$\frac{\partial(\rho\vec{v})}{\partial t} + \nabla(\rho\vec{v} * \vec{v}) = -\nabla p + \nabla[\mu(\nabla\vec{v} + \nabla\vec{v}^{T})]\rho\vec{g}\vec{F} \quad (11)$$

where ∇ is Laplace operator, p represents operating pressure, μ represents dynamic viscosity of mixed-phase, $\vec{F}$ represents a continuum surface force model. Surface tension force considered as a body force, which plays a significant role in two-phase flow. It calculated by Eq. (12)

$$\vec{F} = 2\sigma\frac{k_l\alpha_l \nabla\alpha_l\rho_l + k_v\alpha_v\nabla\alpha_v\rho_v}{(\rho_v + \rho_l)} \quad (12)$$

where σ represents the coefficient of surface tension and k_1, k_v represent the curvature of the liquid–vapor phase medium, respectively.

3 Model Geometry

A schematic configuration of 2-D model geometry is shown in Fig. 1. For a grid independence study shown in Fig. 2, the average temperature of the adiabatic wall is considered. In the numerical analysis, four different grid elements (3688, 14,280, 35,064, 71,560) considered and simulated carried out at 310 K T_{evap} with FR27. The average adiabatic temperature remains approximately constant after grid element 14,280. So, grid size having 14,280 cells taken for further analysis.

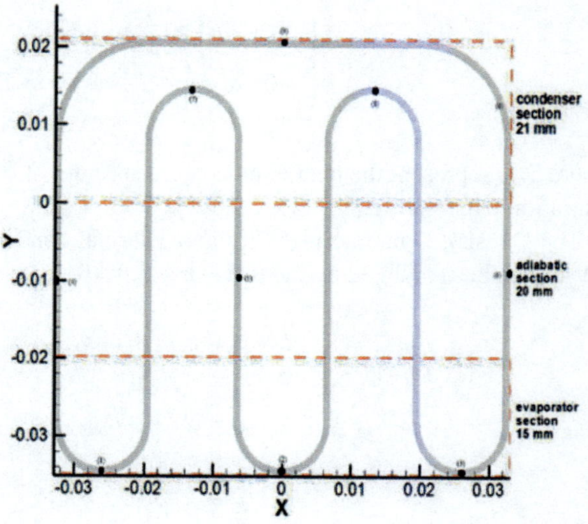

Fig. 1 Schematic configuration of 2-D Model geometry

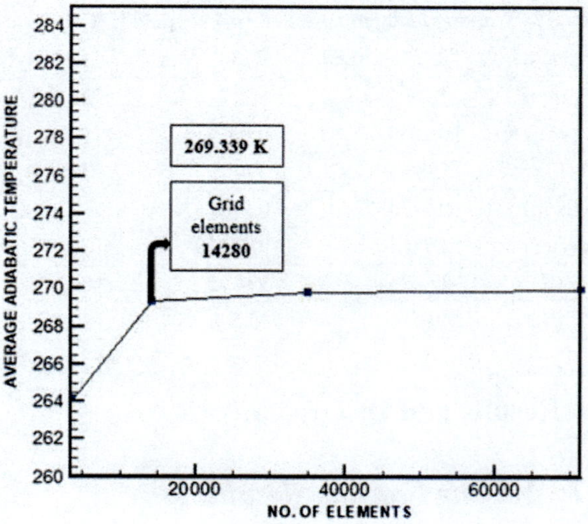

Fig. 2 Grid independence study

3.1 Numerical Simulation Parameters

In this present numerical investigation, constant temperature boundary conditions adopted for the evaporator and condenser section. Here, evaporator temperatures are selected to maintain an equal temperature difference between the evaporator and condenser section for both working fluids. The condenser section temperature for R134a and R1234yf is 246.9 K and 243.70 K, respectively. The validation of our current model is carried out with water as a working fluid and presented in our earlier article [5]. Numerical simulation parameters adopted for present simulation shown in Table 1.

Properties of R134a and R1234yf are at atmospheric pressure and temperature 298.15 K, obtained from REFPRO 8.0 and shown in Table 2.

Table 1 Numerical Simulation Parameters for present Numerical Investigation

Solution method	Selection
Simulation type	2D, Unsteady
Solver	Pressure Based Non-Iterative Time Advancement Scheme (NITA)
Multiphase model	Volume of Fluid (VOF)
Viscous model	Laminar
Pressure–velocity coupling	Pressure-Implicit with Splitting of Operators (PISO)
Momentum and Energy	Second-order upwind
Time step (sec)	10^{-4}

Table 2 Working fluid properties of working fluid HFC-134a and HFO-1234yf

Properties	R134a (liquid)	R134a (vapor)	R1234yf (liquid)	R1234yf (vapor)
ρ (kg/m^3)	1376.7	5.2581	1263.1	5.9812
k (W/m K)	0.10391	0.009313	0.08437	0.009263
C (J/kg K)	1280.5	794.19	1189.9	812.85
σ (N/m)	0.015437	0.015437	0.013808	0.013808
μ (Pa s)	$3.78*10^{-4}$	$9.77*10^{-6}$	$3.06*10^{-4}$	$9.0*10^{-6}$
T_C (K)	374.25	374.25	367.85	367.85
P_C (Bar)	40.5	40.5	33.8	33.8

4 Results and Discussion

4.1 Volume Fraction Results

Volume fraction contours of the CLPHP with HFC-134a and HFO-1234yf obtained with the time are shown in Figs. 3 and 4. The blue colour represents the liquid phase in the volume fraction contours, while the red colour represents the vapour phase. The development and growth of the process of the flow inside the CLPHP exhibited through Figs. 3 and 4a–c. Figures 3 and 4d displays the volume fraction contours after the beginning of working fluid circulation inside the CLPHP. The flow inside CLPHP initiates in any one direction (either clockwise or anti-clockwise) after a few initial oscillation motions of slug-plug.

The thermophysical and heat transport properties of both the working fluids are well suited for CLPHP operation. Still, the factors which affect the start-up mechanism inside the CLPHP are a combination of latent heat of vaporization, surface tension, and density of the working fluid. HFC-134a has all the value superior as compared to HFO-1234yf. So, start-up time is less in the case of the HFC-134a. Adding to this, HFC-134a have slightly higher specific heat, so more heat transfer occurs from the evaporator to the condenser section. HFO-1234yf has a lower dynamic viscosity value than HFC-134a, which offers less shear stress while moving inside CLPHP. Here, the effect of alteration in FRs examines qualitatively. The pressure inside the CLPHP increases as the FR increases. When the pressure increases, the corresponding saturation temperature of the working fluid increases, resulting in more time required for initiation of the boiling and consequently more time required for bubble generation. The liquid slug length increases as the FR increases when comparing the volume fraction results for individual working fluid (either HFC-134a or HFO-1234yf), but liquid slug length increases when we compare for both fluids is more for working fluid HFO-1234yf.

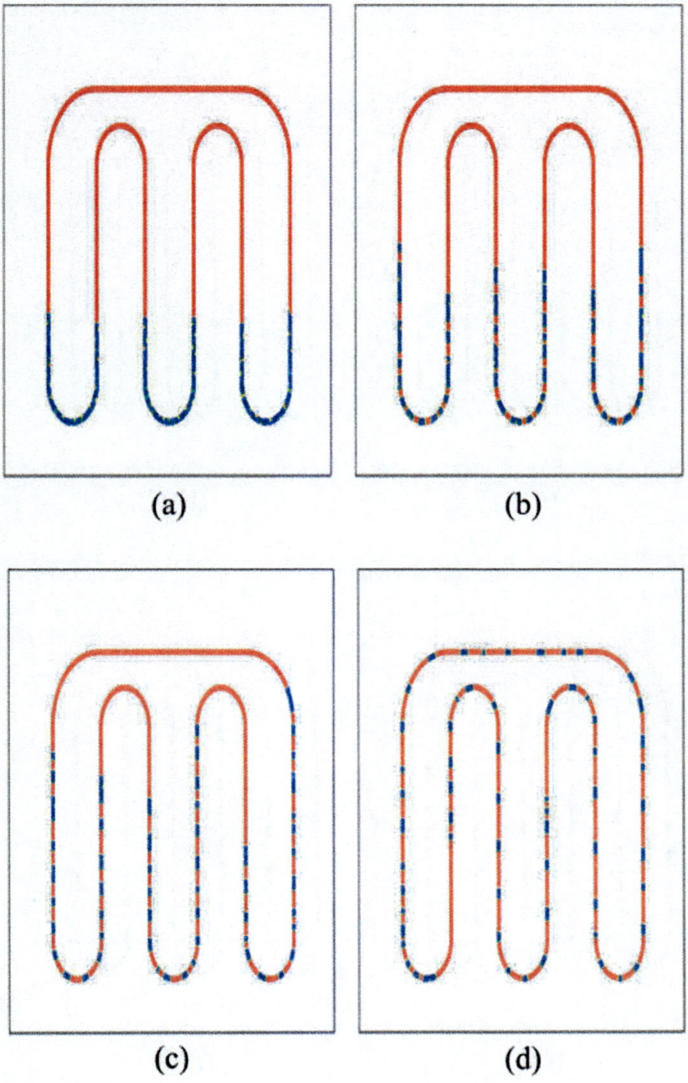

Fig. 3 Volume fraction results at evaporator temperature 285 K with FR 27 **a** $t = 0.10$ s **b** $t = 0.50$ s **c** $t = 1.06$ s **d** $t = 3.50$ s (HFC-134a)

4.2 Time Series Analysis

The time series analysis of CLPHP is a significant and noticeable indication to recognize the flow behaviour of fluid in the numerical investigation. The temperature of the adiabatic section is one of the other criteria that have studied to recognize the chaos inside CLPHP. Here, time series of temperature for different operating conditions tested.

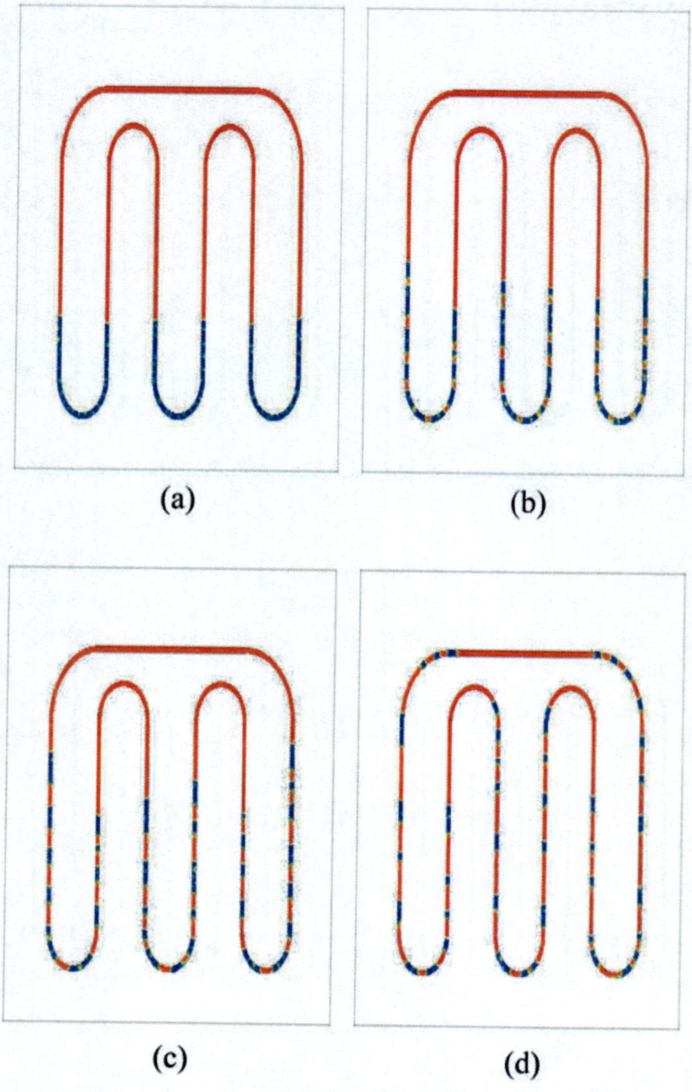

Fig. 4 Volume fraction results at evaporator temperature 281.8 K with FR 27 **a** $t = 0.10$ s **b** $t = 0.50$ s **c** $t = 1.06$ s **d** $t = 3.50$ s (HF0-1234yf)

The different points are located at the CLPHP wall to collect the temperature data concerning time. It shows in geometry Fig. 1. Graphs of the average adiabatic section temperature vs time plotted for different FRs with both working fluids are shown by Figs. 5 and 6. It shows that the fluctuation of the temperature of the adiabatic section decreases as the FRs increases. The reason behind this reduction is the increase in the volume of working fluid and resulting in more time to form a slug-plug flow.

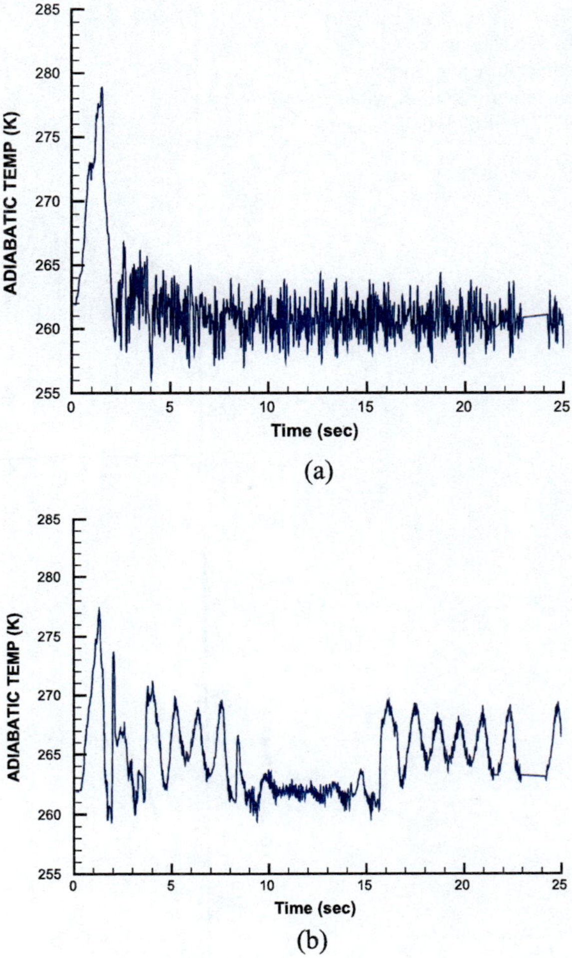

Fig. 5 Time series analysis for adiabatic wall temperature at evaporator temperature 285 K with **a** FR27 **b** FR62 (HFC-134a)

It also observed that the fluctuation inside the adiabatic section is less at the initial stage when the evaporator temperature is close to the condenser temperature.

As the temperature of the evaporator increases, the fluctuations of the adiabatic section increases rapidly, even at lower FR. Furthermore, as evaporator temperature increases, the start-up time for the beginning of fluctuations decreases, and it increases as an increase in FR from lower to higher It displayed in Figs. 5 and 6a, b.

When HFO-1234yf used as a working fluid, the fluctuation inside the adiabatic section is less for all FR at a given evaporator temperature. This fluctuation is directly proportional to slug plug formation in the adiabatic section, so less heat transfer from the evaporator to the condenser section compared to HFC-134a. It observed that the average adiabatic temperature is higher and close to the evaporator in HFC-134a, which shows that the higher heat transfer rate compared to HFO-1234yf.

Fig. 6 Time series analysis for adiabatic wall temperature at evaporator temperature 281.8 K with **a** FR27 **b** FR62 (HFO-1234yf)

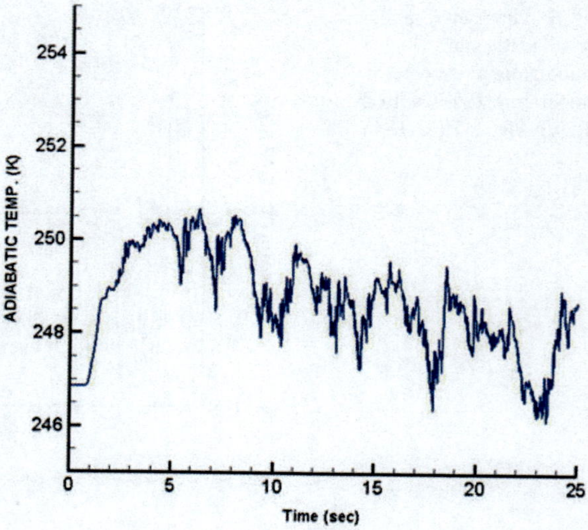

(a)

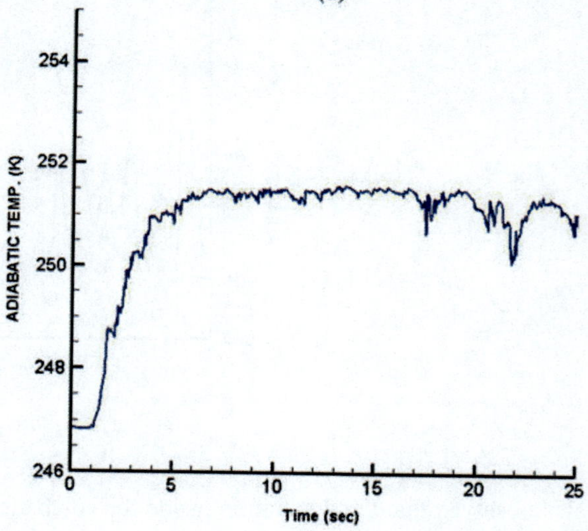

(b)

An average temperature difference of the working fluid into evaporator and condenser section versus evaporator temperature plotted for different FRs is showed in Fig. 7. It is observed that at lower evaporator temperature, the difference is negligible for all FRs. Now, for the lower FR, i.e., 27, if the temperature of the evaporator increases, the amount of heat transfer decreases. The large amount of vapour plug inside the CLPHP at lower FR and vapour plug have low heat carrying capacity so that it transmits less amount of heat from evaporator to condenser section and ultimately lower average temperature of the condenser section. A low average condenser temperature expresses less heat transfer from the evaporator to the condenser section.

Fig. 7 Temperature difference between evaporator and condenser at evaporator temperatures with different FRs **a** HFC-134a **b** HFC-1234yf

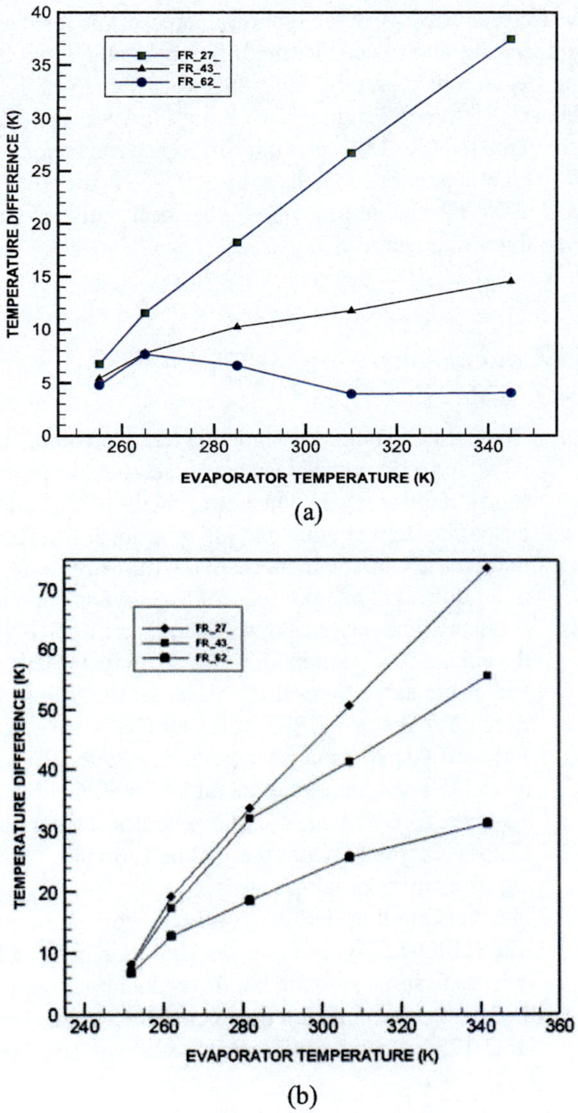

When the FRs increases, there is more amount of liquid inside CLPHP at a particular evaporator temperature. Liquid has a higher heat carrying capacity, so the average temperature of the condenser is higher, and the difference is lower, so a high amount of heat transfer from the evaporator to the condenser section.

When the numerical investigation performed with lower evaporator temperatures for FR 27, 43, and 62, the variation of the temperature difference between the condenser and evaporator section is quite similar, such as the heat transfer rate for HFC-134a and HFO-1234yf. Now, as the temperature of the evaporator increasing the difference between the sections increase, and it shows in fig. At higher evaporator temperature and lower FR, the difference is high, so heat transfer is inferior. Heat transfer rate increased by 29%-89% for HFC-134a. It is observed from the fig that the temperature difference of working fluid between the section is relatively higher compared to HFC-134a. and this difference can be reduced with a further increase in FR. Heat transfer rate increased by 9.5%-57% for HF0-1234yf, as the FRs increased from 27 to 62. Stable flow and higher heat transfer rate obtain with higher FR and evaporator temperature.

5 Conclusions

(1) The used CFD model exhibited effective volume fraction results of FRs (27%, 43%, and 62%) with constant wall temperature condition with both the working fluids. It observed that the performance of CLPHP altered with the variation of evaporator temperature and filling ratio, and the length of the liquid slug found to increase with the increase in the filling ratio. It also observed that the length of a liquid slug is more for HFO-1234yf as compared to HFC-134a.

(2) From the time series analysis of temperature, it observed that fluctuation inside the adiabatic section increases with an increase in the evaporator temperature, and at the same time, it decreases as the filling ratio increases. Fluctuations were more inside CLPHP when HFC-134a used as a working fluid.

(3) The thermal performance increased by 29%-89% as we increase FRs from 27 to 62 at 345 K evaporator temperature for HFC-134a. Similarly, it is increased by 9.5–57% for HFO-1234yf at evaporator temperature 341.8 K. Hence, and the CLPHP exhibited better thermal performance with a higher FR and relatively higher evaporator temperature.

(4) The working fluid HFO-1234yf can work satisfactorily with a CLPHP. Now, when HFO-1234yf use as a working fluid inside CLPHP, the heat transfer rate reduces to some amount, but this reduction is not too significant. When we see the harsh environmental impact of HFC-134a, HFO-1234yf is superior. Thus, HFO-1234yf can be utilized in heat transfer systems as an alternative to the HFC-134a.

References

1. Nazaria M, Ahmadib M, Ghasempoura R, Shafii M, Mahiand O, Kalogirouf S, Omchai S (2018) A review on pulsating heat pipes: From Solar to Cryogenic applications. Appl Energy 222:475–484
2. Bohringer C (2003) The Kyoto protocol: a review and perspectives. Oxf Rev Econ Policy 19:451–466

3. Jiansheng W, Ma H, Zhu Q, Dong Y, Yue K (2016) Numerical and experimental investigation of pulsating heat pipes with corrugated configuration. Appl Therm Eng 102:158–166
4. Sagar K, Naik H, Mehta H (2020) CFD Analysis of cryogenic pulsating heat pipe with near critical diameter under varying gravity conditions. Theor Found Chem Eng 54:64–74
5. Sagar K, Naik H, Mehta H (2021) Numerical study of liquid nitrogen-based pulsating heat pipe for cooling superconductors. Int J Refrigeration 122:33–46

CFD Simulation for Condensation of Humid Air Over Vertical Plate with Eulerian Wall Film Approach

Harshal Narkhede, P. R. Dhamangaonkar, and K. Parashar

Nomenclature

C_{phase}	Condensation constant
S_n	Mass source
S_h	Energy source
$\vec{F}$	External body force
h	Film height
$\vec{D}_T$	Differential advection of temperature
T_S	Film surface temperature
T_f	Average film temperature
T_w	Wall temperature
P_{gas}	Gas flow pressure
P_h	Gravity component normal to wall surface
P_σ	Component due to surface tension

1 Introduction

Condensation of water vapour in air over a surface is common phenomena encountered in various application such as HVAC, surface condensers and nuclear safety

H. Narkhede (✉)
Thermal Engineering, College of Engineering, Pune, India
e-mail: narkhedehs19.mech@coep.ac.in

P. R. Dhamangaonkar
Mechanical Engineering, College of Engineering, Pune, India

K. Parashar
Whirlpool of India Ltd, Pune, India

J. Banerjee et al. (eds.), *Recent Advances in Fluid Dynamics*, Lecture Notes in Mechanical Engineering, https://doi.org/10.1007/978-981-19-3379-0_36

devices. Presence of non-condensable gas results in low condensation rates than condensation of pure steam due to additional heat and mass transfer resistances. The non-condensable component gets piled up near liquid film from which a vapour has to diffuse in order to get condensed onto a film. This layer of non-condensable air reduces the heat and mass transfer in the condensation process. In such condensation, a primary governing factor for condensation rate is diffusion of water vapour through piled up non-condensable gas, i.e. air over liquid film. This diffusion controls the amount of mass transfer from vapour phase to liquid phase.

Earliest attempt to model condensation process is done by Nusselt [1] popularly known as Nusselt film condensation theory. It is simplified form of actual condensation phenomena to formulate mathematical equations using some assumptions. Nusselt proposed theory specifically for film condensation of stagnant pure vapour over vertical flat plate. Siow [2] studied the effects of the presence of a non-condensable gas on the process of internal-flow, laminar film condensation using the full form of the governing conservation equations. An important experimental study was done by Amborsini et al. [3] to study heat and mass transfer rates for condensation of water vapour in the presence of non-condensable gas on vertical flat plate. This benchmarking activity for testing new models and codes for numerically simulating condensation in presence of non-condensable gas is popularly known as CONAN experimental tests. Vyskocil [4] developed a condensation model for commercial CFD code Fluent that could be used for simulations of air–steam flow with condensation and effect of heat transfer through liquid film is studied on condensate rate. It was found that effect of liquid film heat transfer coefficient is very low on condensation rate. O. Patil [5] studied the effect of mass fraction of steam, operating pressure and mass flow rate of mixture on CONAN isothermal wall cases using wall condensation model through Ansys CFX solver. Q. Yi [6] studied influence of the non-condensable gas on distributions of the velocity, temperature, gas concentration and interface characteristics, as well as the heat transfer coefficients. Y. Yang [7] used Eulerian wall film model in Ansys Fluent to simulate defogging process of truck cabin. Eulerian wall film model has a capability to couple the species transport equation with phase change which can be used to simulate diffusion-based condensation.

2 Numerical Methodology

2.1 Conservation Equations

Continuity Equation

$$\frac{\partial \rho}{\partial t} + \nabla \cdot (\rho \vec{v}) = S_m \tag{1}$$

Momentum Equation

$$\frac{\partial}{\partial t}(\rho\vec{v}) + \nabla\cdot(\rho\vec{v}\vec{v}) = -\nabla p + \nabla\cdot\left(\overline{\overline{\tau}}\right) + \rho\vec{g} + \vec{F} \tag{2}$$

Energy Equation

$$\frac{\partial}{\partial t}(\rho E) + \nabla\cdot\left(\rho v\left(h + \frac{v^2}{2}\right)\right) = \nabla\cdot\left(k_{\text{eff}}\nabla T - \sum_j h_j \vec{J_j}\right) + S_h \tag{3}$$

Turbulence Model $k-\omega$ turbulence model is most commonly used model to capture effect of turbulence in flow.$k-\omega$ SST turbulence model is used for these simulations.

Eulerian Wall Film Model EWF model in Ansys Fluent coupled with species transport is used to simulate condensation of water vapour in presence of non-condensable gases. EWF is a multiphase model which is capable of simulations involving thin liquid films forming on surfaces. There are various physical phenomenon regarding liquid films that need to be modelled. Various applications such as film condensation, film evaporation, liquid droplet collection, spraying, film condensation and evaporation, fuel evaporation can be simulated using EWF model. Governing equations for EWF model are as follows:

Film Mass Conservation

$$\frac{\partial\rho_l h}{\partial t} + \nabla_s\cdot\left(\rho_l h\vec{V_l}\right) = \dot{m}_s \tag{4}$$

Film Momentum Conservation

$$\frac{\partial\rho_l h\vec{V_l}}{\partial t} + \nabla_s\cdot\left(\rho_l h\vec{V_l}\vec{V_l} + \vec{D}_V\right) = -h\nabla_s P_L + \rho_l h\vec{g}_\tau + \frac{3}{2}\vec{\tau}_{fs} - \frac{3\mu_l}{h}\vec{V_l} + \vec{\tau}_{\theta_w} \tag{5}$$

where

$$P_L = P_{\text{gas}} + P_h + P_\sigma$$

Second term in the right-hand side of the Eq. 5 includes the effect of gravity in the direction parallel to the film. Third and fourth terms include the net viscous shear force on the gas-film and film-wall interfaces. Fifth term includes the effect of surface tension parallel to the wall. The velocity profile is assumed to be quadratic across the liquid film [8]

Film Energy Conservation

$$\frac{\partial \rho_l h T_f}{\partial t} + \nabla_s \cdot \left(\rho_l h T_f \vec{V}_l + \vec{D}_T \right) = \frac{1}{C_p}\left[\frac{2k_f}{h}(T_s + T_w - 2T_m) + \dot{m}_{\text{phase}} L \right] \tag{6}$$

L is the latent heat associated with the phase change. It is a function of saturation temperature. A piece-wise linear temperature profile has been assumed, i.e. the film temperature varies from T_w to T_f in the lower half of the film and from T_f to T_s in the upper half. T_m is film half depth temperature [8]

Assumptions in EWF model Assumptions involved in Eulerian wall film model are:

1. The EWF model assumes that film always flows parallel to the surface so the normal component of film velocity is zero.
2. The film is assumed to have a parabolic velocity profile and a bilinear temperature profile across its depth.
3. Formation of fog (volumetric condensation if local temperature drops below saturation) is neglected.
4. Film is assumed to be very thin and will not be resolved into a mesh. EWF model uses a virtual 2D film on the surface, which flows parallel to the wall and affects the temperature and species concentration of the core flow but flow momentum coupling is also possible by sharing the same velocity at the interface without slip [8]

Coupling with Species Transport Equation Coupling between EWF and species transport to simulate phase change in mixture is achieved by using source term based on diffusion balance model [8]. Species transport equation is given as

$$\frac{\partial}{\partial t}(\rho Y_i) + \nabla \cdot (\rho \vec{v} Y_i) = -\nabla \cdot \vec{J}_i + S_i \tag{7}$$

Rate of phase change is given by Eq. 8

$$S_i = \dot{m}_{\text{phase}} = \frac{(\rho D/\delta)}{\rho D/\delta + C_{\text{phase}}} C_{\text{phase}} \left(y_{\text{sat}} - y_i \right) \tag{8}$$

y_{sat} is calculated as follows

$$y_{\text{sat}} = \frac{P_{\text{sat}}}{P} \frac{Mi}{M} \tag{9}$$

P_{sat} is saturation pressure. The saturation pressure P_{sat} is function of temperature and computed by using Eq. 10

$$\begin{aligned}\log_{10} P_{\text{sat}} = & -2.1794 + 0.02953(T - 273.17) \\ & - 9.1837e^{-5}(T - 273.17)^2 \\ & + 1.4454e^{-7}(T - 273.17)^3\end{aligned} \tag{10}$$

3 CONAN Benchmark Test

The flat vertical plate encountering condensation of humid air is taken for study. The data for validating results of simulation is a benchmark data available from a CONAN (condensation with aerosols and non-condensable gases) experimental facility at University of Pisa, Italy [3] used by G. Zschaeck [8]. This data is widely accepted for research in film condensation, validation of new models and codes. The facility consists of the primary loop of 2 m long side channel having square cross section of 0.34 m by 0.34 m. Mixture of steam and air is passed through the primary loop at atmospheric pressure with different velocities, temperature and mass fraction of water vapour. Water is passed in the counter direction in a secondary loop which acts as a coolant for the plate. Its inlet and outlet temperature as well as mass flow rate are measured. Along vertical centreline of the aluminium plate, local heat fluxes are recorded. The condensate was collected, and its amount is measured as a function of time. The system is operated on atmospheric pressure. The empirical relation derived from heat and mass transfer analogy can be used to verify the mass transfer process. Simultaneous heat and mass transfer over a vertical flat plate with turbulent flow can be characterized by following empirical relation [3].

$$\text{Nu}_z = 0.0296\text{Re}_z^{0.8}\text{Pr}^{0.33} \tag{11}$$

The corresponding relation for mass transfer can be deduced using heat and mass transfer analogy. It states that under same geometrical conditions, the relations for heat transfer can be used for mass transfer by replacing the non-dimensional numbers in heat transfer with that of mass transfer. The Nusselt number which is non-dimensionalized heat transfer coefficient can be replaced by corresponding non-dimensionalized form which is Sherwood number. Prandtl number which compares relative thickness of velocity and thermal boundary layer can be replaced by Schmidt number. Schmidt number compares relative thickness of velocity and concentration boundary layer. Sherwood number can be expressed as follows

$$\text{Sh}_z = \frac{m''_{\text{cond}}}{\frac{\rho D}{z} ln\left(\frac{x_{\text{nc,bulk}}}{x_{\text{nc,wall}}}\right)} \tag{12}$$

Where Schmidt number is given as

$$\text{Sc} = \frac{\mu}{\rho D} \tag{13}$$

The empirical relation for mass transfer by replacing terms in Eq. 9 can be written as

$$\text{Sh}_z = 0.0296\text{Re}_z^{0.8}\text{Sc}^{0.33} \tag{14}$$

4 Numerical Setup

A numerical investigation is carried out in Ansys Fluent commercial software. Flow is assumed to be incompressible, and pressure-based solver is used. Both steady state and transient solvers are used. Steady state simulations are performed using pseudo-transient adaptive time stepping. Dimensions of geometry are same as CONAN test geometry as shown in Fig. 1. Three zones in which two are fluid zones (humid air and water zone) and one solid zone (AL plate) are used. Thermohydraulic conditions of CONAN test cases are given in Table 1. As the geometry is vertical plate and simple in nature hexahedral meshing is used for discretization of fluid and solid zones. One of the limitations for using EWF model is that the mesh should be conformal at shared boundaries between solid and fluid zone. Conformal mesh means the cells on both zones should share common nodes which also helps in reducing additional interpolation required while performing conjugate heat transfer analysis. So, shared topology is used between zones to ensure conformal meshes. The first layer thickness of inflation layers required for capturing boundary layer, and turbulence effect on a surface is an important parameter while generating the mesh. y + over a condensing wall is kept below 3. Mesh has 0.0001 m first layer height with 20 inflation layers and mesh count of 1,664,476.

The region where condensation is expected is refined to capture steep concentration gradients. The bulk region is significantly large and it do not affect local hydrodynamics of condensing wall. Hence, relatively coarse mesh is used in bulk region.

5 Results and Discussion

For both steady state and transient cases, the converged solution is obtained when condensate rate becomes steady and film mass flow rate at outlet is reaching constant value once the flow stabilizes. Convergence criteria is 1e-3 for continuity and momentum and 1e-6 for energy equations.

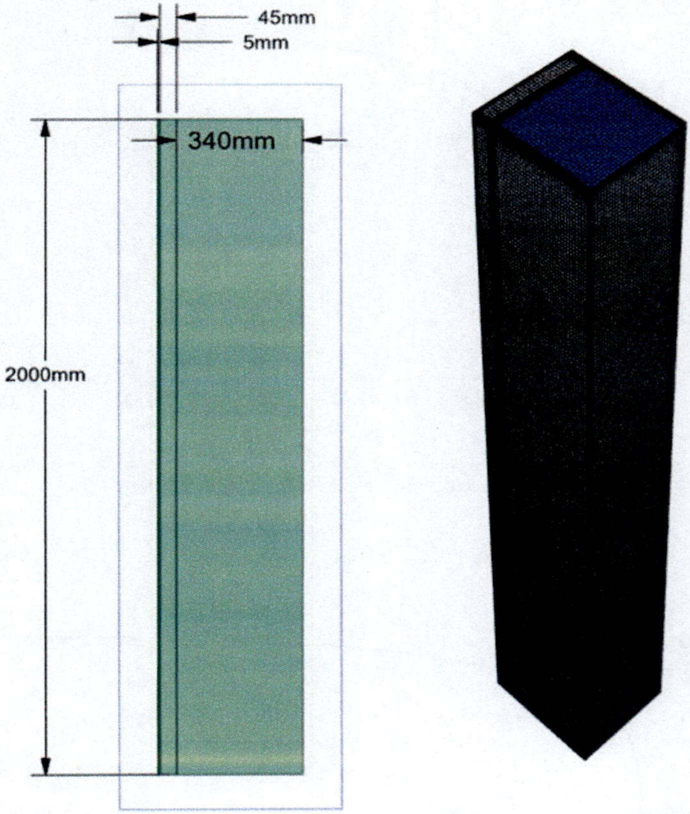

Fig. 1 Dimensions of CAD geometry and mesh

Table 1 Thermohydraulic conditions for CONAN test case [8]

Test name	Inlet velocity air (m/s)	Inlet temperature air (K)	Inlet WV mass fraction of air	Inlet mass flow rate water (kg/s)	Inlet water temp (K)	Condensation rate (gm/s) (Exp)
P10-T30-V15	1.46	355.81	0.4041	1.217	304.39	2.292
P10-T30-V20	2.02	353.76	0.3652	1.217	304.25	2.474
P10-T30-V25	2.52	352.28	0.3307	1.216	304.22	2.604
P10-T30-V30	3.01	351.88	0.2837	1.216	304.06	2.656

Figure 2 shows the mass flow rate difference between air inlet and air outlet has reached steady state. Figure 3 shows that the mass flow rate of condensed water liquid from the outlet has also reached steady state. Mesh independence test is carried out to obtain a solution which is not dependent on mesh size or mesh count. Figure 4

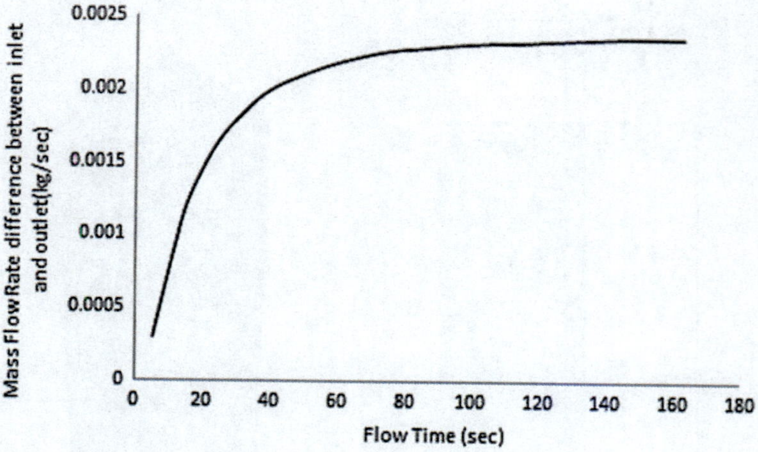

Fig. 2 Convergence of mass flow rate difference across air inlet and outlet

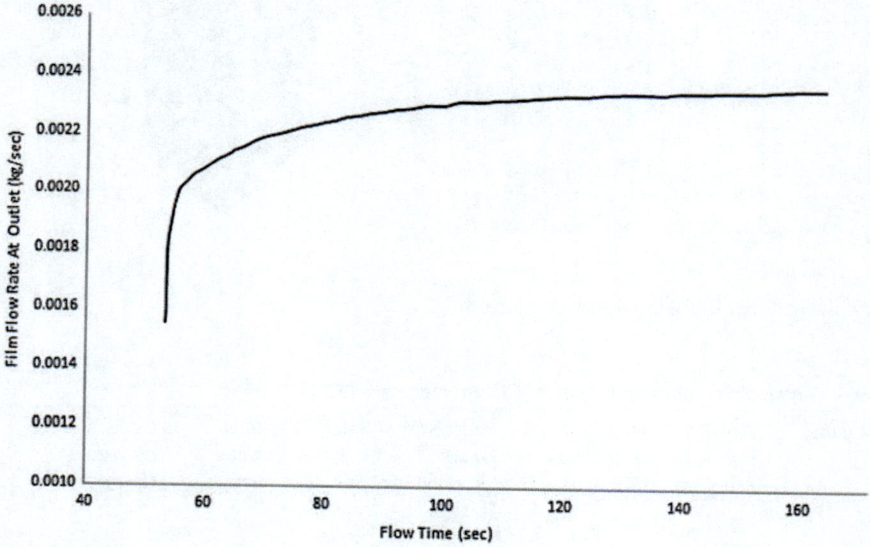

Fig. 3 Convergence of water liquid mass flow rate at air outlet

shows the condensation rate becomes grid independent as mesh count increases.

Figure 5 shows the impact of first layer thickness on condensation rate. First layer thickness of mesh has high impact on condensation rate estimations. Figure 6 shows variation of number of inflation layers on condensation rate. As the EWF-diffusion balance model is a concentration gradient-based model, the mesh resolution near the wall holds importance to capture steep gradients encountered at the wall. The

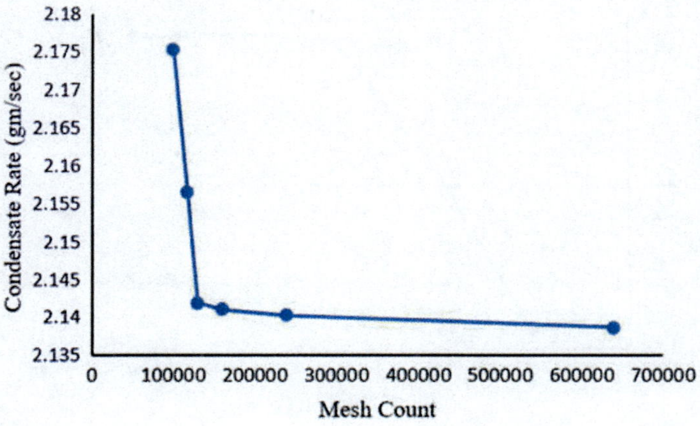

Fig. 4 Mesh independence test

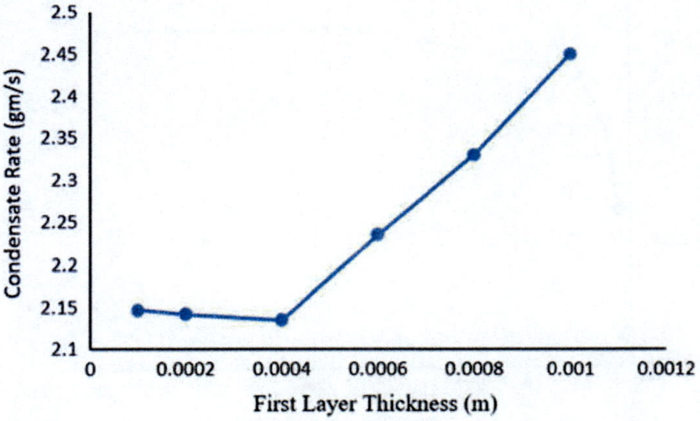

Fig. 5 Variation of condensation rate with first layer thickness

first layer thickness on the wall and number of inflation layers are significant factors affecting the solution.

Figure 7 shows condensation constant (C_{phase}) in the diffusion balance model does not affect the solution much. Condensation constant is a stabilizing parameter for small values involved in computations. Figures 8 and 9 show the results of simulations found to be in good agreement with the experimental set of values of condensation rate for both steady state and transient simulations with maximum variation of 11%. Figures 10 and 11 show heat flux on condensing walls is found to be in good agreement with experimental data except at inlet due to flow development and turbulence effect. An empirical relation based on heat and mass transfer analogy (Eq. 12) is validated on two cases.

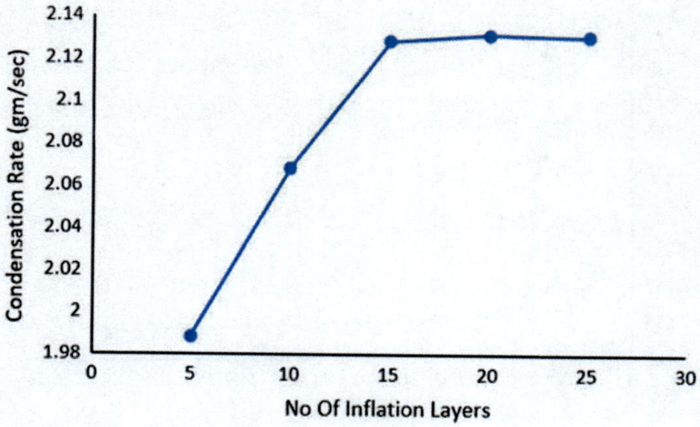

Fig. 6 Variation of condensation rate with number of inflation layers

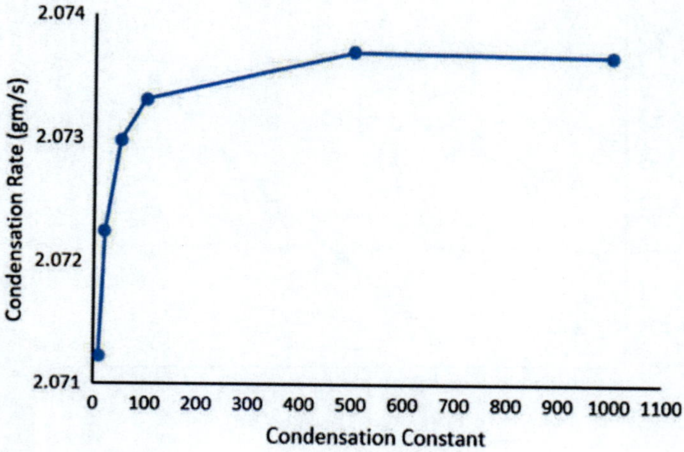

Fig. 7 Variation of condensation rate with condensation constant

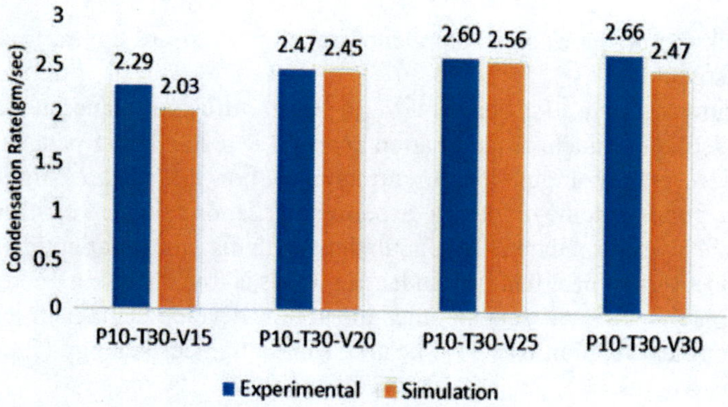

Fig. 8 Comparison of experimental and simulation condensate rate for transient case

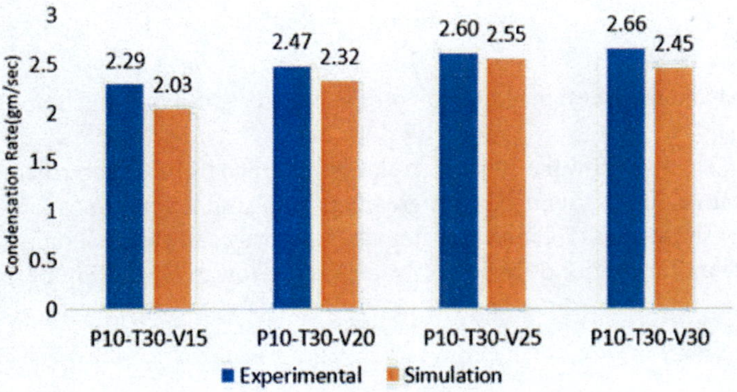

Fig. 9 Comparison of experimental and simulation condensate rate for steady case

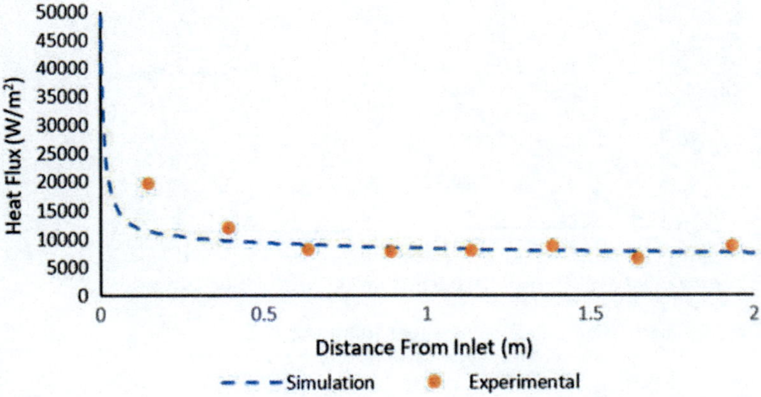

Fig. 10 Validation of surface heat flux for P10-T30-V25 [3]

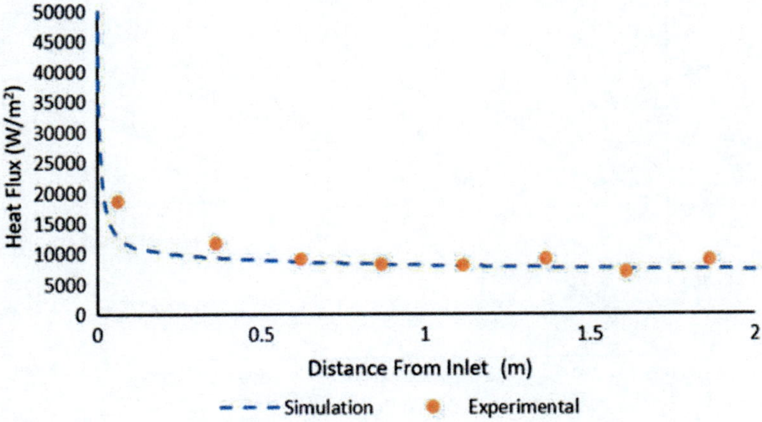

Fig. 11 Validation of surface heat flux for P10-T30-V30 [3]

$$\frac{Sh_z}{Sc^{0.33}} = 0.0296Re_z^{0.8}$$

The left-hand side term is plotted on ordinate, and right-hand side term is plotted on abscissa.

Figures 12 and 13 show the empirical relation obtained from heat and mass transfer analogy (orange line) is found to be in agreement with simulation values (blue curve).

Figure 14 shows mass fraction of water vapour drops near wall generating concentration gradient for further diffusion of water vapour towards wall. Figure 15 shows

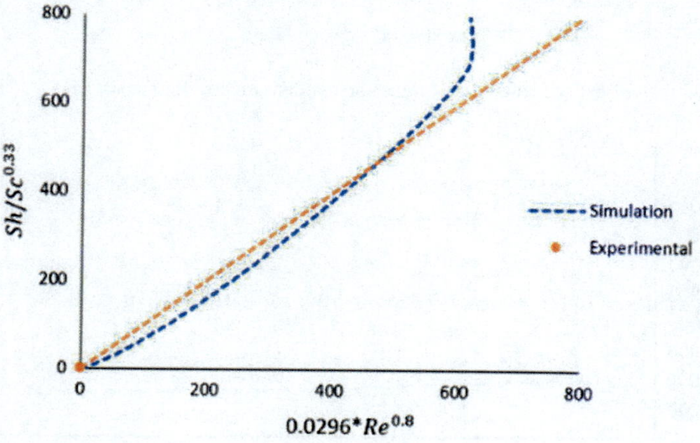

Fig. 12 Empirical correlation validation for case P10-T30-V25

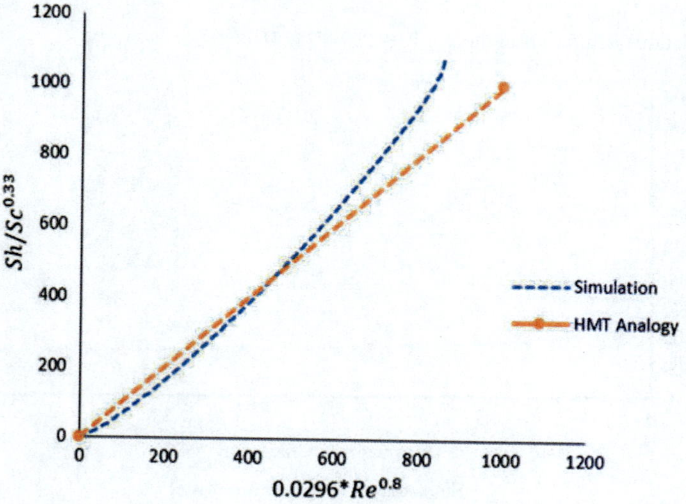

Fig. 13 Empirical correlation validation for case P10-T30-V30

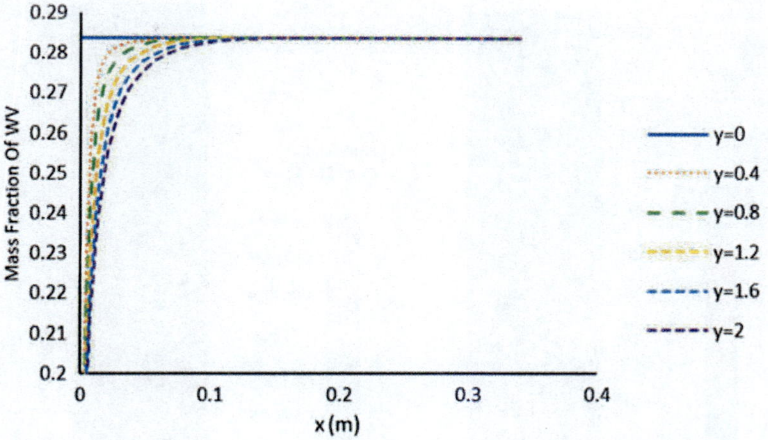

Fig. 14 Water vapour mass fraction variation for case P10-T30-V30

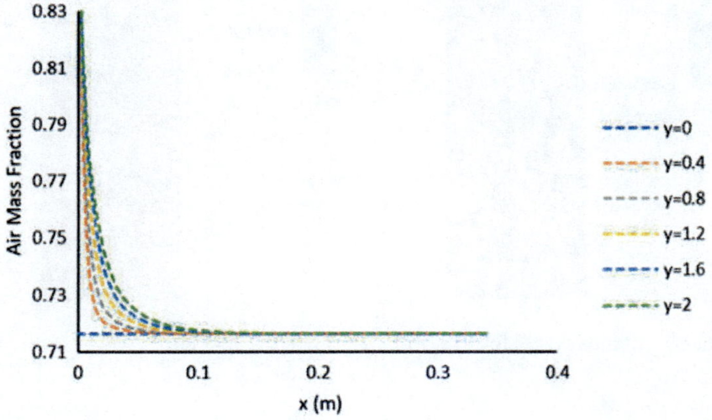

Fig. 15 Air mass fraction variation for case P10-T30-V30

due to condensation of water vapour near the wall, and non-condensable air accumulates which increases its mass fraction and creates further resistance due to cross diffusion. Figure 16 shows contours of film thickness on the condensing wall.

6 Conclusions

The Eulerian wall film model is validated on CONAN benchmark tests. Results are found to be in good agreement with benchmarking set of conditions with maximum variation of 11%. Both steady state and transient approaches are applied and found to be predicting condensation rate with good accuracy. Surface heat fluxes and

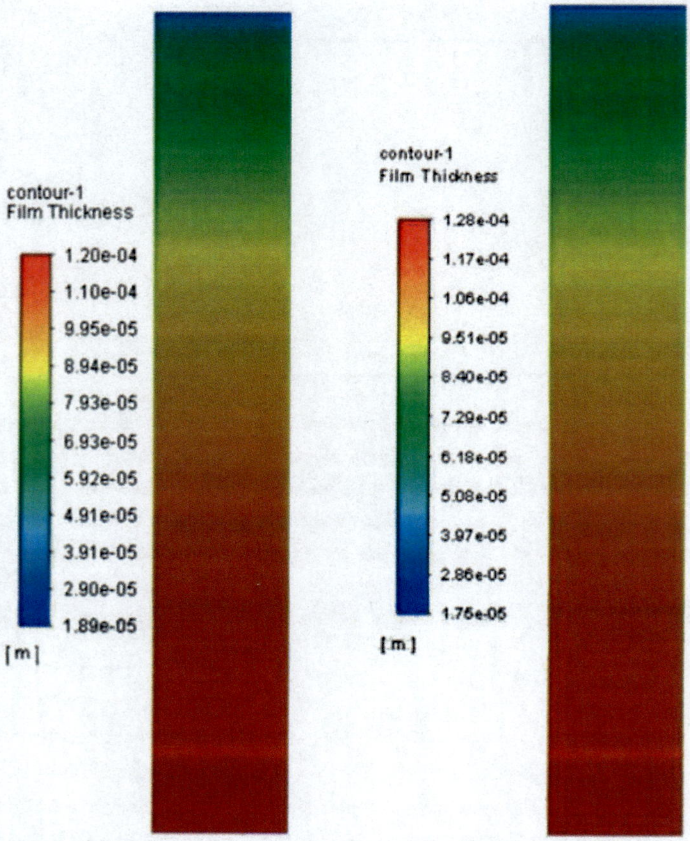

Fig. 16 Contours of film thickness for P10-T30-V15 and P10-T30-V20

correlation based on heat and mass transfer analogy are validated with simulation results.

Acknowledgement Authors would like to thank Whirlpool of India Ltd. Pune for their support and access to Ansys Fluent software.

References

1. Nusselt W (1916) Die oberflächenkondensation des wasserdampfes. Zeitschrift des Vereines Deutscher Ingenieure 60(27):541–546 (in German)
2. Siow EC, Ormiston SJ, Soliman HM (2002) Fully coupled solution of a two-phase model for laminar film condensation of vapour–gas mixtures in horizontal channels. Int J Heat Mass Transfer 45(18):3689–3702

3. Ambrosini W, et al, (2008) Comparison and analysis of the condensation benchmark results. In: The 3rd European Review Meeting on Severe Accident Research (ERMSAR-2008), Nesseber, Bulgaria, pp 23–25
4. Vyskocil L, Schmid J, Macek J (2014) CFD simulation of air–steam flow with condensation. Nuclear Engineering and Design, pp 147–157. http://doi.org/101016/jnucengdes201402014
5. Patil O, Maurya RS (2016) Film condensation behaviour of steam on isothermal walls in presence of non-condensable gases-a numerical investigation. Int J Comp Eng Res 6(5)
6. Yi Q, Tian M, Fang D (2015) CFD simulation of air-steam condensation on an isothermal vertical plate. Int J Heat Technol 33:25–32
7. Yang Y, Huang Y, Zhao J (2020) Optimization of the automotive air conditioning strategy based on the study of dewing phenomenon and defogging progress. Applied Thermal Engineering, vol 169. ISSN 1359-4311. https://doi.org/10.1016/j.applthermaleng.2020.114932
8. Zschaeck G, Frank T, Burns AD (2014) CFD modelling and validation of wall condensation in the presence of non-condensable gases. Nuclear Engineering and Design, pp 137–146. https://doi.org/:101016/jnucengdes201403007

Small-Scale and Pico Hydro Power Generation Techniques Review

Shashikant Mali, Shridhar Motale, Ravindra Adhal, Rushabh Barde, and Sudesh Powar

1 Introduction

Small Pico turbines are one of the promising technologies to generate electricity which could be implemented using different techniques like Titus and Ayalur [1] explained the need for clean energy without the emission of carbon dioxide with the design and fabrication of a turbine to regenerate electricity from sewage water. Nfah and Ngundam [2] discussed the off-grid options in which they concern about renewable energy resources could be a suitable alternative for rural areas, where low electricity was generated (range from 10 to 50KW). Pico hydro turbine is an option for electricity generation in Kenya for 90% of the households, in which they found out it was less costly as compared to the solar system. Maher et al. [3]. Electricity plays an important role in the sociological and economic development of any progressing country. The energy demand significantly increases in India past few decades because of the globalization of the Indian market in 1991 and the rapid industrial revolution. There are too many people who depend on traditional biomass for cooking purposes and they also do not have access to electricity. According to UN-Energy, this gap causes a continuous effect on environmental sustainability laghari and Mokhlis [4]. Briefly explained, energy resources contribution fulfills requirement 78.3% fossil fuel, 2.5% nuclear, remaining around 19% by all renewable energy. The burning of fuel is mainly responsible for global warming, air pollution and environment degradation, so many government policies shifting toward renewable energy generation and also support private players. Pico and mini-microhydro power plant is one best option for renewable energy generation, and the Asia continent consists of around 65% small-scale hydro lower plant in this china account (Fig. 1).

59% then Japan, the USA, Italy, Brazil, then rest of the world. China, Malaysia, Japan unitize their capacity focusing on small-scale hydro power plant and different

S. Mali · S. Motale (✉) · R. Adhal · R. Barde · S. Powar
MIT Academy of Engineering, Pune, India
e-mail: ssmotale@mitaoe.ac.in

J. Banerjee et al. (eds.), *Recent Advances in Fluid Dynamics*, Lecture Notes in Mechanical Engineering, https://doi.org/10.1007/978-981-19-3379-0_37

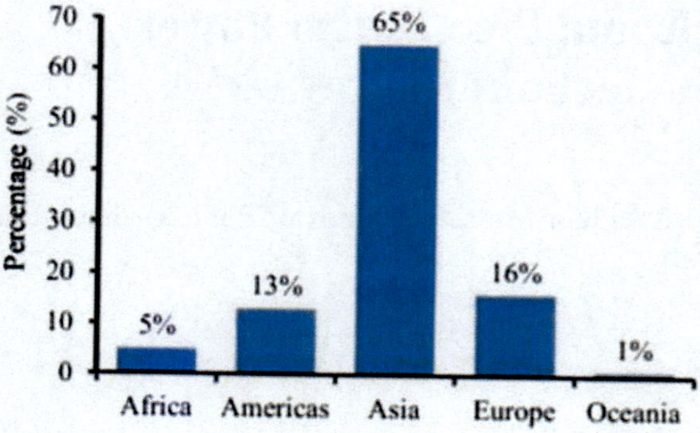

Fig. 1 Small hydro power potential by continents [5]

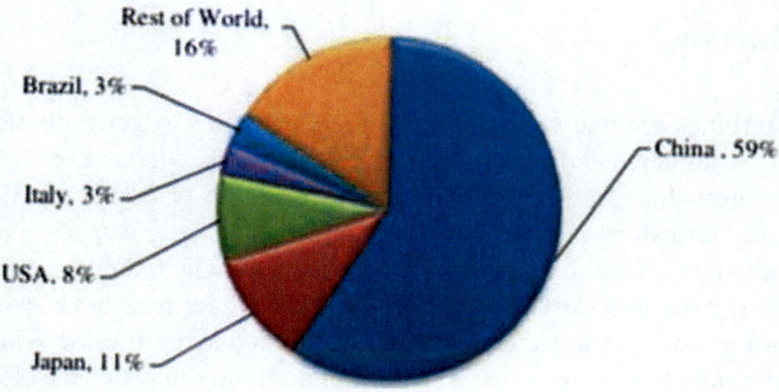

Fig. 2 Pi diagram representation [5]

techniques [5]. The purpose of this paper study small-scale hydro and Pico hydro power working worldwide and it could be an economical option to regenerate electricity in developing countries. This methodology could be applied to a remote region were the head available is very low (Fig. 2).

2 Methods of Small-Scale Hydro-Energy Generation

2.1 PATs as Turbine

De Marchisa et al. [6] presented the case study on 17 distribution networks of Palermo city, where they used pump as a turbine (PAT) as a water distribution network. The main purpose was to reduce the cost of installation of the turbine and maximizes the efficiency of the PAT turbine; a numerical model is a preliminary analysis which they discussed in the light of the capital payback period, enhancing the four different positions of installation in the particular section. The analysis on the economy has been carried out and the result shows different capital payback period in different cases got 2.03 years this very attractive solution and the maximum turbine efficiency that can be reached is the PATs are installed in the main pipe and without affected by the cost of installation and maintenance. Tracey et al.[7] analyze the selection and design process for suitable PAT that depends on practical hydraulic conditions and with the performance of operating pump in reverse as an output maximizes the power production potential of the PAT turbine. One's site is selected and created prototype similar to PAT so that behaviors and efficiency predicted easily. Identifies PAT BEP for machines similarly design as prototype and corresponding as per site design points. It is suitable for the selection process of the turbine once PAT BEP was converted to pump conditions. They repeated this process for each site and the overall selection of the turbine was done. The estimation of the recovery of the water network was done by using the pump as a turbine (PAT) with the following result.

- Required parameter like pressure control was considered for PATs.
- The process developed to pre-prediction of the turbine selection and its performance.
- Compare to single, double PAT was very advantageous in parallel position.

Barbarellia et al. [8] Working on the performance curve of a PAT, the explanation of this curve was found out by using the statics and numerical model. The main parameters are used for the estimation of output were head and discharge. The performance curve consists of potential head vs capacity and turbine efficiency vs capacity. In the section of the calculation, they included efficiency, specific speed, discharge, and head. It included the PAT selection procedure. The case study was included taken input as a different head and found out the key parameters.

Jain and Patel [9] PAT used as a microturbine was discussed. The turbine selection process was discussed by considering the same parameters as head, capacity, back pressure, turbine inlet–outlet velocity, etc. Referring to many literature surveys, they gave theoretical studies and experimental studies for performance prediction for the microturbine. The section of calculation explained force analysis of PAT, loss distribution in PAT, performance improvement in PAT. Ali Maleki et al. [10] A case study on the PAT is included considering various reasons like the generation of the artificial head and scaling effects. Details formation provided about the cost estimation and life cycle of PAT turbine with limitations. The numerical method was used for multistage PAT to determine the changes in the result with the change of fluid viscosity and its patterns. For commercial problems, they went through ANSYS CFX. The impact of fluid viscosity on the hydraulic parameters of the single as well multistage centrifugal pumps in reverse mode was estimated by computational fluid dynamics (CFD) simulations. According to the result of the numerical model, the experimental data for the single-stage PAT turbine was affected by fluid viscosity as it was calculated in different conditions. The higher flow rate was found on higher viscosity.

Menelaos Patelisa et al. [5] the energy recovery and pressure management capabilities using PATs. The hydraulic turbine model of kozani city was performed first and validated to accurate results. In the case of implementation of a PAT in the software, there are certain operational details which are required from the turbine manufacturer. An absolute PAT was found and characteristics points such as the operating curve were found. As excessive pressure reduction effect on the implementation of PAT turbine to achieve significant energy production, combination of PAT turbine installation inside the water distribution network with bigger PAT turbine presents in the water supply chain, so that it can able to generate sufficient energy with enough water supply. Even pressure control does not impact much on turbine performance. Tao Wang et al. [11] the improvement in the effective performance of PAT turbine by changing its design of the blades where the different inlet and outlet angle was going to change. In the section of a theoretical calculation, they found various parameters. It included the experimental setup and result in the analysis. The numerical investigation contains a numerical model on ANSYS software in which hydraulic loss distribution analysis and analysis on internal flow characteristics and static pressure distribution. It compares the result to verify the output of the experiment (Fig. 3).

Fig. 3 Site view of the pump as a turbine [4]

2.2 Pico Hydro Power Generation

Budiarso et al. [12] Main objectives is to developed spoon-based turbo turbine which could be used in the pipeline to increase the electrification ratio. Setup includes dynamometer pulley, tachometer, etc. To calculate RPM and torque to find power output. The ratio of wheel diameter with jet and an optimum number of blades calculations carried out. Turgo turbine and Pelton wheel were analyzed based upon different flow rate regulation and power is calculated and concluded that a sufficient amount of power can be generated using a spoon turgo turbine at low cost. Ibadullah Safdara et al. [13] calculated the efficiency of the generator and turbine, and performance was studied on impulse turbine. Losses due to installation of battery and other components were analyzed. Setup an involved alternator, laser tachometer, battery, rectifier, inverter, transducer, etc., to record required data. Mathematical calculations of shaft power and overall efficiency are completed, and flow rate and turbine efficiency was plotted. It is found that the power generated is less than expected due to lower alternator efficiency. Titus and Ayalur [1] explained the need for clean energy without emission of carbon dioxide, the main aim of review paper is to design and fabricate turbine which will be used fully to regenerate energy from sewage water pipeline without affecting or loss of mass flow rate. Practical experiment was carried out with the head of 14 m and discharge of 9li/sec. In a study, it is found that 212 W electrical power is generated. Titus explained theoretical calculation of power availability, CAD modeling as well as analysis using computational fluid dynamics. In CFD, changes in velocity and pressure in the turbine assembly was studied.

A study by Uchiyama et al. [14] aims at verification of power generation with help of sewage water pipes. Pico hydraulic turbine was practically tested in a closed test rig made with a transparent resin called acrylic to observe the fluid flow and foreign particles and the pump was also used to create discharge. Setup involved guide vane just before runner to enhance efficiency. A different particle that was going to heat blades were tested and accordingly, efficiency was calculated. Gallego et al. [15] paper gives a brief idea about the advantages of the second turbine and its functionalities, i.e., relatively cheap, sustainable, and applicable to a medium head turbine. It could be an economical option to regenerate electricity in developing countries. Different geometric specification, dimensions (of a turbine), and operating conditions were studied using response surface methodology. The second regression plot is obtained and validated. It is found that nozzle diameter affects the efficiency of the turbine and maximum efficiency found to be 93.7% at optimum condition. Ishola et al. [16].

Pelton wheel turbine was designed to generate power from water harvested from a rooftop and stored in a tank. Head of 10 m and other design specifications were fixed and three materials were considered, i.e., steel, aluminum alloy, and plastic, and their performance were majored using Autodesk inventor. Materials tested based on most conservative theory, i.e., von mises theory. Alson 1st principal stresses and 2nd principal stresses calculated and plotted. It is found that aluminum earmarked wheel to be more optimum as compare to other to materials. Williamson et al. [17]

explained turbine selection using multi-criteria methods involving qualitative and quantitative analysis of parameters. Detail selection methodology is discussed by author.13 turbine systems were studied during this research. Bhanbhane village in Nepal was considered for the implementations of this method. This methodology could be applied to a remote region were the head available is very low.

2.2.1 Pico Hydro: A Case Study

Nfah and Ngundam [2] studied the feasibility of Pico hydro together with photovoltaic hybrid power system in Cameroon village incorporating with biogas generator and simulated results concerning HOMERS and hostels of Cameroon village, off-grid options in which they concern about renewable energy resources could be a suitable alternative for rural areas, where low electricity was generated (range from 10 to 50KW) in Cameroon. Thus, electricity supply of the most rural villages in Cameroon may be fulfill with renewable energy options which contain biogas generators.

Maher et al. [3] used Pico hydro turbine as option for off-grid electrification in Kenya country for 90% of households and implement Pico hydro plant in an off-grid area then cost analysis of this find out the continuous electric power generation available is less than 15% of that solar household system, which is calculated by cost per KWh. A coordinated view is required to access the rapid increase in electricity in rural areas. Starts with the closest existing grid present in that place with less cost investment for electricity connections and load distributors. Secondly, in those areas where suitable sites already exist for Pico hydro turbine due to the advantage of the solar system grid connection. To encourage and funding for turbine particular for less income household. Prabir Sarkar et al. [18] studied electric energy generation from greywater in huge raised buildings considered 20th-floor building and greywater collected on 10th-floor tank 76% water use for electricity generation. A pipe is attached to the tank hydraulic turbine Pelton installed into it along with gear and generator responsible for power generation. Calculated different results for different head and cost analysis payback period from investments also calculated.

3 Micro Hydro Power: Case Study Overview

Mutiara Ayu Saria et al. [19] provided a detailed review of the recent available hydro turbine which is suitable for various applications.

This paper gave information about soft water and wastewater utilization in the turbine. It avoids water wastage in the water supply chain and conduit hydro power into an existing one. Studied evolution hydro turbines. Comparison between them considering different parameters like head, hydraulic site, and capacity.

António and Henriques [20] explained design and experimental investigation on small-scale Archimedes screw turbine at hydro power plant test station the research provide detail data with the on the design of Archimedes turbine with experimental data related to torque, efficiency, and power and later, novel contribution is obtained by experimental testing of the plant by using AC and DC generators. But earlier the efficiency of the generator was not estimated by them. The efficiency of the generator was calculated by the ratio of the power output and the input of some DC generator losses. Because of the different inclination angles, available water flows, and rotational speed was power delivered, and the performance of the AC and DC generators. The possibilities and design of test stations and inclination angle from 20° to 40° which having a flow rate of 10 l/s to 12.8 l/s. These results get by considering the different losses of the AC and DC generators. The percentage of measure losses is also large. The difference can be decreased if coefficients are taken from series of various measurements. Elie [21] feasibility study of the micro hydro power plant in Cameroon feasibility check-in in term of the site and material study proper method. Site recognition and topographic analysis, electricity demand in the rural area, the electrical power, and the choice of the technical components of the installation study the best combination from all of that. Calculated cost analysis and payback period power plant. It is visible that the improvement of the KEMKEN MHPP is one of the examples of the contribution of renewable energy for the better sustainable development of the locality. It is capable of 320 KW with a maximum payback period of 7 years and a total investment of 212 486 656 FCFA. This implementation of the project is suitable for local rural areas.

Strategy for successful operation and performance obtained by using the new design micro-level hydro turbine, which is capable of processing under a wide range of fluctuation in stream flow. Sufficient energy producing WWTP is the aim of the vision to reduce the GHG emission of wastewater, as a result of the increase in pressure. This case study by Kyu-Jung Chae et al. [22] states that wastewater treatment can be improved through the implementation of the fluid variables into existing WWTPs. Whether fluid flow through the pipe is sufficient to convert water energy to electrical energy. They included the experimental flowchart for designing the small hydro power by considering the feasible studies, recognition of the need, and output from the experimentation. The flowchart includes the conceptual design stage, the preliminary design stage, data analysis, and results. In the section of results, they show the graphical comparison of electricity generation between the prototype analysis and conceptual design analysis. At the end of the paper, they concludes that fluid flow in pecan to generate electricity.

3.1 Different Technique's

3.1.1 Water Head Generated Using Solar and Wind Energy

Bahadur Singh Pali and Vadhera [23] focuses on novel pumped hydro-energy method with wind energy for electric power generation at a constant voltage output in remote areas. A concept of small electric power generation by using pumped hydroenergy with the help of wind as a primary input. It is proposed for rural and remote areas in which wells are available. The wind turbine does not use for the rotation of the hydraulic turbine but in this case, it rotates the pump which is used to suck water from the store to the upper reservoir. The main purpose of this process is to create a maximum water head and utilized it for Pico hydro turbine for electricity generation. The main advantage of this process maintains continuity in electricity generation at constant voltage output with the change of wind power. Bahadur Singh Pali and Shelly Vadhera [23] purpose a new power generation technology which is better than the solar photovoltaic system, Pico hydro generation, and pump hydro savage system. This idea works only in that area in which open wells and continue sources of water are available. The photovoltaic system is installed on an open well and connected to the solar water pump via electric cables. Water pump which pumps the water and creates potential head and which after utilizing for rotation of the propeller turbine. The generator rotates at constant speed by controlling water flows through an isolated solar system using renewable energy sources. This process is working on a novel operating technique for continuous generation at constant voltage.

Hrvoje Dedic-Jandrek and Nizetic [24] power generation method which replaceable to the PDCV method via a hydraulic turbine. This system converts the remaining differential pressure with a district heating pipe and generates a corresponding amount of electricity. This new method is also called a differential pressure electric power generation system. The system generates power of about 10–20 MWh and also reduces carbon dioxide emission by about 7 tons per year.

3.1.2 Water Hammer Effect

Roberts et al. [25] discussed providing renewable source with the help of mechanical power pressure hammer effect, and hydraulic ram pumps are used in water hammer effect to accelerate water in a sustainable and renewable manner without using fuel in which cam and follower mechanism responsible for rotating effect to shaft connected generator. Positioning a piston crank mechanism upstream to a periodically closing valve, furthermore using a cam mechanism to ensure that on the same time valve closes at the correct crank angle, it can be possible to generate sufficient amount of torque to turn a crankshaft.

3.1.3 Oscillating-Water-Column Wave Energy Converters

António and Henriques [20] focuses on model testing based on rules of thermodynamic, fluid mechanics, and aerodynamic similarity. The testing contains wave energy conversion of oscillating water column type, and compressibility effect on the air in the air chamber and aerodynamics on the air turbine. It states that the necessary volume scale ratio for the air chamber is much higher than identical to the volume ratio for the submerged part of the converter.

It should take compressible flow in thermodynamics through air turbine for turbine simulator. It concludes that if it manage to take approximately volume scale ratio near to dynamic similarity then it can be achieved. But it requires the real hydro turbine practically demonstrate in the laboratory by the phase difference between flow rate and pressure head.

3.2 Turbine Selection

Leguizamón and Avellan [26] explains computational parametric analysis cross-flow low turbine. Design a turbine in steel pipe to reduce the initial cost of hydro projects. In concern with computational modeling and CFD. It has explained estimating equations and different scheme for analysis with the framework of finite volume particle techniques.

Irene Samora et al. [27] has prepared a propeller-type model for microhydro power generation as this concept can be applied parallel to water supply chain systems, small irrigation channels, small rivers and savage water treatment plants, or water drainage systems, and the paper explains that propeller turbine is axial a type of turbine and most suitable for low head and high flow rate, hence this is designed and analyzed with fluent. Concerning design implemented over a 90-degree angle pipe and the efficiency or performance of the turbine shown variation with respect to change in speed. Influence of cavitation factor was also investigated. Jiyun et al. [28] has gained that how water leakage problem and detecting is an important concern and different electrical devices are involved in it. For this purpose, electric energy can be generated by a developing inline vertical cross-flow turbine positioned inside the urban pipeline. Block shapes of water turbine were tested using numerical models and an optimal solution was obtained (Table 1).

Also, Jiyun et al. [28] losses due to tip clearance between the blade tip and pipe surface was analyzed, and self-adjustable blades designed to reduce heat loss as well as negative torque on the blade. The further actual prototype was fabricated and tested for a month which gave a result that electricity generation per day to be 600 W/hr. which could provide a sufficient amount of electricity to the leakage monitoring system. It has explained the selection criteria of the turbine depending upon orientation. Practical test result shown that output power to be 69.1 W which having water head 2.62 m in the pipeline.

Mbelek and Spork [31] Modeling and perform analysis in the MATLAB software and also word on simulation software. It includes the calculation of the turbine model in which they find out the mechanical, electric energy, and pressure difference in the pipe. It gave a brief explanation and calculation of the drive train model generator model. The Simulink block diagram of the electric hydro power generation system. It contains a practical setup description. The result contains various charging currents at different flow rates. Talluri et al. [32]. The development of the design procedure of the Tesla turbine. The paper gave a brief explanation of the Tesla turbine which consists of several components like stator, bladeless rotor with the fixed disk, and

Table 1 Literature review

Method of study, objective and author name and year of publication	Findings			
	Ref. nos.	Objective	Methods of study	Findings
De Marchisa et al. (2013)	[6]	Method of Characteristics used to predict the water head and the flow rate in water distribution network,	Case study and experimental analysis	Economic feasibility of PAT installation in a WDN and cost in additional capital payback period analysis, the different payback period which is 2.03, 2.36, 3.12, and 5.78 years for different ways to install PAT, and the results of four tests are analyzed
Barbarellia et al. (2017)	[8]	Performance curves, head versus strength, efficiency compared to strength. The procedure has been used for the purpose of selecting a PAT once Models to select a pump that acts as a turbine for micro hydro plants	Statistical and numerical models	CQ defined as ratio between the capacity of pump running as turbine and that of pump at best efficiency point found
Warjito Budiarso et al. (2019)	[12]	To study and investigate the performance of a low-cost spoon-based turgo turbine under different operating conditions and parameters	Experimental study	Rural electricity needs a solid, low-cost solution. And a low-cost spoon based on a Pico scale turgo turbine produces an acceptable amount of electrical energy and efficiency

(continued)

Table 1 (continued)

Method of study, objective and author name and year of publication	Findings			
	Ref. nos.	Objective	Methods of study	Findings
Ibadullah Safdara et al. (2020)	[13]	To create a Pico hydro system that can properly generate electricity at varying degrees of water flow by maintaining frequency	Experimental setup	Generator efficiency is recorded at 15.4 gpm and the battery–inverter circuit resides in the Pico hydro control system usually
Titus and Ayalur (2019)	[1]	Design and fabricate turbine which will be usefull to regenerate energy from sewage water pipeline without affecting or loss of mass flow rate	Experimental study	Practical experiment were carried out with head of 14 m and discharge of 9li/sec. In study, it is found that 212 W electrical power is generated
Uchiyama et al. (2016)	[14]	Aims at verification of power generation using sewage water pipes	Experimental analysis	A Pico hydraulic turbine can be used to generate electricity from sewage flowing into pipes
Gallego et al. (2020)	[15]	Making a turgo turbine for low heads to improve its design focuses on increasing the efficiency of the turbine	Numerical analysis (response surface methodology and regressions plot)	It is found that nozzle diameter affects efficiency of turbine. And maximum efficiency found to be 93.7% at optimum condition
Haidar et al. (2012)	[29]	Power supply in remote and hilly areas where the grid system is not economically unequal	Experimental testing and simulated using the MATLAB Simulink block	The speed of the turbine and alternator depends on the water pressure. In this operation, a 1.05 kW alternator is used to charge the battery and the power from DC is converted to 220 V, 50 Hz

(continued)

Table 1 (continued)

Method of study, objective and author name and year of publication	Findings			
	Ref. nos.	Objective	Methods of study	Findings
Kiyarash Rahbara et al. (2017)	[30]	Focuses on the development of the small-scale radial flow turbine and study blade angles and performance	CFD analysis (1D and 3D)	The number of rotor blades has changed and turbine efficiency has improved from 81.3% obtained by mean-line modeling and 84.5% obtained by CFD

toroidal plenum chamber. For calculation purpose, basic rotor model, upgraded rotor model. For analysis, it gave the ANSYS fluent operation. The paper includes the calculation process for each part of the turbine.

Kiyarash Rahbara et al. [30] focuses on the development of the small-scale radial flow turbine. With 1D, 3D, and structural analysis as well as CFD analysis on the ANSYS software. In the calculation, inlet/outlet velocity and angles were found. Also found out the incident losses and power generation. Flow characteristics also were done in the paper. The paper gave a brief explanation about the effect of blade angles, quantity, and thickness on the performance of the Pico turbine. To measure the performance of the individual parts used the various instrument like thermocouples, pressure transducers and gave the result of this. Helena et al. [33]: paper gave importance to the analysis part which included the low-cost solutions where viability analysis and turbine power are considered. The calculation included the various parameters for design conception. Considering the cost-efficient turbine, they worked on the PAT for experimentation. The paper included a brief explanation about the CFD analysis of PAT in which flow characteristics and static analysis were done on the turbine.

4 Result and Discussion

Comparative study of different methods and techniques of power generation as well system implementation is done. PAT technology can be a sustainable option to Pico energy generation as 40% of gross power could be recovered at valve pressure and usage of PAT in parallel has shown a significant increase in power. Although maintenance and implementation cost is high, maximum efficiency can be reached. Pico hydro generation system can be applied to different domains including urban water pipeline, sewage water pipeline, greywater in high raised buildings, and water harvested from the rooftop. And for this purpose, the turgo turbine can be used as it is relatively cheap, sustainable, and applicable to medium head with 93.7% efficiency. It is found that spoon-based turgo turbines can produce a sufficient amount of power at a low cost to increase the electrification ratio.

Losses due to the addition of electrical components found that lower alternator efficiency aluminum leads to lower power output. Considering runner material, the number of materials selected and aluminum found to be the optimal solution. A suitable turbine can be selected using a multi-criterion selection technique based on qualitative and quantitative parameters. In the case of studies, it is found that renewable energy resources is suitable for electrification as the cost of Pico hydro power is 15% less than the solar home system. Elie Bertrand Kengne studied the feasibility of a micro-hydro power plant with the result as 320KWhr capacity and 7 years of the payback period concept to use wind and Pico energy generation by Bahadur Singh Pali, explained the technique by which remote areas having wells can be electrified. This concept involves the usage of wind energy to pump water at the high head as a primary step. And to use this head of water for Pico hydro turbine and generator assembly to generate electricity. This method gives more productivity due to lowered fluctuations in the runner shaft by maintaining the constant flow of water.

Method to use unused pressure in the district heating pipe to generate electricity leads to 10 to 20 Mhr power generation and up to a 7-ton reduction in Co_2 emissions. In turbine selection, it is found that propeller turbines can be implemented easily as their orientation is an axial type of turbine hence can be positioned in a pipe with ease. As well as it is most suitable for low-potential head and higher flow rate. Inline vertical cross-flow aluminum also resembles a similar orientation advantage of propeller turbine is having. Reducing tip clearance with a pipe wall could reduce mass flow losses and backflow problems. Inline vertical cross-flow turbine performance has shown power generation up to 600 W/hr.

5 Conclusion

Different setup for Pico turbine, PATs and micro hydro power plant currently working around the world have been provided established Pico and small-scale hydro power plant. Investment and probable returns in terms of energy regeneration show variation according to technical, economical, and geographical parameters.

A brief study shows PATs more effective water distribution network of metropolitan cities because it acts as PDCV as well for power generation and PATs payback period is also attractive for the investor.

Despite challenges at the technical, economical, and institutional level for Pico hydro power generation but it efficient for urban building with a high water head of tank available for a small amount of energy creation.

Computer-added manufacturing, i.e., 3D printing for manufacturing blades of turbine and ANSYS software for simulation flows effective in terms saving time and alternative for cost of experiment and testing of model for research purpose.

6 Future Scope

- The possibility of cavitation at the throat section under different setups and types of turbine used is a big concern.
- Study related to consumption or losses in electrical equipment used like generator has future scope for increasing power regeneration of Pico turbine research in minimizing losses.
- Effective design of Pico turbine for domestic and greywater pipelines required.
- Methods to reduce minor losses as well as practical losses in a huge system of Pico turbine are needed in near future for a highly reliable and efficient system.

References

1. Titus J, Ayalur B (2019) Design and fabrication of in-line turbine for Pico hydro energy recovery in treated sewage water distribution line. Energy Procedia 156:133–138
2. Nfah EM, Ngundam JM (2009) Feasibility of pico-hydro and photovoltaic hybrid power systems for remote villages in Cameroon. Renewable Energy 34(6):1445–1450
3. Maher P, Smith NPA, Williams A (2003) Assessment of pico hydro as an option for off-grid electrification in Kenya. Renewable Energy 28(9):1357–1369
4. https://doi.org/10.1016/j.rser.2012.12.002
5. Patelisa M, Kanakoudisa V, Gonelasa K (2016) Pressure management and energy recovery capabilities using PATs
6. https://doi.org/10.1016/j.proeng.2014.02.049
7. Lydon T, Coughlan P, McNabola A (2017) Pressure management and energy recovery in water distribution networks. Renewable Energy S0960–1481(17):30749–30758
8. Barbarellia S, Amelioa M, Florioa G, Scornaienchia NM (2017) Procedure selecting pumps running as turbines in micro hydro plants. In: 72nd Conference of the Italian Thermal Machines Engineering Association, ATI2017, 6–8 September 2017, Lecce, Italyia 126(201709), 549–556
9. Jain SV, Patel RN (2014) Investigations on pump running in turbine mode: A review of the state-of-the-art. Renewable Sust Energy Rev 30:841–868
10. Maleki A, Ghorani MM, Haghighi MHS, Riasi A (2019) Numerical study on the effect of viscosity on a multistage pump running in reverse mode
11. Wang T, Kong CWF, Goua Q, Yang S (2017) Theoretical experi mental, and numerical study of special impeller used in turbine mode of centrifugal pump as turbine
12. Budiarso W, Lubis MN, Adanta D (2019) Performance of a low-cost spoon-based Turgo turbine for Pico hydro installation. Energy Procedia 156:447–451
13. Safdara I, Sultana S, Razaa HA, Umerb M, Ali M (2020) Empirical analysis of turbine and generator efficiency of a pico hydro system. Sustainable Energy Technologies and Assessments 37
14. Uchiyama T, Honda S, Okayama T, Degawa T (2016) A feasibility study of power generation from sewage using a hollowed Pico-hydraulic turbine. Engineering 2(4):510–517
15. Gallego E, Rubio-Clemente A, Pineda J, Velásquez L (2020) Experimental analysis on the performance of a pico-hydro Turgo turbine
16. Ishola FA, Azeta J, Agbi G, Olatunji OO, Oyawale F (2019) Simulation for material selection for a Pico Pelton turbine's wheel and buckets. Procedia Manufacturing 35:1172–1177
17. Williamson SJ, Stark BH, Booker JD (2014) Low head pico hydro turbine selection using a multi-criteria analysis. Renewable Energy 61:43–50

18. Sarkar P, Sharma B, Malik U (2014) Energy generation from grey water in high raised buildings: The case of India
19. Saria MA, Badruzzamana M, Cherchia C, Swindleb M, Ajamic N, Jacangelo JG (2018) Recent innovations and trends in in-conduit hydro power technologies and their applications in water distribution systems
20. Falcão AFO, Henriques JCC (2014) Model-prototype similarity of oscillating-water-column wave energy converters
21. Elie Bertrand Kengne S (2017) Methodology feasibility studies of micro- hydro power plant in the cameroon: case of micro-hydro kemken
22. Chae K-J, Kim I-S, Ren X, Cheon K-H (2015) Reliable energy recovery in an existing municipal wastewater treatment plant with a flow-variable micro-hydro power system (2015)
23. Pali BS, Vadhera S (2018) A novel pumped hydro-energy storage scheme with wind energy for power generation at constant voltage in rural areas Renewable Energy 127:802–810
24. Dedic-Jandrek H, Nizetic S (2019) Small scale Archimedes hydro power plant test station: Design and experimental investigation
25. Roberts, Thomas B, Sewell P, Hoare E (2018) Generating renewable power from water hammer pressure surges
26. Leguizamón S, Avellan F (2020) Computational parametric analysis of the design of cross-flow turbines under constraints
27. Samora I, Hasmatuchi V, Münch- Alligné C, Franca MJ, Schleiss AJ, Ramos HM (2016) Experimental characterization of a five blade tubular propeller turbine for pipe inline installation
28. Jiyun D, Hongxing Y, Zhichen S, Xiaodong G (2018) Development of an inline vertical cross-flow turbine for hydro power harvesting in urban water supply pipes
29. Haidar AMA, Senan MFM, Noman A, Radman T (2012) Utilization of pico hydro generation in domestic and commercial loads. Renew Sustain Energy Rev 16(1):518–524
30. Rahbara K, Mahmouda S, Dadaha RKA, Moazamia N, Mirhadizadehb SA (2017) Development and experimental study of a small-scale compressed air radial inflow turbine for distributed power generation
31. Mbelek LN, Spork K (2019) Simulations and experimental validation of Pico conduit pressure hydro power systems with battery storage
32. Talluri L, Fiaschi D, Neri G, Ciappi L (2018) Design and optimization of a Tesla turbine for ORC applications
33. Ramos HM, Borga A, Simao M (2009) New design solutions for low-power energy production in water pipe systems
34. https://doi.org/10.1016/j.sbspro.2014.03.667

Numerical Analysis of Vapor Bubble Influence on the Flow and Temperature Field in Slug Flow Regime of Microchannel

Nirav Chaudhari, Nishant Shah, and Jyotirmay Banerjee

1 Introduction

Flow boiling heat transfer in microchannel has a high ability to dissipate heat transfer in a small area and is one of the most suitable mechanisms for thermal management of high-power microelectronics. However, it is not easy to implement in practical application due to the complex flow boiling convective heat transfer mechanism. Typically, in flow boiling, the microchannel is divided into three sections depending upon the flow pattern, single-phase subcooled section, multi-phase saturated section, and superheated dry-wall section. Different sections of microchannel have unique heat transfer mechanisms [2]. In particular, the slug (elongated bubble) flow regime has been established as an optimal operating condition for micro-heat exchangers, thanks to its efficient heat transfer mechanisms. Due to the small liquid film associated with vapor slug flow and flow circulation inside the slugs, the evaporative convective heat-mass transfer from the wall to the fluid increases significantly. As vapor bubbles grow and elongate, a large liquid–vapor interface generates, and it is responsible for higher heat transfer due to interfacial mass transfer from liquid to vapor and thin liquid film covered by vapor bubbles. Such type of flow gives remarkable thermal performance and the most efficient heat transfer mechanism.

Various experimental studies on flow boiling in a microchannel can be found in the literature, but they have limitations while explaining the dynamics of the boiling flow and the influence of the vapor slug on flow and thermal boundary layer. With the advancements in the numerics of multi-phase flow in the past decade, interface capturing methods can be helpful to predict and visualize the behavior of the vapor slug in a microchannel. The numerical approach to study the growth of vapor bubble was first introduced by Mukherjee and Kandlikar [7]. They used the level set method and presented an interface evaporation phase change model to study the growth of

N. Chaudhari · N. Shah · J. Banerjee (✉)
Department of Mechanical Engineering, SVNIT, Surat, India
e-mail: jbaner@med.svnit.ac.in

J. Banerjee et al. (eds.), *Recent Advances in Fluid Dynamics*, Lecture Notes in Mechanical Engineering, https://doi.org/10.1007/978-981-19-3379-0_38

vapor slug. Magnini et al. [6] presented multiple bubble simulations to capture the essential characteristics of the slug flow. Thermal and hydrodynamic interactions of leading and sequential bubbles have been seen in their work using Fluent. Ferrari et al. [3] used OpenFOAM libraries and presented the impact of liquid film thickness on the heat transfer performance of the microchannel. Using Fluent, Liu et al. [5] examined the transition between slug to annular flow under high heat flux conditions. Okajima and Stephan [8] studied the expansion process of a single vapor bubble in a microchannel using OpenFOAM libraries and presented heat flux variation within the different sections of the vapor bubble. Although multiple numerical studies for multi-phase flow in a microchannel, very few focus on evaporating bubbles and the related heat transfer performance of microchannel. The direct numerical simulations of flow boiling in a microchannel for slug flow regime are complicated even with the available commercial software. So, the present study aims to provide the influence of an evaporating vapor bubble on the flow and temperature field and an understanding of a mechanism that enhances the performance of the microchannel with accurate numerical simulations.

2 Numerical Method

Present work involves two fluids phases which are incompressible, immiscible, and Newtonian in nature, and we have adopted a method in which the problem is reformulated and solved considering both the phase as a single-fluid mixture. Single sets of governing equations are considered for both phases, which share unique velocity, pressure, and temperature. The volume of fluid (VOF) method is adopted in this work to differentiate two phases based on Rudman [10]. To identify each phase, a variable named volume fraction c is used, which is the ratio of the volume of primary fluid in the cell to the volume of the cell. So volume fraction 1 stands for primary fluid (liquid), 0 for secondary fluid (vapor) and $0 < c < 1$ is for interfacial cells. While considering the properties of the fluid like density, viscosity, and thermal conductivity as constant, the properties of the mixture are calculated with the help of volume fraction as shown in Eq. 1. φ stands for any thermophysical property.

$$\varphi = \varphi_{\text{liq}}(c) + \varphi_{\text{vap}}(1 - c) \tag{1}$$

Governing equations The governing equations of VOF transport, continuity equation, momentum conservation equation, and energy conservation equation for single-fluid mixture based solver can be written as

$$\frac{\partial c}{\partial t} + \nabla \cdot (\vec{u}c) = c\nabla \cdot \vec{u} + \frac{\dot{\rho}}{\rho} \tag{2}$$

$$\nabla \cdot \vec{u} = \frac{\dot{\rho}}{\rho} \tag{3}$$

$$\frac{\partial(\rho\vec{u})}{\partial t} + \nabla\cdot(\rho\vec{u}\vec{u}) = -\nabla p + \mu\nabla\vec{u} + \sigma K\nabla C \tag{4}$$

$$\frac{\partial\left(\rho c_p T\right)}{\partial t} + \nabla\cdot\left(\rho c_p T\vec{u}\right) = \nabla\cdot(k\nabla T) + \dot{\rho}_0\cdot h_{lv} + \dot{\rho}c_p T \tag{5}$$

where $\vec{u}$ is the velocity of the fluid, T the temperature of fluid, ρ the density of the mixture, μ the viscosity of the mixture, p the pressure, σ the surface tension force, K the curvature of the interface, C the mollified volume fraction, c_p the heat capacity of the fluid, and k conductivity of the fluid. $\dot{\rho}$ and $\dot{\rho}_0$ are source terms field due to the evaporation and will be explained in the next section.

The last term in the momentum conservation equation represents the source term due to the surface tension force. The formulation is adapted from the continuum surface force method proposed by Brackbill et al. [1]. In the surface tension formulation, interface curvature K is calculated with the convolution of force method, and mollified volume fraction C to reduce spurious currents in the lighter fluid is formulated with the 7×7 Gaussian smoothing kernel. In the energy equation formulation, the heat capacity is assumed the linear function of temperature only. The viscous heating is neglected. The problem involves small temperature jumps (of order $\Delta T = 5$ K), so all the fluid properties are assumed to be constant at the saturation temperature of the fluid.

Phase change model The phase change model estimates the liquid evaporation rate at the interface with the help of local temperature field and volume fraction data. The task of the interfacial phase change model can be divided into two sub-modules,

Calculation of the local evaporation rate: In this step, evaporating mass flux transferred through a liquid–vapor interface is calculated with the help of a local temperature field. With this step complete, the evaporating mass flux is known in each cell that contains the interface.

Calculations of source terms for the conservation equations: The evaporating mass flux calculated in the previous step has to be incorporated into the conservation equations. The stability of the numerical procedure will strongly depend upon this step.

Based on the kinetic theory of gases [11] obtained the relationship between local evaporation rate and local temperature field for microscale boiling phenomena. For a small temperature and pressure jump across the interface, the interfacial mass flux can be written as

$$\dot{m} = \frac{2\gamma}{2-\gamma}\frac{1}{\sqrt{2\pi R}}\left(\frac{P_{\text{vap}}}{\sqrt{T_{\text{vap}}}} - \frac{P_{\text{liq}}}{\sqrt{T_{\text{liq}}}}\right) \tag{6}$$

Later on, Tanasawa [12] simplified this relationship with an assumption that for low superheating temperature values above the saturation temperature, this local interfacial mass flux depends linearly on the local temperature jump across the interface. The simplified equation can be written as

$$\dot{m} = \frac{2\gamma}{2-\gamma}\left(\frac{M}{2\pi R_g}\right)^2 \frac{\rho_{\text{vap}}\, h_{lv}}{T_{\text{vap}}^{3/2}}\left(T_{\text{sat}} - T_{\text{vap}}\right) \tag{7}$$

where M is the molecular mass of fluid, R_g is the gas constant for the fluid, γ is the evaporation coefficient or accommodation coefficient, and h_{lv} is the latent heat of evaporation. The present work involves such a small temperature jump and assumes the vapor phase will be at the saturation temperature all the time, so the above equation can be applied to calculate the interphase mass flux per unit interface area as

$$\dot{m} = \emptyset(T - T_{\text{sat}}) \tag{8}$$

$$\emptyset = \frac{2\gamma}{2-\gamma}\left(\frac{M}{2\pi R_g}\right)^2 \frac{\rho_v\, h_{lv}}{T_v^{3/2}} \tag{9}$$

where $\emptyset$ is kinetic mobility and will be constant throughout the simulation. In some literature, the inverse of $\emptyset$ is used, which is term as interfacial mass transfer resistance. The accommodation coefficient or evaporation coefficient γ varies from 0.001 to 1 according to the situation and type of fluid used. Based on some popular literature, the value of this constant is taken as 1 in this work.

With the calculation of local interfacial mass flux, for each interfacial cell superheated above the saturation temperature, an amount of mass disappears from the nearest liquid cells, and the same amount of mass reappears in the nearest vapor cells to complete the evaporation process. The amount of mass that transfers from each superheated cell is $\dot{m}\delta_s$. So, the local evaporation rate is given by

$$\dot{\rho} = \dot{m}\delta_s = \emptyset(T - T_{\text{sat}})\frac{\left|\overrightarrow{S_{\text{int}}}\right|}{V_{\text{cell}}} \tag{10}$$

where $\left|\overrightarrow{S_{\text{int}}}\right|$ is the area of the interface and V_{cell} is the volume of the cell. The term δ_s is the interface density, representing the surface area of the interface in the cell relative to the volume of the cell.

Theoretically, the source term presented in the Eqs. (2), (4), and (5) can be implemented in the governing equations directly. However, that results in a dense source term field resulting near the interface in both the phases. Due to the discontinues and sensitive nature of the source term field, this may result in numerical instability if the evaporation rate is high. Hence, following the procedure of Hardt and Wondra [4], the sharp source term field is first made smooth and continuous across the interface cells with the help of a simple diffusion equation, and then this smooth source term field is redistributed in the liquid and vapor phase such that the same amount mass that disappears from the liquid phase reappears in the vapor phase. So to remove the instability arising from the addition of the source term and to ensure the global evaporation rate, this method of Hardt and Wondra [4] is used in this work.

3 Validation

In this section, implemented two-phase non-isothermal flow solver with phase change is tested against the benchmark test problem to confirm the validity of the solver.

The growth of spherical bubble This test case is selected in order to validate the phase change model. This test case is widely used to validate different evaporation models in the literature. According to the study of Plesset and Zwick [9], the growth of the vapor bubble in a superheated liquid reaches an optimum state when the growth of the vapor bubble is only controlled by the heat transfer at the liquid–vapor interface. The vapor bubble will be large enough to assume that the vapor temperature equals to the saturation temperature. With this assumption and neglecting the effect of viscous force, Scriven derived the analytical solution for the bubble radius with respect to time

$$R(t) = 2\beta\sqrt{\alpha_{\text{liq}} t} \tag{11}$$

where β is growth constant, as mentioned in Plesset and Zwick [9] and α_{liq} is thermal diffusivity of the liquid.

The spherical bubble of the radius 0.1 mm is initialized in the 2D square domain of size 0.8 mm. The surrounding liquid is superheated above the saturation temperature of the fluid. The test case is symmetric in both x and y directions, so only a quarter of the domain is considered for the simulation. Three different uniform mesh sizes are used, 200×200 ($\Delta x = 4$ μm), 400×400 ($\Delta x = 2$ μm), and 800×800 ($\Delta x = 1$ μm). The fluid refrigerant is HFE-7100, and the properties of the fluid are taken at the saturation temperature. Two different levels of superheating were applied, 5 K and 10 K. At the starting point, the bubble radius is 0.1 mm. The initial temperature profile is shifted to a radius of 0.11 mm for the energy equation's stability and faster convergence rate. This allows the solver to create a proper thermal boundary layer on the liquid side of the bubble before the heat transfer-controlled bubble growth stage.

Figure 1 depicts the growth of the bubble radius over time for various grid sizes and superheating levels. The numerical model nicely predicts the growth. The result converges to the analytical solution as the grid becomes finer. The mesh size of 800×800 ($\Delta \text{x} = 1$ μm) shows an error in the bubble radius at the end of the simulation less than 2% for both cases. Even with the coarsest grid, the error in the final bubble radius at the end of the simulation is less than 6%, as shown in Table 1.

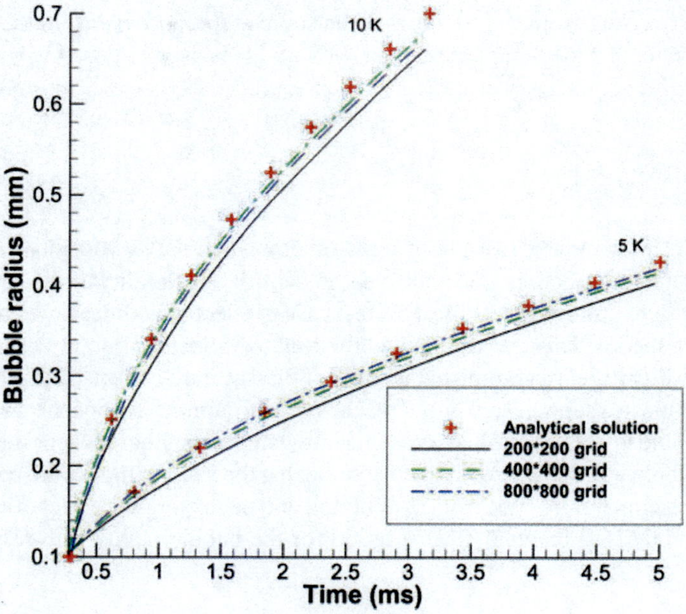

Fig. 1 Validation results for growth of spherical bubble test

Table 1 Error in radius of the vapor bubble at the end of the simulation

	% error
5 K superheating	
200 × 200 ($\Delta x = 4$ μm)	5.49
400 × 400 ($\Delta x = 2$ μm)	2.75
800 × 800 ($\Delta x = 1$ μm)	1.83
10 K superheating	
200 × 200 ($\Delta x = 4$ μm)	5.43
400 × 400 ($\Delta x = 2$ μm)	3.82
800 × 800 ($\Delta x = 1$ μm)	1.99

4 Results and Discussion

With the help of developed in-house code, first, the results are collected for different microchannel flow configurations to establish the superiority of the boiling flow over other flow configurations. Then, the contour plots of velocity and temperature field induced by the vapor bubble in a flow boiling condition are analyzed and explained to justify the superiority of the boiling flow.

Heat transfer analysis of different flow configurations The 2 mm length of the microchannel is divided into 1 mm of adiabatic section and 1 mm of heated section. The adiabatic section is required to establish a developed flow and thermal boundary layer in the heated section before the vapor bubble reaches the heated section. The width of the microchannel is 0.1 mm. A fine uniform mesh of a size 1.5 micron is selected; it is sufficient to capture the interface evaporation phenomena and its effect on the heat transfer rate. The inlet mass flux is 900 kg/m^2 s, which is equivalent to the Reynolds number of 180. Pressure outlet boundary condition is applied at the outlet section, and both side walls are at no-slip boundary condition. The heated section walls are superheated to 4 degrees above the saturation temperature, and fluid inside the domain and at the inlet is at the saturation temperature. The initial elongated vapor bubble is 90 μ m in diameters and has a length of 200 μm. The superiority of the boiling flow is demonstrated with a comparison of single-phase flow, subcooled two-phase flow, and boiling flow under the same simulation conditions.

Figure 2 shows temperature contours with wall heat flux variation along with the length of the microchannel. The interface between the liquid and vapor phase is represented as a thick red line in the contour plots. All of the figures are at the same time step of 1.85 ms to compare the local heat transfer performance of the microchannel.

As depicted from the first figure, as the thermal boundary is developed with time in the case of single-phase flow, the microchannel delivers a constant amount of heat flux. The average value of the local wall heat flux in this case is 5 KW/m^2. However, when we compare the same microchannel section with the subcooled two-phase flow, it gives a wall heat flux of 20 KW/m^2. The main reason behind this improvement is the generation of thin liquid film between the vapor bubble and microchannel wall.

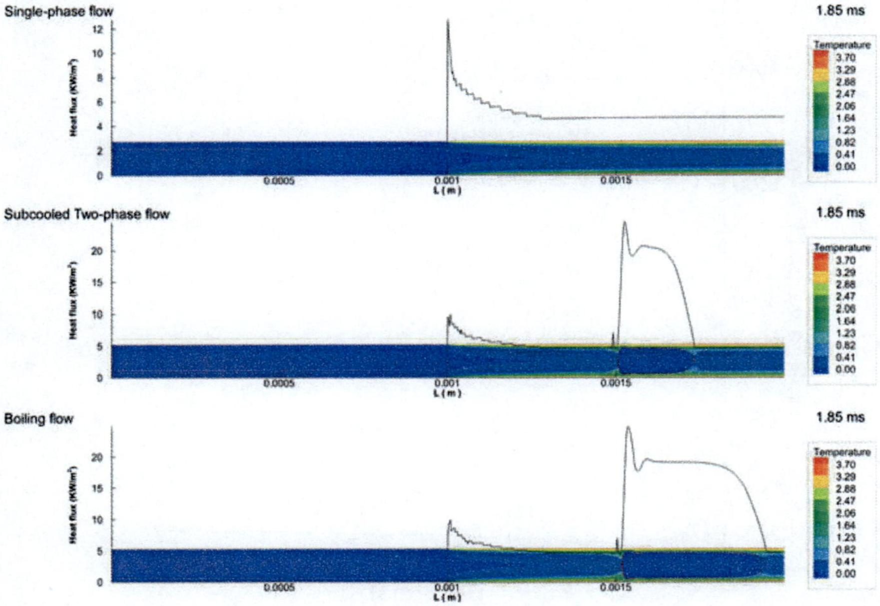

Fig. 2 Temperature contour plot and local wall heat flux variation along the length of the microchannel for different flow configurations

The vapor bubble squeezes the thermal boundary layer near the microchannel wall, which results in a dense temperature profile. So the thinner the liquid film higher will be the heat transfer rate. As depicted from Fig. 2, as we move to the head of the vapor bubble, the thickness of this liquid film gradually increases, and with that value of local wall heat flux gradually decreases. In the case of boiling flow, the volume of the vapor bubble increases due to the evaporation process. This increment in the vapor content results in longitudinal elongation of the vapor bubble so the long vapor bubble provides increased length in the liquid film. The length of the liquid film in the case of subcooled two-phase flow is 0.12 mm, while it is 0.3 mm in the case of boiling flow. So the elevated wall heat flux of 20 KW/m^2 is delivered for a longer length of the microchannel section.

The graph in Fig. 3 shows the total wall heat flux as the simulation progresses for all three microchannel configurations. The graph represents that two-phase flow gives higher heat transfer compared to single-phase flow due to the formation of thin liquid film surrounding the bubble. Boiling flow shows further improvement in heat transfer rate mainly due to two reasons; first, the energy consumed by the evaporation process; second, the elongated liquid film due to the evaporation. As a result, the two-phase flow increases the heat transfer rate up to 50–60% compared to single-phase flow, and the boiling flow further increases this value up to 30–40%.

The vapor bubble induced flow and temperature field analysis The main objective behind this analysis is to relate the flow and temperature field to the dynamics of the evaporating bubble and explain them in reference to heat transfer rate. With

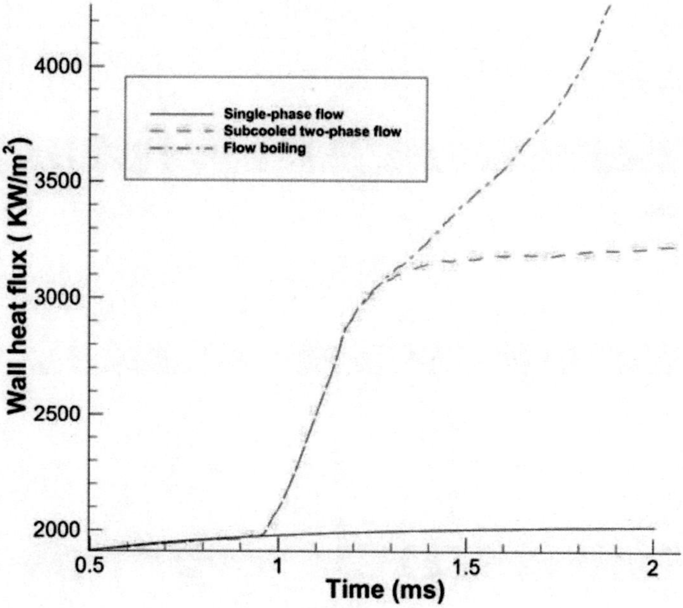

Fig. 3 Comparison of total wall heat flux for different flow configurations

the help of boiling flow simulation, the flow is captured at the time of 1.8 ms. At this instant, the velocity of the head of the bubble is 1.3 m/s, and the tail of the bubble moves with 0.65 m/s. The axial location of the head of the bubble is 1.87 mm, and the axial location of the tail is 1.48 mm. The figures presented in this section are isolines of temperature and velocity field where the vapor bubble is stationed.

The first section is the wake region which is the upstream section of the microchannel from the tail of the vapor bubble. As seen in Fig. 4, the velocity isolines are parallel with the microchannel length up to 1.45 mm. We can see a slight disturbance in the velocity field between 1.45 mm and the vapor bubble tail. In the same way, the temperature isolines are also parallel up to a length of 1.45 mm and show slight disturbance after that. The vapor bubble has just passed from this region, squeezing the thermal boundary layer, and disturbing the flow field. The velocity field and temperature field try to reach steady-state situation holding before the bubble caused the disturbance. Therefore, the local wall heat transfer rate is slightly enhanced by the transient heat convection mechanism in this region. This can be depicted from the wall heat flux graph Fig. 5; the wall heat transfer rate in flow boiling is slightly higher than the heat transfer in single-phase flow.

At 1.47 mm, near the upstream of the bubble tail, the bubble disturbance on the liquid flow becomes remarkable. The liquid velocity at the centerline remains the same as the single-phase case, but the liquid velocity near the wall increases due to the motion of the vapor bubble tail. This prevents the development of the thermal boundary layer. Furthermore, due to the flow field concentrated at the centerline, the temperature field shows a significant spread at the bubble tail ($x = 1.48$ mm). Because of this temperature spread, the local wall heat flux shows a slight drop, as seen in the wall heat flux graph at $x = 1.48$ mm. Coming to the bubble's tail ($x = 1.48$ mm to 1.51 mm), the wavy profile at the bubble's tail generates liquid recirculation. The generated vortices on the bubble interface end up at the bubble tail. This kind of flow field leads to a small rise in temperature near the microchannel wall. The small spike

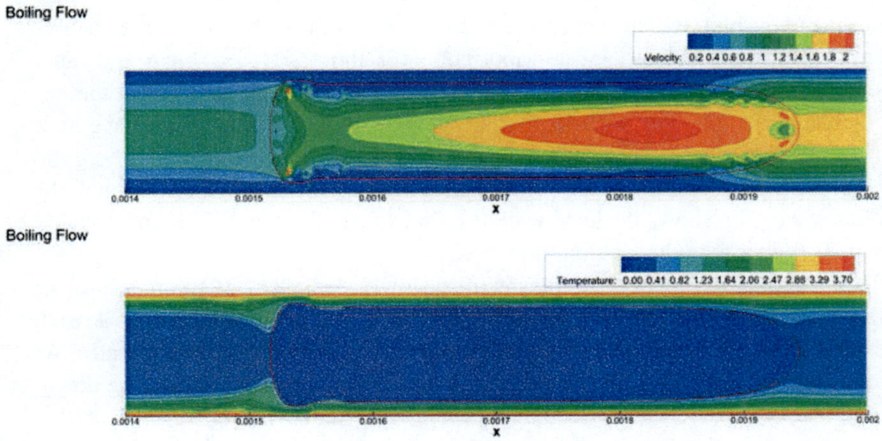

Fig. 4 Velocity and temperature isolines plot for demonstration of vapor bubble influence

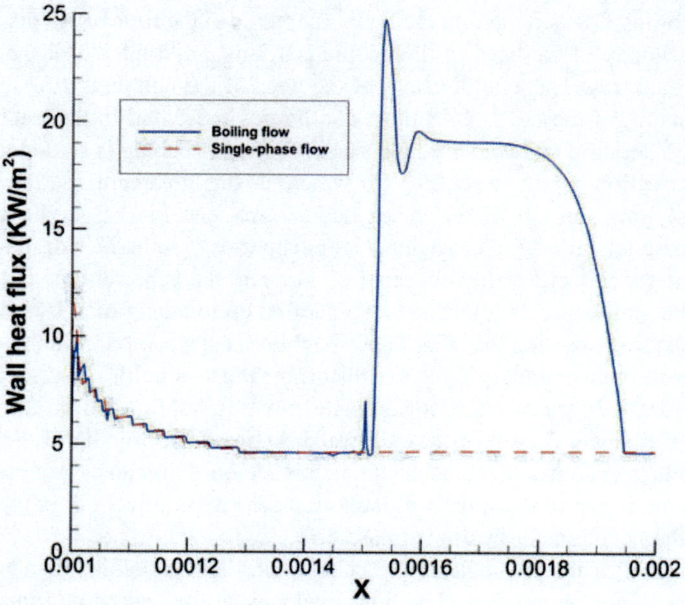

Fig. 5 Comparison of local wall heat flux for single-phase flow and boiling flow

in the local wall heat flux Fig. 5 at $x = 1.49$ mm is a result of that. Additionally, this kind of flow pattern thicken the thermal boundary layer behind the bubble; plus, the bump squeezes the thermal boundary layer to the minimum value. The major spike depicted in the local wall heat flux at $x = 1.50$ mm results from a reduction in the thermal boundary layer.

In the liquid film region at the middle of the bubble ($x = 1.52$ mm to $x = 1.72$ mm), the velocity contours show almost nil velocity in the liquid phase (depicted from Fig. 4). However, the vapor bubble shows an increase in velocity as we move to the centerline of the bubble. The velocity reduces as we move from bubble front to bubble end. The maximum velocity in the vapor bubble is about 2.1 m/s. Due to the vapor bubble, the thermal boundary layer is squeezed, and temperature isolines parallel the microchannel length. The vortices that are generated on the bubble interface have no significant effect on the temperature field. These vortices divert their energy to the centerline, and as a result, the velocity keeps increasing inside the vapor bubble. This bubble region shows the local wall heat flux value of about 20 KW/m^2, roughly five times higher than the single-phase flow.

This value of the local heat transfer rate slowly decreases as we move to the bubble front. The main reason is an increase in the liquid film thickness. As a result of evaporation, the liquid between the bubble and the channel wall accelerates. As the evaporation progresses, the liquid in front of the bubble is forced to move downstream, and as a result, the velocity field develops, as shown in Fig. 4. Compared to single-phase flow, the velocity at the centerline is two times higher (around 1.6 m/s). However, the local wall heat transfer rate does not show any significant changes. The thermal boundary layer formed in front of the vapor bubble closely resembles the same as a single-phase flow. So, the wall heat transfer rate in this region is the same as single-phase flow, as depicted in Fig. 5.

5 Conclusion

We have developed and tested a two-phase flow solver for immiscible, incompressible, and non-isothermal fluid undergoing the evaporation process. The developed solver enables the study of local heat flux variation for different sections of the microchannel and how the flow and temperature field changes with the evaporation. Due to the small liquid film associated with vapor slug and, an increase in the liquid film through interfacial mass transfer improves the performance of the microchannel up to 50–60% compared to single-phase flow. The detailed analysis of the temperature and flow field suggests that the main reason behind this improvement is film evaporation. However, the wake region also shows significant improvement in performance due to transient heat convection current generated by disturbance provided by the vapor bubble.

References

1. Brackbill JU, Kothe DB, Zemach C (1992) A continuum method for modeling surface tension. J Comp Physics 6(100):335–354
2. Consolini L, Thome JR (2010) A heat transfer model for evaporation of coalescing bubbles in micro-channel flow. Int J Heat Fluid Flow 2(31):115–125
3. Ferrari A, Magnini M, Thome JR (2018) Numerical analysis of slug flow boiling in square microchannels. Int J Heat Mass Transfer 8(123):928–944
4. Hardt S, Wondra F (2008) Evaporation model for interfacial flows based on a continuum-field representation of the source terms. J Comp Physics 5(227):5871–5895
5. Liu Q, Wang W, Palm B (2017) A numerical study of the transition from slug to annular flow in micro-channel convective boiling. Appl Thermal Eng 2(112):73–81
6. Magnini M, Pulvirenti B, Thome JR (2013) Numerical investigation of the influence of leading and sequential bubbles on slug flow boiling within a microchannel. Int J Thermal Sciences 9(71):36–52
7. Mukherjee A, Kandlikar SG (2005) Numerical simulation of growth of a vapor bubble during flow boiling of water in a microchannel. Microfluidics Nanofluidics 3(1):137–145
8. Okajima J, Stephan P (2019) Numerical simulation of liquid film formation and its heat transfer through vapor bubble expansion in a microchannel. Int J Heat Mass Trans 6(136):1241–1249
9. Plesset MS, Zwick SA (1954) The growth of vapor bubbles in superheated liquids. J Appl Phy 4(25):493–500
10. Rudman M (1997) Volume-tracking methods for interfacial flow calculation. Int J Numer Meth Fluids 24:671–691
11. Schrage RW (1953) A theoretical study of interphase mass transfer. Columbia University Press, s.l.
12. Tanasawa I (1991) Advances in condensation heat transfer. In: Advances in Heat Transfer Volume 21. s.l.:Elsevier, pp 55–139

Comparison Between Ultra-High-Temperature Thermal Battery and Li-Ion Battery

Alok Kumar Ray, Sagar Vashisht, Jibin M. Joy, and Dibakar Rakshit

Nomenclature

LHS	Latent heat storage
Li-ion	Lithium ion
PCM	Phase change medium
SHS	Sensible heat storage
TCHS	Thermochemical heat storage
TES	Thermal energy storage
T_{ref}, ρ_{ref}	Reference temperature and density

Greek Symbols

δ	Melting fraction
γ	Mushy zone constant
β	Coefficient of thermal expansion
μ	Dynamic viscosity

1 Introduction

Increase in environmental pollution due to fossil fuels, growing energy demand due to urbanization and depletion of fuel reserves around the world have shifted attention

A. K. Ray · S. Vashisht · J. M. Joy · D. Rakshit (✉)
Department of Energy Science and Engineering, IIT Delhi, New Delhi 110016, India
e-mail: dibakar@iitd.ac.in

J. Banerjee et al. (eds.), *Recent Advances in Fluid Dynamics*, Lecture Notes in Mechanical Engineering, https://doi.org/10.1007/978-981-19-3379-0_39

of researchers towards renewable energy sources [1]. However, the intermittency of renewable sources (solar and wind) stands as a major bottleneck for their wide-scale utilization. In this context, energy storage technology can not only address the intermittency of solar and wind power but also enhance the grid stability. Moreover, the storage technology can increase the efficiency of the energy system which eventually reduce the cost of power generation [2]. The energy system used in grid applications plays a significant role in providing electrical energy to different large- and small-scale systems. Electrical energy consumption fluctuates on a daily, seasonal and annual basis. Hence, vacant energy and power generated during peak load must be stored for peak shaving and load levelling.

There are different energy storage systems used for renewable energy as shown in Fig. 1. (1) mechanical, (2) chemical, (3) electrochemical, (4) electromagnetic and (5) thermal energy storage [3, 4]. Electrochemical storage technology involves storage of high-grade electrical energy in chemical form using a reaction in between cathode and anode. There are many different types of batteries based on this technology. Li-ion battery is one of the most commercialized storage methods in market.

In the early 1990s, Sony had been commercially launched lithium-ion batteries [5]. Till now, the development of lithium-ion batteries has dominated the battery technology market. The lithium-ion battery has basic three main working components: anode, cathode and separator. During the discharging process, the lithium

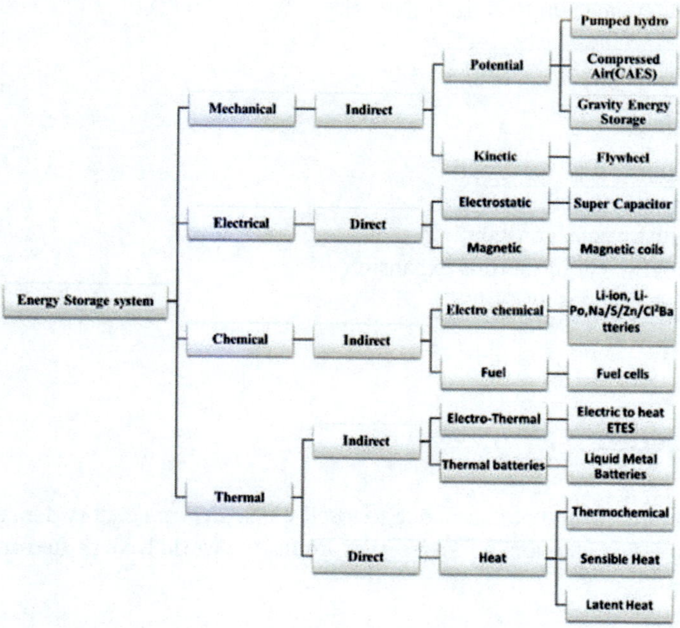

Fig. 1 Classification of energy storage system showing energy conversion method before storing energy

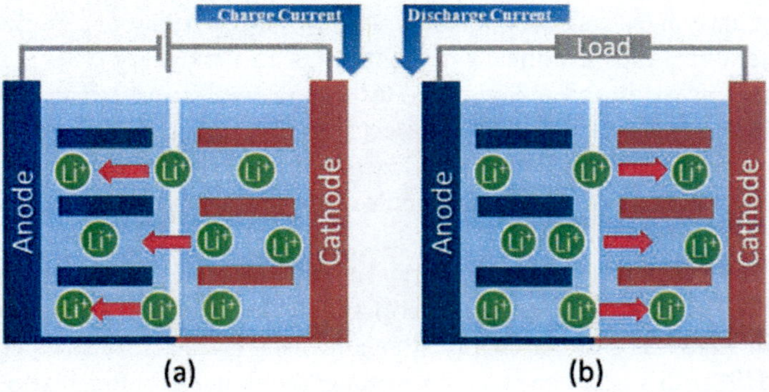

Fig. 2 Charging and discharging of battery

ion moves from anode to cathode layer via separator and this process can also be called oxidation and reduction at the two electrodes. While during charging process, the lithium ion moves from cathode to anode layer via separator. The movement of electron is in the outside circuit and movement of lithium ion is inside the cell. The charging and discharging process is shown in Fig. 2.

In general, the material used for the anode layer is graphite due to its relatively lower price and easy accessibility of carbon. Furthermore, this is commercially used in the lithium-ion battery due to its stability to hold the lithium ions. On the other hand, the cathode layer is made up of lithium metal oxides such as $LiMn_2O_4$, $LiCoO_2$ and $LiFePO_4$. Most Li-ion battery names are based on their cathode chemistry, which is the crucial component of the battery's performance [5].

One of the most excellent solutions for storing energy for a short period with high efficiency is to store electrical energy in electrochemical batteries [6]. The electrochemical storage, i.e. battery must satisfy the complex and large-scale application of energy storage. Hence, capacity, power, energy densities, energy efficiency and lifetime are the requirement for an energy storage application. Lithium-ion batteries dominate the present interest storage application among the various secondary battery technologies available due to their high energy density (up to 200 Wh/kg), increased energy efficiency (>95%) and long cycle life (3000 cycles at a 80% depth of discharge) [7–9].

Lithium-ion batteries are currently used in 77% of running electrical power storage systems in the United States. In recent years, the performance of lithium-ion batteries has successfully developed, and a striking improvement in energy density and performance [5, 10–12]. Li-ion batteries are an attractive candidate for merging with renewable energy sources in grid-level energy storage systems due to their high energy density. The generated electrical energy is stored in lithium-ion batteries and can be used later at a low cost when the end-users demand. Lithium-ion batteries are a viable energy storage solution. However, the several technical challenges associated with lithium-ion batteries such as degradation of capacity with time, heat generation inside battery, the dependence on critical material, high cost, short life span, underperformance in extreme temperatures, safety issues, toxicity of chemicals and recycling after usage are the critical shortcomings of Li-ion battery [13]

TES technology can capture low-grade thermal energy from various renewable sources, which can generate both heat and electricity [14]. Direct solar energy, excess energy from wind and PV power plants and industrial waste heat are potential energy sources. As a result, the TES system can handle renewable energy intermittency while also providing dispatchability [15].

There are three methods for storing thermal energy in a medium. (1) Sensible heat storage (SHS), (2) latent heat storage (LHS) and (3) thermochemical heat storage (TCHS) [16]. LHS has a larger energy storage density than SHS and more maturity than TCHS. LHS system involves state change (melting/solidification) of phase change medium (PCM) nearly at a constant temperature. LHS can store a significant amount of energy in both sensible and latent form. However, the poor thermal conductivity of traditional PCMs decreases the charging (melting) and discharging (solidification) rates of LHS [17]. Metallic PCM like silicon can address this limitation of small storage density of existing high-temperature inorganic salts and conventional Li-ion battery [18, 19]. Hence, metallic PCM-based LHS can be depicted as a thermal battery which can perform like Li-ion battery.

Storage of heat at high- temperature has more exergy content than storage at low-temperature according to second law of thermodynamics. Moreover, high temperatures (>900 °C) can be a viable solution for the co-generation of heat and electricity from storage systems [20]. Furthermore, advanced power cycles such as the supercritical CO_2 cycle, which operates at temperatures higher than standard Rankine cycles, can minimize the cost of power generation from concentrated solar power (CSP). [21]. Hence, a high-temperature thermal battery can be a potential alternative to overcome the challenges of Li-ion battery.

From the preceding literature survey, it can be concluded that there is no quantitative comparison between thermal battery with Li-ion battery even though thermal battery has several benefits over Li-ion battery theoretically. The objective of the present study is to illustrate the merits of thermal battery qualitatively and compare the thermal and economic performance of a specific Li-ion cell with high-temperature thermal battery using silicon as PCM.

2 Problem Description

2.1 *Physical and Computational Domain*

Thermal battery presented in this article is a single tube single pass high-temperature shell and tube heat exchanger. Shell and tube heat exchanger is selected to compare with a cylindrical Li-ion battery with accuracy. The Li-ion battery selected for the comparison is LG 18650HG2. Moreover, shell and tube LHS have the highest efficiency to volume ratio. The high-temperature PCM (silicon) is present inside the

Fig. 3 Physical domain of two batteries. **a** Thermal battery and **b** Li-ion battery

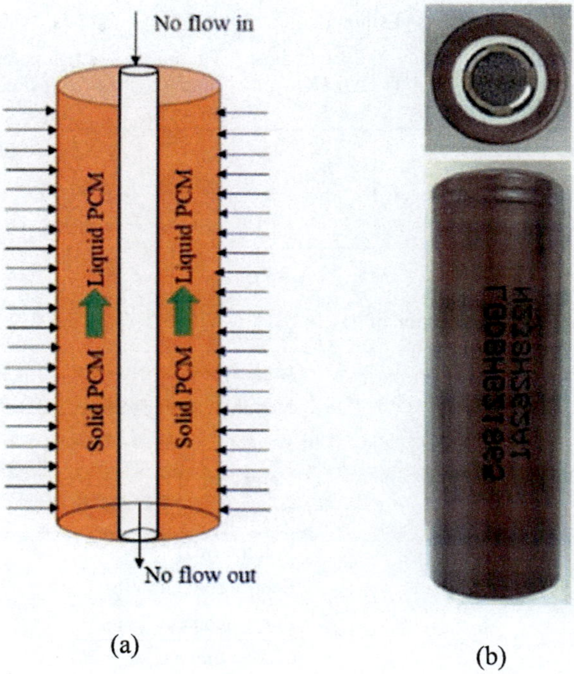

annulus, and uniform heat flux is given on the outer wall of shell. The charging of the thermal battery is performed with uniform heat flux boundary condition which can be analogous to constant current charging of selected Li-ion battery. The height of Li-ion battery is 65 mm and diameter is 18 mm. The comparison has been performed maintaining same storage capacity for both batteries. The storage capacity for LG 18650HG2 is selected as the reference. Physical domain of two batteries for the charging process is illustrated in Fig. 3.

The 3D computational domain of the thermal battery has 140 mm length with shell and tube diameter 28 mm and 26.7 mm, respectively. The charging has been performed by providing uniform heat flux to the outer wall of shell. The lithium-ion cell's 3D model was created using the actual dimensions of a cell. The dimensions of the cell are 18 mm in diameter and 65 mm in height. The nominal capacity and nominal voltage of the cell LG 18650HG2 are 3.0 Ah and 3.60 V, respectively. The thermophysical properties of the silicon and specification of lithium-ion battery are given in Tables 1 and 2, respectively.

The cell's geometry was divided into three sections: the active, the positive and the negative tab zone. The external thin layer is not evaluated independently, but its thermal conductivity is taken into account when determining the active zone's overall thermal conductivity.

The thermal behaviour of the battery cell is analysed by creating the mesh of the geometry. For the evaluation of heat generation inside the battery cell, the energy and MSMD models were activated. At 25 °C, the charging behaviour of the battery cell

Table 1 Properties of silicon

Properties	Density (kg/m^3)	Specific heat (J/kgK)	Thermal conductivity (W/mK)	Latent heat of fusion (J/kg)	Melting temperature (K)	Viscosity (*Pa.s*)
Silicon	$\rho_s =$ 2330 $\rho_l =$ 2570	$c_{p,s} = c_{p,l}$ = 1040	$k_s = 25$ $k_l = 50$	1,800,000	$T_s = 1686$ $T_l = 1688$	$\mu =$ 0.0008

Table 2 Specification of lithium-ion battery

Chemistry	NMC ($LiNiMnCoO_2$)
Capacity (Ah)	3
Nominal voltage (V)	3.6
Weight (g)	48
Height (mm)	65.2
Diameter (mm)	18.5
Volume (m^3)	1.75259×10^{-05}
Energy content (Wh)	10.8
Energy content (kJ)	38.88
Charge time (1.5A @CC-CV)	165 min
Specific energy (Gravimetric) Wh/kg	225
Energy density (Volumetric) Wh/ltr	616.22

was studied at three different C rates (0.5C, 1C and 1.33C). The lowest and highest cut-off voltages were 2.5 V and 4.2 V, respectively. For the anode and cathode tab material, copper and aluminium were selected, respectively. The properties for active zone material used in this case are density is 1703.5 kg/m^3, specific heat is 962.72 J/Kg. K and radial, tangential, axial thermal conductivity is 3.6, 25.84, 25.84 W/m.K., respectively. The UDS diffusivity is used for the active zone material, and the value for uds0 and uds1 are 1,190,000 S/m and 983,000 S/m [22, 23]. The heat transfer coefficient was set to 5 W/m^{2}K, and the free stream temperature was 298 K.

2.2 *Numerical Formulation*

The charging of thermal battery is modelled using finite volume based fixed grid enthalpy-porosity technique with help of Ansys Fluent. During melting, the interface is represented as a mushy zone that separates the solid and liquid phases. The mushy zone is considered a pseudo-porous region with porosity ranging from 0 to 1. Zero signifies solid phase, and one signifies liquid phase. The continuity, momentum and energy equations for the model are expressed considering the following assumptions:

1. The theory assumes the molten PCM to be Newtonian and incompressible in nature.
2. No consideration is given to the volume change that occurs during melting.
3. For both solid and liquid phases, the thermophysical characteristics are considered uniform.
4. The buoyancy force is modelled using the Boussinesq approximation.
5. Viscous heating and radiation loss are neglected.

The energy conservation equation is represented in terms of volumetric enthalpy and temperature.

$$\frac{\partial(\rho H)}{\partial t} + \nabla \cdot (\rho v H) = \nabla \cdot (k \nabla T) + \mathrm{S} \tag{1}$$

$$H = h + \delta\, \mathrm{h_{sl}} \tag{2}$$

$$h{=}h_{\text{ref}} + \int_{T_{\text{ref}}}^{T} c_p dT \tag{3}$$

where H is total enthalpy, h is sensible enthalpy, δ is melting fraction, S is energy dissipation, h_{ref} is sensible enthalpy at T_{ref} and h_{sl} is latent heat of fusion.

$$\delta = \begin{cases} 0 & T < T_{\text{sol}} \\ \dfrac{T - T_{\text{sol}}}{T_{\text{Liq}} - T_{\text{sol}}} & T_{sol} \leq T \leq T_{\text{Liq}} \\ 1 & T > T_{\text{Liq}} \end{cases} \tag{4}$$

where T_{sol} = Solidus temperature of PCM
T_{liq} = Liquidus temperature of PCM

The momentum equation to include natural convection can be expressed as

$$\frac{\partial(\rho v)}{\partial t} + \nabla \cdot (\rho v v) = -\nabla p + \nabla.(\mu \nabla v) + \rho g + M v \tag{5}$$

$$M = \frac{(1-\delta)^2}{\delta^3 + \varepsilon} \gamma \tag{6}$$

where M = porosity function and γ = Mushy zone constant which reflects the morphology of mushy zone. ε is a constant having small value (0.001) to avoid the division by zero.

The fluid density is assumed to be constant in the momentum equation using the Boussinesq approximation, except for the buoyancy force, which would induce natural convection. This body force is modelled using reference density (ρ_{ref}) and reference temperature (T_{ref}). Hence, the momentum equation can be expressed as

$$\frac{\partial(\rho_{\text{ref}}v)}{\partial t} + \nabla\cdot(\rho_{\text{ref}}vv) = -\nabla p + \nabla\cdot(\mu\nabla v) + (\rho-\rho_{\text{ref}})g + \frac{(1-\delta)^2}{\delta^3+\varepsilon}Cv \quad (7)$$

$$(\rho-\rho_{\text{ref}})g = -\rho_{\text{ref}}\beta(T-T_{\text{ref}}) \quad (8)$$

β is the coefficient of thermal expansion of molten PCM.

The continuity equation can be expressed as

$$\frac{\partial(\rho u)}{\partial x} + \frac{\partial(\rho v)}{\partial y} = 0 \quad (9)$$

Similarly, for lithium-ion battery Ansys Fluent package is used for the analysis of the thermal and electrochemical behaviour of a battery cell. The Ansys Fluent battery model is used to carry out the thermal analysis of a battery cell as the battery cell produces heat while operating. The heat generation rate is calculated using a coupled thermal-electrochemical simulation. The governing physics inside the battery cell is the lithium-ion transfer happening inside the anode-separator-cathode sandwich layers. Ansys Fluent uses multi-scale multi-domain (MSMD) approach to model the different physics involved in the multi-solution domain or the active zone (anode-separator-cathode sandwich layers). In this approach, the battery thermal and electrical field in the CFD domain is solved using the following differential equations [24]:

$$\frac{\partial\rho C_p T}{\partial t} - \nabla\cdot(k\nabla T) = \dot{q} \quad (10)$$

$$\dot{q} = \sigma_+|\nabla\varphi_+|^2 + \sigma_-|\nabla\varphi_-|^2 + \dot{q}_{\text{ECh}}$$

$$\dot{q}_{ECh} = j_{ECh}\left[U - (\varphi_+ - \varphi_-) - T\frac{dU}{dT}\right]$$

$$\nabla\cdot(\sigma_+\nabla\varphi_+) = -j_{ECh} \quad (11)$$

$$\nabla\cdot(\sigma_-\nabla\varphi_-) = j_{ECh} \quad (12)$$

where T, σ and φ are the temperature, effective electric conductivities for the electrodes and phase potential for the electrodes respectively, + and − denotes the positive and negative electrodes respectively, $\dot{q}$ is the heat generation rate during the battery operation, j_{ECh} and $\dot{q}_{ECh}$ are the volumetric current transfer rate and the electrochemical reaction heat due to electrochemical reactions, respectively, C_p is the heat capacity, and U is the open-circuit voltage of the battery.

The equations to obtain the source terms, j_{ECh} and $\dot{q}_{ECh}$ depends on the sub-model adopted, i.e. NTGK model for this case. The default parameters given in the Ansys are used for the NTGK model.

Table 3 Thermal-electrical analogy

Thermal battery	Heat (Joule)	Heat flux (W/m^2)	Temperature (K)	Thermal capacity (mc)
Li-ion	Charge (Coulomb)	Current (Ampere)	Voltage (Volt)	Capacitance (Faraday)

2.3 *Analogy Between Thermal and Electrical Parameters*

The selected Li-ion cell has electrical energy storage capacity 38.8 kJ found from technical specification sheet. Corresponding thermal storage capacity of thermal battery has been evaluated considering power conversion efficiency of supercritical CO_2 Brayton cycle at melting point of silicon. The equivalent thermal energy storage capacity is 64.8 kJ. Considering the latent heat of fusion as 1800 kJ/kg and ambient temperature 300 K, the mass and volume of PCM required for thermal battery is 0.02 kg and 0.0078 ltr, respectively. The analogy between different parameters of thermal and Li-ion batteries is established in Table 3.

3 Results and Discussion

The thermal battery of the same energy capacity as that of lithium-ion battery was charged by providing four different heat fluxes. The uniform heat flux used to charge the thermal battery is analogous to the constant current charging in lithium-ion battery. The charging time for thermal battery decreases as the constant flux increases. Table 4 shows the charging time of thermal batteries with different energy fluxes which establishes significantly faster rate of charging than Li-ion cell even for lowest heat flux magnitude. The gravimetric and volumetric energy storage density of thermal battery is also compared with the given Li-ion cell in Table 5.

The volume average temperature and liquid fraction of PCM domain has been evaluated during charging of the thermal battery subjected to uniform heat flux. The temperature increases linearly due to sensible heating and remains constant during phase as evident from Fig. 4. Figure 4 reports the temperature variation till

Table 4 Charging time of thermal battery and Li-ion cell

Thermal battery		Li-ion cell
Flux (W/m^2)	Charging time (minutes)	Charging time (minutes)
625	142.76	165 (1.5A @CC-CV)
1250	71.08	
2500	35.33	
5000	17.675	

Table 5 Energy density comparison between two batteries

Energy density	Thermal battery	Li-ion battery
Gravimetric density (kJ/kg)	3240.4	810
Volumetric density (kJ/m^3)	8327828	2218680

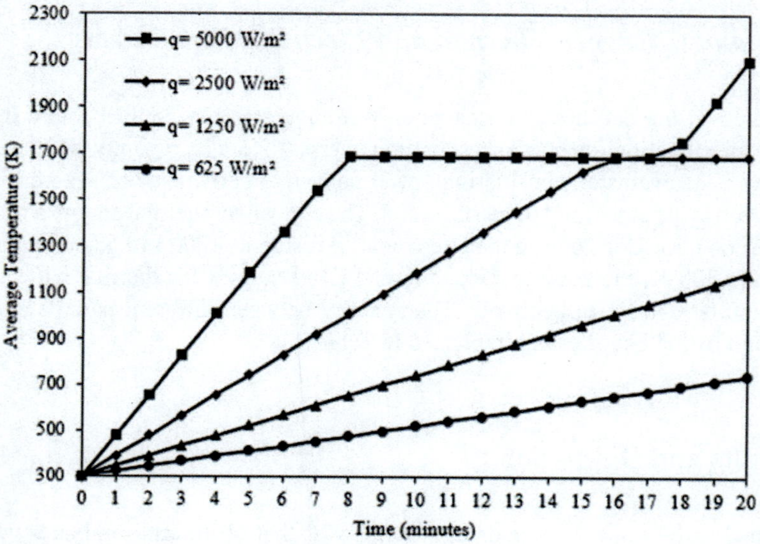

Fig. 4 Temporal variation of average temperature

the complete charging duration of thermal battery for heat flux of 5000 W/m^2. The reduction in charging duration is virtually proportional to the increase in magnitude of heat flux as visible from Fig. 5. Figure 6 represents the melting fraction contour at different instants during charging for thermal battery.

The charging for Li-ion battery was simulated with constant current and the charging time decreases linearly as the C rating increases as shown in Fig. 7. The constant current charging time at 0.5C rating is 110 min.

4 Conclusions

Enthalpy-porosity and MSMD numerical methodology have been successfully implemented to analyse the thermal battery and Li-ion cell, respectively. The conceptual latent storage-based thermal battery performed significantly better compared to the selected Li-ion cell. The charging duration was found to be 9.3 times and 1.15 times smaller than Li-ion cell for highest (5000 W/m^2) and smallest (625 W/m^2) heat flux. The energy storage density was observed to be nearly four

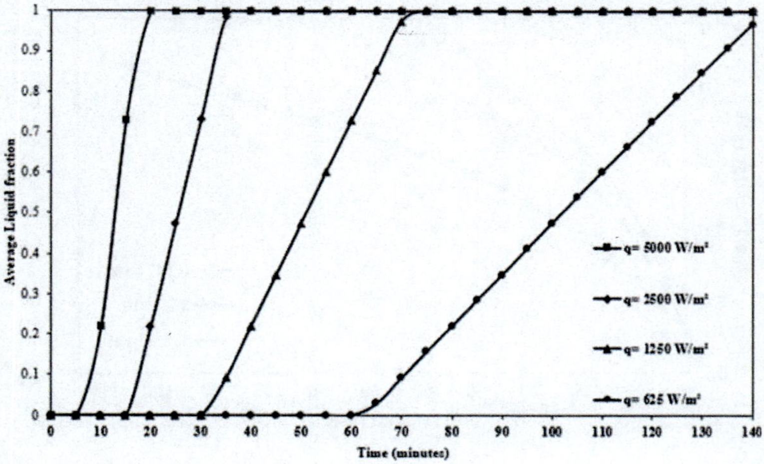

Fig. 5 Temporal variation of average liquid fraction

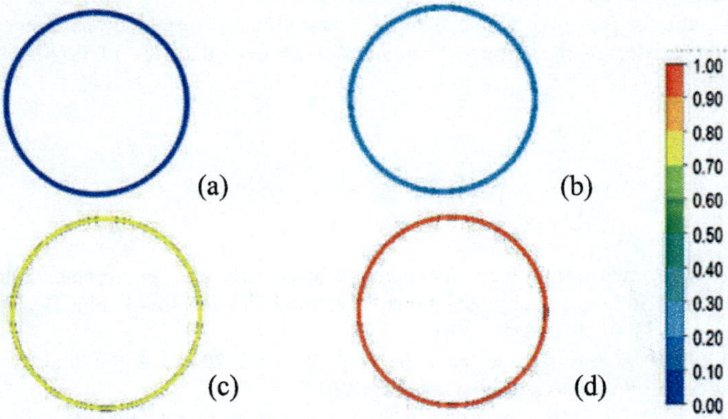

Fig. 6 Melting fraction contour during charging of thermal battery **a** $t = 2400$ s, **b** $t = 4800$ s, **c** $t = 7200$ s and **d** $t = 8400$ s

times higher for thermal battery. Hence, the study proposes to explore the potential of high-temperature thermal battery based on latent storage.

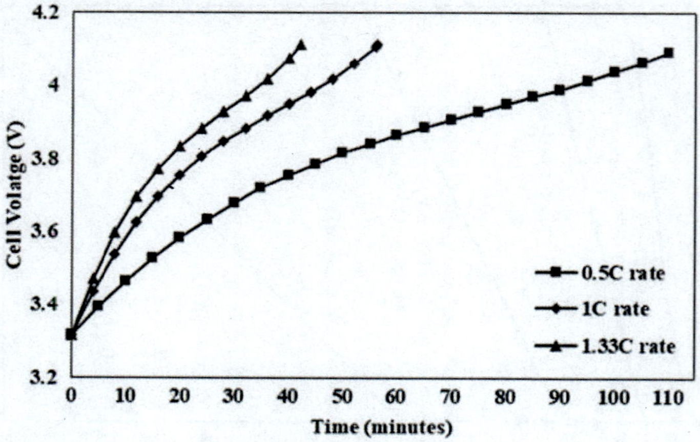

Fig. 7 Temporal variation of cell voltage

Acknowledgements The authors would like to thank Department of Science and Technology, Government of India for providing financial support through project entitled "Different Energy Vector Integration for Storage of Energy"—Grant number-TMD/CERI/MICALL19/2020/03(G).

References

1. Perera F (2018) Pollution from fossil-fuel combustion is the leading environmental threat to global pediatric health and equity: Solutions exist. Int J Environ Res Public Health 15(1). https://doi.org/10.3390/IJERPH15010016
2. Zablocki A (2019) Energy storage: Fact sheet (2019). Eesi, 2040:1–8 [Online]. Available: https://www.eesi.org/papers/view/energy-storage-2019
3. Aneke M, Wang M (2016) Energy storage technologies and real life applications—A state of the art review. Appl Energy 179:350–377. https://doi.org/10.1016/j.apenergy.2016.06.097
4. Ray AK, Rakshit D, Ravikumar K (2021) High-temperature latent thermal storage system for solar power: Materials, concepts, and challenges. Clean. Eng. Technol. 4:100155. https://doi.org/10.1016/j.clet.2021.100155
5. Chen T, Jin Y, Lv H, Yang A, Liu M, Chen B, Xie Y, Chen Q (2020) Applications of lithium-ion batteries in grid-scale energy storage systems. Trans Tianjin Univ 26(3):208–217. https://doi.org/10.1007/S12209-020-00236-W
6. Tetteh S, Yazdani MR, Santasalo-Aarnio A (2021) Cost-effective electro-thermal energy storage to balance small scale renewable energy systems. J Energy Storage 41:102829. https://doi.org/10.1016/J.EST.2021.102829
7. Purvins A, Sumner M (2013) Optimal management of stationary lithium-ion battery system in electricity distribution grids. J Power Sources 242:742–755. https://doi.org/10.1016/J.JPOWSOUR.2013.05.097
8. Valant C, Gaustad G, Nenadic N (2019) Characterizing large-scale, electric-vehicle lithium ion transportation batteries for secondary uses in grid applications. Batter 5(1):8. https://doi.org/10.3390/BATTERIES5010008
9. Crawford AJ, Huang Q, Kintner-Meyer MCW, Zhang JG, Reed DM, Sprenkle VL, Viswanathan VV, Choi D (2018) Lifecycle comparison of selected Li-ion battery chemistries under grid and

electric vehicle duty cycle combinations. J Power Sources 380:185–193. https://doi.org/10.1016/J.JPOWSOUR.2018.01.080
10. Guo L, Zhang S, Xie J, Zhen D, Jin Y, Wan K, Zhuang D, Zheng W, Zhao X (2020) Controlled synthesis of nanosized Si by magnesiothermic reduction from diatomite as anode material for Li-ion batteries. Int J Miner Metall Mater 27(4):515–525. https://doi.org/10.1007/S12613-019-1900-Z
11. Yuan C, Zhang L, Li H, Guo R, Zhao M, Yang L (2018) Highly selective lithium ion adsorbents: polymeric porous microsphere with crown ether groups. Trans Tianjin Univ . 25(2):101–109. https://doi.org/10.1007/S12209-018-0147-5
12. Cao Z, Liu H, Huang W, Chen P, Liu Y, Yu Y, Shan Z, Meng S (2019) Hydrogen bonding-assisted synthesis of silica/oxidized mesocarbon microbeads encapsulated in amorphous carbon as stable anode for optimized/enhanced lithium storage. Trans Tianjin Univ 26(1):13–21. https://doi.org/10.1007/S12209-019-00200-3
13. Einstein A, Cheap-Solar installation um ion battery advantages & disadvantages
14 Luft W (1985) High-temperature solar thermal energy storage. Int. J. Sol. Energy 3(1):25–40. https://doi.org/10.1080/01425918408914381
15. Zhang H, Baeyens J, Cáceres G, Degrève J, Lv Y (2016) Thermal energy storage: Recent developments and practical aspects. Prog Energy Combust Sci 53:1–40. https://doi.org/10.1016/j.pecs.2015.10.003
16. Alva G, Lin Y, Fang G (2018) An overview of thermal energy storage systems. Energy 144:341–378. https://doi.org/10.1016/j.energy.2017.12.037
17. Nazir H, Batool M, Bolivar Osorio FJ, Isaza-Ruiz M, Xu X, Vignarooban K, Phelan P, Inamuddin, Kannan AM (2019) Recent developments in phase change materials for energy storage applications: A review. Int J Heat Mass Transf 129:491–523. https://doi.org/10.1016/j.ijheatmasstransfer.2018.09.126
18. Zeneli M, Malgarinos I, Nikolopoulos A, Nikolopoulos N, Grammelis P, Karellas S, Kakaras E (2019) Numerical simulation of a silicon-based latent heat thermal energy storage system operating at ultra-high temperatures. Appl Energy 242(March):837–853. https://doi.org/10.1016/j.apenergy.2019.03.147
19. Ray AK, Rakshit D, Ravi Kumar K, Gurgenci H (2020) Silicon as high-temperature phase change medium for latent heat storage: A thermo-hydraulic study. Sustain Energy Technol Assessments 46:101249. https://doi.org/10.1016/j.seta.2021.101249
20. Robinson A (2017) Ultra-high temperature thermal energy storage. part 1: concepts. J. Energy Storage 13:277–286. https://doi.org/10.1016/j.est.2017.07.020
21. Dunham MT, Iverson B, Iverson B (2014) High-efficiency thermodynamic power cycles for concentrated solar power systems BYU scholarsarchive citation. Accessed August 23, 2019. Available: https://scholarsarchive.byu.edu/facpub://scholarsarchive.byu.edu/facpub/1585
22. Kim US, Shin CB, Kim CS (2008) Effect of electrode configuration on the thermal behavior of a lithium-polymer battery. J Power Sources 180(2):909–916. https://doi.org/10.1016/J.JPOWSOUR.2007.09.054
23. Estevez MAP, Caligiuri C, Renzi M (2021) A CFD thermal analysis and validation of a Li-ion pouch cell under different temperatures conditions. E3S Web Conf 238:09003.https://doi.org/10.1051/E3SCONF/202123809003
24. ANSYS Inc. (2021) ANSYS fluent theory guide 2021R1, January

The Effect of Thermal Interaction Between Boiling Parallel Microchannels on Flow Distribution

Ankur Miglani, Janmejai Sharma, Shravan Kumar Subramanian, and Pavan Kumar Kankar

Nomenclature

C_{lat}	Lateral thermal conductance (W/K)
H	Height of microchannel (mm)
i	Channel index ($i = 1$ or 2)
K_{wall}	Thermal conductivity of channel wall (W/m-K)
L	Length of microchannel (mm)
n	Number of channels ($n = 2$)
$P_{T,in}$	Total input power (W)
p_{out}	Output pressure (MPa)
$T_{fl,in}$	Fluid inlet temperature (°C)
W	Channel width (mm)
W_T	Total mass flow rate (mg/s)

Greek Symbols

θ	Streamwise non-uniformity parameter

A. Miglani (✉) · J. Sharma · P. K. Kankar
Department of Mechanical Engineering, Microfluidics and Droplet Dynamics Lab, IIT, Indore, India
e-mail: amiglani@iiti.ac.in

S. K. Subramanian
Department of Mechanical Engineering, Swami Vivekananda Institute of Technology, Jawaharlal Nehru Technological University, Hyderabad, Telangana 500085, India

J. Banerjee et al. (eds.), *Recent Advances in Fluid Dynamics*, Lecture Notes in Mechanical Engineering, https://doi.org/10.1007/978-981-19-3379-0_40

1 Introduction

The continued miniaturization of electronic packages, which offer diverse functionalities, has dramatically increased the waste heat load. In particular, the packaging of electronic components associated with high-performance computing systems, HEV/EVs and avionics is highly space constrained, resulting in high-power densities (*e.g.* ~200 W/cm^2 or more), which needs to be dissipated to ensure safe working temperatures and increase system reliability. The high-performance electronic devices feature high-power densities where the utilization of intermediate heat spreaders to smear out the uneven heat flux profiles is not effective. Instead, the ultimate heat sink often experiences the uneven heat flux generated from the temperature variations in the package due to multiple components and devices [1]. A key hydrodynamic implication of uneven heating on the flow boiling in microchannels is the tendency of the flow to distribute unequally (or maldistributed) between the channels. Flow maldistribution is undesirable in heat sinks as the channels that are starved of the flow (compared to uniform flow distribution) may undergo an untimely dry-out and impair its heat transfer performance.

Flow distribution between individual microchannels is strongly affected by their load curves in a parallel channel system. The amount of heat input is one of the key factors that influences the channels characteristic load curve because it governs the thermodynamic state of the working fluid. If the channels have different load curves due to uneven heating, then they must have different flow rates to satisfy the uniform pressure drop boundary condition. In this way, uneven heating can induce flow maldistribution in a parallel channel system [2, 3]. A few studies have explored the thermal implications of flow maldistribution resulting from uneven heating conditions [4–8]. For instance, Cho et al. [6] demonstrated that when a hotspot is located close to the heat sink inlet, a large temperature variation is induced across the heat sink in the transverse direction due to the flow maldistribution. Based on the channel wall temperature and overall pressure drop measurements, they concluded that the maldistribution resulted from an increase in the local pressure drop due to boiling, which rerouted the inlet sub-cooled liquid flow to other locations. Flynn et al. [7, 8] studied the thermal response of flow maldistribution in thermally coupled and thermally isolated parallel microchannels etched on a silicon substrate. The channels were subject to both uniform and non-uniform or uneven heating by independently varying their heat input. In thermally isolated case, uniform heating caused both the channels to either receive single-phase liquid flow or the boiling flow (depending on the heating level), resulting in small temperature difference between channels. However, at high levels of uneven heating, the channel subjected to a lower heat load remained in single-phase liquid regime, while the other channel with higher heat load underwent boiling, resulting in a significant temperature difference between the channels. The existence of flow maldistribution was concluded based on the flow visualizations and temperature measurements.

While the thermal implications of uneven heating have been reported by a few studies, less is known about the hydrodynamic implications of uneven heating, which can explain the observed temperature signatures and degradation in thermal performance. Further, the effects of operational parameters such as the increasing degree of uneven heating and increasing heat load on the flow maldistribution have not been investigated. Other previous attempts [9–11] to capture the flow maldistribution caused by uneven heating conditions were largely motivated by its occurrence in parallel evaporator channels in large-scale steam generation systems, and therefore, focused on long channels with large diameters, which are not suitable for electronics cooling applications. While recent studies have made great strides in performance prediction capabilities, the state-of-the-art models still assume uniform flow distribution between microchannels. Since microchannel flow boiling is uniquely susceptible to flow maldistribution under uneven heating, it is imperative to develop models that account for flow distribution and enable an accurate prediction of the thermal–hydraulic performance of boiling parallel microchannels, for electronic cooling systems having coolant loops with branching flow paths.

2 Two-Phase Flow Distribution Model

The results presented in this study are obtained using the two-phase flow distribution model as described in our previous studies [12, 13]. The modelling approach enables the two-phase flow distribution to be predicted for a system with multiple parallel microchannels for range of operating parameters including the input power, inlet subcooling, channel dimensions (in x, y and z directions), thermal interaction between the channels, number of channels up to 200 and outlet pressure. The model can also capture the flow distribution behaviour under different heat flux boundary conditions such as uniform heating, lateral uneven heating or channel-to-channel differential heating, axial uneven heating and, finally, varying heating profiles along the channel length.

The model is developed to integrate the pump curve with the thermal–hydraulic model [13] for each channel into the flow network equations of the system. The heat conduction is incorporated in the form of convective heat transfer within the channels, the streamwise and lateral heat conduction via the channel walls and the substrate and heat lost to the ambient. Since the thermal connectedness between the channels can have significant effect on the flow distribution between them, it was accounted for in the model via the lateral thermal conductance C_{lat}, which quantifies the level of thermal connectedness between the channels. The lateral thermal conductance (C_{lat}) was determined assuming 1D heat transfer across the vertical mid-plane of the channels as $C_{lat} \cong K_{wall} HL/S$ (where S is the channel pitch). With this definition, a value of $C_{lat} = 0$ W/K signifies no thermal interaction between the channels, while higher values of $C_{lat} = 100$–1000 W/K indicates a strong thermal interaction between them. In the model, the lateral thermal conductance is varied over a range: $C_{lat} = 0.1$–1000 W/K to determine its influence on the flow distribution characteristics under axial uneven heating.

3 Test Cases

A schematic representation of the heating conditions for which the flow distribution behaviour is investigated is shown in Fig. 1. The geometry consists of two parallel microchannels that are placed adjacent to each other in solid copper blocks. The physical dimensions as well as the operating and boundary conditions of this parallel microchannel configuration are adapted from our previous studies [14, 15] and enlisted in Table 1. Two different heating conditions are investigated. First is the baseline case where both the microchannels are heated uniformly and subjected to an equal input power (Fig. 1a). In the second case, the channels are subject to an identical input power, but this is distributed non-uniformly along their lengths. Specifically, the upstream and downstream halves of the channels receive unequal power, thus creating a streamwise uneven heating condition. For both these cases, the flow distribution characteristics are investigated as a function of the total input

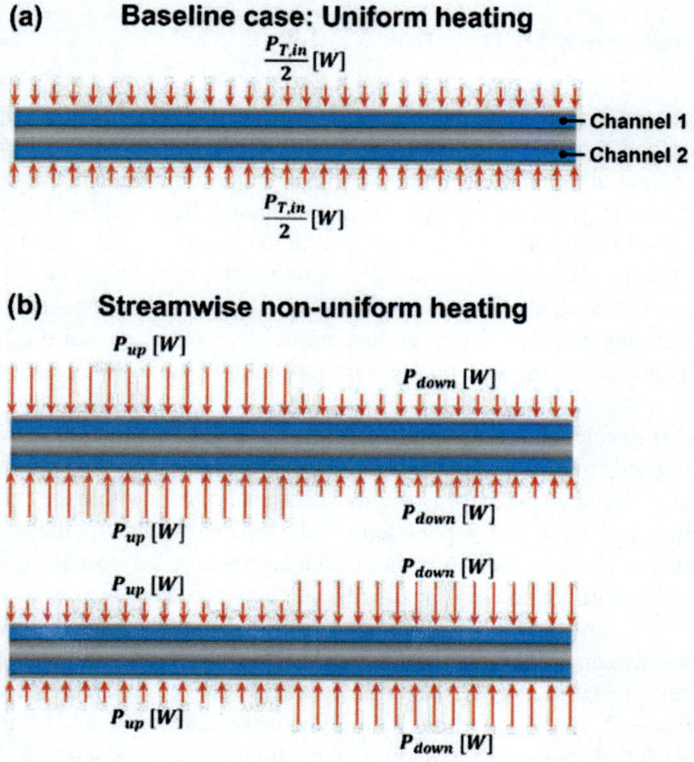

Fig. 1 Schematic diagram of a parallel microchannel configuration showing two different heating conditions investigated in this study: **a** axial uniform heating and **b** axial uneven heating. Note that in the top view of the channels shown here, the arrows of input power drawn on the sides are only for the purpose of representation. In the mode, power input into the channels is from the bottom

Table 1 System parameters used as inputs to the model

Parameters	Magnitude
Channel dimensions: $L, W, H(mm)$	55, 1, 1
Number of channels	2
Inlet mass flow rate $W_T(mg/s)$	0 to 300
Fluid inlet temperature $T_{fl,\text{in}}$ (°C)	88.5
Outlet pressure $p_{\text{out}}(MPa)$	0.105
Input power $P_{T,\text{in}}$(W)	2.2 W, 4.4 W and 8.8 W
Fluid	DI Water
Thermal conductivity of channel wall, copper $K_{\text{wall}}(W/m-K)$	385
Coupling factor or lateral thermal conductance $C_{\text{lat}}(W/K)$	0 to 1000
Streamwise non-uniformity parameter (θ)	– 1 to 1

power and the different degree of thermal connectedness (or thermal conductance) between the channels.

4 Results

4.1 Flow Distribution Diagram

The flow distribution between the channels is pictured as a plot of relative flow rate fraction of each channel versus the total flow rate (see Fig. 2). The fraction of flow rate in each channel is plotted simultaneously on the same diagram, and the sum of the flow rate fractions is equal to unity. The horizontal line with a flow rate fraction of 0.5 represents a uniform flow distribution, where both channels receive equal flow rate. Any deviation from a flow rate fraction of 0.5 indicates flow maldistribution. In this condition, one of the channels receives excess flow, while the other is starved of the flow compared to the uniform flow distribution. For example, in Fig. 2, the total flow rate ranging between 50 mg/s to 190 mg/s experiences flow maldistribution, where channel 1 receives more flow while channel 2 receives less flow. In comparison with the uniform flow conditions, flow maldistribution state is undesirable because the shortage of fluid in the flow starved channel may lead to its premature dry-out, which may severely deteriorate the heat transfer performance of the microchannel system. It is noteworthy to mention that while in Fig. 2, the flow rate in channel 1 is shown to be always higher compared to channel 2, in practice, any one of the channels can undergo boiling, which will then receive reduced flow due to its increased flow resistance, while the other one will receive excess flow.

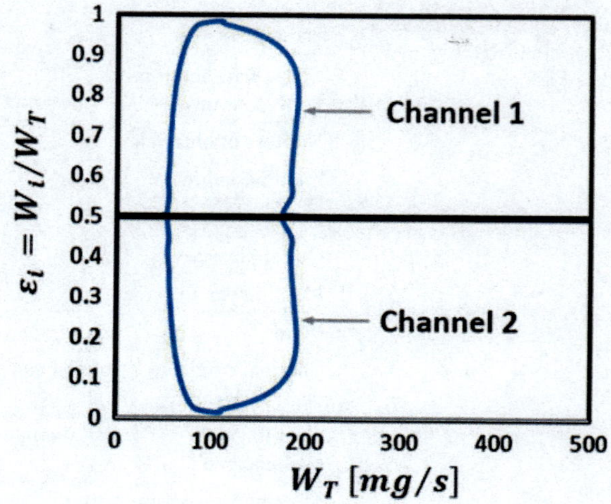

Fig. 2 Representative flow distribution curve showing the flow rate fraction in each channel with increasing total flow rate. Note that it is a single continuous loop, where the top half represents the microchannel receiving excess flow, while the bottom half represents the microchannel that is flow starved

The flow distribution between the channels is visualized in terms of the relative flow rate fraction of each channel versus the total flow rate, as shown in Fig. 3. In Fig. 3, each row represents the effect of increasing thermal conductance between the channels from left to right at a given total input power, while each column represents the effect of increasing input power from top to bottom at a given value of thermal conductance. The range of total flow rate on x-axis is same for each plot. The black curve in each of the plots represents the baseline case with axial uniform heating of the channels, while the blue curves represent the flow distribution characteristics at different degrees of axial uneven heating.

4.2 *Effect of Thermal Interaction Between Channels on Axial Uneven Heating*

For in-phase streamwise non-uniform or uneven heating case, the degree of non-uniformity is quantified by comparing the amount of heating of upstream half of the channels to the heating of downstream half of the channels through the streamwise non-uniformity or unevenness parameter:

$$\theta = \frac{P_{\text{up}} - P_{\text{down}}}{P_{T,in}/2}, -1 \leq \theta \leq 1 \tag{1}$$

where $P_{T,in}/2(= P_{\text{up}} + P_{\text{down}})$ refers to the total power input to each channel heater. The subscripts up and down refer to power input to the upstream half and downstream half of each channel, respectively. Note that the non-uniformity parameter is defined based on a single channel because the streamwise heating profile is identical for

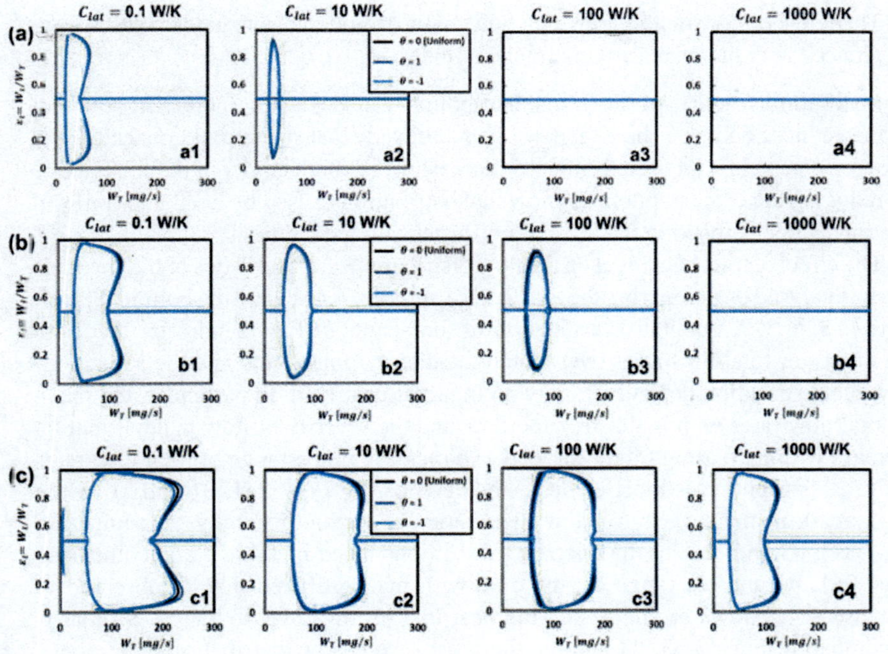

Fig. 3 Relative flow rate distribution, shown as flow rate fraction, versus total flow rate W, for two identical parallel microchannels (parameters in Table 1) subjected to varying degree of axial non-uniform heating at a total heat load of $P_{T,in} = 2.2$ W (top row), 4.4 W (middle row) and 8.8 W (bottom row). The results are for a range of thermal conductance values between the channels: $C_{lat} = 0.1$ W/m-K (leftmost), $C_{lat} = 10$ W/m-K (middle) and $C_{lat} = 100$ W/m-K (middle; third column) and $C_{lat} = 1000$ W/m-K (rightmost)

both the channels. Based on Eq. (1), $\theta = 0$ corresponds to uniform power input to upstream and downstream halves of the channels while extreme uneven heating conditions with entire power input in upstream or downstream halves is given by $\theta = 1$ and $\theta = -1$, respectively.

In comparing the uniform vs. axial non-uniform/uneven heating cases, it is apparent that flow maldistribution is independent of the degree of streamwise uneven heating. This is due to the dominant axial conduction in channel blocks in the present experiments; channels have thick walls with a high thermal conductivity (oxygen-free copper). Due to dominant axial conduction along the channel block, the in-phase axial non-uniform case reduces trivially to that of uniformly heated channels. Therefore, the qualitative features of flow distribution diagrams are also identical to those observed under uniform heating conditions:

- The range of flow rates over which flow maldistribution occurs increases with increasing total heat load.
- The range of flow rates over which the flow maldistribution is most severe is significantly lower compared to that at higher total heat loads (*i.e.* 8.8 W).

- The range of flow rates over which flow maldistribution occurs and is most severe reduces with the increasing thermal coupling.

Maximum severity of flow maldistribution reduces with increasing thermal connectedness between the channels. This indicates that if thermal connectedness between channels is increased either by increasing the conductivity of the substrate or by reducing the channel pitch, the flow maldistribution caused by axial non-uniform heating can be damped to the level of uniform heating conditions.

The effect of total heat input on the flow distribution characteristics of the channels can be analysed by comparing the top (2.2 W), middle (4.4 W) and the bottom (8.8 W) rows in Fig. 3. A qualitative evaluation of the results of Fig. 3 indicates that both the ranges of total flow rates over which maldistribution occurs and the severity of flow maldistribution depend strongly on the total heat load. In particular, the range of total flow rates with maldistributed flow and the severity of flow maldistribution increase with increasing total heat load. An increase in the range of total flow rates with maldistribution is most evident when comparing Figs. 3a1, b1 and c1, while an increase in the severity of flow distribution is visible distinctly on comparing Figs. 3a3, b3 and c3. For the operating conditions listed in Table 1 and comparing Figs. 3a1, b1 and c1, range of flow rate with maldistributed flow is observed to increase ~ 4.2 times on increasing the heat load from 2.2 W to 8.8 W. Similarly, in comparing Figs. 3 a3, b3 and c3, the most severe flow distribution deteriorates significantly on increasing the heat load from 2.2 W to 8.8 W; at 8.8 W, the flow rate fraction can be as low as 0.02 compared to 0.5 at 2.2 W, *i.e.* no flow maldistribution.

The extent to which the total heat load affects the three flow distribution metrics, *i.e.* the range of total flow rates with flow maldistribution, the most severe flow distribution, and the range of total flow rate with most severe flow distribution depends on the strength of thermal connectedness between the channels. Figure 3 shows the effect of increasing the thermal connectedness between the channels by four orders of magnitude: ranging from $C_{\text{lat}} = 0.1$ W/mK (leftmost column) to $C_{\text{lat}} = 1000$ W/K (rightmost column). This covers the range of thermal conductance values that are typical of parallel microchannel heat sinks as would be deployed for thermal management of power dense electronics. It is apparent from Fig. 3 that the increased thermal conductance mitigates the flow maldistribution between the channels. In addition, the strength of thermal interaction between the channels is increased the deterioration in flow maldistribution with an increase in the heat load reduces. The positive influence of thermal conductance in dampening flow maldistribution is due to the possibility of heat relocation between the channels via the channel fins and the substrate of the heat sink containing the channels. Our previous study [15] reported that channel-to-channel heat conduction minimizes the temperature difference between the channels, which leads to redistribution of the heat flux into the channels. Specifically, the channel starved of the flow is at higher temperature compared to the channel with excess flow (at lower temperature), which causes the heat to flow from flow starved channel to the channel with excess flow. This way, the total heat load is shared between the channels such that the channel receiving more flow also carries a larger portion of the heat load, which helps in minimizing the difference in the vapour quality between the channels. This in turn equalizes their hydraulic resistances, which helps in dampening the flow maldistribution. In contrast, when the thermal conductance between the channels is either weak or absent (in case thermally isolated channels), even when a severe maldistribution is present between the channels, the channels must share their respective heat loads individually because the channel-to-channel heat exchange is either restricted (weak thermal interaction) or absent altogether (in thermally isolated channels). Consequently, the severity of flow maldistribution increases with an increase in the total heat load if thermal interaction between the channels is restricted.

It is interesting to note that for a given total heat load, there exists a threshold value of thermal conductance at which the flow maldistribution between the channels vanishes. This is marked by a flat line indicating a relative flow rate fraction of 0.5, meaning equal flow in both the channels. Further, this threshold value increases with an increase in total heat load. For a given total heat load, this threshold value is determined by increasing the thermal conductance value in small steps and identifying the value at which flow maldistribution is absent. For instance, this on the order of $C_{\text{lat}} = 100$ W/K at 2.2 W, and $C_{\text{lat}} = 1000$ W/K at 4.4 W. However, with further increase in heat load to 8.8 W, this threshold value ceases to exist. This means that up to certain heat loads, increasing the thermal conductance between the channels is an effective method of eliminating flow maldistribution. At higher heat loads (specific to operating conditions and heat sink design), it may not be possible to eliminate flow maldistribution completely, but it can be damped to a significant extent in terms of reducing the three flow distribution metrics.

5 Conclusions

In this study, the effect of thermal interaction between boiling parallel microchannels on their flow distribution behaviour is investigated under streamwise uneven heating conditions and as a function of increasing magnitude of total heat load. This aids in quantifying the degree of flow maldistribution that may occur under streamwise uneven heating conditions for microchannel heat sinks (for electronics cooling applications) with varying degree of thermal interaction between the channels. The following important conclusions are inferred from this study:

- Streamwise uneven heating induces flow maldistribution between the channels which worsens with increasing input power in terms of the following three metrics:
  - The range of total flow rates with flow maldistribution
  - The worst flow maldistribution where the flow rate imbalance between the channels is maximum
  - The range of total flow rates with this worst flow maldistribution

- Increasing the thermal conductance between the channels (either by increasing substrate conductivity of decreasing channel pitch or wall thickness) aids in dampening flow maldistribution under non-uniform heating conditions by allowing channel-to-channel heat conduction.
- By increasing the thermal conductance, the flow maldistribution between the channels can be dampened but only up to a certain level. This threshold value of thermal conductance beyond which no reduction is observed increases with increasing head load.
- Compared to the existing methods of dampening flow maldistribution such as inlet restrictors, pressure drop elements, varying channel cross sections, increasing the thermal conductance between microchannels is simple, easy to implement and an effective method for mitigating flow maldistribution. However, it may not be suitable for eliminating flow maldistribution, especially at high heat loads.

References

1. Hamann HF, Weger A, Lacey JA, Hu Z, Bose P, Cohen E, Wakil J (2006) Hotspot-limited microprocessors: Direct temperature and power distribution measurements. IEEE J Solid-State Circuits 42(1):56–65
2. Cho ES, Choi JW, Yoon JS, Kim MS (2010) Experimental study on microchannel heat sinks considering mass flow distribution with non-uniform heat flux conditions. Int J Heat Mass Transf 53(9–10):2159–2168
3. Costa-Patry E (2011) Cooling high heat flux micro-electronic systems using refrigerants in high aspect ratio multi-microchannel evaporators. EPFL, Lausanne, Ph.D. Thesis, 230
4. Alam T, Lee PS, Yap CR, Jin L (2013) A comparative study of flow boiling heat transfer and pressure drop characteristics in microgap and microchannel heat sink and an evaluation of microgap heat sink for hotspot mitigation. Int J Heat Mass Transf 58(1–2):335–347
5. Bogojevic D, Sefiane K, Walton AJ, Lin H, Cummins G, Kenning DBR, Karayiannis TG (2011) Experimental investigation of non-uniform heating effect on flow boiling instabilities in a microchannel-based heat sink. Int J Therm Sci 50(3):309–324
6. Ritchey SN, Weibel JA, Garimella SV (2014) Local measurement of flow boiling heat transfer in an array of non-uniformly heated microchannels. Int J Heat Mass Transf 71:206–216
7. Sarangi RK, Bhattacharya A, Prasher RS (2009) Numerical modelling of boiling heat transfer in microchannels. Appl Therm Eng 29(2–3):300–309
8. Revellin R, Thome JR (2008) A theoretical model for the prediction of the critical heat flux in heated microchannels. Int J Heat Mass Transf 51(5–6):1216–1225
9. Revellin R, QuibÉn JM, Bonjour J, Thome JR (2008) Effect of local hot spots on the maximum dissipation rates during flow boiling in a microchannel. IEEE Trans Compon Packag Technol 31(2):407–416
10. Miler JL, Flynn R, Refai-Ahmed G, Touzelbaev M, David M, Steinbrenner J, Goodson KE (2009) Effects of transient heating on two-phase flow response in microchannel heat exchangers. Int Elect Packaging Technical Conference Exhibition 43604:563–569
11. Kingston TA, Weibel JA, Garimella SV (2020) Time-resolved characterization of microchannel flow boiling during transient heating: Part 1–Dynamic response to a single heat flux pulse. Int J Heat Mass Transf 154:119643
12. Van Oevelen T, Weibel JA, Garimella SV (2018) The effect of lateral thermal coupling between parallel microchannels on two-phase flow distribution. Int J Heat Mass Transf 124:769–781

13. Van Oevelen T, Weibel JA, Garimella SV (2017) Predicting two-phase flow distribution and stability in systems with many parallel heated channels. Int J Heat Mass Transf 107:557–571
14. Miglani A, Weibel JA, Garimella SV (2021) Measurement of flow maldistribution induced by the Ledinegg instability during boiling in thermally isolated parallel microchannels. Int J Multiph Flow 139:103644
15. Miglani A, Weibel JA, Garimella SV (2021) An experimental investigation of the effect of thermal coupling between parallel microchannels undergoing boiling on the Ledinegg instability-induced flow maldistribution. Int J Multiph Flow 139:103536

Study on Direct Contact Condensation in Stagnant and Flowing Media

Deepak Kumar Agarwal, Vishnu Viswanath, Anant Singhal, Jophy Peter, T. John Tharakan, and S. Sunil Kumar

Nomenclature

B	Condensation driving potential
ΔT_{sub}	Pool subcooling
D	Injector exit diameter
G	Steam mass flux
α	Volume fraction
τ	Tress tensor
Γ	Interphase mass transfer

Subscripts

i	Vapour–liquid interface
q	Qth phase
p	Pth phase
m	Mixture

D. K. Agarwal (✉) · V. Viswanath · A. Singhal · J. Peter · T. J. Tharakan · S. Sunil Kumar
Liquid Propulsion Systems Centre, ISRO, Trivandrum, Kerala 695547, India
e-mail: dagarwal.iitk@gmail.com

J. Banerjee et al. (eds.), *Recent Advances in Fluid Dynamics*, Lecture Notes in Mechanical Engineering, https://doi.org/10.1007/978-981-19-3379-0_41

1 Introduction

Direct contact condensation (DCC) is a homogeneous mode of condensation heat transfer which occurs when a condensable gas is injected into its liquid at subcooled conditions. DCC phenomenon is an intricate phenomenon as it involves hydrodynamics, thermodynamics, bubble dynamics, turbulent mixing, natural convection and also the instabilities depending upon the flow regime. Besides, the gas condensation results in pressure fluctuations that create pressure loads on the walls of the containment chamber. This makes the problem of DCC complex, and its study requires experimental, numerical and theoretical methods for complete elucidation.

DCC is encountered in the booster pump exit line of liquid oxygen-kerosene rocket engine. The hot turbine drive gas, which is mostly gaseous oxygen, gets mixed with liquid oxygen at the exit of booster pump and subsequently condenses downstream. Gaseous oxygen has to be fully condensed before it reaches pump inlet to avoid performance degradation of the main pump. Understanding the interfacial heat and mass transfer during this, DCC phase is pivotal for the design of the system. This paper consists of the details of experimental study that has been carried out to investigate the DCC of steam in water, and results were reported by Viswanath et al. [18]. Experiments were also performed for steam condensation inflowing water. As part of the complete study, numerical simulations of gaseous oxygen condensation were also conducted.

2 Literature Review and Objectives

Earliest study reported on DCC of steam in water focussed on the change in shape of the steam plume with variation in bath temperature, Boehm et al. [2]. Later on, investigators were principally interested in developing empirical correlations for penetration length as in [9, 19]. Chan et al. [3] published a detailed regime map for inversely buoyant steam jet with steam mass flux up to 175 kg/m^2s. Chun et al. [4], Kim et al. [10], Aya et al. [1] proposed empirical correlations for heat transfer coefficient for direct condensation of vapour jets in subcooled water pool. Liang et al. [11] calculated the product of interfacial heat transfer coefficient and interfacial area for sub-sonic, positively buoyant steam jets as depending on degree of subcooling, steam mass flux and more importantly turbulent parameters. It was observed that for the sonic condensation jet, the heat transfer coefficient was in the range of 1–3.5 $MW/(m^2\ K)$, whereas for the subsonic jet, it was 1/5 to 1/10 that of the sonic jet as reported by Chun et al. [4]. Wu et al. [20] investigated the horizontal sonic steam injection into subcooled water and calculated heat transfer coefficient based on surface renewal model. The surface area was calculated from the photographs.

Gulawani et al. [6], Gulawani [7] and Dahikar et al. [5] have numerically studied the phenomena of steam injection at supersonic velocity in water. These numerical studies were carried out using the two-fluid modeling approach of ANSYS CFX. The shape of the stable steam jet and the interfacial heat transfer at high velocity were explained. Li et al. (2015) studied DCC of steam occurring at low velocities in the VOF framework. The characteristics of steam jet in oscillatory regime were studied. Patel et al. [13] studied the chugging regimes using a two-fluid model in the CFD packages NEPTUNE CFD and Open FOAM. Song et al. [17] studied unstable steam jet using VOF model of ANSYS Fluent. The turbulent flow field was resolved using LES model. The pressure oscillation during condensation of unstable steam jet was brought out. Jayachandran et al. [8] carried out numerical investigation of unstable DCC of steam in water using a two-fluid modeling approach in ANSYS CFX. In addition to studying the interfacial heat transfer, the bubble oscillation stages were divided into stable and unstable regions.

From literature, it was observed that most of the studies focussed on very high steam mass flux conditions $G > 150$ kg/m^{2}s. The steam was injected at sonic condition, and the subsonic regime was not explored thoroughly. Moreover, the possibility of using shadowgraphy technique for image visualisation was also not explored. Based on this background, experiments were conducted to develop an understanding of the subsonic inversely buoyant steam condensation at low mass fluxes. Experiments were therefore conducted for steam in stagnant and flowing water at low mass flux of steam. Numerical investigation of gaseous oxygen from discrete orifices into cross-flowing liquid oxygen was also carried out. The trajectory of injected gaseous oxygen and the flow behaviour of the condensing gas were studied. Simulations were also carried out for various gas mass flow rates and temperatures to study their effect on the volume fraction of gas at the line exit.

3 Domains of Study

The challenges involved in addressing DCC phenomena in cryogenic liquid rocket engine are manyfold. In this regard, comprehensive experimental and numerical studies were targeted as shown in Fig. 1

Experimental study of steam condensation in stagnant water. The test setup for steam condensation studies in stagnant water is shown in Fig. 2. The setup consists of steam generation system, cuboidal water containment chamber, steam supply lines, injectors and a Z—type Herchellian shadowgraph as in Settles [14] with high-speed camera. The details of the experimental test rig and measurements implemented are detailed in [18].

The steam generation system consists of stainless steel tubes (SS304L), subjected to volumetric heating and wound with two layers of insulation to reduce thermal losses to the ambient. First layer of insulation comprised of insulation cloth ($K =$ 0.0589 W/m K at 198 °C) having thickness of 5 mm. Outer layer was insulation tape ($K = 0.2$ W/mK at 198 °C) with thickness of 0.1 mm. De-mineralized water was fed into the electrical-heated feed system, and once steam was formed, it was admitted into the injector submerged in the pool of water. The injector line was double-walled

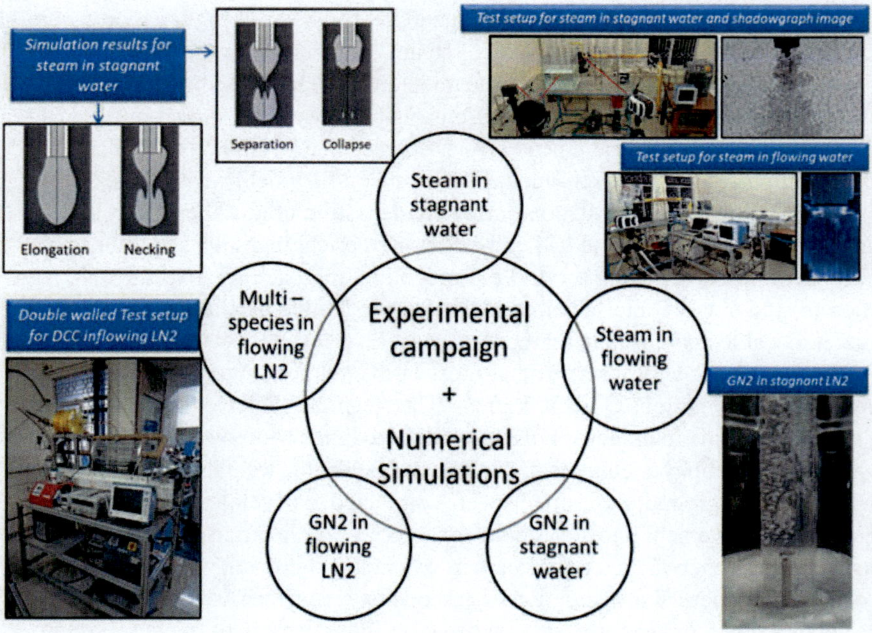

Fig. 1 Domains of study

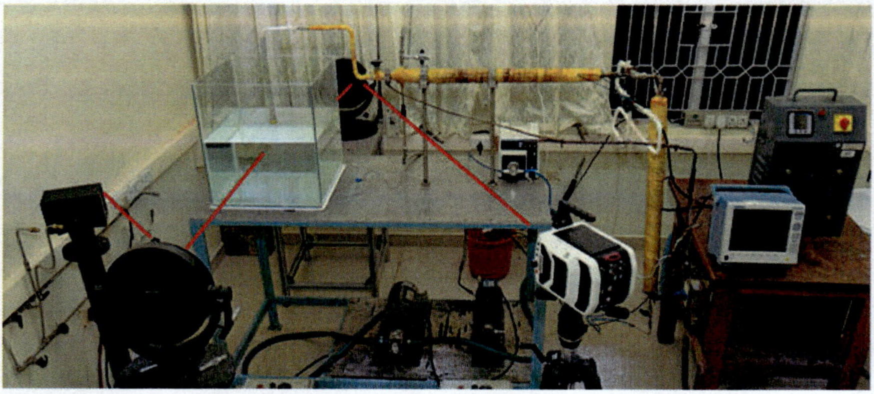

Fig. 2 Test setup with shadowgraph for image visualisation

insulated line and was pre-heated with steam for 30 min before data acquisition was started.

Image Processing Technique. Shadowgraph method was used to distinguish the condensing steam in water. To distinguish steam from hot water, hot water was injected into the pool of water at different temperatures as shown in Fig. 3.

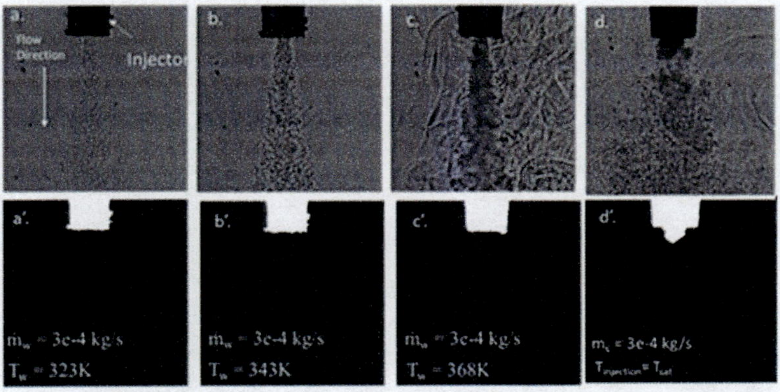

Fig. 3 **a** Shadowgraph image of hot water injection and **a′** corresponding post-processed image. Images **b**, **c** and **d** are the raw images, and **b′**, **c′** and **d′** are the corresponding processed images

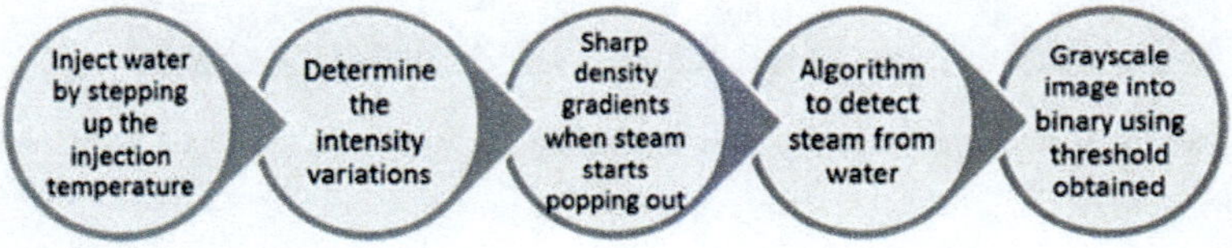

Fig. 4 Image processing methodology

With water, density gradient was not sharp enough to create any detectable intensity changes, whereas for steam injection, there was a visible intensity change near the exit of the injector as shown in Fig. 3d′. The methodology adopted for image processing is shown in Figs. 3 and 4.

Experimental study of steam condensation in flowing water. An experimental test setup with provision for steam injection into cross-flowing water was established as shown in Fig. 5. Water supply to the steam generator was ensured using digital micro-gear pump (Coleparmer make Model No: 75211-35) which maintains a constant flow rate of water to the inlet of the steam generating unit. The 0.1 hp motor can deliver a flow ranging from 0 to 60 ml/min with a flow accuracy of 0.5%. Cross-flow of water was obtained by using a regulated centrifugal pump, with a flow range of 1–50 lpm. The test section consists of a transparent borosilicate tube of 24 mm diameter and 2000 mm length with a flange for injecting steam perpendicularly into the flowing water.

For measuring the temperature of flowing water, three T-type thermocouples (Composition: Copper—Constantan, Range: 23–623 K, 0.4% accuracy) are mounted at the inlet of the feedline just before of the borosilicate tube. Similarly, state of steam was determined with pressure transducers (Strain gauge based, steady-state pressure measuring transducers, 1–5 bar, accuracy ±1% of full scale) and T-type thermocouples. The injected steam was monitored through Phantom V1210 high-speed camera.

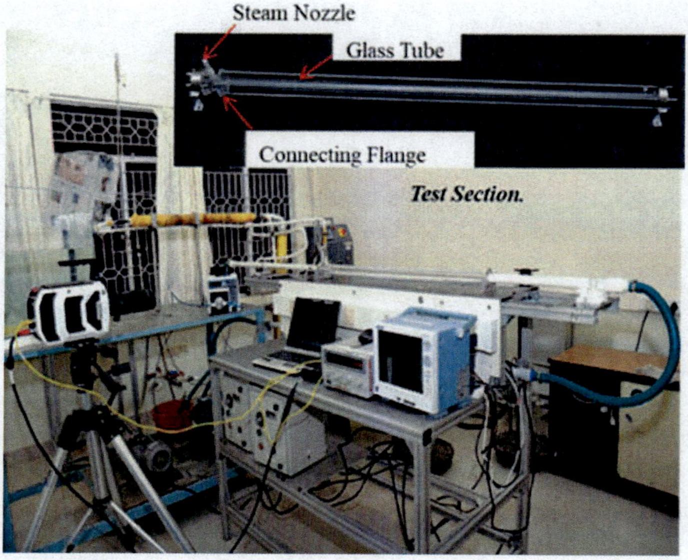

Fig. 5 Test setup for studying steam condensation in flowing water with the elements of the test section shown above

Experimental test setups for DCC in cryogenic media. The experimental test setup for cryogenic media was also established, and preliminary studies were carried out for GN_2 condensation in stagnant LN_2 as shown in Fig. 1. However, more comprehensive works are planned, and a double-walled vacuum-jacketed test section with provision for injecting mixture of condensable gases was made. Work is in progress to ensure leak tightness of the test section. Figure 6 shows the test section for cryogenic experiments.

Experiments planned. In order to simulate the scenario of gaseous mixture condensing in the liquid oxygen stream of the rocket engine, experiments are planned with gaseous nitrogen (GN_2), carbon dioxide (CO_2) and water vapour (H_2O) condensing in liquid nitrogen (LN_2) for several gas to liquid momentum flux ratios and gas compositions.

Numerical study of steam in stagnant water. Computational fluid dynamic study of steam in stagnant water was conducted. The analyses were conducted using pressure-based coupled solver of ANSYS Fluent.

Governing equations. The gas–liquid two-phase flow is formulated in the two-fluid Eulerian framework. A set of volume-averaged continuity, momentum and energy equations are solved for each of the two phases given in Eqs. (1)–(4).

$$\frac{\partial}{\partial t}(\alpha_{\mathrm{q}}\rho_{\mathrm{q}}) + \nabla.\left(\alpha_{\mathrm{q}}\rho_{\mathrm{q}}\vec{V}_{\mathrm{q}}\right) = \sum_{p=1}^{n}(\Gamma_{\mathrm{pq}} - \Gamma_{\mathrm{qp}}) \quad (1)$$

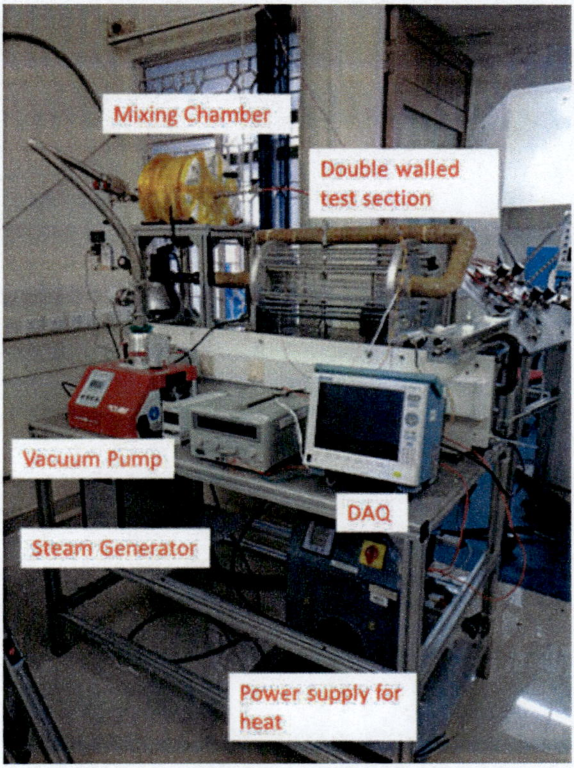

Fig. 6 Test setup for cryogenic experiments

$$\frac{\partial}{\partial t}\left(\alpha_q\rho_q\vec{V}_q\right)+\nabla.\left(\alpha_q\rho_q\vec{V}_q\vec{V}_q\right)$$
$$=-\alpha_q\nabla p+\nabla.\overline{\overline{\tau}}_q+\alpha_q\rho_q\vec{g}$$
$$+\sum_{p=1}^{n}\left(\vec{F}_{Dpq}+\Gamma_{pq}\vec{V}_{pq}-\Gamma_{qp}\vec{V}_{qp}\right) \tag{2}$$

where

$$\overline{\overline{\tau}}_q=\alpha_q\mu_q\left(\nabla\vec{V}_q+\nabla\vec{V}_q^{T}\right)+\alpha_q\left(\daleth_q-\frac{2}{3}\mu_q\right)\nabla.\vec{V}_q\overline{\overline{I}} \tag{3}$$

$$\frac{\partial}{\partial t}\left(\alpha_q\rho_q h_q\right)+\nabla.\left(\alpha_q\rho_q\vec{u}_q h_q\right)$$
$$=\alpha_q\frac{dp_q}{dt}+\overline{\overline{\tau}}_q:\nabla\vec{u}_q-\nabla.\vec{q}_q$$
$$+\sum_{n}^{p=1}\left(Q_{pq}+h_{pq}\Gamma_{qp}-\Gamma_{qp}h_{qp}\right) \tag{4}$$

Interfacial Heat and Mass Transfer. Heat transfer process is considered on both sides of the interface. The liquid–vapour interface is assumed to be at saturation temperature.

$$Q_q = h_q A_i (T_i - T_q) + \Gamma_{pq} H_q \tag{5}$$

$$Q_p = h_p A_i (T_i - T_p) - \Gamma_{pq} H_p \tag{6}$$

Heat transfer coefficient is solved by specifying Ranz–Marshall Nusselt (Nu) correlation at both the liquid and the vapour sides.

$$\mathrm{Nu} = 2.0 + 0.6\mathrm{Re}_p^{1/2} \mathrm{Pr}_p^{1/3} \tag{7}$$

$$\mathrm{Re}_p = \frac{\rho_p \left| \vec{V}_p - \vec{V}_q \right| d_p}{\mu_p} \tag{8}$$

$$\mathrm{Pr}_p = \frac{Cp_p \mu_p}{k_p} \tag{9}$$

$$h_p = \frac{\mathrm{Nu}_p k_p}{d_p} \tag{10}$$

The interphase mass flux is calculated from the heat balance at the interface.

$$Q_p + Q_q = 0 \tag{11}$$

Turbulence Modeling. Turbulence is modelled using standard k-ϵ model. The transport equations for turbulent kinetic energy and turbulent dissipation rate solved are given in Eqs. (12) and (13), respectively.

$$\frac{\partial}{\partial t}(\rho_m k + \nabla.\left(\rho_m \vec{V}_m k\right) = \nabla.\left(\left(\mu_m + \frac{\mu_{t,m}}{\sigma_k}\right)\nabla k\right) + G_{k,m} - \rho_m \varepsilon \tag{12}$$

$$\frac{\partial}{\partial t}(\rho_m \varepsilon + \nabla.\left(\rho_m \vec{V}_m \varepsilon\right) = \nabla.\left(\left(\mu_m + \frac{\mu_{t,m}}{\sigma_\varepsilon}\right)\nabla \varepsilon\right) + \frac{\varepsilon}{k}\left(c_{1\varepsilon} G_{k,m} - C_{2\varepsilon} \rho_m \varepsilon\right) \tag{13}$$

A pressure-based coupled solver of ANSYS Fluent was used to solve the governing equations. The algorithm solves the system of momentum, continuity and energy equations in a closely coupled manner. The equations of turbulence were solved in a sequential manner. The nonlinear governing equations were solved iteratively until the solution converged.

Initially, axi-symmetric CFD simulations were carried out for the reported experimental conditions of Simpson et al. [15]. The computational domain considered for

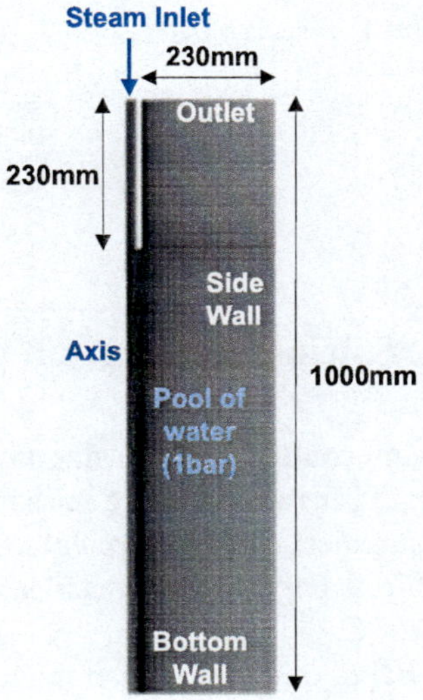

Fig. 7 Computational domain for steam–water simulation

the simulations is shown in Fig. 7. Steam mass flux of 145.2 kg/m^2s at saturated conditions and ambient pressure at outlet were specified as boundary conditions for the analysis.

Numerical study of gaseous oxygen condensation in liquid oxygen. The 3-dimensional computational domain used for the CFD simulations is shown in Fig. 8. It can be seen that symmetric half of the flow line is modelled. The domain consists of the exit line of the booster pump and the orifices (16 nos.) through which gaseous oxygen is injected into the flowing liquid oxygen.

Mass flow rates of liquid oxygen and gaseous oxygen are specified at their respective inlets. Pressure is specified at the exit of the line. The values of boundary conditions specified for the analysis are given in Table 1.

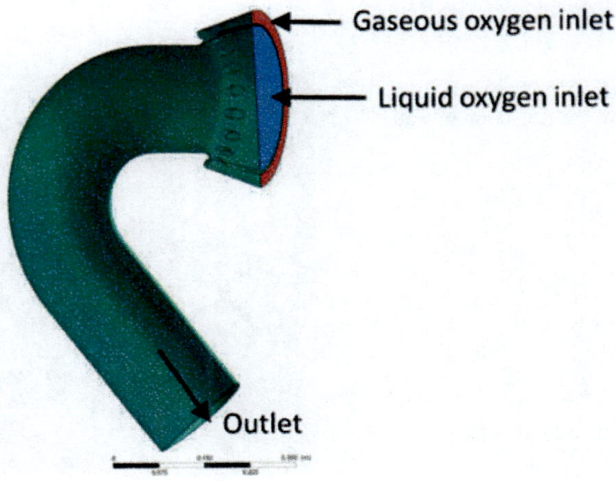

Fig. 8 Computational domain

Table 1 Boundary conditions

	Mass flow rate (kg/s)	Temperature (K)	Pressure (MPa)
Liquid inlet	430	95	–
Gas inlet	7.5	400	–
Outlet	–	–	2.0

4 Saleint Results from the Studies

Steam condensation in stagnant. When steam condenses in stagnant pool of water, there forms a vapour core and a bubbly two-phase zone having an interface dominated by turbulent eddies. A typical scenario that was observed as part of the study is shown in Fig. 9. For the results detailed in this section, the water sub-cooling was maintained at 70 °C.

Based on the shadowgraph images, a regime classification was proposed. When mass flux of steam was less than 15 kg/m^{2}s, a chugging behaviour of steam was observed. This regime was characterised by intermittent entry of water into the injector inlet as was evidenced by the thermocouple readings which showed rapid fluctuations. The steam condensation was violent as shown in Fig. 10.

When the steam mass flux was in the range of 17 32–32 kg/m^{2}s, entry of water into the injector ceased and the condensation of steam became less violent. This regime was called transition bubbly regime as there was a visible chug of vapour separating out from the vapour core. A typical transition bubbly vapour condensation is shown in Fig. 11.

Finally, when the steam mass flux was significantly higher but less than 128 kg/m^{2}s, an unstable jetting regime was observed. Here, the condensation of steam was smooth, but the interface was still rugged and wavy. A typical unstable jet is shown in Fig. 12. Detailed analysis of the experimental results pertaining to dimensionless penetration length and heat transfer coefficient estimation is given in [18].

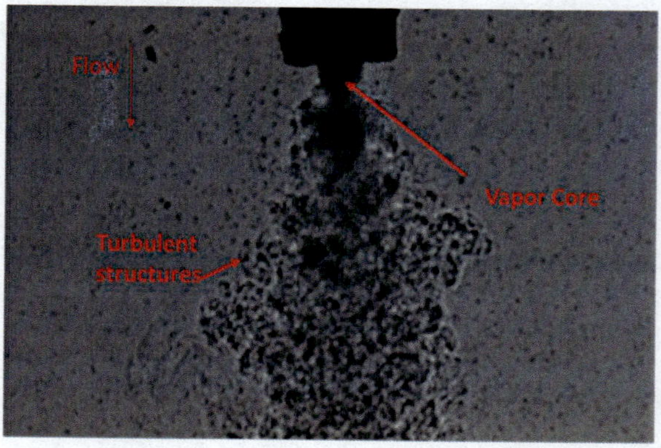

Fig. 9 Shadowgraphy of steam condensation in water

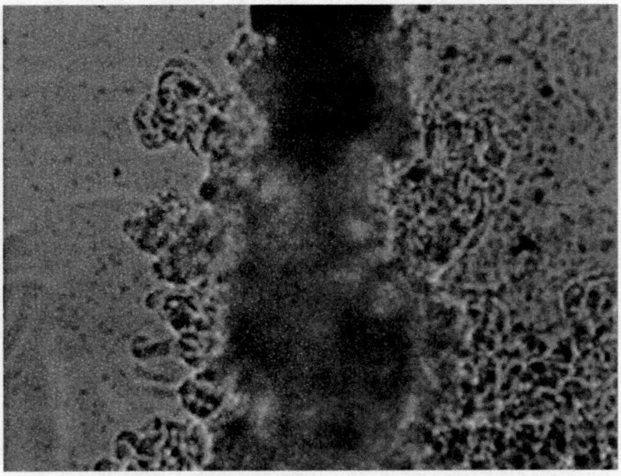

Fig. 10 Chugging regime of steam condensation

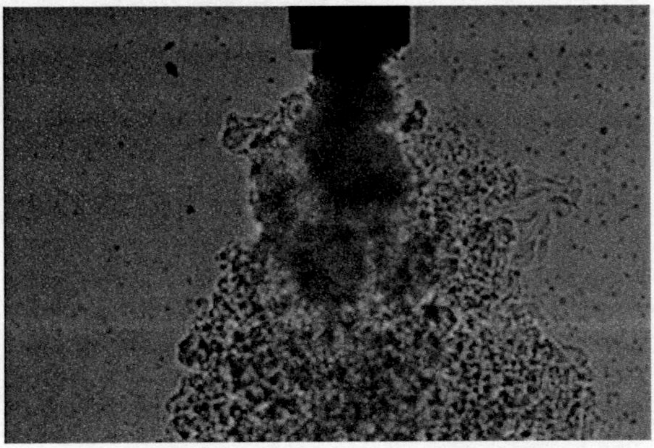

Fig. 11 Transition bubbly vapour condensation

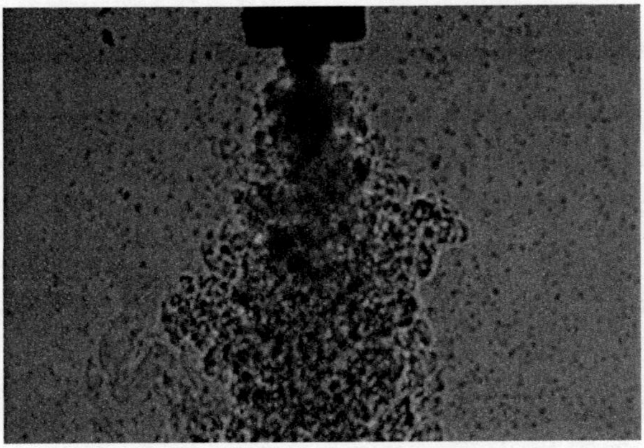

Fig. 12 Unstable jet steam condensation

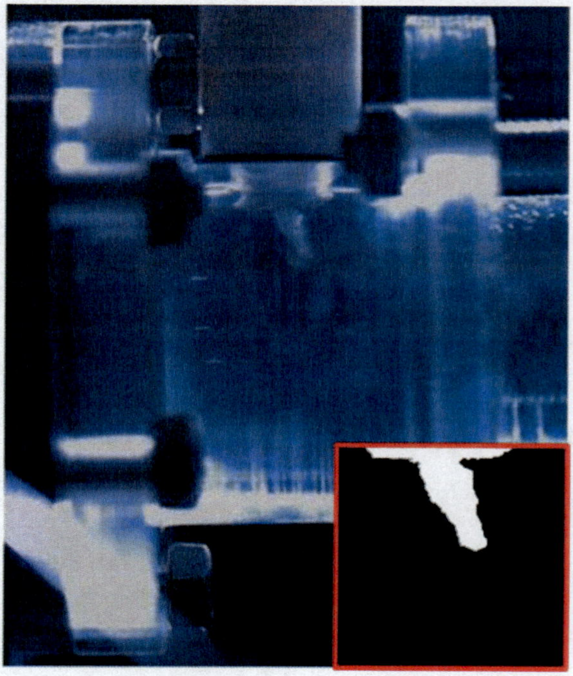

Fig. 13 Raw and post-processed image of steam condensation in flowing water

Steam condensation in flowing water. Based on the experience gained from the steam in stagnant water experiments, steam was injected at higher Reynolds number into cross-flowing stream of water. The experiments were conducted for Reynolds number of steam in the range of 12,400–24,900, with a cross-flow water Reynolds number of 14,000, 43,000 and 74,400. The water subcooling was maintained at 70 °C. The raw and the post-processed images are shown in Fig. 13. The variation of nondimensional penetration length calculated along the axis of the condensing steam is shown in Fig. 14.

Numerical study of steam in stagnant water. The pressure oscillations resulting from the bubble motion and interfacial condensation are shown in Fig. 15.

It can be seen that the oscillatory bubble motion results in the periodic pressure variation between 105 and 102 kPa. Figure 16 shows the pressure spike that was obtained at the time of secondary bubble separation. A second pressure peak, which is lower than the first, is observed when the secondary bubble collapses due to condensation. These numerical results are in-line with the experimental observations reported by Simpson and Chan [15].

Numerical study of gaseous oxygen condensation in liquid oxygen. The results of the CFD simulations were post-processed in CFD-post once the solution converged. Figure 17 shows the contours of gaseous oxygen volume fraction along the walls of the domain. It can be seen that the gas streams from the discrete orifices expand and merge downstream of the injection location. The merged stream of gas condenses as it flows downstream along the line.

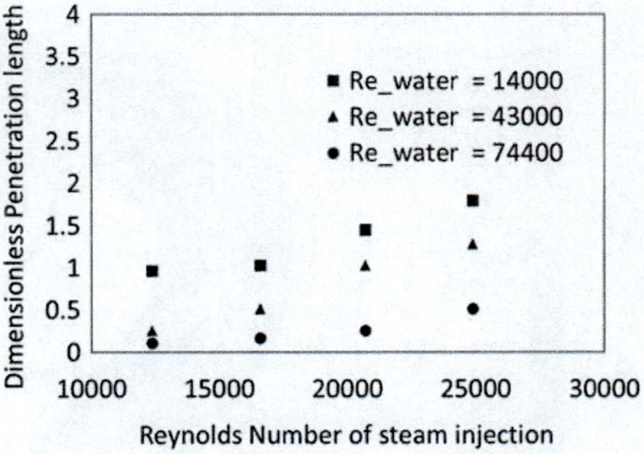

Fig. 14 Dimensionless penetration length for different Reynolds number of steam injection and water flow

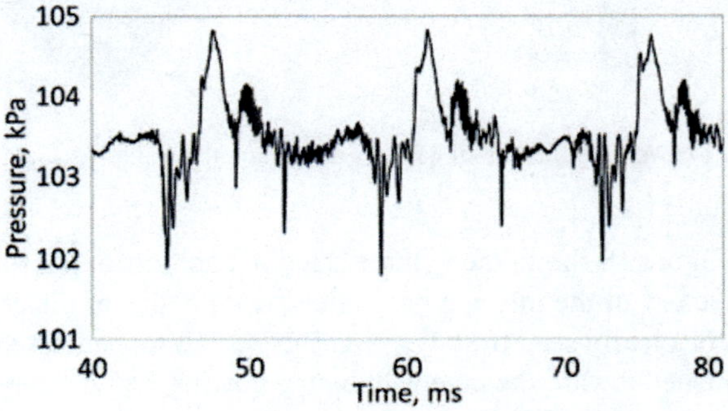

Fig. 15 Pressure oscillations from the computational model

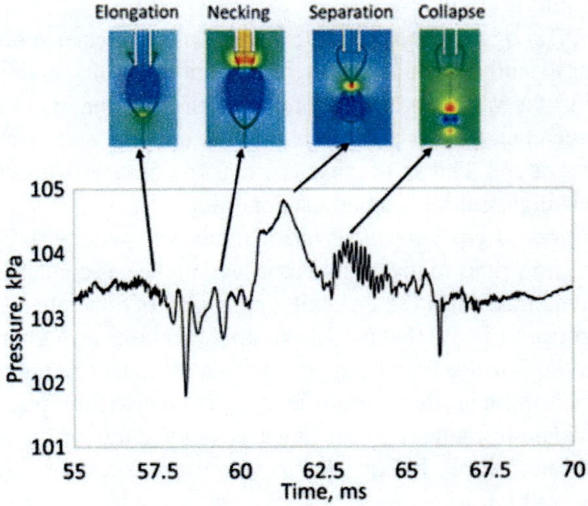

Fig. 16 Pressure variation during to different phase of steam condensation

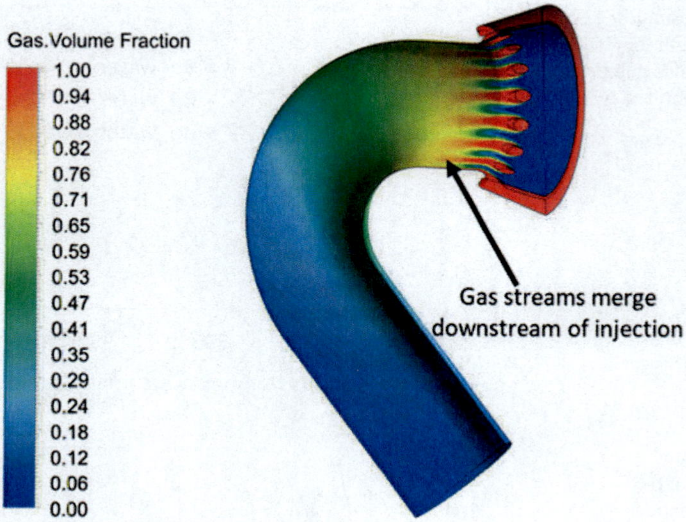

Fig. 17 Volume fraction of gaseous oxygen on the walls

Figure 18 shows the volume fraction contours of gas on the symmetric plane. The trajectory of the injected gas under the influence of the cross-flowing liquid oxygen can be clearly seen from Fig. 18. The gas initially enters the flow line at an angle and is pushed toward the pipe wall by the flowing liquid oxygen. The volume fraction of gas at the exit of the line is obtained as 0.0057 for the operating conditions reported in Table 1.

Effect of gas temperature. Simulations were carried out for various gas temperatures to study its influence on the amount of gas that condenses in the line. Figure 19 shows the variation of gas volume fraction at line exit with gas temperature. It can be seen that the gas volume fraction at the pipe exit increases nonlinearly with gas temperature. This is because the sensible heat transfer component for the injected gas is higher at high superheat condition.

Effect of gas mass flow rate. Simulations were also carried out for various gas mass flow rates to investigate its effect on the gas condensation. The variation of gas volume fraction at the line exit with gas mass flow rate is shown in Fig. 20. It can be seen from Fig. 20 that the gas volume increases as higher amount of gas is injected into the flowing liquid oxygen. This is attributed to higher gas velocities at higher mass flow rates which results in lower residence time available for heat transfer. This is evident from the contours of gas velocity at the symmetry plane, for 4 and 7.5 kg/s flow rates, shown in Fig. 21. It is seen that the gas flows at higher velocity along the pipe wall for higher mass flow rates resulting in lower residence time.

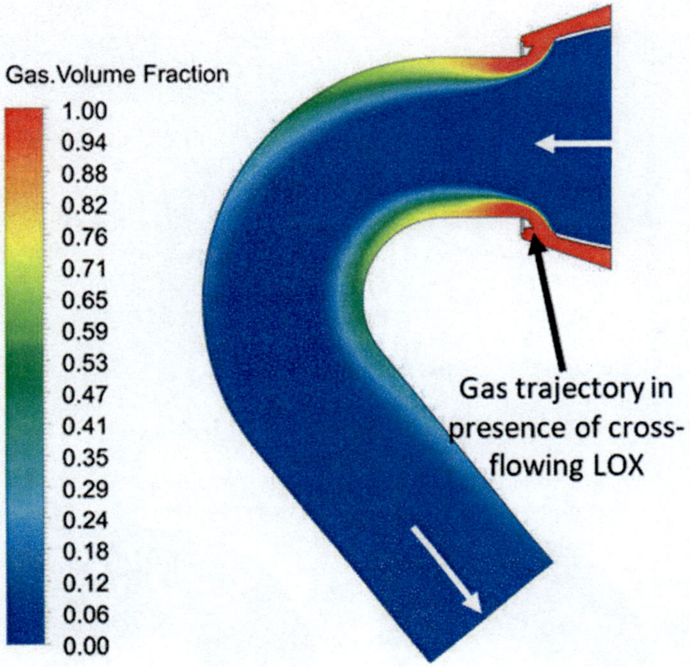

Fig. 18 Volume fraction of gaseous oxygen on the symmetry plane

Fig. 19 Gas volume fraction at line exit for various gas temperature

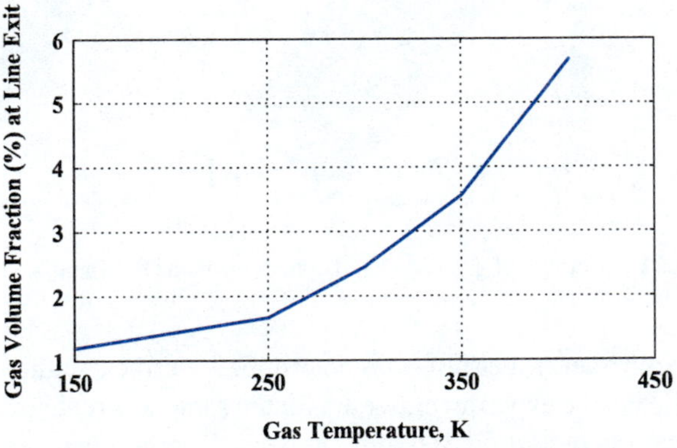

5 Conclusions

The phenomena of DCC was studied by injecting subsonic steam at various steam injection Reynolds numbers and pool subcooling (3000 < Re < 20,000, 50 < Pool subcooling < 70). Steam behaviour was analysed, and a regime classification was proposed. Steam condensation in cross-flowing water was also studied. The trend of variation of dimensionless penetration length for different Reynolds number of steam

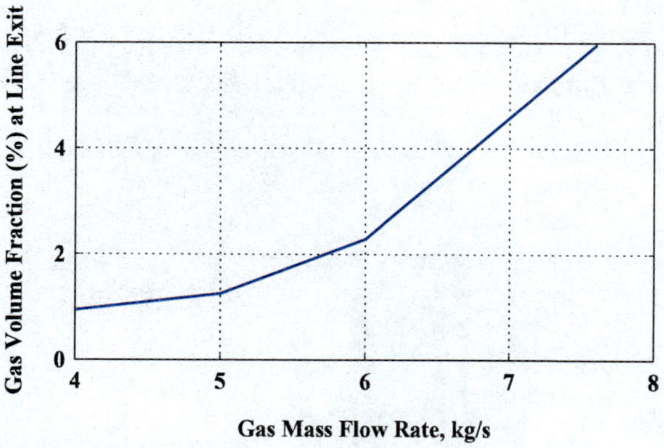

Fig. 20 Gas volume fraction at line exit for various gas mass flow rates

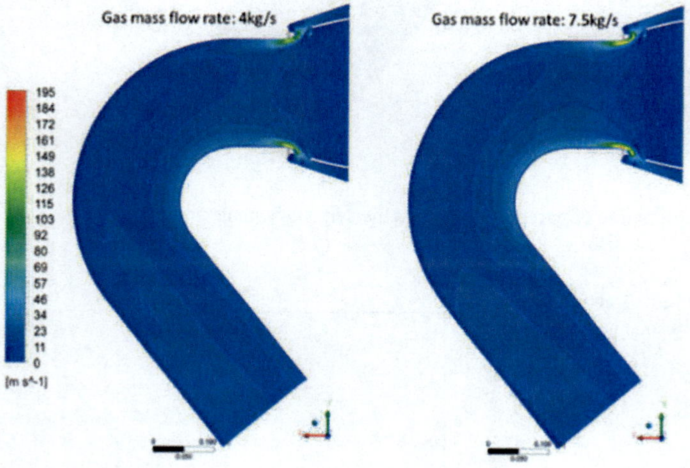

Fig. 21 Contours of gas velocity at symmetry plane for 4 and 7.5 kg/s gas mass flow rates

injection and water flow was determined. Further, studies are planned with GN_2 and LN_2, and the experimental setup for the same was realised. To decipher the intricacies of gas condensation in liquid, numerical studies were also conducted. The different events that occur during condensation were captured by the numerical model, and their relation with pressure oscillations was also determined. A parametric study of gaseous oxygen condensation in flowing LOX was carried out to determine the effect of injected gas temperature and mass flow rate on condensation characteristics.

Acknowledgements The authors thank Prof. Prathap C., Indian Institute of Space Science and Technology, Valiamala for his consultation and guidance during the test programme. The authors would also like to thank Prof. M.D. Atrey and Prof. Atul Shrivastava, IIT Bombay for their continued support and guidance.

References

1. Aya, Nariai H (1991) Evaluation of heat-transfer coefficient at direct-contact condensation of cold water and steam. Nucl Eng Design 131(1):17–24
2. Boehm J, Gesundh I (1938) pp 591–595
3. Chan C, Lee C (1982) A regime map for direct contact condensation. Int J Multiph Flow 8(1):11–20
4. Chun MH, Kim YS, Park JW (1996) An investigation of direct condensation of steam jet in subcooled water. Int Commun Heat Mass Transfer 23(7):947–958
5. Dahikar SK, Sathe MJ, Joshi JB (2010) Investigation of flow and temperature patterns in direct contact condensation using PIV, PLIF and CFD. Chem Eng Sci 65(16):4606–4620
6. Gulawani SS, Joshi JB, Shah MS, RamaPrasad CS, Shukla DS (2006) CFD analysis of flow pattern and heat transfer in direct contact steam condensation. Chem Eng Sci 61(16):5204–5220
7. Gulawani SS (2009) Analysis of flow pattern and heat transfer in direct contact condensation. Chem Eng Sci 64(8):1719–1738
8. Jayachandran KN, Roy A, Ghosh P (2020) Numerical investigation on unstable direct contact condensation of steam in subcooled water. Heat Transf Eng 1–21
9. Kerney P, Faeth G, Olson D (1972) Penetration characteristics of a submerged steam jet. AIChE J 18(3):548–553
10. Kim HY, Bae YY, Song CH, Park JK, Choi SM (2001) Experimental study on stable steam condensation in a quenching tank. Int J Energy Res 25(3):239–252
11. Liang KS, Griffith P (1994) Experimental and analytical study of direct contact condensation of steam in water. Nucl Eng Des 147(3):425–435
12. Moffat RJ (1988) Describing the uncertainties in experimental results. Exp Thermal Fluid Sci 1(1):3–17
13. Patel G, Tanskanen V, Kyrki-Rajamäki R (2014) Numerical modelling of low-Reynolds number direct contact condensation in a suppression pool test facility. Ann Nucl Energy 71:376–387
14. Settles GS (2012) Schlieren and shadowgraph techniques: visualizing phenomena in transparent media. Springer Science & Business Media
15. Simpson ME, Chan CK (1982) Hydraulics of a subsonic vapour jet in sub-cooled liquid. J Heat Transfer 104:271–278
16. Sonin AA (1984) Suppression pool dynamics research at MIT (NUREG/CP--0048-Vol3). United States
17. Song S, Yue X, Zhao Q, Chong D, Chen W, Yan J (2020) Numerical study on mechanism of condensation oscillation of unstable steam jet. Chem Eng Sci 211:115303
18. Viswanath V, Peter J, Agarwal DK, Tharakan TJ, Kumar SS, Vasudevan MK, Prathap C (2020) Direct contact condensation of subsonic, inversely buoyant steam jet in a stagnant pool of water. In: Proceedings of the 8th international and 47th national conference on fluid mechanics and fluid power (FMFP), IIT Guwahati, Guwahati-781039, Assam, India
19. Weimer J, Faeth G, Olson D (1973) Penetration of vapor jets submerged in subcooled liquids. AIChE J 19(3):552–558
20. Wu XZ, Yan JJ, Li WJ, Pan DD, Chong DT (2009) Experimental study on sonic steam jet condensation in quiescent subcooled water. Chem Eng Sci 64(23):5002–5012

Unsteady Free Convection of Fluid with Variable Viscosity in a Partially Porous Cube Under an Influence of Energy Source

M. S. Astanina and M. A. Sheremet

Nomenclature

c	Thermal capacity
g	Gravity acceleration
h	Size of the porous layer
k	Heat conductivity
K	Permeability of porous medium
L	Size of the chamber
Nu	Nusselt number
Pr	Prandtl number
Ra	Rayleigh number
T_c	Cold border temperature (K)
T_h	Heater temperature
u, v, w	Non-dimensional velocity projections
x, y, z	Non-dimensional coordinates

Greek Symbols

α	Thermal diffusivity
β	Expansion parameter
ε	Porosity of the solid structure
ΔT	Temperature drop
θ	Non-dimensional temperature

M. S. Astanina (✉) · M. A. Sheremet
Laboratory on Convective Heat and Mass Transfer, Tomsk State University, 634050 Tomsk, Russia
e-mail: astanina.marina@bk.ru

J. Banerjee et al. (eds.), *Recent Advances in Fluid Dynamics*, Lecture Notes in Mechanical Engineering, https://doi.org/10.1007/978-981-19-3379-0_42

μ_0	Reference dynamic viscosity
$\mu = \exp(-\xi\theta)$	Dimensionless dynamic viscosity
ξ	Viscosity varying characteristic
ρ	Density
τ	dimensionless time
$\psi_x,\ \psi_y,\ \psi_z$	Dimensionless vector potential functions
$\omega_x,\ \omega_y,\ \omega_z$	Dimensionless projections of vorticity vector

1 Introduction

The convective heat transfer is a main process in different engineering devices. There are a lot of engineering applications where study of natural convection is a basis for design of an electronic equipment. The important area of research is a study of two- and three-dimensional fluid flow and heat transfer inside cavities with different heating elements, working fluids, and other additions including porous insertions. For example, Zhu et al. [1] have studied the heat transfer inside closed differentially heated cube filled with an anisotropic porous medium. The obtained results have demonstrated that the fluid flow inside cavity depends on the buoyancy ratio, the anisotropy ratio, and porous Rayleigh number. Raizah et al. [2] have conducted unsteady convective flow of nanofluid in an E-cavity in the presence of homogeneous/heterogeneous porous layers. The authors have showed that the growth of Rayleigh number leads to increase in heat transfer ratio. In addition, the influence of the location of the porous layer on the characteristics of convective flows has been analyzed. Miles and Bessaïh [3] have reported results of numerical simulation of laminar convection inside a porous layer between two cylinders filled with a nanofluid. It was illustrated that a variation of porous medium characteristics leads to an increase in the convective fluid flow strength and Nusselt number. Kramer et al. [4] have studied numerically 3D free convection inside porous cube under the heating/cooling effect from vertical walls. As a result, authors have noted that the reduction of Darcy number and rise of Rayleigh number are a good way for an intensification of heat transfer inside the cavity. Zhang et al. [5] have demonstrated results of transient convection melting inside closed cubical chamber with a cylindrical heated element. It has been found an optimal location of the inner cylinder and thermal boundary conditions for decrease of full melting time.

In the present research, the numerical simulation of natural convection within partially porous cubical cavity with a local heated element has been worked out.

2 Problem Description

Figure 1 illustrates the scheme of the research region. We have a closed cube of size L filled with a fluid of variable viscosity in the presence of the porous insertion of height h and a local energy source of constant temperature T_{h} at the bottom wall. The side surfaces are kept at constant temperature T_{c} ($T_{\mathrm{h}} > T_{\mathrm{c}}$). It is supposed that the working fluid is Newtonian, heat-conducting, and the Boussinesq model is valid. In addition, the viscosity of the liquid depends on the temperature according to the exponential law. The porous medium is homogenous and isotropic. In addition, it is expected that the temperature of the porous medium is equal to the temperature of the working fluid, and the LTE technique is used for simulation.

The mathematical model has been formulated using dimensionless non-primitive variables "vector potential functions and vorticity vector" as follows

- for the clear fluid zone:

$$\nabla^2\psi_x = -\omega_x, \nabla^2\psi_y = -\omega_y, \nabla^2\psi_z = -\omega_z \tag{1}$$

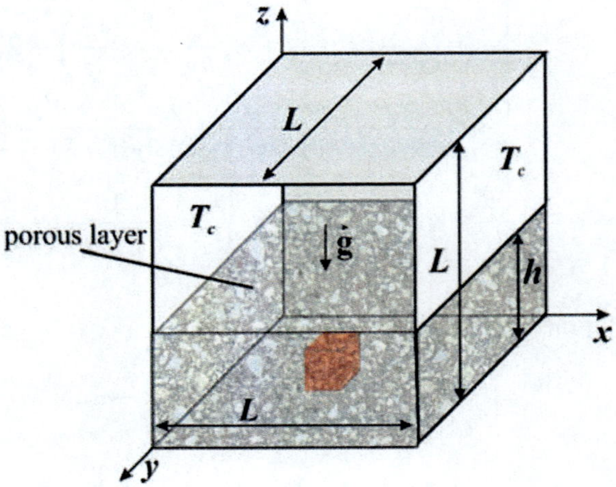

Fig. 1 Physical model of the considered problem

$$\left|\begin{array}{l}\dfrac{\partial \omega_x}{\partial \tau}+u\dfrac{\partial \omega_x}{\partial x}+v\dfrac{\partial \omega_x}{\partial y}+w\dfrac{\partial \omega_x}{\partial z}-\omega_x\dfrac{\partial u}{\partial x}-\omega_y\dfrac{\partial u}{\partial y}-\omega_z\dfrac{\partial u}{\partial z}\\ =\sqrt{\dfrac{\mathrm{Pr}}{\mathrm{Ra}}}\left(\dfrac{\partial^2(\mu\omega_x)}{\partial x^2}+\dfrac{\partial^2(\mu\omega_x)}{\partial y^2}+\dfrac{\partial^2(\mu\omega_x)}{\partial z^2}\right)\\ -\sqrt{\dfrac{\mathrm{Pr}}{\mathrm{Ra}}}\dfrac{\partial}{\partial x}\left(\omega_x\dfrac{\partial \mu}{\partial x}+\omega_y\dfrac{\partial \mu}{\partial y}+\omega_z\dfrac{\partial \mu}{\partial z}\right)+2\sqrt{\dfrac{\mathrm{Pr}}{\mathrm{Ra}}}\left[\dfrac{\partial v}{\partial z}\dfrac{\partial^2 \mu}{\partial y^2}\right.\\ \left.+\dfrac{\partial u}{\partial z}\dfrac{\partial^2 \mu}{\partial x \partial y}-\dfrac{\partial w}{\partial y}\dfrac{\partial^2 \mu}{\partial z^2}-\dfrac{\partial u}{\partial y}\dfrac{\partial^2 \mu}{\partial x \partial z}+\dfrac{\partial^2 \mu}{\partial y \partial z}\left(\dfrac{\partial w}{\partial z}-\dfrac{\partial v}{\partial y}\right)\right]+\dfrac{\partial \theta}{\partial y}\end{array}\right. \tag{2}$$

$$\left|\begin{array}{l}\dfrac{\partial \omega_y}{\partial \tau}+u\dfrac{\partial \omega_y}{\partial x}+v\dfrac{\partial \omega_y}{\partial y}+w\dfrac{\partial \omega_y}{\partial z}-\omega_x\dfrac{\partial v}{\partial x}-\omega_y\dfrac{\partial v}{\partial y}-\omega_z\dfrac{\partial v}{\partial z}\\ =\sqrt{\dfrac{\mathrm{Pr}}{\mathrm{Ra}}}\left(\dfrac{\partial^2(\mu\omega_y)}{\partial x^2}+\dfrac{\partial^2(\mu\omega_y)}{\partial y^2}+\dfrac{\partial^2(\mu\omega_y)}{\partial z^2}\right)\\ -\sqrt{\dfrac{\mathrm{Pr}}{\mathrm{Ra}}}\dfrac{\partial}{\partial y}\left(\omega_x\dfrac{\partial \mu}{\partial x}+\omega_y\dfrac{\partial \mu}{\partial y}+\omega_z\dfrac{\partial \mu}{\partial z}\right)+2\sqrt{\dfrac{\mathrm{Pr}}{\mathrm{Ra}}}\left[\dfrac{\partial v}{\partial x}\dfrac{\partial^2 \mu}{\partial y \partial z}\right.\\ \left.-\dfrac{\partial v}{\partial z}\dfrac{\partial^2 \mu}{\partial x \partial y}-\dfrac{\partial u}{\partial z}\dfrac{\partial^2 \mu}{\partial x^2}+\dfrac{\partial w}{\partial x}\dfrac{\partial^2 \mu}{\partial z^2}+\dfrac{\partial^2 \mu}{\partial x \partial z}\left(\dfrac{\partial u}{\partial x}-\dfrac{\partial w}{\partial z}\right)\right]-\dfrac{\partial \theta}{\partial x}\end{array}\right. \tag{3}$$

$$\left|\begin{array}{l}\dfrac{\partial \omega_z}{\partial \tau}+u\dfrac{\partial \omega_z}{\partial x}+v\dfrac{\partial \omega_z}{\partial y}+w\dfrac{\partial \omega_z}{\partial z}-\omega_x\dfrac{\partial w}{\partial x}-\omega_y\dfrac{\partial w}{\partial y}-\omega_z\dfrac{\partial w}{\partial z}\\ =\sqrt{\dfrac{\mathrm{Pr}}{\mathrm{Ra}}}\left(\dfrac{\partial^2(\mu\omega_z)}{\partial x^2}+\dfrac{\partial^2(\mu\omega_z)}{\partial y^2}+\dfrac{\partial^2(\mu\omega_z)}{\partial z^2}\right)\\ -\sqrt{\dfrac{\mathrm{Pr}}{\mathrm{Ra}}}\dfrac{\partial}{\partial z}\left(\omega_x\dfrac{\partial \mu}{\partial x}+\omega_y\dfrac{\partial \mu}{\partial y}+\omega_z\dfrac{\partial \mu}{\partial z}\right)+2\sqrt{\dfrac{\mathrm{Pr}}{\mathrm{Ra}}}\left[\dfrac{\partial w}{\partial y}\dfrac{\partial^2 \mu}{\partial x \partial z}\right.\\ \left.-\dfrac{\partial w}{\partial x}\dfrac{\partial^2 \mu}{\partial y \partial z}+\dfrac{\partial u}{\partial y}\dfrac{\partial^2 \mu}{\partial x^2}-\dfrac{\partial v}{\partial x}\dfrac{\partial^2 \mu}{\partial y^2}+\dfrac{\partial^2 \mu}{\partial x \partial y}\left(\dfrac{\partial v}{\partial y}-\dfrac{\partial u}{\partial x}\right)\right]\end{array}\right. \tag{4}$$

$$\frac{\partial \theta}{\partial \tau}+u\frac{\partial \theta}{\partial x}+v\frac{\partial \theta}{\partial y}+w\frac{\partial \theta}{\partial z}=\frac{1}{\sqrt{\mathrm{Ra}\cdot\mathrm{Pr}}}\left(\frac{\partial^2 \theta}{\partial x^2}+\frac{\partial^2 \theta}{\partial y^2}+\frac{\partial^2 \theta}{\partial z^2}\right) \tag{5}$$

- for the porous zone:

$$\nabla^2\psi_x=-\omega_x, \nabla^2\psi_y=-\omega_y, \nabla^2\psi_z=-\omega_z \tag{6}$$

$$
\begin{aligned}
&\varepsilon\frac{\partial\omega_x}{\partial\tau}+u\frac{\partial\omega_x}{\partial x}+v\frac{\partial\omega_x}{\partial y}+w\frac{\partial\omega_x}{\partial z}-\omega_x\frac{\partial u}{\partial x}-\omega_y\frac{\partial u}{\partial y}-\omega_z\frac{\partial u}{\partial z}\\
&=\varepsilon\sqrt{\frac{\Pr}{\mathrm{Ra}}}\left(\frac{\partial^2(\mu\omega_x)}{\partial x^2}+\frac{\partial^2(\mu\omega_x)}{\partial y^2}+\frac{\partial^2(\mu\omega_x)}{\partial z^2}-\varepsilon\frac{\mu\omega_x}{Da}\right)\\
&-\varepsilon\sqrt{\frac{\Pr}{\mathrm{Ra}}}\frac{\partial}{\partial x}\left(\omega_x\frac{\partial\mu}{\partial x}+\omega_y\frac{\partial\mu}{\partial y}+\omega_z\frac{\partial\mu}{\partial z}\right)\\
&+2\varepsilon\sqrt{\frac{\Pr}{\mathrm{Ra}}}\left[\frac{\partial v}{\partial z}\frac{\partial^2\mu}{\partial y^2}+\frac{\partial u}{\partial z}\frac{\partial^2\mu}{\partial x\partial y}-\frac{\partial w}{\partial y}\frac{\partial^2\mu}{\partial z^2}-\frac{\partial u}{\partial y}\frac{\partial^2\mu}{\partial x\partial z}\right.\\
&\left.+\frac{\partial^2\mu}{\partial y\partial z}\left(\frac{\partial w}{\partial z}-\frac{\partial v}{\partial y}\right)+\frac{\varepsilon v}{2Da}\frac{\partial\mu}{\partial z}-\frac{\varepsilon w}{2Da}\frac{\partial\mu}{\partial y}\right]+\varepsilon^2\frac{\partial\theta}{\partial y}
\end{aligned}
\tag{7}
$$

$$
\begin{aligned}
&\varepsilon\frac{\partial\omega_y}{\partial\tau}+u\frac{\partial\omega_y}{\partial x}+v\frac{\partial\omega_y}{\partial y}+w\frac{\partial\omega_y}{\partial z}-\omega_x\frac{\partial v}{\partial x}-\omega_y\frac{\partial v}{\partial y}-\omega_z\frac{\partial v}{\partial z}\\
&=\varepsilon\sqrt{\frac{\Pr}{\mathrm{Ra}}}\left(\frac{\partial^2(\mu\omega_y)}{\partial x^2}+\frac{\partial^2(\mu\omega_y)}{\partial y^2}+\frac{\partial^2(\mu\omega_y)}{\partial z^2}-\varepsilon\frac{\mu\omega_y}{Da}\right)\\
&-\varepsilon\sqrt{\frac{\Pr}{\mathrm{Ra}}}\frac{\partial}{\partial y}\left(\omega_x\frac{\partial\mu}{\partial x}+\omega_y\frac{\partial\mu}{\partial y}+\omega_z\frac{\partial\mu}{\partial z}\right)\\
&+2\varepsilon\sqrt{\frac{\Pr}{\mathrm{Ra}}}\left[\frac{\partial v}{\partial x}\frac{\partial^2\mu}{\partial y\partial z}-\frac{\partial v}{\partial z}\frac{\partial^2\mu}{\partial x\partial y}-\frac{\partial u}{\partial z}\frac{\partial^2\mu}{\partial x^2}+\frac{\partial w}{\partial x}\frac{\partial^2\mu}{\partial z^2}\right.\\
&\left.+\frac{\partial^2\mu}{\partial x\partial z}\left(\frac{\partial u}{\partial x}-\frac{\partial w}{\partial z}\right)+\frac{\varepsilon w}{2Da}\frac{\partial\mu}{\partial x}-\frac{\varepsilon u}{2Da}\frac{\partial\mu}{\partial z}\right]-\varepsilon^2\frac{\partial\theta}{\partial x}
\end{aligned}
\tag{8}
$$

$$
\begin{aligned}
&\varepsilon\frac{\partial\omega_z}{\partial\tau}+u\frac{\partial\omega_z}{\partial x}+v\frac{\partial\omega_z}{\partial y}+w\frac{\partial\omega_z}{\partial z}-\omega_x\frac{\partial w}{\partial x}-\omega_y\frac{\partial w}{\partial y}\\
&-\omega_z\frac{\partial w}{\partial z}=\varepsilon\sqrt{\frac{\Pr}{\mathrm{Ra}}}\left(\frac{\partial^2(\mu\omega_z)}{\partial x^2}+\frac{\partial^2(\mu\omega_z)}{\partial y^2}\right.\\
&\left.+\frac{\partial^2(\mu\omega_z)}{\partial z^2}-\varepsilon\frac{\mu\omega_z}{Da}\right)-\varepsilon\sqrt{\frac{\Pr}{Ra}}\frac{\partial}{\partial z}\left(\omega_x\frac{\partial\mu}{\partial x}+\omega_y\frac{\partial\mu}{\partial y}\right.\\
&\left.+\omega_z\frac{\partial\mu}{\partial z}\right)+2\varepsilon\sqrt{\frac{\Pr}{\mathrm{Ra}}}\left[\frac{\partial w}{\partial y}\frac{\partial^2\mu}{\partial x\partial z}-\frac{\partial w}{\partial x}\frac{\partial^2\mu}{\partial y\partial z}+\frac{\partial u}{\partial y}\frac{\partial^2\mu}{\partial x^2}\right.\\
&\left.-\frac{\partial v}{\partial x}\frac{\partial^2\mu}{\partial y^2}+\frac{\partial^2\mu}{\partial x\partial y}\left(\frac{\partial v}{\partial y}-\frac{\partial u}{\partial x}\right)+\frac{\varepsilon u}{2Da}\frac{\partial\mu}{\partial y}-\frac{\varepsilon v}{2Da}\frac{\partial\mu}{\partial x}\right]
\end{aligned}
\tag{9}
$$

$$\eta\frac{\partial\theta}{\partial\tau}+u\frac{\partial\theta}{\partial x}+v\frac{\partial\theta}{\partial y}+w\frac{\partial\theta}{\partial z}=\frac{\alpha_{\mathrm{pm}}/\alpha_{\mathrm{f}}}{\sqrt{\mathrm{Ra}\cdot\mathrm{Pr}}}\left(\frac{\partial^2\theta}{\partial x^2}+\frac{\partial^2\theta}{\partial y^2}+\frac{\partial^2\theta}{\partial z^2}\right) \tag{10}$$

The initial and boundary conditions for the governing Eqs. (1)–(10) are written as follows:

$$\tau=0:\begin{cases}\psi_x=0,\\ \psi_y=0,\\ \psi_z=0\end{cases}\begin{cases}\omega_x=0,\\ \omega_y=0,\ \theta=0\text{ at}\\ \omega_z=0\end{cases}\left|\begin{array}{l}0<x\le 1,\\ 0\le y\le 1,\\ 0\le z\le 1\end{array}\right.$$

$$\tau>0:\begin{cases}\dfrac{\partial\psi_x}{\partial x}=0,\\ \psi_y=0,\\ \psi_z=0\end{cases}\begin{cases}\omega_x=0,\\ \omega_y=-\dfrac{\partial w}{\partial x},\ \theta=0\text{ at}\\ \omega_z=\dfrac{\partial v}{\partial x}\end{cases}\left|\begin{array}{l}x=0\text{ and }x=1,\\ 0\le y\le 1,\\ 0\le z\le 1\end{array}\right.$$

$$\begin{cases}\psi_x=0,\\ \dfrac{\partial\psi_y}{\partial y}=0,\\ \psi_z=0\end{cases}\begin{cases}\omega_x=\dfrac{\partial w}{\partial y},\\ \omega_y=0,\quad\dfrac{\partial\theta}{\partial y}=0\text{ at}\\ \omega_z=-\dfrac{\partial u}{\partial y}\end{cases}\left|\begin{array}{l}y=0\text{ and }y=1,\\ 0\le x\le 1,\\ 0\le z\le 1\end{array}\right.\qquad\text{for the heater }\theta$$

$$\begin{cases}\psi_x=0,\\ \psi_y=0,\\ \dfrac{\partial\psi_z}{\partial z}=0\end{cases}\begin{cases}\omega_x=-\dfrac{\partial v}{\partial z},\\ \omega_y=\dfrac{\partial u}{\partial z},\quad\dfrac{\partial\theta}{\partial z}=0\text{ at}\\ \omega_z=0\end{cases}\left|\begin{array}{l}z=0\text{ and }z=1,\\ 0\le x\le 1,\\ 0\le y\le 1\end{array}\right.$$

$= 1$;

for the internal fluid/porous interface:

$$\psi_{\mathrm{pm}}=\psi_f,\quad\frac{\partial\psi_{\mathrm{pm}}}{\partial z}=\frac{\partial\psi_{\mathrm{f}}}{\partial z},$$

$$\omega_{\mathrm{pm}}=\omega_{\mathrm{f}},\quad\frac{\partial\omega_{\mathrm{pm}}}{\partial z}=\frac{\partial\omega_{\mathrm{f}}}{\partial z},$$

$$\left.\frac{\partial\theta_{\mathrm{f}}}{\partial z}\right|_{\mathrm{cl.f}}=\left.\frac{\lambda_{\mathrm{s}}}{\lambda_{\mathrm{f}}}\frac{\partial\theta_{\mathrm{s}}}{\partial z}\right|_{\mathrm{pm}},\ \theta_{\mathrm{f}}|_{\mathrm{cl.f}}=\theta_{\mathrm{f}}|_{\mathrm{pm}}=\theta_{\mathrm{s}}|_{\mathrm{pm}}$$

The definable integral parameter is the average Nusselt number at surface of the local heated element: $\overline{\mathrm{Nu}}=-\frac{1}{l}\int\limits_0^1\frac{\partial\theta}{\partial n}\mathrm{d}l$.

Table 1 Values of Nu_{min} for different grid sizes

Present results	Present results	Data of Bessonov et al. [6]	Data of Fusegi et al. [7]
100 × 100 × 100	50 × 50 × 50	86 × 65 × 65	62 × 62 × 62
0.727	0.7869	0.675	0.7867

The system of governing Eqs. (1)–(10) with initial and boundary conditions has been implemented using the finite difference method and a uniform grid. The Poisson Eqs. (1) and (6) have been worked out using the central differences at the first step and then successive over-relaxation technique. The other parabolic equations have been solved using locally one-dimensional method of Samarskii. The last step of simulations is solving of all equations using C++ language code.

The worked out computational method has been tested employing the results from article of other authors [6, 7]. Table 1 demonstrates the comparison of obtained data for Nu_{min} at middle-cross $y = 0.5$ with results from literature for different meshes.

The mathematical modeling has been performed for following range of the dimensionless parameters: $Pr = 7.0$, $Da = 10^{-4}$–10^{-2}, $\varepsilon = 0.3$–0.9, $Ra = 10^4$–10^6, $\xi = 0.0$–2.0.

Figure 2 shows the 3D patterns of temperature inside the cavity for different values of viscosity parameter ξ. Anyway, the obtained data clearly demonstrate the area of heating of the cube from the heater at bottom surface. A minimum temperature is located on the vertical borders of the chamber $x = 0$ and $x = 1$. A thermal plume is formed above the isothermal source, and then, it expands near the upper adiabatic boundary. The jump from a working fluid of a constant viscosity (Fig. 2a) to a liquid with a variable viscosity (Fig. 2b, c) is visible in the growth in convective fluid flow within chamber.

3 Conclusions

Convective heat and mass transfer in a partially porous cubic chamber filled with the fluid of variable viscosity has been worked out numerically. The governing system of equations has been written using the dimensionless variables "vector potential functions—vorticity vector—temperature." The results have demonstrated that use of the fluid with variable viscosity leads to an intensification of convective fluid flow.

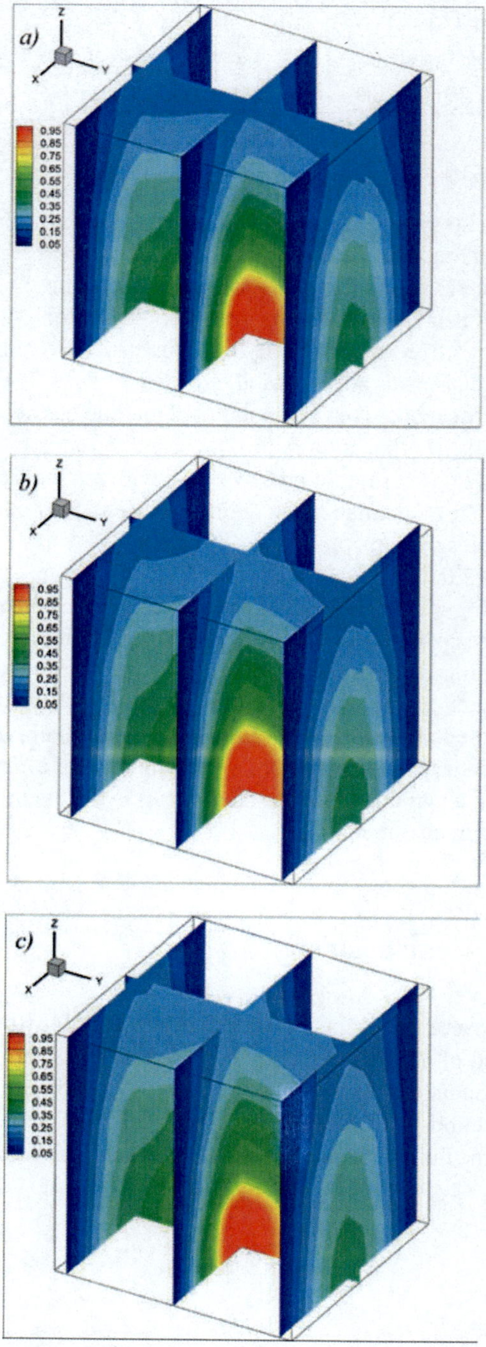

Fig. 2 Three-dimensional temperature fields for Da = 10^{-3}, Ra = 10^4, Pr = 7.0, ε = 0.6, $h = 0.5$ L and different values of ξ: **a** ξ = 0, **b** $\xi = 1.0$, **c** $\xi = 2.0$

Acknowledgements The reported study was funded by RFBR, project number 20-31-90080.

References

1. Zhu QY, Zhuang YJ, Yu HZ (2017) Three-dimensional numerical investigation on thermosolutal convection of power-law fluids in anisotropic porous media. Int J Heat Mass Transf 104:897–917
2. Raizah ZAS, Ahmed SE, Aly AM (2020) ISPH simulations of natural convection flow in E-enclosure filled with a nanofluid including homogeneous/heterogeneous porous media and solid particles. Int J Heat Mass Transf 160:120153
3. Miles A, Bessaïh R (2021) Heat transfer and entropy generation analysis of three-dimensional nanofluids flow in a cylindrical annulus filled with porous media. Int J Heat Mass Transf 105240
4. Kramer J, Ravnik J, Jecl R, Škerget L (2011) Simulation of 3D flow in porous media by boundary element method. Eng Anal Bound Elem 35:1256–1264
5. Zhang X, Wang L, Li D (2021) Lattice Boltzmann simulation of natural convection melting in a cubic cavity with an internal cylindrical heat source. Int J Therm Sci 165:106917
6. Bessonov OA, Brailovskay VA, Nikitin SA, Polezhaev VI (1997) Three-dimensional natural convection in a cubical enclosure: a benchmark numerical solution, In: de Vahl Davis G, Leonardi E (eds) International symposium on advances in computational heat transferring, 26–30 May 1997 Cesme. Begell House, Inc., pp 157–165
7. Fusegi T, Hyin JM, Kuwahara K (1991) A numerical study of 3D natural convection in a differently heated cubical enclosure. Int J Heat Mass Transf 34:1543–1557